The Arthropoda

HABITS, FUNCTIONAL MORPHOLOGY, AND EVOLUTION

S. M. MANTON

CLARENDON PRESS · OXFORD
1977

Oxford University Press, Walton Street, Oxford OX2 6DP

OXFORD LONDON GLASGOW NEW YORK
TORONTO MELBOURNE WELLINGTON CAPE TOWN
IBADAN NAIROBI DAR ES SALAAM LUSAKA ADDIS ABABA
KUALA LUMPUR SINGAPORE JAKARTA HONG KONG TOKYO
DELHI BOMBAY CALCUTTA MADRAS KARACHI

© OXFORD UNIVERSITY PRESS 1977

All rights reserved. No part of this publication may be reproduced, stored in a retrieval system, or transmitted, in any form or by any means, electronic, mechanical, photocopying, recording or otherwise, without the prior permission of Oxford University Press.

British Library Cataloguing in Publication Data

Manton, Sidnie Milana
  The arthropoda
  1. Arthropoda
  I. Title
  595′.2 QL434

ISBN 0-19-857391-X

PRINTED IN GREAT BRITAIN
BY THOMSON LITHO LTD., EAST KILBRIDE, SCOTLAND

TO

J. GORDON BLOWER

Preface

DURING recent decades much work has been published on the habits and the correlated functional morphology of arthropods, but the reading of some of the original papers in the scientific journals, which of necessity embrace detailed proof, can be arduous for those who are not particularly familiar with the field. Since enough work has now been completed to give a new and overall picture of arthropod evolution and relationships, the time has come to present a review of the essentials of this new evidence in a form suitable for the more general reader. This work is correlated in an essential manner with the newer comparative embryological studies on annelids and arthropods with its functional approach, a field of work indispensable for any modern assessment of the Arthropoda. Included too are the most recent and beautiful of the palaeontological evidences concerning hitherto unknown details of the anatomy of trilobites and of other early arthropods.

An abundance of drawings are here reproduced from the original papers, together with some new ones. Accuracy and not schematic simplification is the keynote of the illustrations concerning functional morphology. In modern comparative functional embryology, on the contrary, schematic fate maps are essential. Colour is used to distinguish endoskeletal tendons from apodemes on the figures throughout Chapter 3, but the colours are omitted from a few comparable figures in Chapters 1, 8, and 9, where the characteristics of the endoskeleton are shown in either black or white.

A brief technical Appendix summarizes the methods employed in obtaining the data, which may be of service to those pursuing this type of work.

I owe much, over the years, to the stimulating discussions of zoological problems which I have had with, in particular, Dr. W. T. Calman, F.R.S., Professor H. Graham Cannon, F.R.S., Professor O. W. Tiegs, F.R.S., Professor G. O.

Evans, Mr. J. Gordon Blower, Professor D. T. Anderson, F.R.S., and many others. Professor H. B. Whittington, F.R.S., has given me much of his time and assistance concerning fossil arthropods; and I am grateful to several palaeontologists for supplying me with copies of their drawings in advance of publication for use in this book.

I have travelled widely in search of animals in their natural environments and I have had much assistance in obtaining live material from marine stations and from zoologists in many parts of the world. In particular I should like to record my thanks to the following: to Dr. E. Batham, Professor V. V. Hickman in Australasia; to Professor T. Thomson Flynn, who helped me establish myself on a then almost uninhabited part of Mount Wellington, Tasmania for some weeks work on living Syncarida, to Dr. R. F. Lawrence for assisting my animal hunting in South Africa and for subsequently sending me much live material; also from Africa, to Dr. A. J. Alexanser for attending to my wants on her own expedition to Trinidad; to Dr. N. B. Causey and Dr. H. L. Sanders, U.S.A. for hunting and sending me live animals; to Dr. O. Schubart in South America for similar services; and in Britain to Dr. Eason, Mr. J. Colman, Professor H. E. Hinton, F.R.S., Mrs. J. Froud, and Mr. J. Sankey of Juniper Hall Field Station, whose knowledge of old flint walls and their fauna has been invaluable. But above all I am indebted to Mr. J. Gordon Blower for his indefatigable collecting and identifications in this country, for accompanying me on a myriapod hunt in Sicily, for his diligent reading of the manuscript of this book, and for all his helpful suggestions. My daughter, Mrs. E. A. Clifford, has come to the rescue with some of the new drawings owing to my present arthritic disabilities. The photography on the plates, except for one fossil, is the work of Dr. J. P. Harding and all are of living animals doing something of evolutionary significance to them. Many hundreds of fine photographs have been prepared in order to make the selection shown on the eight plates. Finally my thanks are due for the kind hospitality I have received over many years at the British Museum (Natural History), for the cooperation of the staff of the Zoological Department and of the libraries and for the access I have had to the collections of preserved material.

London S. M. M.
May 1975

Acknowledgements

Thanks are due to the following authors and publishers who have given permission for the use of illustrations:

Professor D. T. Anderson for Fig. 1.15(g); Dr. J. Bartlet for Fig. 9.10; Dr. J. Gordon Blower for Fig. 7.14; Dr. L. A. Borrodaile for Fig. 4.9; Professor F. Brioli Figs. 1.8, 6.6; Dr. W. T. Calman for Fig. 4.8(a); Professor H. Graham Cannon for Figs. 2.9–11, 3.7, 3.9(a), 4.8(c), 4.13–17; Professor J. L. Cisne for Figs. 1.3–5, 4.2; Dr. R. J. Daniel for Fig. 4.20; Dr. M. A. Edwards for Fig. 6.11(b); Dr. M. E. G. Evans for Figs. 7.14, 7.15; Dr. J. W. Folsom for Fig. 6.8(e)–(g); Professor N. M. Hanken for Fig. 1.12; Dr. J. P. Harding for Pls. 1(b)–(f), 2, 3(a)–(e), (g)–(i), 4–8; Professor J. W. Hedgpeth for Figs. 1.16(c), 10.9(e); Professor R. R. Hessler for Fig. 3.1(b), (f); Professor C. A. Horridge for Fig. 6.11; Dr. C. P. Hughes for Fig. 1.7; Professor G. E. Hutchinson for Fig. 1.17(a); Dr. I. K. Ingham for Figs. 1.18, 5.19; Dr. E. Marcus for Figs. 1.19, 6.15; Dr. C. Mettam for Fig. 4.21(d)–(e); Dr. E. Morgan for Fig. 10.9(a)–(d); Professor C. F. A. Pantin for Fig. 4.1; Miss J. C. Perryman for Fig. 4.19; Dr. H. Prell for Fig. 5.12(b), (c); Professor J. W. S. Pringle for Figs. 9.15, 9.17(c)–(d); Professor G. Ramazotti for Figs. 1.19, 6.15; Professor P. E. Raymond for Fig. 1.14(a); Drs. R. and E. Richter for Fig. 1.14(b); Dr. W. D. I. Rolfe for Figs. 1.18, 5.19; Dr. P. Ruck for Fig. 6.11(b); Dr. H. L. Sanders for Figs. 1.15(a)–(e), 2.7, 4.6–7, 4.8(b); Professor G. O. Sars for Figs 1.16(a), 3.1(g); Mr. D. J. Scourfield for Fig. 6.7; Dr. A. Serfaty for Fig. 1.13(b); Dr. A. G. Sharov for Fig. 1.17(b); Dr. R. E. Snodgrass for Fig. 9.15(a); Professor L. Størmer for Figs. 1.9–12, 1.14(c)–(d), 4.3(c), 6.2, 10.6; Professor W. Sturmer for Pl. 1 (a); Sir James Stubblefield for Fig. 1.2(g)–(k); Professor O. W. Tiegs for Figs. 9.16, 9.17(c)–(d); Professor M. Vachon for Fig. 1.13; Professor H. Weber for Fig. 9.17(a)–(b); Professor H. B. Whittington for Figs. 1.1, 1.2(a)–(f), 1.6, 4.3(a)–(b), 6.16; *Annals and Magazine of Natural History* for

Pl. 8(d); *Biological Reviews* for Figs. 1.8–11, 1.14, 1.15(f), 1.16–17, 4.3(c), 6.2, 6.6–7; *Bulletin of the Geological Survey of Canada* for Fig. 1.6; *Bulletin du Muséum national d'histoire naturelle* for Fig. 1.13(b); *Cambridge Monographs on Experimental Biology* for Figs. 9.15, 9.17; *Endeavour* for Fig. 1.13(a); *Fossils and Strata* for Figs. 1.1, 1.2(a)–(b), 1.3–5, 1.7, 1.12, 4.2, 4.3(a)–(b); *Journal of Experimental Biology* for Fig. 10.9 (d); *Journal of the Linnean Society (Zoology)* for Figs. 1.2(g)–(k), 3.22, 4.9, 4.27–9, 5.1–14, 5.16–18, 6.13, 7.1, 7.4–7, 7.11, 7.16, 8.1(a)–(b), 8.2–17, A.1, Pls. 1(f), 4(a)–(n), 5(a)–(c), 6(a), (c)–(i), (j)–(m); 7(c)–(i); *Journal of Morphology* for Fig. 6.11(b); *Journal of Natural History* for Figs. 4.21(d)–(e), 4.25–6; *Journal of Zoology, London* for Figs. 2.1, 4.21(d)–(e), 4.24, 6.12, 6.14, 7.15; *Memoirs of the Connecticut Academy of Arts and Science* for Figs. 1.15(a)–(c), 2.7, 3.1(b), 4.6–7; *Memorie delgi Instituto Italiano Idrobiologia* for Figs. 1.9, 6.15; *Nature* for Fig. 7.14; *Philosophical Transactions of the Royal Society*, Series B, for Figs. 2.6, 2.10, 2.12, 3.1(c)–(e), 3.2, 3.4–6, 3.11–14, 3.15(a)–(c), 3.16–21, 3.23–31, 4.14–16, 6.4, 6.5, 6.8–9, 6.16; *Proceedings of the Linnean Society of London* for Fig. 4.1; *Proceedings of the Zoological Society of London* for Pl. 3(f); *Quarterly Journal of Microscopical Science* for Fig. 8.1(c); Liverpool University Press for Fig. 4.20, from *Lancashire Sea Fisheries Laboratories Reports*; *Scottish Journal of Geology* for Fig. 1.18; *Senckenbergiana, Lethaia* for Fig. 1.14(d); *Transactions of the Royal Society of Edinburgh* for Figs. 2.9, 3.9(b), 3.10, 4.13, 4.17; the Geological Society of America for Figs. 1.2(c)–(f), 2.2, 3.3, 4.5, 4.8, from the *Treatise on Invertebrate Paleontology*; *Zoological Journal of the Linnean Society* for Figs. 5.15, 7.2–3, 7.8–10, 7.12–13, 9.1–14, 10.1–5, 10.7–8, Pls. 1(b)–(e), 7(a)–(b), 8(a)–(c), (e)–(j); *Zoologica, Stuttgart* for Fig. 5.12(b), (c); the Pergamon Press for Fig. 1.15(g), from *Embryology and phylogeny in annelids*.

Contents

	LIST OF PLATES	xix
1.	**INTRODUCTION**	**1**
	1. A review of the Arthropoda, living and extinct	1
	A. Trilobita	2
	B. Some early non-trilobites	8
	C. Chelicerata	12
	(i) Merostomata, Xiphosura	12
	(ii) Merostomata, Eurypterida	13
	(iii) Arachnida	15
	D. Early merostome-like animals	18
	E. Crustacea	18
	F. Pycnogonida	22
	G. Uniramia (Onychophora, Myriapoda, Hexapoda)	24
	H. Tardigrada	29
	2. What this book is about	30
	A. Functional morphology and habits	30
	B. Types of evolutionary change	33
	(i) Adaptive radiation	33
	(ii) Advances	33
	C. The significance of animal shapes and the functional approach	34
2.	**HABIT DIVERGENCIES AND ARTHROPODAN FOOD-GATHERING IN THE PALAEOZOIC, AND SOME BASIC LEG MOVEMENTS**	**37**
	1. Introduction	37
	2. Basic limb movements employed in walking	39
	A. The promotor–remotor swing	39
	B. The adductor–abductor (or levator–depressor) movements	42
	C. Leg movements on and off the ground	42
	D. The integration of movements of successive legs, metachronal rhythm, used for walking and burrowing	44
	3. The collection of food and its transport forwards towards the mouth	49

A. Food transport by mechanical means in Crustacea and in *Limulus* — 49
B. Food transport by mechanical means in trilobites and other early arthropods — 51
4. Hydraulic systems formed by the limbs of aquatic arthropods — 53
   A. The transfer of food forwards towards the mouth by currents in Crustacea — 57
5. The transport of food from below upwards towards the mouth in the Uniramia (Onychophora, Myriapoda, Hexapoda) — 59

3. THE EVOLUTION OF ARTHROPODAN JAWS — 63
1. Introduction — 63
2. Crustacean mandibles — 66
   A. Mandibles serving simple raptatory and sometimes filter-feeding in Copepoda — 68
   B. Basic form of crustacean mandibles used for primitive grinding movements — 69
      (i) Promotor and remotor muscles — 72
      (ii) Muscles to the transverse mandibular tendon — 72
      (iii) Posterior adductor muscles — 73
   C. Crustacean mandibles used for cutting large food — 73
      (i) Cutting incisor processes — 74
      (ii) Large, hard food — 77
      (iii) Transverse cutting of large food by the mandibles of Isopoda — 77
      (iv) Transverse biting by the mandibles of the crayfish and crab — 80
      (v) Mandibles used for large-food feeding in certain Peracarida and Syncarida — 83
   D. Conclusions on the jaw mechanisms of Crustacea — 87
   E. Endoskeleton supporting mandibular movements in Crustacea — 87
      (i) Basement membranes — 87
      (ii) Transverse mandibular tendons — 87
      (iii) Apodemes — 89
3. Gnathobases of Chelicerata and Trilobita — 90
   A. Xiphosura — 92
   B. Arachnida — 94
   C. Trilobita — 95
4. Conclusions concerning the jaws of the primarily aquatic arthropods — 96
5. Jaws and mandibles of the Uniramia (Onychophora, Myriapoda, Hexapoda) — 97
   A. Basic form of uniramian mandibles and the jaws of the Onychophora — 97
      (i) Preoral cavity — 101
   B. Mandibles of the Myriapoda — 103
      (i) Diplopoda — 103

(ii) Symphyla	108
(iii) Chilopoda	111
(iv) Conclusions on the jaw mechanisms of Myriapoda	113
c. Mandibles of the Hexapoda	114
(i) The tentorial apodemes	114
(ii) The basic movements and musculature of hexapod mandibles	115
(iii) Mandibular skeleto-musculature of Thysanura and Pterygota	116
(i) Protrusible mandibles suiting fine-food feeding in hexapods	120
(v) Collembola	122
(vi) Diplura	127
(vii) Conclusions on the jaw mechanisms of hexapods	132

4. THE EVOLUTION OF ARTHROPODAN TYPES OF LIMBS AND THE EFFECTS OF HABITS ON THE LIMBS OF *LIMULUS* AND ON EVOLUTION WITHIN THE CRUSTACEA ... 135

1. Introduction	135
2. The biramous limbs of trilobites, of other fossil arthropods, and of *Limulus*	137
3. The biramous limbs of Crustacea	142
A. Basic plan	142
(i) Exites	144
(ii) Endites	144
(iii) Exopods	144
(iv) Endopods	146
(v) Stenopodia and Phylopodia	147
B. Serial differentiation of crustacean limbs	148
c. Changes in crustacean limbs during ontogeny	153
D. Special habits of Crustacea; the correlated limb structure; and crustacean evaluation	153
(i) The raptatory habit and its derivatives	155
(ii) Filter and suspension feeding in: barnacles, copepods, branchiopods, cephalocarids, malacostracans, leptostracans, euphausids, and ostracods	156
(iii) Escape jumping through the water in copepods and shrimps	169
(iv) Conclusions concerning crustacean limbs and habits	176
4. The limbs of annelids	176
5. The uniramous lobopodial limbs and respiratory system of the Onychophora	182
6. The basic construction of the uniramous trunk limbs of myriapods and hexapods	191
A. Hinge joints	192
B. Pivot joints	192
c. Extrinsic leg muscles	194

	D. Intrinsic leg muscles	194
	E. Derivation of a simple walking leg	196

5. SPECIAL REQUIREMENTS OF LAND ARTHROPODS: STEPPING; LEG-ROCKING; AND JOINTS IN THE UNIRAMIA — 199

1. Introduction — 199
2. Hanging stance and leg length — 200
3. Fields of movement, leg numbers, and the evolution of hexapody — 201
4. Leg structure and stepping suited to produce (i) strong movements and (ii) the mutually exclusive fast movements — 204
   A. Legs suiting strong, slow movements — 204
   B. Legs suiting fast movements — 204
   C. Muscle numbers — 205
   D. Sliding sternites — 206
   E. Muscle leverage — 209
   F. Positions of axis of swing — 209
5. Leg-rocking in myriapods and in hexapods — 212
   A. Myriapod methods of leg-rocking — 212
   B. Hexapod methods of leg-rocking — 215
      (i) The Diplura — 215
      (ii) The Protura — 216
      (iii) The Collembola — 218
      (iv) The Pterygota — 220
      (v) The Thysanura — 222
   C. Conclusions concerning the coxa–body joint and the leg-rocking mechanism of myriapods and hexapods — 224
6. Particular leg joints of certain myriapods and hexapods — 225
   A. Joints giving levator–depressor movements — 228
   B. Joints providing active flexor movements of the leg — 228
   C. Skeleto-musculature associated with pushing by the dorsal surface in the Diplopoda — 229
   D. Synovial cavities in myriapodan joints — 230
7. The legs of Carboniferous and modern Myriapoda — 233

6. THE BROAD SUBDIVISIONS OF THE ARTHROPODA — 236

1. Introduction — 236
2. Similarities and differences between arthropdan taxa of phylogenetic significance — 238
   A. Body regions, or tagmata, of arthropods — 238
      (i) The head, cephalon, and prosoma — 238
      (ii) The trunk; including thorax; abdomen; pygidium; opisthosoma; mesosoma and metasoma — 240
   B. Types of arthropodan limbs — 241
      (i) The limbs of Trilobita and of Crustacea — 241
      (ii) The limbs of Chelicerata; Merostomata; Arachnida; and early merostome-like arthropods — 242

	(iii) The limbs of the Pycnogonida	244
	(iv) The limbs of the Uniramia	244
c.	The cephalon, prosoma, and the heads of arthropods	244
	(i) Types of anterior end of the body	245
	(ii) Ontogenetic development of a head	247
	(iii) Definitions of a segment; the acron; and the telson	251
	(iv) Further composition of arthropodan heads	252
d.	The jaws of arthropods	253
	(i) Types of jaws	254
	(ii) Jaws of the Uniramia	254
	(iii) Trilobite gnathobases	256
e.	Conclusions concerning the relationships of the larger groups of Arthropoda	257
	(i) Chelicerata, Crustacea, Uniramia, and embryology	257
	(ii) Trilobite internal structure	258
	(iii) Trilobite and crustacean modes of development and the relationships between these two taxa	260
	(iv) The status of the Trilobita	261
	(v) The status of the Pycnogonida	261
	(vi) Crustacean evolution	266
3.	Convergent similarities among arthropods	270
a.	Transversely biting mandibles	270
b.	Entognathy	270
c.	Organs of sight	273
4.	Grades of uniramian evolution: Monognatha, Dignatha, and Trignatha, and their phylogenetic groupings	279
5.	The evolution of the Uniramia: Onychophora, Myriapoda, and Hexapoda	281
a.	Heads and limbs	281
c.	Series of gaits of the Uniramia	282
d.	Differentiation of habits and uniramian evolution	284
6.	The haemocoel	287
7.	Other metamerically segmented invertebrates	288
a.	The Tardigrada	288
b.	*Opabinia regalis* Walcott	289
8.	The origin of the Arthropoda	291

7. LOCOMOTORY MECHANISMS — 293

1.	General features of locomotory mechanisms	293
2.	The uses of trunk musculature	295
3.	Speeds of progression	298
a.	The frequency of stepping	298
b.	The relative duration of the forward and backward strokes	298
c.	The angle of swing of the leg	302
d.	Leg length	304
4.	The phase differences between successive and paired legs	309
a.	Metachronal waves and phase differences between successive legs	310

B. The phase differences between paired legs — 312
C. The effects of the phase differences between successive and paired legs on leg disposition — 313
5. The determinants of segment numbers; the order of footfalls; and the loading on the legs — 314
6. The relationships between the gaits used by different Uniramia for walking and running — 319
  A. The advantages of the retention of a lobopodium; of narrow, pointed limb-tips; and of leg-rocking — 319
    (i) The lobopodium — 319
    (ii) The narrow, pointed tarsus — 319
    (iii) Leg-rocking — 320
  B. The gaits of Onychophora — 320
  C. The gaits of Myriapoda and Hexapoda — 320
    (i) The gaits of Symphyla and of epimorphic Chilopoda — 321
    (ii) The gaits of anamorphic Chilopoda — 325
    (iii) Divergence in feeding habits of epimorphic and anamorphic Chilopoda as determinants of choice of gaits — 325
    (iv) The gaits of Pauropoda — 327
    (v) The gaits of Diplopoda — 327
    (vi) The effects of body size on locomotory forces — 328
    (vii) Conclusions concerning gaits and facilitating morphology of the myriapod classes — 330
    (viii) The gaits of Hexapoda — 330
7. Jumping in Myriapoda and Hexapoda — 331
  A. Jumping gaits of Thysanura — 332
  B. Jumping millipedes — 335
  C. High jumping and escape reactions of hexapods — 336
    (i) High jumping by Machilidae — 337
    (ii) High jumping by Collembola — 338
8. Neural co-ordination of arthropod movement — 339

8. HABITS AND THE EVOLUTION OF THE MYRIAPODA TOGETHER WITH THE BASIC SKELETO-MUSCULATURE OF THE UNIRAMIA — 344

1. Introduction — 344
2. Basic form of the skeleto-musculature — 345
  A. The skeleton — 345
  B. The principal myriapodan divergencies in exoskeleton — 348
  C. Muscular systems and their functions — 348
    (i) Superficial muscles — 348
    (ii) Longitudinal muscles — 349
      Dorsal longitudinal muscles — 349
      Lateral longitudinal muscles — 349
      Sternal longitudinal muscles — 350
    (iii) Deep oblique muscles — 350
    (iv) Deep dorso-ventral muscles — 350
    (v) Conclusions concerning trunk musculature — 351

3. The Diplopoda	351
A. Diplopodan burrowing techniques	352
(i) The iuliform type of head on pushing and the evolution of diplosegments	352
(ii) The polydesmoidean type of dorsal-surface or 'flat-back' pushing	356
(iii) The nematophoran and colobognathan type of wedge-pushing	360
B. Diplopodan spiralling techniques	364
C. Secondary diplopodan achievements	368
(i) The Lysiopetaloidea	368
(ii) The Pselaphognatha	370
D. Conclusions concerning the Diplopoda	372
4. The Symphyla	372
A. Extra tergites of Symphyla	373
B. Intercalary tergites of Symphyla	374
C. Musculature and other features of Symphyla	374
D. Conclusions concerning Symphyla	376
5. The Chilopoda	376
A. Structural features correlated with basic habits common to all Chilopoda	377
B. Structural features of the Geophilomorpha associated with burrowing	377
C. Facilitation of fast running in Chilopoda by cuticular and muscular features	380
(i) Tergite heteronomy in Scolopendromorpha and Anamorpha	381
(ii) Musculature and heteronomy	382
D. A secondary chilopodan achievement: *Craterostigmus*	387
E. Conclusions concerning Chilopoda	394
6. The Pauropoda	396
7. General conclusions on myriapodan evolution	396

9. HABITS AND THE EVOLUTION OF THE HEXAPOD CLASSES — 398

1. Introduction	398
2. The Collembola	401
A. The collembolan jump	402
(i) Collembolan musculature associated with jumping	403
(ii) Exoskeletal and endoskeletal features associated with the collembolan jumping mechanism	408
(iii) Features of collembolan legs associated with running and abdominal jumping	410
D. Conclusions concerning Collembola	414
3. The Diplura	415
A. Structural facilitations to crevice-living in Japygidae	416
B. Structural facilitations to crevice-living in Campodeidae	420
C. Conclusions concerning Diplura	421

    4. The Protura   422
    5. The Thysanura   423
        A. Structural features associated with the jumping gaits of Thysanura   423
            (i) Coxal movements and associated structure in Thysanura   424
            (ii) Telopod movements and associated structure in Thysanura   432
        B. The mechanism of the high escape jumps of *Petrobius*   433
        C. Conclusions concerning the Thysanura   434
    6. The Pterygota   438
        A. The origin and evolution of the Pterygota   438
            (i) The origin of flight mechanisms   439
            (ii) Thoracic musculature   440
            (iii) Wing movements and musculature   441
            (iv) Coxa–body articulation and wing evolution   445
    7. Conclusions concerning the hexapod classes   447

10. LOCOMOTORY HABITS AND THE EVOLUTION OF THE CHELICERATA AND PYCNOGONIDA   449
    1. Introduction   449
    2. The Merostomata   449
    3. The invasion of the land   450
    4. The Arachnida   451
    5. Arachnid locomotory mechanisms associated with terrestrial life   452
        A. Arachnid coxae   453
        B. Arachnid leg positions   453
        C. Arachnid leg movements: the promotor–remotor swing: leg-rocking: and levator–depressor movements   458
        D. Arachnid leg mechanisms   461
            (i) Stepping movements and leg mechanism of scorpions   461
            (ii) Leg mechanism of spiders   464
            (iii) Leg mechanism of an amblypygid (Pedipalpi)   467
            (iv) Leg mechanism of Solifugae   468
        E. Conclusions concerning leg mechanisms of Arachnida   469
    6. Arachnid gaits   472
    7. Hexapodous stepping in Arachnida   477
    8. Hydrostatic pressure and arachnid jumping   479
    9. The leg mechanism of Pycnogonida   479
    10. Conclusions concerning locomotory habits and the evolution of Arachnida and Pycnogonida   483

11. A FURTHER WORD ON POLYPHYLY   487

CLASSIFICATION OF THE ARTHROPODS   495

TECHNICAL APPENDIX   501

REFERENCES   508

INDEX   515

List of plates
(*between* pp. 466–7)

PLATE 1. (a) *Cheloniellon calmani*, Broili, X-ray; (b) *Hermonia hystrix* (Savigny), walking; (c) *Neanthes fucata* (Savigny), crawling; (d, e) *Alentia gelatinosa* (Sars), walking; (f) *Scolopendra morsitans*, running.

PLATE 2. (a) *Triops cancriformis*, swimming; (b) *Sida crystallina*, swimming; (c) *Balanus* sp., 'fishing'.

PLATE 3. (a) *Cypria ophthalmica;* (b) *Cyclops vicinus;* (c) *Cyclops viridis;* (d) *Armadillidium* sp. enrolled and walking; (e) *Glomeris marginata*, enrolled and walking; (f) *Anaspides tasmaniae*, progressing slowly; (g) *Phytisica marina*, holding on to a hydroid stem; (h) *Macropipus puber*, swimming; (i) *Carcinus maenas*, running.

PLATE 4. (a, b) *Peripatopsis moseleyi*, walking and resting; (c) *Peripatopis novaezealandiae;* (d, e) *Scutigerella immaculata*, crawling and walking; (f) *Callipus longobardius*, walking; (g) *Polyzonium germanicum*, walking; (h) *Craspedosoma rawlinsi*, harnessed to a sledge; (i) *Polydesmus angustus*, running; (j, k) *Scutigera coleoptrata*, running and resting; (l, m) *Pauropus gracilis*, first instar and adult walking; (n) *Craterostigmus tasmanianus*, first instar, starting to walk; (o) *Blaniulus guttulatus*, enrolled.

PLATE 5. (a, b) *Haplophilus subterraneus*, extended and contracted; (c) large geophilomorph centipede, progressing slowly; (d) *Cryptops hortensis*, running; (e) *Lithobius forficatus*, running; (f) *Polydesmus angustus*, running; (g) *Lithobius variegatus*, running; (h) *Lithobius forficatus*, second instar; (i) slit passable to specimen in (a) and (b).

PLATE 6. (a) *Tomocerus longicornis;* (b) T.S. of spines of *Polyxenus lagarus;* (c) *Neanura muscorum;* (d) *Tomocerus longicornis*, walking; (e) *Campodea staphylina*, running; (f) *Petrobius brevistylis*, running; (g) *Thermobia domestica*, running; (h) dipluran japygid sp.; (i) *Cylindronotus laevioctostriatus*, walking; (j) *Polyxenus lagarus;* (k) *Acerentomon nemorale*, walking; (l) *Polyxenus lagurus*, walking; (m) *Petrobius brevistylis*, progressing along a narrow ledge; (n) *Glomeris marginata*, walking.

PLATE 7. (a) *Euscorpius carpathicus*, walking; (b) *Buthus australis*, walking; (c, d) *Ophistreptus guineensis*, pushing and running; (e) iuliform diplopod, slow gait; (f) *Siphonophora hartii*, walking; (g) *Dolistenus savii*, walking; (h) *Polydesmus angustus*, first instar; (i) *Craterostigmus tasmanianus*, running slowly.

PLATE 8. (a, b) Amblypygid sp; (c) *Galeodes arabs*, running; (d) *Trochosa ruricola*, running; (e) *Allothrombium fulginosum*, walking; (f) *Ixodes hexagonus*, walking; (g, h) *Opiliones* sp., running and walking; (i, j) *Chelifer* sp., running; (k) *Anisopus* moving by undulations.

THE EVOLUTION OF ANIMAL LIFE: SOME OUTSTANDING EVENTS AND EPISODES

Approximate time durations

Period	Events	Age	Era	Duration
PLEISTOCENE & RECENT	Rise of modern man.	Age of Man	QUATERNARY	2 million years
PLIOCENE	Man-apes in Africa at the close of the period. Waning of larger mammals.			5 million years
MIOCENE	Acme of mammals. Unspecialized apes abundant in Africa and in the tropical forests of the Old World. Supposed man-apes in India and Africa.		TERTIARY	19 million years
OLIGOCENE	Rise of modern mammals.	Age of Mammals		12 million years
EOCENE & PALAEOCENE	Archaic mammals Rise of snakes. Rise of new large Foraminifera. Rise of modern lamellibranchs and gastropods.			27·5 million years
CRETACEOUS	At the close of the period extinction of many large reptiles (dinosaurs, marine reptiles, pterosaurs). Last ammonites and almost last belemnites. Extreme specialization of the reptiles on land, in the sea, and in the air. First true birds. Mammals small, shrew-like, and probably nocturnal.	Age of reptiles	MESOZOIC	71 million years
JURASSIC	Ammonites and belemnites abundant. Spread of the reptiles to land, sea, and sky. Dinosaurs abundant. Toothed birds evolve. Small primitive mammals in existence.			57 million years
TRIASSIC	Acme of ammonites. Belemnites and first scleractiniid corals. First dinosaurs and large marine reptiles. First mammals (small and primitive).			32 million years
PERMIAN	Life in shallow seas reduced in numbers and species. Last trilobites. Later rise and spread of ceratitic ammonites and reptiles.			55 million years
CARBONIFEROUS	Spread of amphibians and ancient sharks. Rise of reptiles before the close of the period. Insects rapidly evolving and abundant. In Britain non-marine lamellibranchs abundant in 'deltaic' facies, corals, brachiopods, crinoids, and goniatites being abundant in marine facies. First true amphibians.	Age of amphibians		65 million years
DEVONIAN	Amphibians evolving from air-breathing fishes at the end of the period. Bony fishes, including air-breathing forms, very abundant, notably in fresh water. Goniatites evolve and spread. First known insects and spiders. Last graptolites.	Age of fishes		50 million years
SILURIAN	Armoured jawless fishes very abundant towards the end of the period. Acme of eurypterids. Incoming of scorpions. Waning of graptolites. Rich coral-reef faunas locally in Britain with crinoids and brachiopods abundant.		PALAEOZOIC	40 million years
ORDOVICIAN	Acme of graptolites. Appearance and increase of rugose and tabulate corals. Brachiopods and cystids abundant. Spread of molluscs, notably straight cephalopods. First armoured jawless fishes.			65 million years
CAMBRIAN	First graptolites in late Cambrian times. Dominance of trilobites. Brachiopods and molluscs primitive and thin-shelled. First corals and crinoid-like cystids. Sudden appearance of representatives of nearly all the invertebrate phyla.	Age of invertebrates		70 million years
PRE-CAMBRIAN	Traces of life rare, except locally in very late Pre-Cambrian times when exceptional conditions have very occasionally permitted the preservation of soft-bodied animals. For instance, in the Ediacara Hills area of South Australia there have been found supposed representatives of annelids, possible fore-runners of the trilobites, 'worms', and medusoid and colonial coelenterates ('jelly-fish' and 'sea pens' respectively). Possible 'sea pens' have also been found in Charnwood Forest, Leicestershire (*Charnia*), and in South Africa. Sponge spicules have been claimed present in the Pre-Cambrian, and there are markings which may be interpreted as worm burrows and tracks.			

From R. M. C. Eagar, Manchester Museum 1968.

1 Introduction

1. A review of the Arthropoda, living and extinct

THERE are many invertebrate phyla whose bodies are covered by cuticle, but the Arthropoda exploit the use of cuticle to the full. Often it is thick and stiff, forming sclerites or surface armour of various kinds, but this is not always so; the cuticle may be thick and flexible. Serially repeated joints are present on body and legs where the cuticle is stiff. There are endless elaborations of the cuticle in the form of articulations and other skeletal features which are just as striking as is the form of the vertebrate skeleton.

Cuvier called these animals the Articulata, but their name has been changed several times and the term 'Arthropoda' was adopted by von Siebold in 1848. The view that the Arthropoda are monophyletic was not seriously questioned until 1958 (Tiegs and Manton). The Arthropoda do not appear to have originated from one primeval group and certainly not from one known annelid-like ancestor. There is no evidence that the modern annelid parapodium and arthropodan limbs are homologous. Evidence will be presented below to substantiate the view that the arthropods are polyphyletic, the various groups being evolved from several different, as yet unknown, ancestors.

In number of species the Arthropoda exceeds the whole of the rest of the animal kingdom put together. In size they range from mites, a fraction of a millimetre in length, to a Japanese crab *Macrocheira* with a leg span of 3·5 m. Their habitats extend from the cold darkness of the deepest oceans to the even colder mountains and to the Antarctic continent, as well as to all more congenial environments. Yet we know nothing of the origin of the primarily aquatic arthropods which flourished long before there was any land fauna.

The earliest rocks bearing recognizable fossil remains of animals provide evidence of abundant and diversified marine faunas of armoured arthropods with jointed bodies and jointed, often long, legs. They lived in many ways: on the

Introduction

bottom, swimming, and burrowing. Habits of shallow burrowing probably gave some protection from predators. The trilobites (Figs. 1.1–5) dominated the Palaeozoic seas but became extinct at least 300 million years ago. Most of the great group called the Merostomata disappeared also; only *Limulus* (Fig. 1.9) and its related genera, representing the Xiphosura, persist to the present day. The Eurypterida, which form the other component group of the Merostomata, were wholly Palaeozoic, reaching their prime from the Ordovician to the Devonian (Figs. 1.10–12). Some of them attained gigantic size for an arthropod, being 1·0 to 1·8 m in length. Others of smaller size had a scorpion-like body. The later eurypterids appear to have been inhabitants of brackish or fresh water. The fossil scorpion-like eurypterids (Figs. 1.11, 1.12) give no precise evidence of the origin of terrestrial arachnids but are suggestive of such an origin (Fig. 1.14(d)).

A. *Trilobita*

The trilobites comprise many of the earliest known fossil arthropods, besides being the most abundant in their time. They persisted from the Cambrian to the Permian, a span of at least 250 million years. Some 2000 different species have been recognized, but only recently has it been possible to reconstruct their bodies at all fully because most fossils reveal only the hard, sclerotized dorsal surface. By cutting the rock away from the fossils (Whittington 1974, and others), by cutting serial sections of a trilobite in the rock and polishing the surface of each slice (Størmer 1939), and from X-ray photographs of well-preserved fossils in suitable rocks (Cisne 1975; Størmer 1974) we at last know both the ventral as well as the dorsal aspect, the number of legs on the cephalon, the form of the legs all along the body, quite a lot about the musculature, the endoskeletal and digestive systems; information which was never expected from a fossil has been obtained. Some of Cisne's reconstructions of *Triarthrus* and those of Whittington for *Olenoides* are reproduced in Figs. 1.1–1.5.

The trilobed appearance of the dorsal face of a trilobite is due to pleural expansions overhanging the legs laterally on either side. The carapace of *Limulus* (Fig. 3.12, lateral groove) is demarcated from the central region of the body on the fused prosomal segments, a shape that confers mechanical strength to the exoskeleton and prevents bending. The flattened cephalon of *Triarthrus*, uniting several segments,

Introduction

bore dorso-laterally a pair of compound eyes. The structure and modes of action of these eyes have recently received much study, in particular by Clarkson and Levi-Setti (1975), showing that about 500 million years ago the trilobites had met some of the optical needs for acute vision in water which were first appreciated in the seventeenth century by Des-

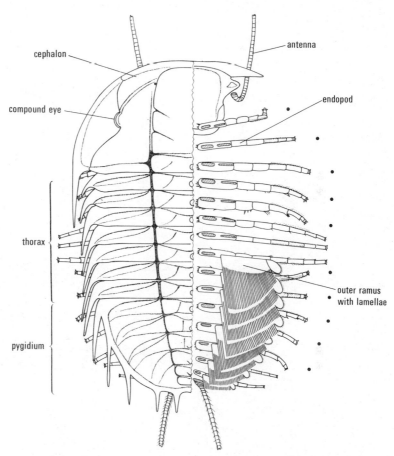

FIG. 1.1. The trilobite *Olenoides serratus* Meek, 7 cm, Middle Cambrian, reconstructed in dorsal view. On the right the dorsal exoskeleton and ventral trunk cuticle is removed to show the limbs and the inner view of the rostral plate and hypostome (ventral exoskeleton of the head). The outer ramus of limbs I–VII bearing the gill lamellae are removed and their areas of attachment to the coxa are diagonally ruled and lie to the right of the hatched areas indicating the attachment of the leg to the body. The biramous limbs are shown at one moment during the performance of a probable gait (see Fig. 2.5(a), (b)). The tips of the endopods opposite the black spots are on the ground, performing the propulsive backstroke, while the outstretched endopods are off the ground and swinging forwards. (After Whittington 1975.)

Introduction

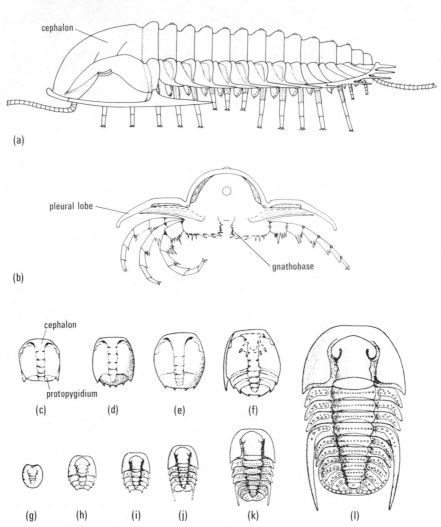

FIG. 1.2. The trilobite *Olenoides serratus* Meek. (a) Lateral view of the animal progressing by a probable gait. The ventral parts of the coxal are visible and the legs then rise below the pleural arch, becoming visible again as they descend to the ground in groups of five propulsive legs, the legs swinging forwards being off the ground and seen somewhat end on and foreshortened. (b) Transverse section of the body bearing an outstretched endopod on the right; on the left is shown the manner in which the leg could flex, bringing the tip towards the middle line. The outer ramus bearing gill lamellae is shown arising from the distal face of the coxa and projecting backwards, see Fig. 1.1. (After Whittington 1975.) (c)–(l) Larval stages of trilobites: (c)–(f) protaspis stages of *Sao hirsuta* Barrande, Middle Cambrian (after Whittington 1957); (g)–(k) meraspis stages of *Shumardia pusilla* (Sars); (l) holaspis stage of *S. pusilla*. (After Stubblefield 1926.)

Introduction

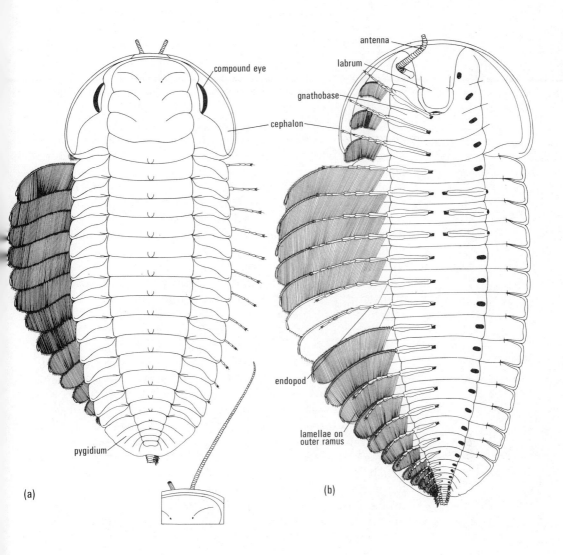

FIG. 1.3. The trilobite *Triarthrus eatoni* (Hall), 30 mm, Upper Ordovisian. (a) Dorsal, (b) ventral view. The endopods of the limb are shown entire on the left in (b) and projecting on the right in (a). The outer ramus of the limbs, bearing gill lamellae, are seen on the left in (a) and (b). The gnathobases from the coxae of all legs are shown on the left in (b) and the leg insertion, where legs are removed, are seen on the right in (b). Note the free trunk segments forming most of the body behind the head; the fused pygideal segments, five in number, at the hinder end. (After Cisne 1975.)

Introduction

cartes and Huygens. Present-day work is making progress in revealing the full functional significance of the structure of these eyes (see also Chapter 6 §3C). Ventrally a pair of antennae arose in front of the labrum or hypostoma and three pairs of cephalic postoral limbs followed. No other group of arthropods has a cephalon or a head of this composition. The mouth was backwardly directed and a narrow oesophagus passed forwards to a capacious crop occupying the anterior parts of the cephalic cavity (Fig. 1.5), much as in *Limulus* (Fig. 3.13). Behind the cephalon a series of body segments, articulated with one another, permitted dorso-ventral but no lateral bending of the body; each segment carried a pair of biramous legs. Posteriorly a short pygidium was formed of fused segments, each of which carried a pair of biramous limbs. In *Triarthrus*, according to Cisne (1975), but perhaps not in all trilobites, there was a small post-pygideal abdomen of limb-bearing segments but no pleural expansions and at the posterior end there was a formative zone from which the segments grew just in front of a minute telson bearing the anus (Fig. 1.3(b)). The outer branch of each limb carried a series of thin lamellae which were probably respiratory (see Chapter 4, Figs. 4.2, 4.3(c)), while the inner branch formed a walking endopod. From the coxa, or basal segment of the limb, a huge cusped gnathobase projected towards the mid-line ventrally (Figs. 1.1, 1.2(b), 1.3(b), 1.4(b), 1.5).

The trilobites showed an adaptive radiation of their own, but on the whole they were conservative in their basic organization. Bottom-living was probably a primary habit, with shallow burrowing in search of soft invertebrates, such as worms or molluscs, as an easy derivation. The head shield, which could be raised and lowered on the trunk, suited such a habit, the motive force coming from the endopods of the limbs. No lateral movements in the horizontal plane of the body were possible, but a nodding movement in the vertical plane enabled some trilobites to enroll, just as a woodlouse or

---

FIG. 1.4 (see opposite). The trilobite *Triarthrus eatoni* (Hall). (a) Reconstruction of the head end in dorsal view to show the musculature, the dorsal longitudinal muscles being shown on the right with a part cut away in thoracic segments 2 and 3. (b) Transverse section at the overlap of trunk segments 5 and 6. The lateral pleural lobe overhangs the coxa and its huge gnathobase; the endopod and the outer ramus are removed. The ventral and dorsal longitudinal and some other muscles are shown, together with one transverse endoskeletal bar. (After Cisne 1975.)

Introduction

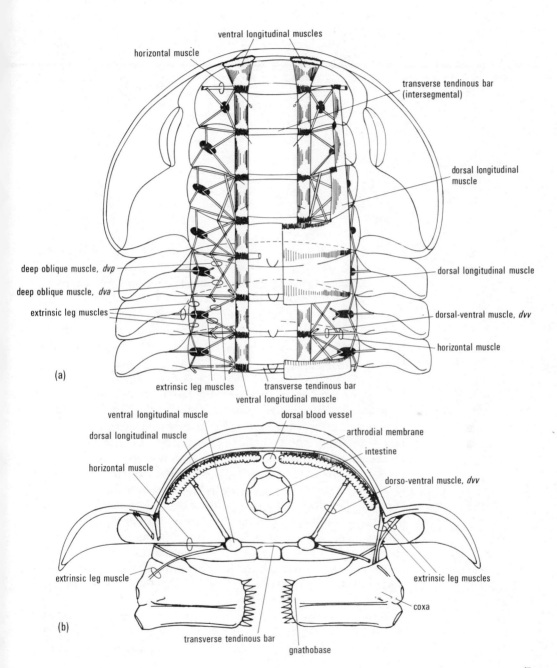

Introduction

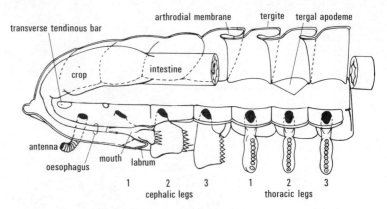

FIG. 1.5. The trilobite *Triarthrus eatoni* (Hall) reconstructed in a longitudinal view of the right half of the body. The head bears the paired antenna and three pairs of limbs whose gnathobases are a little backwardly directed (see Fig. 1.3(b)). The first gnathobase lies far lateral to the preoral cavity. (After Cisne 1975.)

a pill millipede does today (Pl. 3(d), (e)). Since the insertions of the legs on the body were relatively small (see Figs. 1.3(b) and 4.2, 4.3, cf. 3.12) and the muscles slender (Fig. 1.4), no great force could have been exerted by the limbs compared with modern arthropods. Some trilobites probably indulged in jumping along the bottom, as revealed by their tracks (Miller 1973), while others probably could swim (see also Bergström 1973).

B. *Some early non-trilobites*

There are many well-preserved fossil arthropods in the Burgess Shale of the Middle Cambrian whose anatomy differs not only from that of the trilobites but from all other large groups of known arthropods, living or extinct. *Marrella splendens*, so beautifully reconstructed by Whittington (1971), is an example; it is the commonest animal in the Burgess Shale (Fig. 1.6). *Marrella* shows a primitive level of cephalic development. It was blind and there were two pairs of long antennae, but no jaw or coxal endites on the serially uniform legs. A dorsal cephalic shield was present bearing four remarkable long processes. The manner of life of these animals is unknown but the long second antenna, with setose distal podomeres, or segments, might have been used to sweep up detritus of food value.

There are other equally well-preserved fossils, such as *Burgessia* and *Cheloniellon* (Figs. 1.7, 1.8) which do not easily fit into our larger taxa of arthropods. *Burgessia* possessed a

Introduction

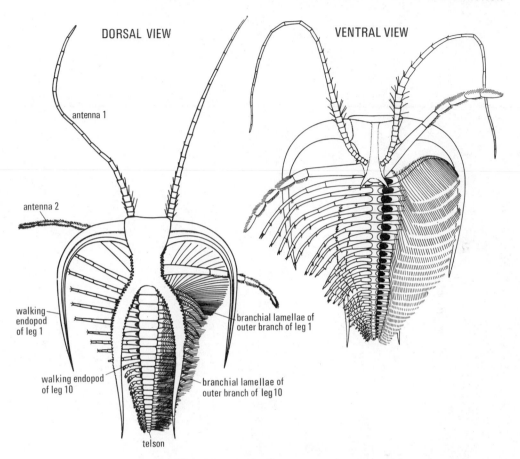

Fig. 1.6. *Marrella splendens* Walcott, 15 mm, Middle Cambrian. (a) Restoration in dorsal view, gill branches removed on the left side. (b) Restoration in ventral view, left walking endopods and right gill branches removed. The black areas represent the cut surface of the limb immediately distal to the insertion of the gill branch. (After Whittington 1971.)

flat, roughly circular carapace lacking eyes. Ventrally, one pair of preoral antennae was followed by three cephalic biramous legs, the inner branch being a walking endopod and the outer branch filiform. The first seven pairs of trunk limbs were also biramous, the outer branch being a single thin lamella. The last pair of legs was shorter, uniramous, and spine-like. There were no coxal endites or gnathobases. The absence of a labrum is remarkable, since trilobites and all major extant arthropodan groups possess this structure. A

Introduction

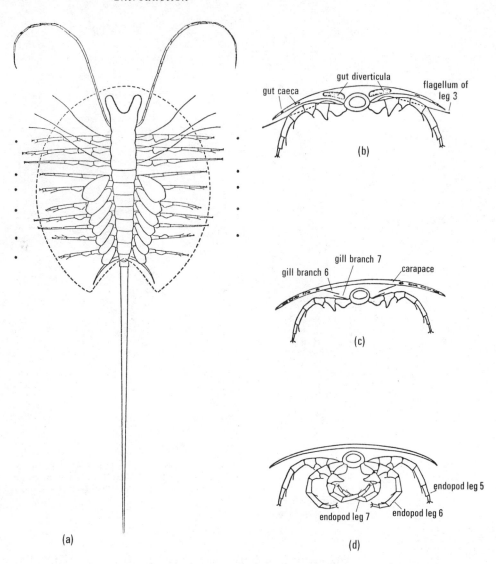

Fig. 1.7. *Burgessia bella* Walcott, Middle Cambrian, 10 mm. (a) Dorsal view of a reconstruction of the animal with the carapace removed but indicated by the dotted line; walking by a possible gait, propulsive limb-tips in contact with the ground marked by black spots alongside. The first three post-antennal limbs each carry a flagelliform outer ramus and are borne on the cephalon. The outer ramus of the following 7 trunk limbs is gill-like. (b) Transverse section through the head at the level of the 3rd biramus limb showing the origin of the flagellum and the carapace united with the head. (c) Transverse section at the origin of the 7th biramus limb showing the carapace free from the body and the origin of the 8th gill branch. (d) Transverse section showing superimposed walking endopods 5–7 in different degrees of ventral flexure. (After Hughes 1975.)

Introduction

long terminal spine of form similar to that of *Limulus* (Figs. 1.7, 1.9) where it arises during larval life, the early larva having no spine, Fig. 1.9(c). The spine may have been used in shallow grubbing, levering the anterior end down into the surface layers of the substratum.

We do not know where to place the well-preserved, but tantalizing, *Cheloniellon* from the Devonian (Fig. 1.8, Pl.

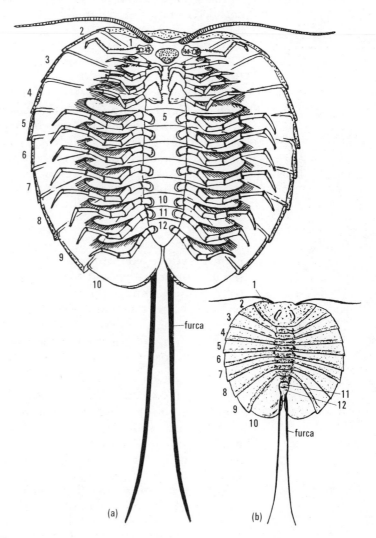

FIG. 1.8. *Cheloniellon calmani* Broili, 105 mm, Devonian. Reconstruction in (a) ventral and (b) dorsal view. (After Broili 1933, from Tiegs and Manton 1958.)

Introduction

1(a)), which possessed long antennules, possibly antennae, no mandibles, and serially similar biramous limbs, much as in trilobites, of which only four pairs behind the mouth carried gnathobases. The body was flat and the tergites laterally expanded to give a shield, much as in *Burgessia* but segmented, and the body terminated in a long furca. The similarity in shape of these two animals suggests that they perhaps had similar habits. Examples of convergence are illustrated by the enrolled isopod and diplopod in Pl. 3(d), (e) and by the pair of flattened swimming limbs of a eurypterid and a swimming crab, Fig. 1.12 and Pl. 3(h).

There are many other well-preserved arthropods in the Burgess Shale, but they do not help in ascertaining the origin of Crustacea or of Chelicerata. The real importance of fossils such as *Marrella* and *Cheloniellon* lies in the evidence they give of a primordial arthropodan fauna, with relics persisting even into the Devonian, which was characterized by serially uniform biramous limbs bearing a respiratory outer branch, by lack of jaws, and by the presence of a large labrum or upper lip. Within this fauna must lie the ancestors of the trilobites, the merostomes, and possibly the crustaceans.

C. *Chelicerata*

This great natural assemblage of animals comprises the mainly extinct Merostomata, with *Limulus* (Fig. 1.9) as a living representative and the modern Arachnida (Pls. 7(a), (b), 8), which comprise the dominant invertebrate carnivores on land. Many merostome-like fossils extended upwards in time from the Cambrian. The presence of pincer-like chelicerae in front of the mouth instead of antennae and the absence of mandibles characterizes these animals.

(i) *Merostomata, Xiphosura*. The 'horseshoe' or 'king-crab', *Limulus*, is no crab but a representative of the Merostomata, Xiphosura (Fig. 1.9). It possesses a carapace, covering the head and bearing a pair of dorsal eyes. Ventrally preoral chelicerae are followed by five pairs of postoral limbs, each of which is provided with a strong gnathobase (see Chapter 3 §3A, Figs. 3.12, 3.13, Chapter 4 §2, Fig. 4.4(a)). Hinged to this prosoma is another dorsal sclerite covering the mesosomal part of the opisthosoma which bears a ventral trough containing six pairs of flattened limbs, comprising the genital operculum and the five branchiate swimming limbs. The expanded outer lobe or branch of these limbs each carry

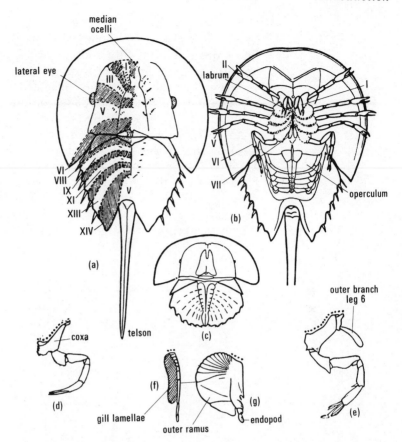

FIG. 1.9. *Limulus polyphemus* L. 450 mm. (a) Dorsal view. (b) Ventral view. (c) Larval stage. (d), (e) Fifth and sixth prosomal limbs respectively, coxa–body junction marked by dots. (f), (g) Mesosomal limb with respiratory gill lamellae, surface view in (g), longitudinal section through outer ramus in (f), coxa–body union marked by dots. (After Størmer 1944, from Tiegs and Manton 1958.)

100–150 gill-lamellae reminiscent of the respiratory lamellae on the outer rami of trilobite limbs. The body terminates in a long, strong spine.

Limulus and related species attain a size of 300 mm or more, excluding the spine. They are active burrowers and in some parts of the world *Limulus* denudes muddy bottoms of their clam population (see Chapter 3). Locally they are regarded as pests where clams are a commercial product, yet *Limulus* is found only in certain parts of the oceans of the world.

(ii) *Merostomata, Eurypterida*. These animals are wholly extinct (Figs. 1.10–12). They had their prime from the

Introduction

Ordovician to the Devonian and many were of spectacular size. The prosoma was narrower than in *Limulus* and the limbs projected laterally and showed considerable differentiation one from another. Individual segments of the mesosoma are evident. Here the coxae of five pairs of limbs are represented by flat coxal plates situated below the body proper and covering gills on their dorsal faces. The limbless metasoma was segmented and usually tapered to a spinous telson. One of the latest reconstructions of *Mixopterus* by Hanken and Størmer (1975) is shown making its tracks on the bottom as it walks (Fig. 1.12). The long limb of *Mixopterus*, corresponding with the pedipalp of a scorpion, is large, but even larger is the next limb, clearly prehensile in function, which corresponds with the simple first walking leg of a

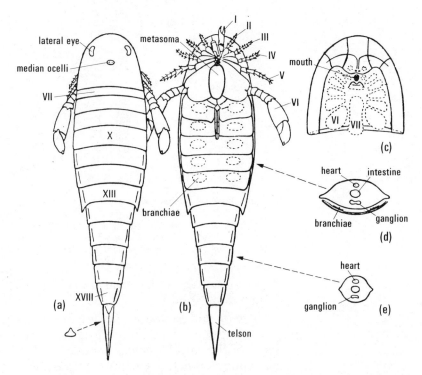

FIG. 1.10. A swimming eurypterid *Hughmilleria norvegica* (Kaier), 100 mm, Silurian. (a) Dorsal view. (b) Ventral view. (c) Ventral view of prosoma with limbs removed. (d) Transverse section of mesosoma. The position of the branchiae appears to have been on the sternal surface and not on the inner side of the covering coxal plates (Waterston 1975). (e) Transverse section of metasoma. (After Størmer 1944, in Tiegs and Manton 1958.)

scorpion or a spider. Eurypterids were also swimmers; compare the flattened 6th prosomal limb of *Hughmilleria* (Fig. 1.10) and the posterior thoracic legs of a swimming crab, Pl. 3(h).

(iii) *Arachnida*. The arachnids today comprise the dominant invertebrate carnivores on land. The preoral chelicerae are followed by a pair of pedipalps situated at the sides of the mouth; these limbs may be long and complex but usually bear a chewing gnathobase flanking a preoral cavity (Fig. 3.14). Four pairs of limbs, customarily labelled I–IV, follow and a united cuticular shield or carapace covers the anterior end or prosoma and prosomal limb bases. The posterior part of the body may be long and segmented (scorpion), or the segments

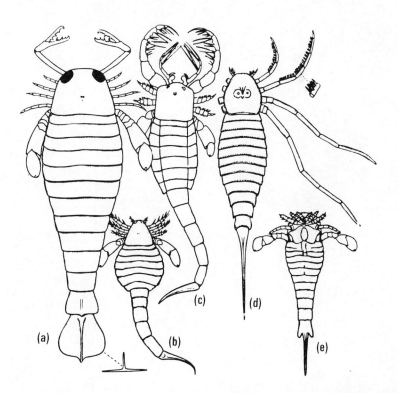

FIG. 1.11. Diversity of form in eurypterids. (a) *Jaekelopterus rhenaniae* (Jaekel), 1·8–1 m, Lower Devonian. (b) *Paracarcinosoma scorpionis* (Grote and Pitt), 1–0·4 m, Upper Silurian. (c) *Mixopterus kiaeri* Størmer, 1–0·67 m, Uppermost Silurian. (d) *Hallipterus excelsior* (Hall), 1·5–1 m, Devonian. (e) *Baltoeurypterus tetragonophthalmus* (Fischer), 1–0·24 m, Upper Silurian. (After Størmer 1944, in Tiegs and Manton 1958.)

Introduction

may be fused together (spider), and there may be a terminal spine or sting, as in some eurypterids (cf. scorpion and *Mixopterus*, Figs. 1.12, 1.13, Pls. 7, 8), the posterior part of the body being curved over the dorsal surface for stability. A

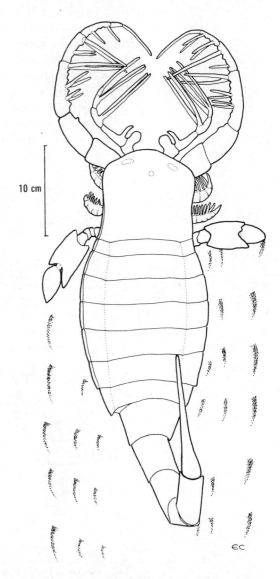

FIG. 1.12. *Mixopterus kiaeri* Størmer, 1–0·67 m, Uppermost Silurian. Drawing of a reconstructed model in a probable natural position walking on the tracks which have been fossilized. (After Størmer and Hanken 1975.)

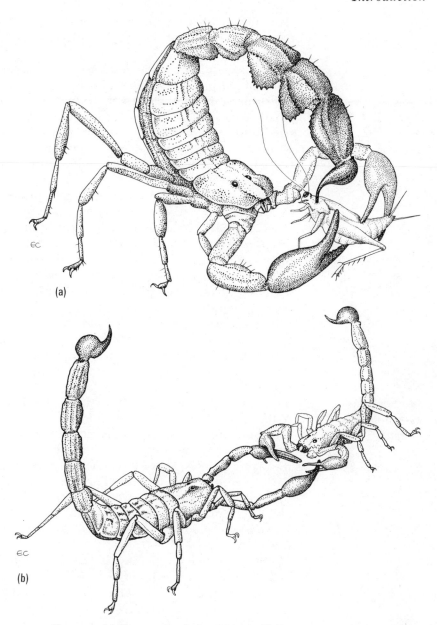

Fig. 1.13. (a) The scorpion *Androctonus australis* L., 95 mm, stinging its prey with the terminal sting and holding the prey with the pedipalpal chelae. The chelicerae overhang the mouth and walking legs I–IV arise on the prosoma composed dorsally of fused segments. (After Vachon 1953.) (b) The scorpion *Butholus alticola* (Puc.). The 'premarital *promenade à deux*', the male holding the female. (After Serfaty and Vachon 1950.)

Introduction

reduction of the opisthosoma, the body region behind the prosoma, is evident in most adult arachnids other than scorpions.

The earliest known arachnids are several species of aquatic Silurian scorpions, but there is no evidence supporting a view that scorpions gave rise to the other arachnid orders (see Chapter 10 §10). There is little doubt that the arachnids arose from some marine chelicerate stock or stocks, but from which we do not know, neither do we know whether the invasion of the land happened once only, or more than once. Størmer (1970) has described a Lower Devonian scorpion, presumably aquatic, with long external gills projecting from the ventral surface and spreading out to the sides of the animal (Fig. 1.14(d)). Otherwise *Waeringoscorpio* is very like a modern terrestrial scorpion (Fig. 1.13). Many other arachnid orders appear as fossils abruptly in the Devonian and Carboniferous. But the earliest known land arachnid, *Alkenia mirabilis*, lived contemporaneously with the aquatic gill-bearing scorpions (Størmer 1970).

The success of modern arachnids can be judged from Bristowe's (1947) estimate that the spiders alone on the arable land of England and Wales consume annually in their insect food more than the weight of the humans living in the same area.

D. *Early merostome-like animals*

The Aglaspida (Fig. 1.14(c)) lived from the Lower Cambrian to the Upper Ordovician and appear to be the most primitive members of the Chelicerata. A rather trilobite-like cephalic shield covered the prosoma, but the limbs do not resemble those of trilobites, a pair of chelicerae being followed by five pairs of simple walking legs. The succeeding free segments, with pleural expansions, were sometimes limbless and the body terminated in a long spine. The relationships of *Weinbergina* and *Euproöps* (Fig. 1.14(a), (b)) probably lie with the Xiphosura.

Whatever their origin may be, the higher merostomes show a dichotomy into crawling types, with a soldering together of opisthosomal segments, which led to the Xiphosura, and bottom-living types capable also of swimming in which the segments remained free and led to the Eurypterida.

E. *Crustacea*

The Crustacea (Pls. 2, 3) are the dominant marine arthropods today. A small proportion of them have spread into fresh water and fewer still onto the land. They possess a calcified

Introduction

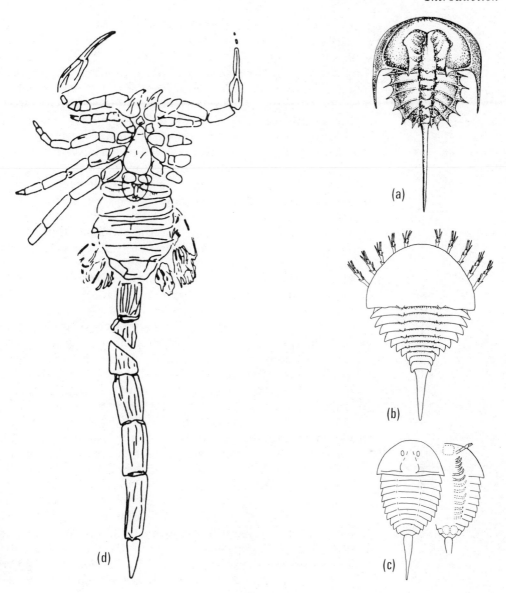

FIG. 1.14. (a)–(c) Merostomata, Aglaspida. (a) *Euproöps thompsoni* Raymond, 44 mm, Carboniferous. (After Raymond 1944, from Tiegs and Manton 1958.) (b) *Weinbergina opitzi* Richter, 96 mm, Devonian. (After R. and E. Richter 1929, from Tiegs and Manton 1958.) (c) *Aglaspella eatoni* (Whiff.), Cambrian. (After Størmer 1944, from Tiegs and Manton 1958.) (d) The aquatic scorpion *Waeringoscorpio hefteri* Størmer, Lower Devonian. (After Størmer 1970.) Note the gills projecting beyond either side of the middle region of the body.

Introduction

and sclerotized exoskeleton and are very various in habit and associated structural modifications. They possess a characteristic head, differing from the chelicerates and trilobites. Two pairs of antennae lie in front of the mouth in the adult; a pair of mandibles flanks the mouth and is followed by a pair of maxillae 1 (maxillules) used in feeding. In most crustaceans a pair of feeding maxillae 2 follows, with from one to three pairs of feeding maxillipedes in certain groups (Figs. 1.15, 3.7, 3.9, 4.10, 4.12, Pls. 3(a)–(c), (f)–(i)). The legs behind are used for swimming or for walking, or for both. Many are biramous. The inner rami may be rounded in section, as in the trilobites, but the outer rami never bear trilobite-like respiratory lamellae; the rami may be flattened, either one or both, according to the habits of the animals. Exites and endites may be present as lobes arising from the medial and lateral face of the leg base (see Chapter 4).

The body behind the head is usually divisible into regions or tagmata which do not correspond with those noted above for other groups. A limb-bearing thorax behind the head usually has a constant number of segments in each major crustacean group, but the number differs from group to group. An abdomen follows which may or may not carry limbs. The limbs here may be reduced or absent, but are well-formed and used for swimming in many Malacostraca (e.g. shrimps and lobsters). In the majority of crustaceans a carapace or fold (Pls. 2(a), (b), 3(a), (h)) grows out from the head and envelops part or all of the body, usually remaining free from it. Thus a single shield covers succeeding segments, but is not derived from the fused tergites of these segments as in Chelicerata (Figs. 1.15(g), Pls. 2(a), 3(a), (h)).

The fossil record tells us nothing about the origin or interrelationships of the five major component groups of the Crustacea. The Branchiopoda are mostly small, swimming or bottom-living in habit, filter or detritus feeding, with an immense adaptive radiation of their own in fresh water (Pl. 2(a), (b)). The Copepoda (Pl. 3(b), (c)) are also mostly small, bottom-living or pelagic, filter or predatory feeders, inhabiting salt and fresh water. The Ostracoda are entirely covered by a bivalved shell, the carapace, and swim, or live on the

---

FIG. 1.15 (see opposite). A primitive crustacean *Hutchinsoniella macracantha* Sanders, 3 mm. (a) In ventral view. (b)–(e) illustrate the extreme flexibility of the body. (After Sanders 1963.) (f) Typical crustacean nauplius larva (from Tiegs and Manton 1958). (g) Typical crustacean metanauplius larva (from Anderson 1973).

Introduction

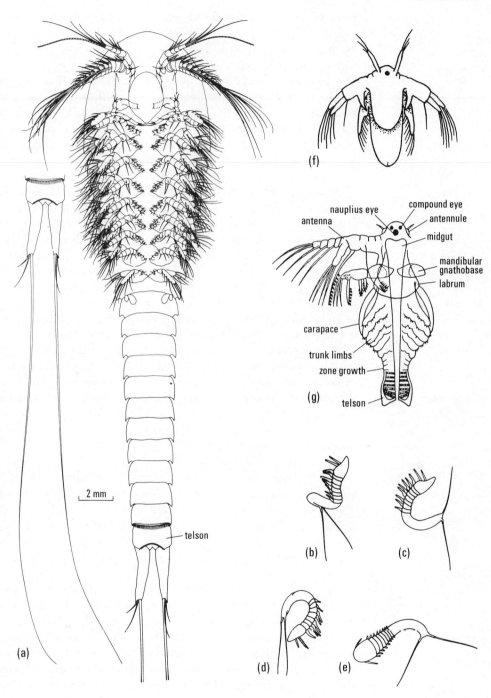

Introduction

bottom, in the sea and fresh water (Pl. 3(a)). The barnacles of many kinds, sessile in the adult stage (fixed to the substratum such as rock, a ship, or a whale) are grouped together as the Cirripedia (Fig. 4.13, Pl. 2(c)). The Malacostraca comprise the large and most conspicuous crustaceans such as shrimps, prawns, lobsters, crabs, woodlice, sandhoppers, etc., and most of them live in the sea (Fig. 4.10, Pls. 3(d), (d1), (f), (h), (i)). River crabs go hundreds of miles up European rivers annually and back to the sea to spawn. There are specialists such as the coconut crabs which climb coconut trees, opening the nuts and feeding on them; these crabs live largely on land.

The fossil Syncarida (Malacostraca) from the Permian are not very different from *Anaspides* living in fresh water today (Pl. 3(f)). Many series of fossil crabs in more recent deposits indicate some of the evolutionary changes which have gone on within the several crab assemblages. There are crustacean-like early remains which are imperfectly fossilized and Lower Cambrian Ostracoda and the malacostracan group the Phyllocarida. We have no sure fossil evidence of the ancestors of most modern Crustacea, which presumably lived alongside the early arthropods but did not arise from any known group. The recently discovered minute Cephalocarida (Fig. 1.15) which live in or on certain flocculent deposits on the sea floor, in some ways are probably the most primitive extant Crustacea (see Chapter 3 and Chapter 6 §2E(vi)). The structure and life-history of the Cephalocarida is of considerable interest in suggesting modes of crustacean evolution (Sanders 1963*a*, *b*; Hessler 1964).

F. *Pycnogonida or Pantopoda*

This small group of marine animals, sometimes called sea spiders, live in shallow water crawling over seaweeds and among hydroids upon which they feed (Fig. 1.16). They can also swim. The species are widely distributed and extend to the greatest depths. The leg span is 3 mm to 50 cm, the abyssal species being the largest. Some 500 species are known.

Pycnogonids possess very long legs relative to a small compact body composed of few segments. Usually four pairs of 8-segmented legs are present, the terminal segment being a claw. No biramous limbs are present. Ten-legged forms exist in the same families comprising the normal octopodous species and are probably polymerous variants of these species. One 12-legged form exists. Anterior to the walking

Introduction

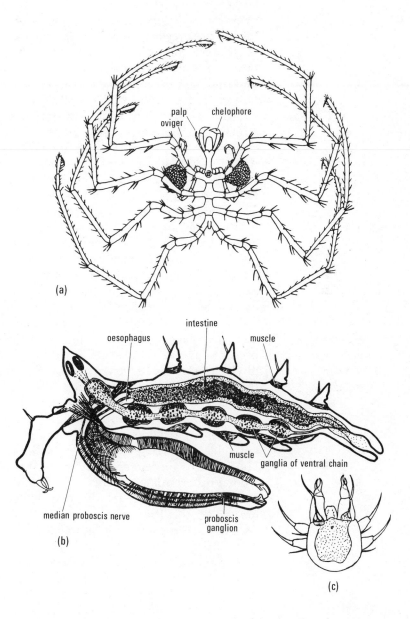

Fig. 1.16. Pycnogonida. (a) Dorsal view of *Nymphon rubrum* Hodge, 2 mm. (After Sars, from Tiegs and Manton 1958.) (b) Diagrammatic section of *Ascorhynchus castelli* (Dohrn). (After Dohrn, from Tiegs and Manton 1958.) (c) Protonymphon larva of *Pentanymphon antarcticum*. (After Hedgpeth 1947, from Tiegs and Manton 1958.)

Introduction

legs there may be a pair of legs, the ovigers, which are present in most males and carry the eggs, but are absent in some females. Anteriorly too there are usually a pair of chelate chelifores and sensory palpi. A proboscis lined by cuticle bears the mouth at its tip. The proboscis may be small or large and may be movable at its base on the rest of the trunk. A narrow oesophagus connects the proboscis cavity with the stomach cavity, from which diverticula extend into the legs, which also house the sex glands. There are no special respiratory or excretory systems, but the usual arthropodan dorsal ostiate heart, supraoesophageal and suboesophageal galglia, and ventral chain of ganglia are present. Pycnogonids hatch into a protonymphon larva which differs from all other arthropodan larvae (Fig. 1.16(c)).

Whatever may be the significance of the polymerous condition of some pycnogonids, it is a phenomenon without parallel among the Arthropoda. Only the notostracan branchiopod *Triops* possesses a slightly different feature where some of the posterior of the trunk limb-bearing segments carry many limbs each (Pl. 2(a)).

G. *Uniramia (Onychophora, Myriapoda, Hexapoda)*

This great assemblage of animals, terrestrial in habit at the present day, we must now regard as a separate arthropodan phylum (Manton 1972, 1973; and Chapter 6). It comprises the Onychophora, soft-bodied animals with one pair of antennae, one pair of jaws, and many legs (Pl. 4(a), (b)); the Myriapoda, divisible into the millipedes, centipedes, symphylans, and pauropodans, all with many legs and one pair of antennae (Pls. 1(f), 4(d)–(k), 5, 6(j), (l), 7(c)–(i)); and the Hexapoda or six-legged animals, the three pairs of legs arising from a differentiated thorax and the abdomen being usually limbless (Pl. 6(a), (c)–(i), (k), (m)). There are five classes of hexapods, the Pterygota, or winged insects, and the four wingless classes: the Diplura, Collembola, Protura, and Thysanura. Every component group within the Uniramia moves about by uniramous legs, in contrast to the biramous limbs of the large marine groups so far considered. The jaw mechanisms of the Uniramia differ fundamentally from those of the Chelicerata and Crustacea (Chapter 3).

The relationship of these three groups rests on morphological and embryological features and not least on the evidence provided by their jaw and locomotory mechanisms. Every component group of the Uniramia moves about by

Introduction

legs, which employ a series of gaits (Fig. 6.13). At different times these gaits change, by alteration of certain variables, in a precise and integrated manner, which serves momentary needs. Comparable series of gaits are not used by Arachnida or by Polychaeta or by vertebrates, where the patterns are more stereotyped and do not appear to be modifiable into a series of variable gaits.

The three large aquatic groups of arthropods, the Trilobita, Chelicerata, and Crustacea, are armoured and jointed and quite distinct from one another as far back as the fossil record goes. Diversification of the Uniramia, by contrast, must have taken place much more recently, and on land, from soft-bodied animals. The comparatively new studies of functional morphology, embracing the whole animal in relation to its habits, has yielded an abundance of evidence concerning the evolution of the Uniramia, as will be shown in subsequent chapters. This evidence tallies with that derived from the newer comparative embryological work employing the concept of fate-maps as used in vertebrate embryology (Anderson 1973).

An animal must be a working whole at all stages in its evolution. The end terms of uniramian evolution differ both structurally and functionally. The paths by which they could have evolved can be appreciated now for the first time. If functional continuity was maintained at all stages in uniramian evolution, as indeed it must, then the Onychophora, Myriapoda, and Hexapoda must have diverged from terrestrial soft-bodied animals with lobopodial, unjointed limbs, not parapodia (see Chapters 6-8).

The evidence concerning the diversification on land of the Uniramia implies the existence of a soft-bodied marine stock of animals which emerged from the sea and then evolved into the modern representatives. This marine stock must have differed profoundly from the contemporary armoured arthropods mentioned above. Heavy sclerotization which could form sclerites was certainly absent on the trunk, but thin surface sclerotization is always present on the flexible parts of the cuticles of all Uniramia, as on the entire body surface of the Onychophora, the pleuron of the Chilopoda, Symphyla, and Pauropoda, and on the arthrodial membranes of all joints. Whether or not sclerites are formed in Uniramia is a matter of degree, not of capacity or incapacity to sclerotize the cuticle.

Introduction

The only lobopodium we can study alive is a leg of an onychophoran. This uniramous limb is worked by antagonistic muscles and by the flow and resistance of haemolymph (Chapter 4 §5). Such a limb could readily have given rise to a jointed unbranched leg of a uniramian. But a polychaete parapodium works in an entirely different manner (coelomic fluid plays a static role); and a parapodium, in spite of its diversity in different annelids, is not a practicable origin for a jointed leg (Chapter 4 §4–6). Thus we are led to the postulate that not only did the Uniramia leave the sea with a soft body and lobopodial uniramous limbs, but that a haemocoel was also present providing part of the mechanism of movement of the lobopodium. Such postulated soft-bodied animals stand in marked contrast to all the annelid groups possessing a large segmental coelom and parapodial limbs which must have lived in the sea alongside the ancestral Uniramia. It is reasonable to suppose that before becoming terrestrial the Uniramia possessed some cephalic sclerotization, and perhaps various incipient types of head differentiation, all in contrast to contemporaneous annelids and armoured arthropods.

The probable derivation of the terrestrial Uniramia, from soft-bodied marine ancestors with short, unjointed uniramous limbs, is so different from the probable origin of the Arachnida from aquatic chelicerates, already highly sclerotized with long, jointed legs representing the endopods of originally biramous limbs, that it is not surprising to find that the general adaptation to land life and the locomotory mechanisms in the two groups are profoundly different (Chapters 7 and 10).

Marine arthropods lacking heavy sclerotization are not readily preserved as recognizable fossils. A possible member of the marine Uniramia is *Aysheaia pedunculata* Walcott from the Middle Cambrian (Fig. 1.17(a)). Its head end has been variously restored by those who have studied the fossils. It superficially resembles the Onychophora in trunk texture and in limbs. There are a few other problematical remains, such as *Xenusion*, which is less precisely known and conceivably might belong to the same line of evolution. The chances of formation of recognizable fossils of soft-bodied animals of Palaeozoic age are more remote than for the highly sclerotized arthropods.

The fossil record helps us little with the earliest terrestrial

Introduction

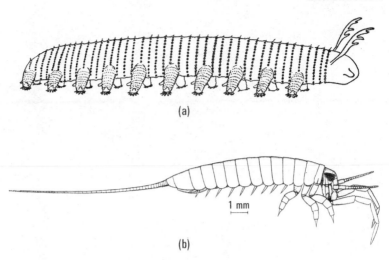

FIG. 1.17. (a) *Aysheaia pedunculata* Walcott, 50 mm, Middle Cambrian, conjectural restoration, doubtful concerning the head end. (After Hutchinson 1930, from Tiegs and Manton 1958.) (b) The monuran *Dasyleptus brongniarti* Sharov, 10 mm, Lower Permian, an apterygote hexapod. (After Sharov 1957, from Tiegs and Manton 1958.)

Uniramia. All the recognizable fossils are too advanced to illuminate the differentiation of the larger taxa. There are plenty of fossil millipedes (Diplopoda) and winged insects (Pterygota), often of large size in the Devonian and Carboniferous, as well as spiders and scorpions. The earliest fossil collembolan, *Rhyniella praecursor* from the Mid-Devonian, is already advanced in type. Some of the fossils do not fit present-day taxonomic categories. An example is the Carboniferous arthropod *Arthropleura* (Fig. 1.18), the leg of which is illustrated in Fig. 5.19. The animal was probably 1·5–1·8 m long; the gut is filled with fossilized plant debris; this arthropod is no close relative of any existing myriapod group.

The only other early fossil Uniramia of note are the beautifully preserved Monura, such as *Dasyleptus brongniarti*, 10 mm long, and other species described by Sharov (1957) from the Lower Permian (Fig. 1.17(b)). This animal bears many resemblances to the Thysanura Machilidae but is presumably more primitive in that pleural rudiments are present from the manidular to the labial segments and are clearly defined in the head capsule; the three pairs of thoracic limbs are followed by nine pairs of short abdominal limbs and the 14th trunk segment is as large as the 1st and bears a long

Introduction

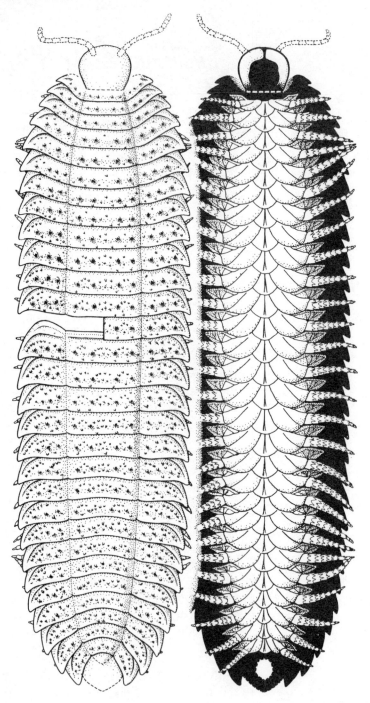

median style. The species of Monura presumably stand close to the Thysanura.

H. *The Tardigrada*

The Tardigrada comprise over 400 described species. These arthropods are minute in size; the smallest are only 50 μm in length and the average is 350–500 μm. They live in the suface film of moisture over mosses and in fresh and salt water. They can withstand considerable periods of desiccation. They possess four pairs of lobopodial limbs projecting ventro-laterally from the soft body wall. The cuticular covering is shed at ecdysis as in other arthropods; Fig. 1.19 shows a tardigrade just before the moult with the old cuticle already separated from the new one formed internal to it. There are

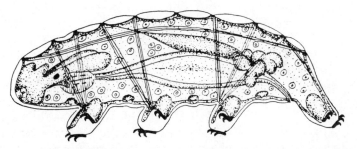

FIG. 1.19. Diagrammatic lateral view of the tardigrade *Macrobiotus* just before the moult, a new set of pedal claws already being formed within the cuticle about to be shed. The organs shown from left to right are: cerebral ganglia; optic lobes and lateral eyes; salivary glands, dorsal to the suctorial muscular pharynx which leads to the oesophagus and midgut; the paired gonads lie dorsal to the midgut; globular glands and malpighian tubes lie at the junction of midgut and hindgut. (After Ramazzotti, from Marcus 1972.)

no feeding limbs. A terminal mouth opens into an eversible buccal apparatus, which leads to a muscular pharynx followed by a narrow oesophagus merging into a straight gut. The cerebral ganglia and ventral chain are as in other arthropods. Terminal claws on the legs are various in different genera, ranging from only two to many more (Fig. 6.15). A few long sensory hairs project from the body, resembling the trichobothria of pauropods (Fig. 5.11). There is very little variation within the group.

FIG. 1.18 (see opposite). The Carboniferous arthropod *Arthropleura armata*, 1·6–2 m long. (After Rolfe and Ingham 1967.) The number of body segments is conjectural and the head details are hardly known.

Introduction

2. What this book is about

The fossil record tells us something about the existence of extinct groups of arthropods; it tells us a little about evolution within the Trilobita but nothing certain as to their relationships; the past history of the chelicerate groups is not entirely obscure; but the record of the rocks reveals very little indeed concerning the origin, the past history, or the relationships of Crustacea and of marine or terrestrial Uniramia (Onychophora, Myriapoda, Hexapoda). By contrast, modern work on functional morphology, carried out on a broad comparative basis, now gives an abundance of sound evidence on this subject, as does the work on functional comparative embryology of arthropods and annelids (Anderson 1973). The evidence derived from the functional morphology and the conclusions reached are the subject of the following pages. The embryological evidence is summarized in Chapter 6 §2E(i) and Chapter 10 §10.

A. *Functional morphology and habits*

The study of functional morphology, meaning an attempt to ascertain in some detail how a whole animal works, is a recent approach towards understanding the evolution of the different shapes and proficiencies of arthropods. It is essential to study habits of life as well as the usefulness of general body shapes, details of the skeleto-musculature, and so on. The yield in sound evidence obtained during some 35 years is so great as to justify the writing of this book to present in outline and as simply as possible the results and implications of modern knowledge of functional morphology of arthropods. This approach has proved to be a most powerful tool, not only in gaining sound understanding of the significance of the different shapes and proficiencies of arthropods, but of the evolution of the groups themselves. This latter and most important outcome of the work was not initially either planned or anticipated; the evolutionary aspects grew along with the progress of functional understanding of structure. A long series of papers recording much of this work has been published in the Linnean Society of London's *Zoological Journal* under the heading 'The evolution of arthropodan locomotory mechanisms: Parts 1–11', although a much better title would have been 'The habits and the evolution of the larger groups of the Arthropoda'. The scope and productivity of the project extended as time and accomplished work progressed.

For very many years there has been active discussion about

Introduction

the origin, evolution, and relationships of the Arthropoda and of the insects in particular. In the absence of direct fossil evidence discussions have been, of necessity, largely speculative.

Until recently the Arthropoda have been considered to be a unified and monophyletic phylum, but now a polyphyletic derivation seems inescapable (Chapter 6). Arthropodization, that is the formation of a head or of a cephalon, of limbs in series which are often diversified, and of sclerotization forming sclerites which necessitated jointing of the body and legs, cannot have been evolved once only. We must now recognize as separate phyla, at least the Chelicerata, the Crustacea, and the Uniramia among extant groups, and the position of the extinct Trilobita and others will be considered in Chapters 6 and 11. If the Arthropoda are polyphyletic, it follows that certain similarities between some of the component taxa have been convergently acquired (Chapter 6). There is indeed no doubt about many of the convergencies which differ in structural detail and show only general resemblances which meet particular needs. Other convergencies do not represent adaptations to particular circumstances at all and are due merely to a limitation in the number of ways in which it is mechanically possible for simple structures such as joint articulations, to be devised (Chapters 5, 6–10).

The importance of the comparative approach to the study of functional morphology cannot be overestimated. One has to start somewhere in attempting to understand the meaning of structure and this necessitates studying one whole animal in considerable detail. Many morphological features will be found to be associated either with special performances and abilities or with particular habits, but one can seldom get very far by considering one animal alone. For it is only by comparing structure and function in many species, both related and distantly placed in our classifications, that a proper appreciation of any one animal can be arrived at and some evidence gained as to how the species may have evolved.

The basic habits with which arthropodan structure is associated may be very easy to recognize, such as differing locomotory, burrowing, or feeding habits, but the details of the achievements in these fields may be most complex. Excellence in performance, either of various special capabilities, or of habits of life, depend upon the differentiation of

Introduction

structure facilitating certain movements, sensory perception, etc. Structural features can be understood only when a sufficient number of comparisons are made between many types of animals, and always on a functional basis, because habits and structure have evolved together. Of the many things an arthropod can do, the habit of particular evolutionary significance may be easy to appreciate. The method of burrowing used by a geophilomorph centipede and its manner of breaking into the encasing armour of a sizeable arthropod and feeding on the soft contents of the body of its prey are the keys to understanding its whole skeleto-musculature and general body form and much of its evolution (Chapters 3 and 8). But in other cases only one habit, or achievement, out of many which are practised by the same animal may have been of paramount evolutionary significance, its greatest survival value being manifest only under exceptional circumstances. The significance of such a habit may be much less easy to comprehend. For example, a curious little diplopod *Polyxenus* is proficient at many things, but the ability to live on glass-smooth flint ceilings of rocky hides under adverse circumstances, or cataclysmal conditions, is made possible by many morphological details of a remarkable nature, which have led to survival (Chapter 8 §3C(ii), Pl. 6(j), (l)) and (Manton 1956). In this example the proficiency of evolutionary significance may not be exceptionally advantageous all the time, but only occasionally. The explanation of the remarkable differences in structure and locomotory mechanisms between *Polyxenus* and all other diplopods is not at all obvious, but a comprehensive functional study of the morphology of many diplopods leaves these conclusions in no doubt whatever.

The diversification of habits and of the associated morphology which facilitate each habit become mutually exclusive. A millipede which burrows like a bulldozer by the motive force of its legs (Pl. 7(c)–(e)) can only excel in those pursuits if it is slow-moving, strongly constructed; and possesses many legs so that force can be transmitted to the head end with the body-line curved in any position; it must also possess a host of other structural features. The same animal can never run fast and catch flies for a meal as can the highly advanced centipede *Scutigera* (Pl. 4(j)) with quite different morphology. A consideration of how these mutually exclusive morphologies could have evolved by continuous

Introduction

series of functionally workable ancestors leads us to a terrestrial soft-bodied uniramian with lobopodial, unjointed limbs (Chapters 6–9).

B. *Types of evolutionary change*

One can recognize two types of evolutionary change which are very different in their extreme manifestations. There are morphological changes associated with assuming a distinctive habit and there are structural specializations suiting a particular habitat. The former gave rise to main phyletic patterns of structure and may be called advances, while the latter led to adaptive radiations within a phylum.

(i) *Adaptive radiation.* Evolutionary specialization towards one particular detailed set of environmental circumstances renders the animal unfit to live in other circumstances. A hermit crab demonstrates the requirements for living in a gastropod shell in very great detail. Some other object may replace the gastropod shell, but the hermit is tied to this sort of life. Swimming and burrowing crabs show other structural features which fit them for selected environments and none of these animals is suited to meet environmental pressure by predators, food supply, or general conditions and circumstances other than the ones normal to them. This kind of evolutionary change, called *adaptive radiation*, fits individuals within large groups, such as malacostracan crustaceans or winged insects, to invade this or that specialized niche to which they are structurally suited. It is probable that such adaptive radiations have had little, if anything, to do with the establishment of large arthropodan taxa, but instead appertain to the recently diversified end-terms of large and successful lines of evolution.

(ii) *Advances.* There is another type of evolutionary change of much greater significance concerning the origin of large groups. This type of change in structure facilitates general habits and enables an animal to live better, or more easily, in the same or in a variety of habitats. These changes we can call *advances*. They concern the assumption of a habit which does not tie the animal down to one set of environmental circumstances. The many lines of mammal-like reptiles show advances towards the mammalian type of organization but they do not show particular structural features which would facilitate living in limited habitats. Within the arthropods a large number of animals of various kinds can be found within

Introduction

the same environment; e.g., a decaying log in South Africa. Within or under it, at least during the daytime, you may find: *Peripatus*; many sorts of millipedes and centipedes; Symphyla; Pauropoda; scorpions; mites and other arachnids; winged and wingless hexapods. These arthropods are all momentarily living in the same environment, the log, but they can live in many other places as well. It is their present and past habits which differ and with these habits are associated their very different structures. The evolutionary changes resulting in differing morphologies in these classes of arthropods have not restricted them to any one set of environmental circumstances as has the morphology of a hermit crab; they can all live within the same log as well as elsewhere.

Since habits and structure have evolved together, habits which have persisted for countless millions of years become of great structural importance. Two divergent habits of fossil swimming reptiles were recognized by D. M. S. Watson (1949). Both groups fed on fish. The one line of reptiles bore down on their prey by fast swimming and became streamlined, like a fish, with shorter and shorter necks as geological time went on, while in the other line the necks became longer and longer and the feeding technique must have been the snatching of a fish from one side or another by neck flexibility, rather than moving the whole body forwards. The important concept is the persistence of a habit over long periods of geological time which fits an animal to live with greater efficiency in the same or in a variety of circumstances, the open sea or the decaying log. Such advantage of structure leads to the evolution of large taxa. The important principles may be expressed as major changes of policy or strategy as opposed to adaptive compromise with the circumstances of the environment; tactics, one might say.

C. *The significance of animal shapes and the functional approach*

It is easy to appreciate something of the significance of animal shapes among vertebrates. The usefulness of the shape of a herring, a dolphin, a giraffe, or a gibbon is obvious in outline, although it may require much specialized knowledge to appreciate all the detailed correlations of structure, both with function and with habit. The very different shapes of arthropods built upon a multitude of basic plans are far less easy to interpret. Without special knowledge the functional significance of, for example, the shape of a flat-backed or of a

Introduction

colobognathan millipede (Pls. 4(g), 5(f), 7(f), (g)), or of any one type of chydorid crustacean (figures in Fryer 1968), are not at all obvious and much functional knowledge is needed before they become so. Studies of functional morphology have illuminated our views on many particular evolutionary changes. We now know which are the more advanced and which are the more primitive members of certain arthropod groups and we can arrange them in a rough morphological series. We know, for example, that the centipede *Scutigera*, which, among all types of centipede, has the most acute eyesight, the most highly advanced mandibles, is the most rapid runner, and quickest seizer of fast-moving prey in the form of spiders and flies, is not the most primitive member of this class, as has sometimes been suggested, but is in fact the most advanced. We can now appreciate the superlative efficiency of its morphological advances because, for the first time, the locomotory and other habits of this and many other types of centipede have been properly studied and correlated with their skeleto-musculature, respiratory mechanisms, and habits (Chapter 8 and Manton 1965). *Scutigera* cannot be regarded as a primitive centipede on any count and its remarkable structure is readily understandible on a functional basis. Similarly the significance of convergencies of arthropodan structure, whereby animals of different ancestries come to resemble one another, some in surprising detail, can often be understood on a functional and evolutionary basis.

The newer functional approach to arthropodan structure is yielding results of the greatest importance. This type of work, using induction based on reliable evidence, is replacing the older speculations which were made because the fossil record provided no evidence. Large edifices of theory, which have been raised upon non-functional bases, can now be set aside: for example the many theories of the origin of insects, the theoretical basic form of pleurites (lateral sclerites) and their evolution in myriapods and apterygotes. The older theories are untenable because they imply either ancestral animals which could not function, or which did not exhibit progressive advances of the right kind. Even the correct anatomy of pleurites has been unknown until recently among myriapods and apterygote hexapods. When these matters are studied in a comparative and functional manner the myth of the insectan sub-coxal segment vanishes and

Introduction

much is learnt from the very different mobile pleurites and leg-bases of myriapods and apterygote hexapod classes. A new concept emerges showing that the leg-bases of the five hexapod classes are functionally and structurally so different from one another that their evolution can only be envisaged as having proceeded independently along five separate lines from soft-bodied uniramian ancestors (Chapters 5, 6, and 9). Moreover this type of evidence is consistent; for example, the Symphyla, once a strongly-favoured near-relative or ancestor of the hexapods, prove to be quite distinct in the head structure, the jaw anatomy, and the limb structure, and can have had no part in the evolution of any hexapod.

The functional and comparative approach has been applied over a wide field; indeed, if this had not been so the progress in our understanding of arthropodan evolution would have been the less. Enough has been done to illuminate relationships of the major arthropodan taxa, as well as to show how evolution has progressed in the details of myriapod and hexapod classes in particular. In the following pages an outline only is presented of the major fields of investigation and conclusions which have been reached. Much work remains to be accomplished but it cannot be done quickly or by considering one animal alone or in part. The sound evidence which emerges from this work concerning the diversification of arthropodan taxa gives a very different picture from the previous speculations on this subject.† The newer approach has yielded fresh light on the puzzle presented by the Pycnogonida (Chapters 6 and 10) and embraces the Tardigrada also (Chapter 6).

---

† No support has been found for the flights of imagination and the erroneous statements by Sharov (1966) concerning basic arthropodan evolution; see Anderson (1966).

2 Habit divergencies and arthropodan food-gathering in the Palaeozoic and some basic leg movements

1. Introduction

An ability to walk about on the sea floor presumably preceded swimming capabilities. A swimming limb can be a flattened paddle with minimal or no jointing, but a walking leg, if the cuticle is well sclerotized, must be jointed, as shown below. The earliest known arthropods possessed well-jointed limbs, as seen in trilobites, merostomes, and others (Figs. 1.1–3, 1.6–12). They were presumably bottom-living in habit. Some of the more advanced forms, such as eurypterids, must have been able to swim and some of the earlier arthropods may have been able to swim to various extents as an alternative to a predominantly bottom-living habit. Food was presumably collected mainly from the bottom or surface layers of the soft substratum by the early arthropods, using a very simple series of limbs. Elaborate prehensile anterior limbs, as possessed by the eurypterid *Mixopterus* (Fig. 1.12), probably indicate a carnivorous animal capable of dealing with large, strong prey, a habit impossible to the early arthropods with a simple series of jointed legs.

The Trilobita, Merostomata, and many other early armoured arthropods, such as *Marrella* (Fig. 1.6), possessed a labrum (upper lip) but no mandibles. Food must have been ingested at the mouth by pharyngeal swallowing movements, as in all arthropods. Suction is easily caused by alternate dilation and contraction of a pharynx or by peristaltic movements along an oesophagus, as is seen today in *Limulus* (Fig. 3.13a) and in the hexapod *Petrobius*, animals very far removed from one another taxonomically. The function of mandibles, or the equivalent gnathobases of the prosomal legs of *Limulus*, or of arachnid pedipalps, is to cut, grind, or squeeze the food as the case may be, but not to push it into the mouth. In the absence of mandibles and of gnathobases food must presumably have consisted of soft invertebrates grabbed on or just below the surface of the substratum. Detritus, dead animals, and small food particles may have been swept towards the mid-ventral surface of the early armoured arthropods by a sweeping action of the legs; the long setose second antennae of *Marrella* were particularly suitable for such movements. The means whereby food might then have been transported forwards towards the mouth is of considerable interest because two divergent habits have been associated with a basic evolutionary dichotomy.

The importance of habits which have persisted for countless millions of years (see Chapter 1) is well illustrated by the two contrasting routes used by arthropods in conveying food to the mouth. By the one route food is passed forwards, close to the body, between a series of paired limbs and by the other food is raised from below directly upwards towards the mouth and does not travel along the ventral face of the body (Fig. 2.1). The former route suits primarily aquatic ar-

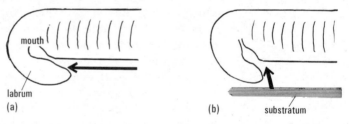

FIG. 2.1. Diagram showing two contrasting routes taken by the food to the mouth. In (a) the food travels forwards close to the body between the bases of the legs to reach the mouth, as in some of the less specialized Crustacea and presumably in the Trilobita and other early armoured arthropods. In (b) the food travels from below upwards to the mouth and not close to the body. This route is seen in the Uniramia. (From Manton 1973a).

thropods, well sclerotized and often with little body differentiation. The latter route characterizes the Uniramia (Onychophora, Myriapoda, Hexapoda Pls. 4–7) and suits land living, but must have existed also in the marine non-armoured ancestors of the Uniramia. Certain Crustacea, specialized for feeding on large food masses or large prey, have secondarily developed the ability to raise food directly upwards towards the mouth, an accomplishment resulting from structural advances of various kinds which have been independently acquired in a parallel manner by several crustacean groups and are not at all primitive (see Chapter 3 §2C). These two basically different routes for the food to the mouth are correlated with the evolution of mutually exclusive types of jaw.

The extant Arthropoda can be classified into a few large categories on the type of jaw they possess (see Chapter 3). The contrasting gnathobasic jaws present in Chelicerata and in Crustacea are formed from a pair of proximal inner lobes, called endites, on the first pair of postoral limbs, the biting being done by the base of each limb; the distal part of the limb either remains, or becomes progressively reduced. In the Uniramia (Onychophora, Myriapoda, Hexapoda) the jaw is a whole limb which bites with the tip and not with the base. These three types of jaw will be considered in detail in Chapter 3. Firstly it is necessary to note some fundamental movements of trunk limbs, both singly and in series with one another, as they exist today. Presumably these movements were similarly used by arthropods before there were any jaws. Basic limb movements provide simple mechanisms of locomotion, food transport, and respiration which are operative irrespective of the presence or absence of paired jaws.

2. Basic limb movements employed in walking

During walking most arthropodan limbs perform two simple movements, which, in animals with jointed limbs, are located at different joints.

A. *The promotor–remotor swing*

This movement of the leg base on the trunk is shown in Fig. 2.2(a), (b) and takes place about an axis of swing lying roughly in the transverse plane of the body. This axis may be ventral, ventro-lateral, or lateral in position. The promotor–remotor movement about this axis forms the basis of stepping and is used in walking on the substratum, in shallow burrowing within it, and in simple swimming. The movement is implemented by extrinsic protractor, retractor, levator, and depressor muscles from the proximal part of the leg which pass into the trunk and insert on exoskeletal and endoskeletal structures of various kinds in the same or in more distant segments. Intrinsic muscles within the leg are used in stepping and comprise levator, depressor, and flexor muscles.

In many of the less advanced Crustacea, such as Cephalocarida and some of the Branchiopoda and others, the base of the leg is wide, its proximal segment, the coxa, being elongated and flattened in the transverse plane (Figs. 2.8–10, 4.4–6). The coxa swings at its anterior and posterior proximal margins, without any particular articulations with the body sclerites; a sufficiency of flexible arthrodial membranes present at the joint permits the movement. The angle through which such a simple leg base can swing is limited, the most suitable being perhaps of the order of 16–27°. Much greater angles of swing result from the presence of narrow coxa–body junctions or from coxa–body articulations of considerable strength or complexity. The narrow coxa–body junction of iuliform diplopods permits a coxal swing of about 50–65°; the range is greater in a chilopod, about 70–85°, where great specializations of both coxa–body and coxa–trochanter joints contribute to this end (Chapter 5). The swimming thoracic legs of copepods have a most remarkable and unique coupler apparatus at the coxa–body joint enabling the leg to swing through some 100–105° (see Chapter 4, Figs. 4.18, 4.19). But the fossil armoured and soft-bodied arthropods have none of these refinements. The extreme position of the promotor–remotor swing of the 5th and 6th prosomal limbs of *Limulus* during walking are shown on the opposite sides of Fig. 2.6(a), (b) recorded from a cinematograph film. Legs of a

FIG. 2.2 (see opposite). Diagrams showing the two basic movements employed by many arthropodan limbs, exemplified by *Polydesmus* (Diplopoda). One or other of these movements are also used by the several types of mandibles. Arthrodial membrane at the coxa–body joint is indicated in black. (a) Ventral view of two successive pairs of legs, those on one side of the body being cut short close to the coxa–sternite articulation. The axis of swing, passing through the dicondylic coxa–sternite articulation, is shown by a heavy line. The legs on the right show the forward and backward positions resulting from the promotor–remotor swing of the coxa about its axis of movement on the body. (b) Lateral view of three successive coxae cut short near the coxa–sternite articulation to show the promotor–remotor swing about the axis (indicated by a black dot). (c) Ventral view of two successive legs from one side of the body in position of abduction, away from the fellow, and adduction, towards the fellow. The coxae cannot participate in this movement. The axes of swing of the leg-joints distal to the coxa are shown by heavy lines, the two proximal joints being dicondylic pivot joints and the three distal joints being hinge joints. The tarsal claw is hinged to the tarsus in the same plane as the other hinge joints. (d) Proximal dicondylic pivot joint on a leg showing two positions of the distal segment; the antagonistic muscles are indicated by arrows within the leg segment. A diagrammatic transverse section through the joint at the level indicated shows the lateral points of locally strengthened cuticle on the two leg podomeres united by very short arthrodial membrane. (e) Distal hinge joint on the leg showing two positions of the distal podomere, the single flexor muscle (adductor or depressor) being indicated by an arrow within the leg podomere. A diagrammatic transverse section through the joint at the level indicated shows the dorsal point of close union between the strengthened cuticle of the two leg segments which forms the hinge. (From Manton 1969.)

Habit divergencies and arthropodan food-gathering

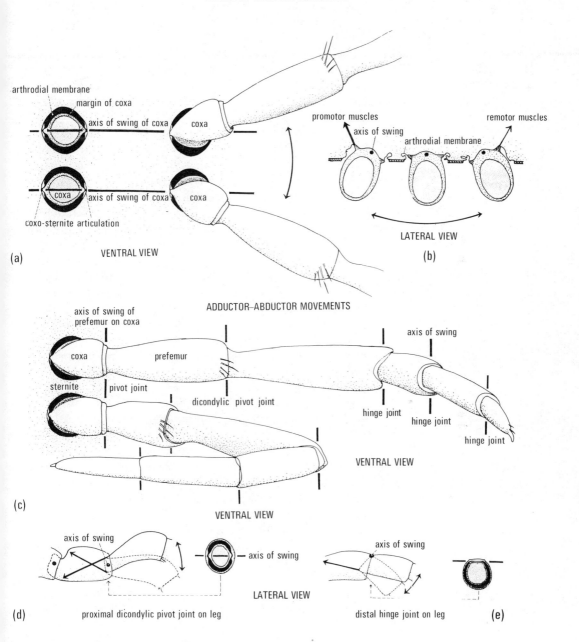

Habit divergencies and arthropodan food-gathering

pair actually step in similar phase (not as shown); that is they step in unison. Outlines of the legs are given as seen in ventral view through a glass-bottomed tank of sea water. The knee joint, stippled on leg 5, projects dorsally away from the substratum (see also Fig. 3.12(b)).

B. *The adductor–abductor (or levator–depressor) movements*

These actions usually take place at right angles to the promotor–remotor swing (Fig. 2.2(c)) and occur distal to the coxa when the leg is jointed. The levator–depressor movement is usually implemented by one or two successive pivot articulations situated between the overlapping cuticles of successive leg-segments or podomeres, just beyond the coxa, and by more distal hinge or pivot articulations at the joints present along the more distal part of the leg. Pivot articulations (Figs. 2.2(d), 4.27(c)–(e), and 4.28) are worked by paired antagonistic muscles within the proximal podomere which can alternatively levate and depress the immediately distal podomere, which swings at the more or less equatorial pivot articulation on the anterior and posterior faces of the leg. Hinge articulations, frequently single and dorsal in position (Figs. 2.2(e), 4.27(a), and 4.28), are usually worked by flexor muscles only, the straightening of the leg, after flexure at a hinge joint, taking place passively by hydrostatic pressure when the leg is doing no outside work, or by strong proximal depressor muscles which press the limb-tip firmly on the ground and indirectly extend the distal part of the leg during part of the propulsive backstroke (see Chapter 5). There are also many examples where the indirect method of leg extension is not used and antagonistic pairs of muscles are present at every joint along the leg, as in many winged insects and malacostracan Crustacea. In the flattened legs of branchiopod Crustacea flexibility is more limited and the joints are ill-defined (Figs. 2.7, Pl. 2 (a)(b)). The details of articulations at joints between podomeres among modern arthropods are not easy to ascertain (Figs. 4.27, 28, 5.15–18, 10.1–5), except in large species with heavily sclerotized or calcified cuticles, and the nature of the inter-podomere joints in fossil species is little known.

C. *Leg movements on and off the ground*

A combination of promotor–remotor with the levator–depressor movements can result in simple stepping (Fig. 2.3(a), (b)). Position HJ shows the outstretched leg with its tip on the substratum at the beginning or end of the backstroke, the ventral face of the body being held clear of the ground. Position HK shows the needed flexure when halfway through the backstroke (see below), a knee rising above the level of the leg base, and position HO indicates full extension during the forward swing, off the ground.

When a leg swings equally forward and backward about its base during stepping, the leg must shorten to HK when half-

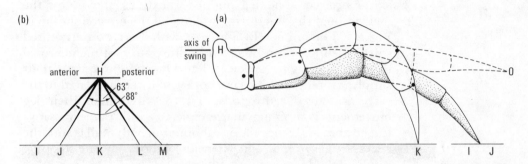

FIG. 2.3. The stepping movements of a jointed leg, articulated to the body at H. (a) The anterior face of the leg in three positions and (b) a lateral view of the stepping movements, the leg swinging through two alternative angles of swing. Leg HJ is outstretched at the beginning of the backstroke, its tip being in contact with the ground. Half-way through the backstroke the tip is at K relative to the body, the distance HK being shorter than HJ. During the latter part of the backstroke the leg elongates to position HL. The body is moved forwards by a distance equal to JL during a backstroke through an angle of 63°. A longer leg, HI, can move through a larger angle of swing, 88°, provided the leg is capable of relatively greater flexures when half-way through the backstroke, HK. The dotted outline HO in (a) shows the outstretched leg performing the forward swing off the ground.

way through the backstroke if the limb-tip is not to slip on the ground. Extension of the leg takes place during the latter part of the backstroke to HJ. The larger the angle of swing of the leg, the greater will be the necessary shortening in position HK and the greater will be the necessary joint specializations to permit this. Longer legs and larger angles of swing, such as 88° (in Fig. 2.3(b) by leg HI), will give a long distance IM provided the leg can shorten to HK when half-way through the backstroke. Large angles of swing and long legs can give long strides and fast running, but the latter could not have been practised by the early arthropods because they lacked the highly specialized joint articulations and the refinements in shape of the leg podomeres which are required for fast running (see Chapter 5).

D. *The integration of movement of successive legs; metachronal rhythm, used for walking and burrowing*

A many-legged animal cannot achieve steady locomotion by moving all of its legs, or all legs on one side of the body, in unison. The four pairs of large flat thoracic legs of copepods are an exception; they are moved almost synchronously and this results in sudden and intermittent jumping through the water (Chapter 4 §3D (iii)), but such movements are rare and having nothing to do with walking or the forward transport of food close to the ventral surface. Usually a phase difference exists between one leg and the next, as shown by the metachronal rhythm during walking of the onychophoran in Fig. 2.4. This animal (Pl. 4(a), (b)), with short, non-armoured soft-walled legs, can be taken in illustration of metachronal rhythm in a living species. The Onychophora with their simple, uniramous limbs are not even distantly related to the early armoured arthropods, but the rhythm of their leg movements is common, in principle, to a wide range of many-legged animals. Since the onychophoran body wall is soft, the necessary shortening and extension of the leg during stepping can be done by muscles, without joints.

Figure 2.4 shows groups of limbs, each consisting of one leg shown in white swinging forwards off the ground, followed by two legs shown in black performing the propulsive backstroke, on segments 1–10 where they form cycles of limb movement or metachronal waves. When leg $n-1$ is put on the ground a little in advance of leg n (earlier in time) the metachronal wave appears to travel forwards over the body. The positions of limb-tips relative to the head, plotted from a cinematograph film, extend to the right of the body on Fig. 2.4, the forward stroke being marked by thin lines and the propulsive backstroke by thick lines. There is considerable uniformity throughout the seven paces performed in just over 5 seconds of the record, but between left legs 13 and 18 and right legs 10 and 15, there are four successive legs and not two in the propulsive phase.

When legs are several times the length of the segment which bears them greater precision in stepping is needed than is shown by the Onychophora. Longer metachronal waves are also required, unless the angle of swing of the leg is very small. Legs six times the length of the segment which bears them are shown in Fig. 2.5. Each metachronal wave consists of eight legs, five in the propulsive backstroke and diverging from one another, and three performing the forward swing, converging and off the ground. The levels of the footprints

Habit divergencies and arthropodan food-gathering

are marked by black spots alongside the propulsive legs. The legs swinging forward off the ground are more fully outstretched and project furthest from the body. Legs of a pair are in similar phase, that is they step in unison. In many

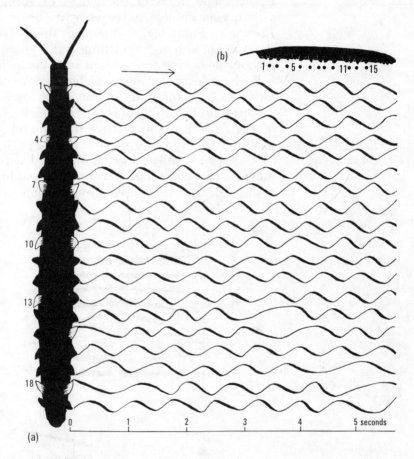

FIG. 2.4. Onychophora walking. Cinematograph records of *Peripatopsis sedgwicki*. (a) Dorsal view of the animal with left legs 1, 4, 7, 10, 13, 18 and right legs 1, 4, 7, 10, 15, 18 off the ground performing the forward swing and shown in white, the intervening legs being on the ground performing the propulsive backstroke. On the right of the animal the positions of the limb-tips relative to the head are shown for 7½ paces, heavy lines indicate the backstroke and thin lines the forward swing. Legs 1–12 step fairly regularly and the legs of a pair move in unison. There is some irregularity in the movement of the next five pairs of legs but the rhythm is almost restored by legs 18–19. (b) Side view moving by a slower gait, legs 1, 5, 11, 15 being off the ground; there is no stepping at all by the more posterior limbs. The propulsive legs are marked by black spots.

Habit divergencies and arthropodan food-gathering

arthropods paired legs move in opposite phase, but it is unlikely that trilobites and other early armoured arthropods stepped thus because they probably had no use for gaits requiring this phase relationship which could not give forward movement to food situated mid-ventrally (see below). The fields of movement of successive legs may overlap considerably, as shown by the heavy vertical lines for *Limulus* in Fig. 2.6(c), but the legs themselves do not touch one another although crowding and even crossing of legs is practicable during the forward recovery swing.

Each metachronal wave of limb movements progresses forwards over the body in the above examples, the most posterior leg in a group of forwardly moving legs, legs 1 and 9 in Fig. 2.5(a), being put on the ground while the most posterior leg in each propulsive group, leg 6 in Fig. 2.5(a) is raised, legs 2–5 still occupying the same footprints. The body moves forwards as shown by the arrowed line between Fig. 2.5(a)–(b). Each propulsive group of legs may be likened to

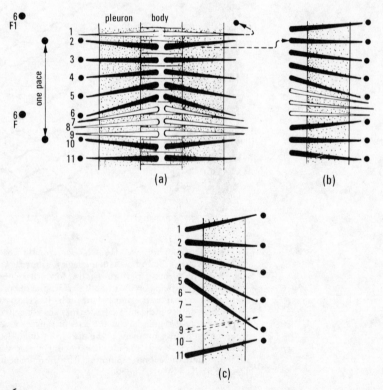

Habit divergencies and arthropodan food-gathering

the progression of a caterpillar wheel, the tread in contact with the ground being always the same length but the surface for ever changing. Stride lengths are indicated on the left, leg 6, raised from the level of spot 6F, will be put down at the level of spot 6F1. The repeat of the waves of movement also gives the stride length, e.g. the marked distance between the footprints of legs 2 and 10.

Limb movements are suited in great detail to the shape and the requirements of an animal. Length of a metachronal wave can change at different times and the length is also suited to the total number of locomotory limbs present. When the legs are long, as in Fig. 2.5, and some 20 pairs are present, a few complete waves, or one and a part of another wave, may be present at any one moment. A small angle of swing of the leg, about $22°$, is shown in Fig. 2.5(a), (b) and contrasts with the much larger angles shown in Fig. 2.3(b). If the angle in Fig. 2.5(a) be increased to $37°$ the leg disposition becomes as shown in Fig. 2.5(c). The footfalls lie further apart and the same

---

FIG. 2.5 (see opposite). Diagrammatic ventral views of practicable and impracticable gaits for the trunk limbs of armoured arthropods; no gnathobases are drawn, but the position of the trilobite pleuron is indicated. The walking endopods in contact with the ground and performing the remotor backstroke are shown in solid black and the endopods off the ground and performing the forward swing are shown by open lines. Footfalls or footprints where the legs are in contact with the ground are marked alongside by black spots. (a), (b) The legs swing through about $22°$, five legs are simultaneously propulsive and are followed by three in the forward swing, so making a metachronal wave of eight leg pairs, probably a suitable length for a total number of legs of about 20 pairs: Leg 1 has just been lifted from the ground, leg 2 has just been placed upon the ground a little in front of the transverse plane of the body. Legs 3–6 show progressive stages in the backstroke, legs 7–9 are swinging forwards and project further from the body because they are outstretched sideways. Leg 10 shows a repeat of leg 2. The stride length is indicated on the left. Leg 6 is raised from its footprint level with the spot 6 F and will be put down level with the spot 6 F1. The distance between the footfalls of legs 2 and 10 also show one stride or pace length, as marked on the left. Leg 1 has nearly completed the forward swing, and it will be put on the ground as indicated on the right by the arrow towards the single footprint marked by a black spot. (b) The appearance of the metachronal wave when leg 1 has been put down on the ground, leg 6 has been raised, and the body has moved forwards as indicated by the arrowed line between (a) and (b). (c) Diagram showing the effect of increasing the angle of the swing of the legs to about $37°$. The same length of the metachronal wave becomes impossible because leg 9, starting its propulsive stroke, would cross leg 5, also propulsive, giving a mechanically impossible situation. At least five legs in the forward swing would be needed to allow leg 11 to repeat leg 1. But metachronal waves of 10 or 11 legs might be too long for the length of the body and the number of propulsive legs would be fewer than those in the forward swing, giving a probably impracticable gait. The gait shown in (a)–(b) is a more likely gait for a trilobite, but others may also have been used.

47

length of metachronal wave becomes impracticable because leg 9, starting its backstroke, would have to cross over leg 5, a movement not performed by arthropods. A minimum of 5 instead of 3 legs in the forward swing would just permit leg 11 to repeat the action of leg 1, but an equal number of propulsive and recovering legs would probably represent a 'faster pattern' (see Chapter 7) of gait than would suit a trilobite, although a diplopod with a more highly organized coxa–body joint can perform this gait. The 'pattern' of a gait (see Chapter 7, §3B), can readily change in the opposite direction, most leg pairs being in a propulsive phase and a minimal single pair at a time swinging forwards.

The limbs of the trilobite *Olenoides* can be eight times the length of the segment which bears them. The fields of movement of these limbs must have overlapped considerably on the forward stroke and accurate stepping by small angles of swing of some 16° would be expected. But the stepping might have been fairly speedily performed. Swimming as well as walking might have been possible for a trilobite because of the flattened form of the respiratory lamellae on the outer ramus (see below), and jumping may also have taken place. If about six successive legs performed the backstroke almost in unison, the animal would leave the bottom, travel forwards and sink down again. The same legs could repeat the jump leaving the interrupted series of fossil tracks described by Miller (1973).

Shallow burrowing as well as walking could result from a trilobite using gaits of the order shown in Fig. 2.5. The overhang of the pleural fold would protect the respiratory lamellae borne by the outer ramus of each limb and metachronal rhythm of the entire limbs would provide both the locomotory force and the necessary water change for respiration. A downward flexure of the cephalic shield would tip its anterior margin into the substratum. If this was soft, metachronal waves of limb movement in which, at any one moment, most of the legs were in the propulsive phase and few in the forward swing, would result in shallow ploughing. Better burrowing is achieved by *Limulus* because its 6th prosomal limbs form the principal digging organ and have a distal fan of large strong setae (Fig. 3.12(b)) with which the substratum can be shovelled away posteriorly. The fields of movement of these 6th limbs in *Limulus* scarcely overlap those of the 5th pair and thus the digging limbs can be used independently of the rest during active excavation (Fig. 2.6(c)).

3. Food collection and its transport towards the mouth

The early armoured arthropods, as in *Limulus* today (Fig. 3.12(a)), could have collected soft-bodied invertebrates from the surface or surface layers of the substratum by grasping adductor–abductor movements of leg pairs. Flexures of the distal podomeres could bring the limb-tips and the prey towards the ventral surface of the body (Figs. 1.2(b), 4.3). The strong spines on the median side of the limbs of *Olenoides* and other trilobites would be helpful for such actions. Simple walking legs of extant arthropods do not carry such heavy armature. *Burgessia* has a pair of pincer-like processes on each leg formed from the ventral face of the coxa and the next podomere (Fig. 1.7). The second-antennal limbs of *Marrella* could serve as admirable sweeps for collecting detritus and small food from the sea floor.

Thus food material could have been collected in plenty and brought to the mid-ventral line between the leg bases. From this position the food could have been shifted forwards to the mouth in an automatic manner (see below). A lobe from the ventral surface of the body projecting downward and backward, the labrum or upper lip, could have prevented food from escaping ingestion by suction at the oral orifice. Before considering fossil arthropods further it is useful to recognize how mechanical means are used by extant arthropods in moving food forwards mid-ventrally.

A. *Food transport by mechanical means in Crustacea and in* Limulus

Two crustacean examples will suffice, the notostracan Branchiopoda, which lived in the Lower Triassic exhibiting much the same body structure as today (Pl. 2(a)), and *Nebalia* among the more primitive of the Malacostraca. The Notostraca are not filter feeders. They eat particles of a variety of sizes which they collect from the bottom by the metachronal promotor–remotor movement of a series of closely set limbs. These limbs bear knobbed, setose, short endites (Fig. 4.16, Pl. 2(a), *Triops*-type) which beat the food backward into the mid-ventral space between the paired legs, a feature unique among Branchiopoda. The proximal endite, or gnathobase, is forwardly directed. The successive limbs are close enough together for metachronal rhythm to allow each pair of gnathobases to push food material to the pair in front at the

Habit divergencies and arthropodan food-gathering

moment when leg n is finishing its backstroke and leg $n+1$ has started to move forwards. In this way a column of food material is pushed forwards to the mouth.

Nebalia is a filter feeder, the process of filtration leaving food material in the middle line between the thoracic legs (see Chapter 4, §3D(ii)). The forward transport of this food material is done mechanically. A horizontal section through the carapace and thoracic legs of *Nebalia* is shown in Fig. 2.11. From the basal part of each leg arises a short row of long brush setae which are forwardly directed. One from each leg is drawn, the others being above and below. Metachronal rhythm of the promotor–remotor swing of the legs enables one pair of brushes to push food material forwards in the

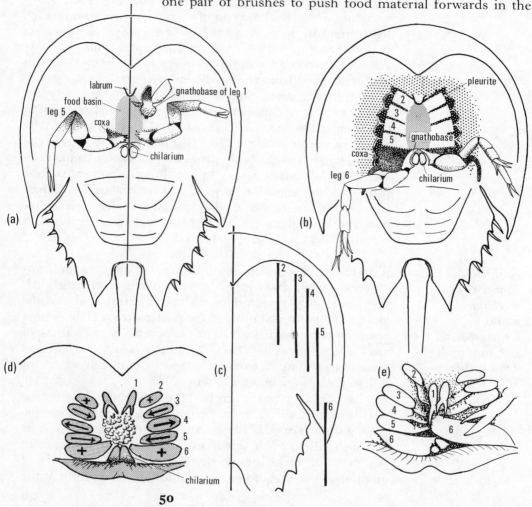

Habit divergencies and arthropodan food-gathering

middle line on to the next anterior pair of brushes when leg n is finishing the backstroke and leg $n+1$ is starting the forward stroke, utilizing exactly that feature of the metachronal rhythm used by *Triops* (below). By this means the food is pushed to the mouth parts and so to the mouth.

Limulus possesses five pairs of prosomal limbs each with a large cusped and setose gnathobase. These are crowded together with little space between one pair and the next (Figs. 2.6, 3.13(a)). Two distinct movements are used, the metachronal locomotory promotor–remotor swing and at right angles to this a biting adductor–abductor movement. Gnathobases 3 to 5 each bear a movable endite directed upwards almost into the mouth and capable of actively swinging anteriorly but not posteriorly (Fig. 3.13(b)). The sixth and most powerful biting gnathobases can shift food forwards to the gnathobases in front, as shown by Fig. 2.6(e). A combination of all movements passes food mid-ventrally forwards into the mouth, the forceps on the 1st postoral leg assisting as required.

B. *Food transport by mechanical means in trilobites and other early arthropods*

The successive legs of the trilobites *Olenoides* and *Triarthrus* differ from those of the Notostraca (e.g. *Triops*), *Nebalia*, and *Limulus* in being set further from one another and in possessing a small coxa–body junction (Figs. 1.1, 1.3). The legs of *Limulus* and of the Crustacea mentioned above are all expanded in the transverse plane at their origin from the body and *Limulus* has a strong dorso-lateral articulation between the coxa and body (Fig. 3.12).

FIG. 2.6 (see opposite). *Limulus polyphemus* in ventral view through glass from a cinematograph film. Arthrodial membrane at the leg joints is shown in black, the knee being much foreshortened and stippled, rising from the substratum. (a), (b) Walking movement. (a) The animal's right prosomal leg 5 is at the end of the remotor backstroke. The animal's left prosomal leg 5 is at the end of its forward swing and the knee is still strongly flexed upward. (b) The extreme forward and backward positions of prosomal leg 6 are shown on opposite sides of the diagram. Legs of a pair move in similar phase, the legs on either side of (a, b) being taken from frames half a cycle of limb movements apart in time. A whole leg viewed from the transverse plane is shown on Fig. 3.12. (c) The heavy vertical lines depict the range of movement of prosomal limb-tips 2–6 relative to the anterior end of the carapace. (d) Feeding movements of prosomal limbs taking place at right angles to the walking promotor–remotor swing. The genital operculum hangs down behind the chilaria. Horizontal sections are shown of the bases of the coxae. Those of legs 2 and 6 are stationary while coxae 3–5 move in opposite phase as indicated by the arrows, legs of a pair moving in the same phase. The food is soft and lies in the middle line. At other moments and when food is harder, the full series of coxae chew, but not always so regularly, (e) showing a momentary strong heave by the left 6th coxa. (From Manton 1964.)

Habit divergencies and arthropodan food-gathering

The reconstructions of *Triarthrus* (Cisne 1975) provide, for the first time, clear evidence of the nature of the coxa–body junction. The head of the coxa is narrow, there is no coxa–body articulation, and the musculature is not massive (Fig. 1.4). Therefore coxal movements on the body could have been varied and less restricted to certain planes than they are in most Crustacea, but the movements must of necessity have been weak. At rest the coxae and their large gnathobases were transverse in position, except for some fanning out at the anterior and posterior ends of the body. The coxa–body joint could have permitted a promotor–remotor movement, an adductor–abductor movement, and a slight twisting so that the axis of swing at the end of the backstroke brought the gnathobase forwards. Suitable combinations of these three basic movements could transport food forwards in the middle line. Adduction would bring the gnathobases together, gripping food by their stout median cusps, but no biting could be accomplished by such weak musculature. The metachronal promotor–remotor swing could have shifted food forwards, an essential aid being a twist at the end of the backstroke giving a final grip and forward push to the food mass. Such a twist is only possible in the absence of an articulated coxa–body junction. The leg bases of *Daphnia* and *Nebalia* are wide in the transverse plane, yet permit a small twist at the end of the backstroke (Cannon 1927, 1933, and see Chapter 4, §3D(ii)).

Hutchinsoniella has a similar wide leg base and the promotor–remotor swing is modified into a decided elliptical movement. The short cusps on the protopodal endites are

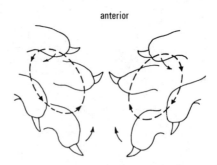

FIG. 2.7. *Hutchinsoniella macracantha*. Ventral view of a protopodal endite moving through its elliptical oscillatory cycle. The movement is not a simple promotor–remotor swing (for whole limb see Fig. 4.6). (After Sanders 1963.)

moved as shown in Fig. 2.7, the whole leg being seen in Figs. 4.6 and 4.7. A pair of endite cusps come together at the end of the backstroke, grip the food and carry it forward, and then release it by an outward movement; the cycle is then repeated. The great length of the gnathobase in *Triarthrus* appears to be related to the distance between the bases of paired and successive legs. Without this remarkable length of gnathobase a small twist of the leg at the end of the backstroke could hardly bring one pair of gnathobases up to the pair in front.

Such leg movements would be suitable for trilobites feeding on soft invertebrates, the median spines on the endopods and the coarse cusps on the median edge of the gnathobases being of importance in gripping prey or dead material. None of these features could have provided a filtratory method of feeding.

Early armoured arthropods, such as *Marrella* and *Burgessia* (Figs. 1.6 and 1.7), lacked gnathobases. Their successive legs were closer together than in *Triarthrus*, particularly so in *Marrella*. These animals could have used the metachronal movement of their legs for food transfer close to the body and towards the mouth, the close approximation of the legs mid-ventrally in *Marrella* probably having been particularly helpful. The presence of gnathobases was not essential for forward movement of food in the middle line but doubtless increased the possible size of soft-bodied prey organisms.

The early arthropods, with many leg pairs used in metachronal rhythm, fulfilled the needs of feeding, surface locomotion, and shallow burrowing. Such movements were a prerequisite for the evolution of gnathobasic mandibles as present in the modern marine Crustacea and the gnathobase of *Limulus*. Mandibles (jaws) are not required for the actual ingestion of food which is done by muscular action of the pharynx and oesophagus. The mandibles serve to prepare the food for swallowing.

4. Hydraulic systems formed by the limbs of aquatic arthropods

Metachronal waves of limb movement are used for swimming and also for respiration, by changing the water surrounding flattened respiratory surfaces, such as gill lamellae of *Limulus*, outer rami of trilobites, and exites in *Anaspides* and *Chirocephalus*.

Two successive flattened legs moving in unison are shown in side view in Fig. 2.8(a); in position B they hang vertically

Habit divergencies and arthropodan food-gathering

from the body, while positions A and C show different degrees of forward and backward inclination. The inter-limb space is greatest in position B. When a couple of legs swing from behind forwards towards the vertical, movement D in Fig. 2.8(b), water is sucked into the inter-limb space (interrupted arrow), while in movement E, where the legs are swinging posteriorly, the water is expelled from this space (interrupted arrow).

The effects of metachronal rhythm of the limbs on water movements is shown in Fig. 2.8(c). In position F leg N has reached the end of its forward stroke, it is momentarily stationary and lies just anterior to the vertical; leg N1 has

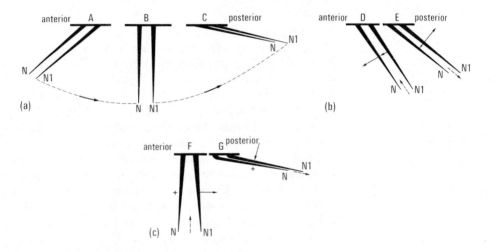

FIG. 2.8. Diagrams to show some features concerning swimming; the legs are seen in side view, appended from the body above (heavy line); solid arrows indicate leg movements; interrupted arrows show water movements. (a) Two consecutive legs, N and N1, moving in similar phase, (A) in a forwardly directed position, (B) in a vertical position, and (C) in an almost extreme posterior position. (b) Two similar consecutive legs moving in similar phase in the direction of the heavy arrows, in (D) the inter-limb space is increasing in volume and water is being sucked in between the legs N and N1, approaching the maximum size of this space shown in (a) position (B) above; in (E) the inter-limb space is decreasing in volume, moving towards position (C) in (a) above and water is being squeezed out of the inter-limb space. (c) A similar couple of successive legs, but moving in metachronal rhythm, leg N1 being slightly in advanced phase on leg N. In (F) leg N is at the end of the promotor swing and just about to start the remotor movement, it is stationary and marked by a cross. Leg N1 has just started to move backwards, water is being sucked into the inter-limb space which is larger than in (a) position (B). In (G) leg N is stationary at the end of the backstroke while leg N1 has started to swing forward, so that the two legs are squeezed tightly together and any intervening water is expelled, cf. (a) position (C).

started to move posteriorly. The inter-limb space is now greater than in Fig. 2.8(a), position B, and gives maximum suction of water into this space. In position G leg N is at the end of its backstroke and stationary while leg N1 has started to move forwards, so giving a maximal squeeze on the inter-limb space from which all water can be expelled. The use of alternate suction and expulsion of water into and out from an inter-limb space serves swimming and other purposes; it is advantageous for the limbs to move forwards only as far as position F and then to employ the remotor swing.

Most effective swimming will result when the limbs are flattened in the transverse plane so that they offer greatest resistance to the water. Respiratory structures such as crustacean exites or trilobite gill lamellae are often flattened and if they are moved as described maximum change of water between the respiratory surfaces will result. Two consecutive legs of a branchiopod crustacean *Chirocephalus* are shown in Fig. 2.9 moving in metachronal rhythm. The lateral and medial edges of the flat limbs are curved posteriorly and overlap the limb behind. In addition, flexures take place, as shown, along the length of the limb which enhance the alternate inter-limb suction and the production of a swimming current. *Branchinella*, Fig. 2.10, moves essentially as in *Chirocephalus* and in *Sida*, Pl. 2(b). Above each trunk limb an arrow shows the direction of swing of the leg, the length of the arrowed line indicating roughly the speed of movement of each limb. The inter-limb water ejected from between legs 1 to 5 form a swimming current; a maximum squeeze on the inter-limb space occurs where legs 5 and 6 are moving towards each other and the maximum suction takes place between legs 10 and 11 which are moving momentarily in opposite directions (see also *Sida* in Pl. 2(b)). The feeding currents and filter plates will be considered later (Chapter 4, §3D(ii)). When legs are used as simple paddles in swimming the leg can be outstretched throughout the propulsive backstroke as well as during the forward swing, except for minor flexures contributing to hydraulic efficiency; well-formed joints permitting the leg to shorten when half-way through the backstroke are not needed as in legs used for walking, because the cuticle is so thin and flexible.

Swimming can result from the use of parts of limbs, the whole limb not serving as a paddle. The copepod *Calanus* uses metachronal movements of the endopods of antenna 1,

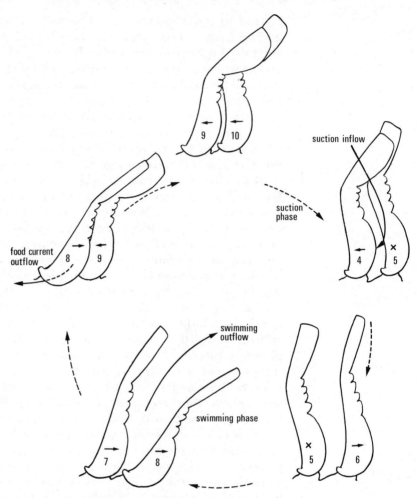

FIG. 2.9. Diagram to show a complete oscillation, divided into five phases, of two consecutive legs from the left side of *Chirocephalus*, viewed from their median aspect. (After Cannon 1928.)

antenna 2, and mandible to create swimming currents which propel the animal gently forwards (see also Pl. 3(c)). The elaborate morphology permitting the rapid escape jumping through the water by the thoracic limbs is considered in Chapter 4, §3B(iii). The thoracic exopods of *Anaspides* (Fig. 4.5(a), Pl. 3(f)) and of *Hemimysis* (Fig. 3.7) are moved by a rotatory, or in some other species by a backward and forward, beat, the long setae spreading out on the backstroke, so

Habit divergencies and arthropodan food-gathering

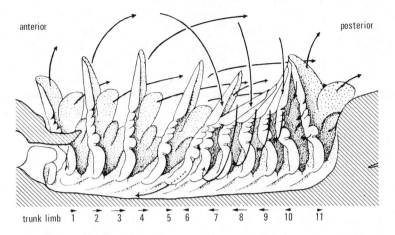

FIG. 2.10. Outline sketch of left half of an anostracan crustacean (based on *Branchinella australiensis*) to show swimming and feeding currents. The arrows below the trunk limbs indicate their relative movements. All setae have been omitted. The animal is on its back, the normal position in which it swims. (After Cannon 1933a). See also *Sida* in Pl. 2(b).

providing resistance to the water, and folding against the axis of the exopod on the forward movement. The pleopods also contribute to swimming (see Pl. 3(f) and legend).

The trilobites *Olenoides* and *Triarthrus*, with their narrow coxal bases and wide separation between one leg pair and the next, can have had no hydraulic system capable of providing a swimming current from the whole leg. But the exopods of *Triarthrus* in general form resemble those of *Anaspides* and presumably could have been used for swimming, the respiratory lamellae providing the resistance to the water in much the same way as the setae of *Anaspides*. In *Olenoides* the exopods are shorter and the respiratory lamellae longer and all are covered by the pleural expansions of the body. Appropriate movements of these exopods might have provided both swimming and respiratory currents.

A. *The transfer of food forwards towards the mouth by currents in Crustacea*

It is doubtful whether the trilobites or other early armoured arthropods could have utilized currents to transfer food forwards in the mid-ventral line. They must have relied entirely upon mechanical rather than hydraulic transport of food forwards to the mouth. The whole limb of *Triarthrus* and of *Olenoides* could provide no effective paddle action because the parts are rounded and the coxal head is narrow.

57

Habit divergencies and arthropodan food-gathering

The exopod movement might have provided an axial flow of water, as in the crustacean *Hemimysis*, but there is no structural arrangement capable of directing part of this flow forwards (see Fig. 3.7 and Chapter 4, §3D(ii)).

The Crustacea show various methods of forward transport of food by currents. In the branchiopod shown in Fig. 2.10 there is a forwardly-directed gulley leading from each inter-limb space to the middle line. At the moment of maximum compression of an inter-limb space (see that between legs 5 and 6, where the endite setae close this space on the median side but are not drawn) the only exit for the last of the inter-limb water is through the gulley to the mid-ventral line; an

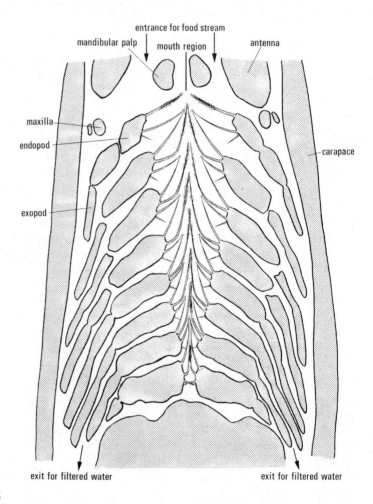

arrow marks this flow. During the metachronal beating of the trunk limbs, paired forwardly-directed jets of water pass into the food groove intermittently and combine to form an orally-directed food current leading particles filtered from the water by filter plates (see Chapter 4, §3D(ii)) to the mouth.

The various methods employed by crustaceans in transporting food to the mouth mid-ventrally by currents are dependent upon much higher levels of morphological differentiation than are seen in trilobites. Insufficient accurate detail is known about other early arthropods, but it is much more likely that their mid-ventral food transfer was effected by mechanical and not by hydraulic means.

The crustacean *Nebalia* (Fig. 2.11) filters an antero-posterior mid-ventral current, which is produced by metachronal rhythm of the flattened thoracic limbs, much as in *Chirocephalus* or *Branchinella*, but the inter-limb water leaves this space in a postero-lateral direction (see Chapter 4, §3D(ii)). Filtered particles left on the filter walls flanking the mid-ventral space are not collected by currents, but are moved mechanically by the metachronal action of sweeping brush setae which push food forward towards the mouth and against the incoming current (see above §3A).

5. The transport of food from below upwards towards the mouth in the Uniramia (Onychophora, Myriapoda, Hexapoda)

The route for the food to the mouth along the mid-ventral line which has just been considered must have been widespread in the early jawless arthropods which were provided with good sclerotization and jointed bodies and it paved the way for the evolution of grinding and biting mandibles in Crustacea and of pedipalpal chewing in Arachnida. A contrasting route for the food to the mouth is exemplified by the terrestrial Uniramia (the Onychophora, Myriapoda, and Hexapoda) and shown diagrammatically in Fig. 2.1(b). There is no transport from behind forwards close to the body, but only directly upwards towards the mouth, which is again

FIG. 2.11 (see opposite). Horizontal section through the thoracic legs of the leptostracan crustacean *Nebalia bipes* to show the food-gathering setae. Hooked setae interlock at their tips from one leg to another; they arise in a row, one above another, and form a lateral filter wall on either side of the middle space. Metachronal rhythm of these limbs, together with their shape and the presence of a carapace, results in a current of water entering the mid-ventral space from in front, passing through the lateral filter walls of hooked setae on either side, and out posteriorly, as shown by the arrows. Metachronal rhythm causes the short spiked setae to shift fine material off the filter plates back into the mid-ventral space and here the long brush setae push the food particles forward from one set of brushes to the next, against the incoming current, and so to the mouth region. (After Cannon 1927.)

Habit divergencies and arthropodan food-gathering

situated above and behind a large labrum. A long series of legs may be present and they move in metachronal rhythm, but are not usually concerned in feeding. Only a few pairs of limbs placed immediately behind the mouth are so used and the postoral pair always manipulates food with the tips and not with the limb bases. The specialized paired uniramian limbs, or mouth parts, are very various in construction, usually only two or three in number and suited to a vast variety of food and methods of obtaining it. Only rarely can very finely divided food be collected on land, as is done in the water, unless the minute size of a uniramian species matches its minute food particles, such as the fungal spores which are eaten by very small Collembola, but it can be done for example by the Machilidae (Hexapoda, Thysanura, (Manton 1964)).

The simplest manipulation of food by uniramian limbs is seen in the anterior two or three pairs of trunk legs of scolopendromorph centipedes and of lysiopetaloid diplopods which at times hold or steady a food mass (Fig. 2.12). The centipede shows no structural specializations for this, but the diplopod possesses rows of strong setae on these legs which help to hold the food (cf. Fig. 2.12(b) with (c)).

In order that a pair of limbs may manipulate food by their tips, terminal setae or claws must be present and there must also be sufficient flexibility all along the leg. In the myriapods the leg joints provide this mobility and in the Onychophora the lobopodial limbs are very flexible all along their length, being capable of elongation, shortening, and bending in any direction (see Chapter 4, §5). In all Uniramia terminal claws are present on the walking legs.

The route for the food from below upwards presumably characterized the earliest terrestrial Uniramia and there is no evidence against the supposition that the tips of a few pairs of undifferentiated postoral limbs may have been used in feeding before the emergence on land. The Uniramia can never have gone through a stage utilizing the feeding methods of the armoured marine arthropods or some evidence of this would be expected to survive.

FIG. 2.12 (see opposite). (a) Ventral view of the head end of the scavenging carnivorous diplopod *Callipus foetidissimus* (Savi), male, to show the three short spinous anterior pairs of legs which are used for holding food but not for walking. (b) and (c) Anterior views of legs 2 and 30 respectively. The spines shown in heavy line are thick and highly sclerotized and absent on legs behind the third pair. (From Manton 1964.)

Habit divergencies and arthropodan food-gathering

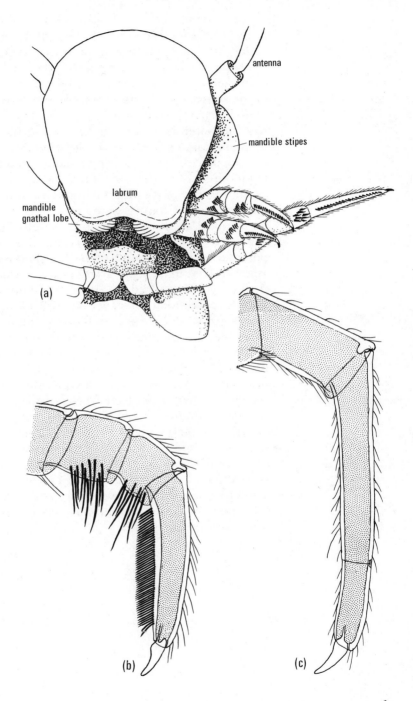

Habit divergencies and arthropodan food-gathering

If an aquatic ancestral uniramian employed a lobopodial limb to help food towards the mouth it could only do so by using the clawed tip. It could not use the soft coxal region of the leg in the manner of the armoured arthropods. That this manipulation of food by limb-tips is a particular potentiality of the supposed ancestral Uniramia is shown by the contrast with the parapodial limbs of polychaetes. No parapodium has the potentiality of transformation into a limb capable of manipulating food with the tip. All parapodia utilize an acicular mechanism for changing the length of the parapodium during stepping and for assisting the promotor–remotor swing taking place in parapodia used for crawling (see Chapter 4, §4). Parapodia are incapable of performing flexures along their length such as described above, an ability which characterizes all Uniramia and makes possible the manipulation of food by their limb-tips. Modern annelids feed in a great variety of ways, but they do not use their parapodia to hold food or to help it towards their mouths because parapodial construction does not permit such movements. Annelids take in their food by a highly muscular, often eversible, and sometimes armoured pharynx and they do not possess a labrum. Ancestral aquatic Uniramia probably relied upon a pharyngeal method of food intake or swallowing, but in this they were not unique. It is possible that the holding of a food mass by the tips of a few pairs of anterior limbs preceded the differentiation of one postoral pair of jaws. Such jaws cut or grind food material prior to ingestion by pharyngeal movement; and they do it with the tips and not with the base of the limbs.

The scene is thus set for the evolution of mandibles in armoured and in soft bodied aquatic arthropods, already possessing two well-organized and different routes for the food towards the mouth.

3 The evolution of arthropodan jaws

1.Introduction

LEG movements probably used in walking, shallow ploughing and burrowing, and in swimming by the early jawless armoured arthropods were considered in Chapter 2. In these animals and also in many modern Crustacea the food material travels forwards mid-ventrally close to the body to reach the mouth. This transfer is usually done mechanically, but sometimes by currents in the Crustacea. Ingestion of food takes place by swallowing movements and the evolution of jaws or mandibles provides mechanisms for dividing the food finely enough for swallowing.

The modern Arthropoda have evolved many ways of preparing their food for swallowing which involve seizing, cutting, and grinding. Variable numbers of postoral limbs are used, the number being constant in each taxon. The lobster catches prey with its large chelae on the 6th postoral limbs. Anterior to these great claws lie three pairs of maxillipeds which cut and shred the food and pass portions on to the paired maxillae 1 and maxillae 2 and so to the paired mandibles. A praying mantis on land is similar in that food, grasped by the prehensile claws of the 1st walking legs, is passed to the mouth parts and mandibles.

An understanding of the habits of life and of the ways in which the jaws of arthropods work, on a broad comparative basis, has only recently been achieved (Manton 1964). From it, much information advances our comprehension of evolution and classification of the Arthropoda. The essentials of the various jaw mechanisms presented in this chapter form part of the basis for the conclusions considered in Chapter 6. Many of the figures are simplified for the present purpose in that some muscles are omitted but other figures are reproduced in their original form, giving not only the features under consideration, but detail which was essential for the original investigation and which here assists in understanding

Evolution of arthropodan jaws

the jaw mechanisms. The habits of life, including certain feeding mechanisms and associated morphology, of some of the groups of animals whose mandibular mechanisms are described below will be considered in later chapters.

The immediately postoral pair of limbs in Crustacea (Fig. 3.1) and Uniramia (Onychophora, Myriapoda, Hexapoda) (Figs. 3.15 and 3.16) are the mandibles. The basic mandibular function is usually one of grinding or fine shredding of the food by a rolling movement (see below) and not of pushing food into the mouth. Crustacean and hexapod mandibles capable of strong cutting in the transverse plane of the body have been independently achieved by long series of evolutionary changes from primitive rolling mandibles. These advances have occured in many taxa and differ from one another in detail, although the end terms may be functionally similar in that strong cutting of the food takes place by movements in the transverse plane of the body.

The Chelicerata and Trilobita, on the other hand, lack a pair of mandibles, limbs used for feeding only. The Xiphosura employ a series of five pairs of gnathobases on their prosomal legs while the Arachnida use one pair on the pedipalps (Figs. 3.12–14). These gnathobases chew, shred, and sometimes execute strong cutting. The feeding movements (adductor–abductor) are always in the transverse plane in the Chelicerata contrasting with the basic mandibular movements (promotor–remotor) of the crustacean gnathobase (see below). The myriapodan mandibles bite together in the transverse plane but by their tips and not by a gnathobase, in contrast to the Chelicerata (Figs. 3.17–22). Myriapod mandibles are usually jointed (Figs. 3.16–18). The Trilobita possess a very large gnathobase projecting from the coxa towards the mid-ventral line (Figs. 1.1, 1.2(b), 1.3(b), 1.4(b), 1.5) on all limbs behind the antennae. This gnathobase must have been used in feeding, but rather differently from those of the Chelicerata.

The crustacean mandibles, formed by elaboration of one pair of gnathobases, show the distal part of the leg reduced to a palp and in many species it disappears entirely, in contrast to the gnathobasic walking legs of trilobites and Xiphosura. Uniramian mandibles lack a palp because they bite with the tips of their uniramous limbs. As shown below, there is nothing in common between the essential functioning of the crustacean mandible and of the chelicerate gnathobase, the two are fundamentally different in their modes of operation.

The contrasting jaws of the Chelicerata, Crustacea, and Uniramia must have evolved independently. Those of the two former groups could have been developed from early armoured arthropods possessing at least some pairs of gnathobases in which metachronal limb movements brought food forwards towards the mouth, close to the body, as considered in Chapter 2, Fig. 2.1(a). The whole-limb mandibles of the Uniramia, on the contrary, appear to have originated from soft lobopodia of early non-armoured arthropods with terminal claws which transported the food upwards directly towards the mouth, as shown in Fig. 2.1(b). The gnathobasic mandibles of Crustacea and the whole-limb mandibles of the Uniramia serve a variety of similar needs and it is not surprising, therefore, to find some convergencies in evolutionary end results. All the evidence points to the resemblances being superimposed upon fundamental differences in jaw evolution. The modes of functioning of these three types of jaws may now be considered. The mandibles of Crustacea have resulted from the utilization of the normal promotor–remotor swing of arthropodan walking legs (see Chapter 2, Fig. 2.2(a), (b)). The gnathobasic jaws of the Chelicerata use the adductor–abductor limb movements taking place at right angles to the promotor–remotor swing at the coxa–body junction (Fig. 2.2(c)). In the Uniramia biting with the tip is done in various ways, by rolling in the Hexapoda, and by transverse biting in the Myriapoda. A summary is given in Table 3.1 and the essential details follow in this chapter.

TABLE 3.1
Summary of jaw mechanisms (see further Chapter 6, Fig. 6.1)

Using	Promotor–remotor swing giving 'ROLLING' JAWS	Prehensile movements in transverse plane giving 'BITING' JAWS
Proximal endite or gnathobase	CRUSTACEA	CHELICERATA
Distal end of a whole limb	HEXAPODA	MYRIAPODA

The endoskeleton supporting mandibular or gnathobasic muscles is shown on the text-figures as pink stipple for connective tissue structures and red for cuticular apodemes.

Evolution of arthropodan jaws

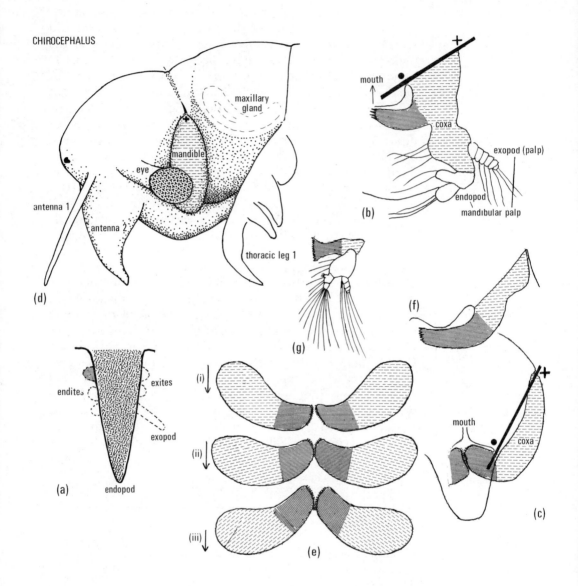

2. Crustacean mandibles

The essential form of crustacean mandibles has been known for a long time, but the way in which these organs work has been fundamentally misunderstood.

An example of a simple mandible among extant Crustacea is seen in the larva of *Hutchinsoniella macracantha*, Fig. 3.1(b).

Evolution of arthropodan jaws

FIG. 3.1 (see opposite). Basic structure and movements of crustacean mandibles. Interrupted ruling indicates the coxal segment of the leg; horizontal ruling the gnathobase or proximal endite; heavy lines show the axis of swing of the mandible on the head, the cross marking the dorsal articulation, or point of close union, with the head, and the black spot the ventral end of this axis where no articulation is present. It should be noted that there is only one transverse dorsal groove across the head; that it is not intersegmental; and that it is a mechanical device strengthening the union between cuticle and suspensory arms of the mandibular tendon (see Fig. 3.2(a)). No support exists for the often repeated drawing by Snodgrass showing three supposed grooves marking supposed intersegmental boundaries across the head, or for the theoretical deductions therefrom. (a) Embryonic simple limb rudiment of a crustacean. The positions in which lobes from the limb will grow are shown by dotted outlines; exites and exopod on the outer side and endites on the inner side of the future coxal region. The proximal endite can form a gnathobase and the distal part of the limb rudiment forms the endopod. (b) Mandible of the larva of *Hutchinsoniella macracantha* Sanders with mandibular palp formed from a reduced endopod and exopod. (After Hessler 1964.) (c) Mandible of *Chirocephalus diaphanus*, viewed from the transverse plane with the outline of the labrum, which lies in front and below the mandible. (After Manton 1964.) (d) Lateral view of the head of *Chirocephalus diaphanus* with the mandible exposed laterally and the posterior end of the labrum obscured by trunk limb 1. (After Manton 1964.) (e) Diagrammatic horizontal section of a pair of mandibles of *Chirocephalus* showing progressive positions (i–iii) during the remotor roll about the axis of movement, to show the changing positions of the molar processes of the mandible. (After Manton 1964.) (f) Mandible of adult *Hutchinsoniella macracantha*, palp now absent. (After Hessler 1964.) (g) Mandible of adult *Calanus finmarchicus*, palp well developed. (After Sars 1903.)

The coxal segment† is large (interrupted lines); it bears an inwardly-directed proximal endite, or gnathobase, which meets its fellow in the middle line (horizontal ruling); the remaining part of the main axis of the leg, the endopod, is reduced and the exopod arises distally on the coxa, as is usual in Crustacea. Together the endopod and exopod form the mandibular palp, here biramous as in many adult Crustacea, e.g. the copepod *Calanus* (Fig. 3.1(g)). A well developed uniramous palp (endopod), lacking an exopod, is present in *Anaspides*, mysids, crayfish, and crab (Figs. 3.4, 3.5(a), 3.6(a), (e), 3.7). The adult *Hutchinsoniella* lacks a mandibular palp (Fig. 3.1(f)).

The axis of swing of the mandible on the body lies obliquely in the transverse plane in the larva of *Hutchinsoniella*, Fig. 3.1(b), in contrast to the horizontal axis of swing of the walking legs of *Anaspides* (Fig. 4.5). A small shift of this axis towards the vertical in the adult *Hutchinsoniella* is carried further in *Chirocephalus*, where the mandible forms

† The simplest terminology is employed in this book. The coxa or the coxal segments here means the protopod, or basal unbranched portion of the leg. The coxa is usually in one piece but it may be divided into podomeres, moveable or fixed on one another; see the united 'coxa' and 'basis' of *Anaspides* (Fig. 4.5) forming parts of the coxa, as usually defined, and not flexible on one another.

much of the side wall of the head (Fig. 3.1(c), (d)).

The simplest mandibular movements consist essentially of the promotor–remotor movements seen in walking legs. The remotor movement is the stronger and it brings together the mesially-directed margins of the pair of mandibles (see below). The movements and the musculature in *Hutchinsoniella* are not simple and result in a postero-medial rotation, which brings the cusped median ends of the gnathobases together in the middle line; a somewhat similar movement is shown by the trunk limbs, Fig. 2.7. The gnathobases are curved so that their cusps reach the mouth opening. On the forward stroke of the mandibular base the mandibular cusps move backward and apart; on the backstroke these cusps rotate forward while the lateral part of the mandible moves backward (see below). Food particles are masticated on the forward stroke of the cusped mandibular margin. The mandibular musculature is rather more elaborate than that described below for *Chirocephalus*.

A. *Mandibles serving simple raptatory and sometimes filter feeding in Copepoda*

The type of mandible present in the bottom-living *Hutchinsoniella* (Fig. 3.1(f)) masticates food particles which are already small. Although filter-feeding copepods are well known, most species are raptatory whether bottom-living or swimming (Pls. 3(b), (c)). Their mandibles are of the *Hutchinsoniella* type but more robust, with stronger, larger cusps on the gnathobase; a bent axis fits the L-shaped gnathal margin well into the oral aperture; and a ball-and-socket articulation dorso-laterally with the head corresponds with the point marked by a cross, Figs. 3.1(b), (c), (d). Stronger masticatory movements are thus possible, the interlocking cusps comminuting larger food. Prehensile postoral limbs hold relatively large food, such as chironomid larvae, in many cyclopoids. The mandibular movement appears to be a slight modification of the promotor–remotor swing of an ambulatory limb (see below). Filtering food from the water appears to be a secondary accomplishment in, for example, certain calanoids; the food particles reach the mandibles from behind and there is little change in the structure of the mandible, apart from the cusps being slightly smaller than in the predators.

Raptatory bottom-living copepods are not far removed from those with epizoic habits or habits of scraping up food from the surface of larger prey. Two free living harpacticid

Evolution of arthropodan jaws

copepods are shown in Figs. 6.10(a), (b). They both possess simple mandibles with very strong terminal cusps which lie right opposite the mouth. In *Amphiascus* (Fig. 6.10(a)) the cusps are situated between the labrum and the paired paragnaths, but in *Idya* (Fig. 6.10(b)) a trunk is formed by apposition of the labrum against the fused paragnaths. The latter wrap round the mandibles distally and greater oesophageal suction now can be exerted, the mandibles and maxillae 1 working at the tip of the trunk. Such anatomy could pave the way for suctorial feeding and parasitism, the mandibles then becoming reduced to stylets (see entognathy in Chapter 6 §3).

B. *Basic form of crustacean mandibles used for primitive grinding movements*

The mandibles of the branchiopod crustacean *Chirocephalus* differ from those of *Hutchinsoniella* in the relatively larger size and in the molar processes which bear fine grinding ridges instead of cusps. The upper and lower ends of the axis of swing on the head are marked by a spot and a cross on Fig. 3.1(c) and on other figures. The essential detail of the simple dorsal articulation of the mandible, (Figs. 3.2(a), 3.11(a)), consists of a short flexible link of lightly sclerotized cuticle uniting strongly sclerotized opposing faces of cuticle forming the actual articulation. There is no precise ventral articulation and laterally the mandible is connected with the head by ample flexible cuticle, the arthrodial membrane, which permits the rolling mandibular movements about the axis.

The route for food particles to the molar lobes of the mandibles is from behind forwards, close to the body, as described in Chapter 2 §3 and Fig. 2.10 for the anostracan *Branchinella*, essentially resembling *Chirocephalus*. A similar route is present in many filter-feeding Malacostraca, some of which live largely on the bottom, for example *Anaspides* (Pl. 3(f)), while others are mainly pelagic, such as *Paranspides* and mysids. The food current lies between the bases of the thoracic limbs of the mysid in Fig. 3.7 and passes forwards to the molar lobes as shown by the heavy arrow on Fig. 3.8(a). The actual mode of filtration is considered in Chapter 4.

The movement and musculature of the mandible of *Chirocephalus* is described below; the type may be taken as basic to the mandibles of many other Crustacea serving particular purposes. In some orders no examples of primitive mandibles exist today.

Evolution of arthropodan jaws

The mandibles of *Chirocephalus* do not bite together in the transverse plane. They execute movements directly comparable to the promotor–remotor swing of a walking leg, but owing to the mandible being short and close to the body, the ambulatory swing becomes a rolling movement whereby the molar processes rub against each other giving a grinding effect. These rolling mandibular movements are shown in Fig. 3.1(e) by horizontal diagrammatic sections of the pair of mandibles at successive moments during a remotor swing or roll about the axis of movement. The medial molar surfaces of the gnathobase are ridged and during the remotor backward roll they come together, first anteriorly (i), then they roll against each other (ii), and finally the posterior corners come together (iii). The gnathobases are curved, so facilitating the rolling movement and a forward shift of the food across the molar ridges (Figs. 3.1(e), 3.2(c)). Usually the two mandibles are slightly asymmetrical in shape, an assistance in the detailed movement and grinding of food particles. A much more elaborate asymmetry of mandibles is seen in the mysid, Fig. 3.10. Sometimes grinding is assisted by the pair of mandibles moving slightly out of phase with one another, i.e. not exactly synchronously, resulting in one set of molar ridges rubbing across the other. These movements take place in the space boxed in between the labrum and the ventral body surface. The finely-divided food material is then sucked into the mouth by muscular movements of the oesophagus.

The movements of the simpler crustacean mandibles are not due to direct transverse adductor muscles. The rolling movements are caused by three categories of muscles: (i) promotor and remotor muscles arising from the proximal rim of the mandible and inserting on the dorsal body wall (Fig.

FIG. 3.2 (see opposite). Skeleto-musculature of the mandible of *Chirocephalus diaphanus*. (a) Thick transverse section, viewed from in front, to show the muscles and endoskeleton (pink); level A passes in front of the anterior margin of the mandible and level B passes through the middle of the mandible, showing its internal space and posterior margin. (b) Transverse section at level C, posterior to level B in (a), at the posterior junction of mandible and head to show the direct posterior remotor (adductor) muscles from the posterior mandibular margin. (c) Horizontal diagrammatic view through the mandibular tendon and mandible to show muscles 5a and 5b (promotors and remotors) attached to the tendon (pink) and the forward position of the molar processes of the mandible, seen as if the body was transparent. Muscles 3 and 5b execute the promotor mandibular roll but become abductors in large-food feeders; muscles 4 and 5a execute the remotor mandibular roll but become adductors in large-food feeders (see Fig. 3.6). (After Manton 1964.)

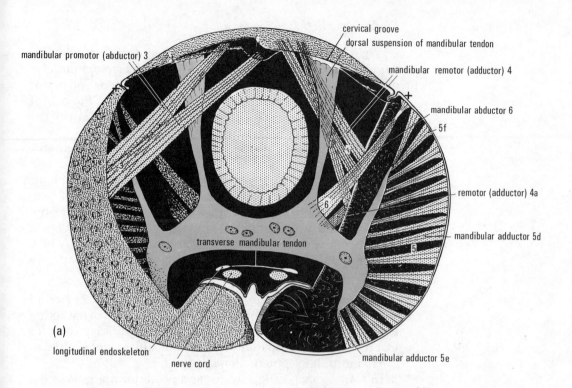
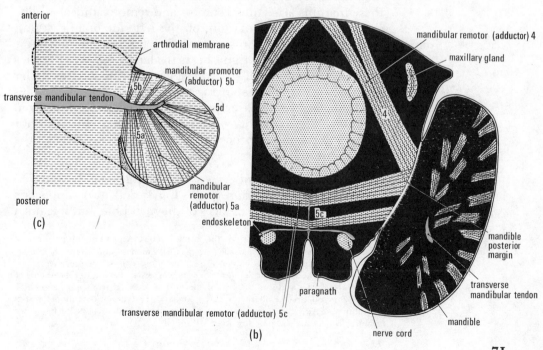

Evolution of arthropodan jaws

3.2(a), (b))†; (ii) muscles arising from the internal mandibular surface and inserting on a substantial transverse mandibular tendon (pink, Figs. 3.2(a), (c)); and (iii) true posterior adductor muscles uniting the posterior margin of the pair of mandibles (these muscles are smaller in bulk than those of the other categories (Fig. 3.2(b))).

In the examples considered below the promotor–remotor rolling muscles 3 and 4 of *Chirocephalus*, mysids, and *Anaspides* (Fig. 3.2(a), 3.5(a)) become secondarily abductor and adductor respectively in function in the isopods (e.g. *Ligia* (Fig. 3.5(b))). In the crayfish muscle 4 remains as the principal adductor but in the crab this muscle is small and a very strong adductor complex is secondarily formed by new muscles, adductors 1 and 2, arising from a remarkably constructed lever (see below).

(i) *Promotor and remotor muscles*. These muscles correspond with the extrinsic promotor and remotor muscles of a walking leg which insert on the tergum. The mandibular muscles insert dorsally as is spatially convenient and are not limited to the mandibular segment alone.

In *Chirocephalus* the anterior and posterior margins of the mandibles are approximately vertical in position and from these margins arise the promotor and remotor muscles. The anterior mandibular margin, on the left side of Fig. 3.2(a), bears the marked promotor muscles 3, while the posterior margin, shown on the right, bears the larger remotor muscles 4. The promotor roll gives a slight parting of the molar processes (gnathobases) owing to the shape of the mandibles and in that sense only are the promotor muscles 3 abductor in function. The remotor roll ultimately brings the posterior corners of the molar processes firmly together and again the remotor muscles 4 here are adductor only in this sense.

(ii) *Muscles to the transverse mandibular tendon*. The transverse mandibular tendon (pink, Fig. 3.2(a), (c)) supports a mass of muscles from all faces of the mandible, but it is not a

---

† Muscles do not actually insert on the cuticle but on basement membrane or subectodermal sheets of connective tissue (Fig. 3.11(a)) which are united with the cuticle by tonofibrils. This qualification concerning insertions will not be repeated in the text. Secondly, the view that the head consists of a series of cylindrical segments is invalid. Much of the dorsal surface of the head is derived from non-segmental embryonic ectoderm (see Chapter 6 §2A(i), Fig. 8.3). Thus arguments about how far mandibular extrinsic muscles pass out of their own segment dorsally has little meaning for *Chirocephalus* or any other arthropod.

Evolution of arthropodan jaws

'transverse adductor tendon' as usually stated in the literature; and neither is the entirety of muscles from the mandible to this tendon adductor in function. The horizontal section in Fig. 3.2(c) shows the muscles marked 5b to be much shorter and less bulky than those marked 5a. The former contribute to the passive promotor roll and the latter to the active grinding remotor movement. The transverse mandibular tendon is formed of densely packed connective-tissue fibres (the nuclei of a few of their formative cells are shown in Fig. 3.2(a)). Processes from the main tendon provide maximum support for the promotor and remotor rolling muscles and the whole tendon system is anchored dorsally to the cervical groove. This groove does not represent an intersegmental boundary† but is a simple mechanical device providing support for the tendon without incurring buckling of the outer cuticle.

(iii) *Posterior adductor muscles*. Some final force is given to the end of the remotor roll by the transverse muscles 5c marked on Fig. 3.2(b) which unite the posterior margins of the pair of mandibles. Contraction of these muscles brings the posterior parts of the molar processes hard together, completing the forward movement of the gnathobases. Only these muscles are truly adductor in function in *Chirocephalus* and they are seen again in the mysid, marked on the left of Fig. 3.11(b), and in other primitive Malacostraca. But in the large-food feeding Crustacea considered below the strongest adductor muscles either correspond with remotor muscle 4, pulling on a strong apodeme from the posterior margin of the mandible (Figs. 3.5(a), 3.6(b), (c), 3.11(c)), or are new developments, but they do not span the head transversely.

C. *Crustacean mandibles used for cutting large food*

The type of grinding mandibles just described characterizes the Branchiopoda and the simpler Malacostraca and presumably was present in many early Crustacea where the gnathobase ended in molar ridges instead of simple cusps, as in *Hutchinsoniella* and many copepods. Grinding molar processes must have been a usual first step in mandibular evolution associated with filter feeding by post-mandibular limbs, but elaborate filter feeding as we now see it, must be the result of much evolutionary progress; see Chapter 4 §3D(i).

† See footnote on p. 72.

Evolution of arthropodan jaws

A later step in mandibular evolution results in the production of organs capable of some cutting of the food, as well as grinding. Ultimately the grinding becomes incompatible with strong cutting and is abandoned. There are many different ways in which crustacean mandibles have become highly modified for biting.

(i) *Cutting incisor processes.* The formation of incisor processes on crustacean mandibles is a widespread advance in many taxa and preceded the evolution of the highly specialized mandibles of contrasting types which can cut very strongly.

Ideally simple, hypothetical mandibles are shown in Figs. 3.3(a), (b) which are symmetrical, with the axis of swing passing through the middle of each (heavy line), the mandible rolling equally in either direction about this axis. The mandibles of *Chirocephalus*, considered above, are only a little different (Figs. 3.3(c), (d)) in that the molar process is directed slightly anteriorly, resulting in a better forward shift of the food.

In *Anaspides* (Figs. 3.3(e), (f), Pl. 3(f)) three changes have occurred. The amount of mandible situated anterior to the axis of swing is reduced while that posterior to it is increased. Cutting cusps lie on an incisor process at the posterior corners of the gnathobase, far from the axis. Thirdly the dorsal end of the axis of swing of the mandible, marked by a cross in Fig. 3.1(c), has shifted posteriorly so that the axis of swing no longer lies in the transverse plane; compare with Fig. 3.3(c) and (e). The outer aspect of the mandible of *Anaspides* with its oblique axis of swing is shown in Fig. 3.5(a). Ventral views of the mandibles of *Anaspides*, Fig. 3.4, show the movement from the extreme promotor roll in (a) to the final limit of the remotor roll in (d). The dorsal end of the axis of swing, marked by a cross, lies far below the plane of the paper. The

---

FIG. 3.3 (see opposite). Diagrams showing the movement of simple crustacean mandibles. Figures on the right show a median aspect of each mandible as seen in a left half of the body with the anterior end to the right. Figures on the left show horizontal sections across the pair of mandibles in progressive positions of the remotor roll. The axis of swing is marked by a heavy line on the right and by a spot (a section of this line), on the left. (a), (b) An ideally simple symmetrical mandible serving grinding only by the molar areas. (c), (d) Mandible of *Chirocephalus diaphanus* with the molar process directed slightly forward. (e), (f) Mandible of *Anaspides tasmaniae* Thompson showing the oblique axis of swing; reduction of the preaxial part of mandible; enlargement of the postaxial part, the posterior ventral corner of which forms a cutting incisor process. The molar areas still grind. (After Manton 1969.)

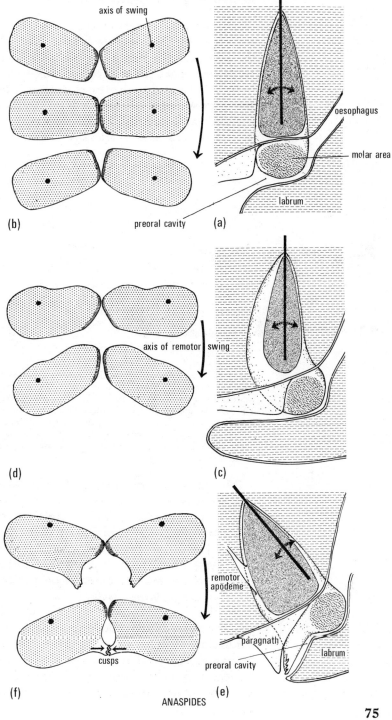

ANASPIDES

Evolution of arthropodan jaws

black spot which bears an arrow is at the end of the promotor roll at (a) and the spot moves round (b)–(d) in the direction of the arrow as the mandible rolls. In (a) the molar grinding ridges are apart. They then roll together, dorsal to the labral margin. The incisor processes can never be further apart than the width of the labrum or food would escape; they come together and cut the food, close against the labrum. The incisor process is seen laterally in Fig. 3.5(a).

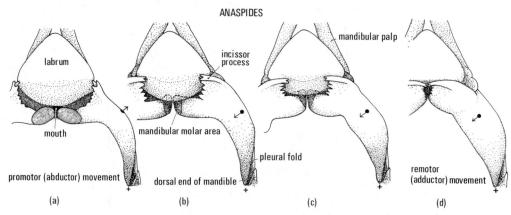

Fig. 3.4. Ventral views of a mandible of *Anaspides tasmaniae* Thompson to show the range of movement about the axis seen in Figs. 3.3(e), (f), 3.5(a). The dorsal articulation of the mandible, marked by a cross, lies far below the plane of the paper. The spot bearing an arrow marks the same place on the mandible in each figure and the arrow shows the direction of mandibular roll. The extreme promotor roll is shown in (a) with the molar and incisor processes parted. In (b)–(d) the molar processes come together and the incisor processes bite just behind the posterior edge of the labrum. (After Manton 1964.)

The musculature causing the mandibular movements of *Anaspides* is essentially as in *Chirocephalus*. Two promotor (abductor) muscles 2 and 3, working at poor mechanical advantage, are drawn in Fig. 3.5(a) arising from the small preaxial part of the mandible. An end-on view of the transverse mandibular tendon is indicated in black in the middle of the mandible and it bears the same category-(ii) muscles as in *Chirocephalus*.

At the posterior edge of the mandible arises a stout posterior mandibular apodeme from which a mass of remotor (adductor) muscles 4 pass upwards, but also laterally, to the carapace, a cervical groove again providing an anti-buckling device. Transverse category-(iii) muscles unite the posterior margins of the mandible, as in *Chirocephalus*. It is the strength of the bulky remotor muscles 4, swinging the mandible posteriorly about its axis of movement (interrupted line ending in a cross and a spot in Fig. 3.5), which gives particular strength to the cutting action of the incisor processes.

A mandible such as that of *Anaspides* is found in the less advanced Malacostraca, for example mysids and *Paranaspides*, where the incisor processes at times can cut into soft pieces of food held by the thoracic endopods. A diagrammatic view from the middle line of such a dual purpose mandible is shown in Fig. 3.8(a) where the route of filtered food to the mandible is shown by the heavy arrow passing directly on to the ridged molar process. The distal incisor process is small. The actual mandible of *Hemimysis* in Fig. 3.9(a) gives the mandibular details, extra teeth being situated above the incisor process (see below). The feeding mechanism is described in Chapter 4 §3D(ii).

(ii) *Large, hard food.* An ability to feed on large, hard food either by scavenging or by catching live prey, may be regarded as an advanced achievement, independently acquired by many Crustacea. Three examples of the mandibles of such Crustacea will be considered below, but there are other solutions to the same problem existing in the various taxa. Many are based upon the primitive grinding mandible and its musculature (e.g. *Chirocephalus*) on which has been superimposed a cutting incisor process, e.g. *Anaspides*, which can eat large, soft food, such as tadpoles, as well as filter.

The various methods of obtaining very strong cutting in the transverse plane are incompatible with mandibular grinding and the dual purpose mandible, with both incisor and molar processes, present in the less advanced Malacostraca, has given way to entirely shearing mandibles in the more advanced members of this group.

(iii) *Transverse cutting of large food by the mandibles of Isopoda*. *Ligia* bites into food masses by the strong incisor processes of the mandibles coming together in the transverse plane, biting also being done by the strong distal setae on the maxillae 1. A backward shift of the dorsal end of the axis of movement of the mandible, beyond its position in Fig. 3.5(a) to that in 3.5(b), brings the axis in *Ligia* to a horizontal position at the side of the head. A hinge-like articulation of the mandible on the head in this position results in

Evolution of arthropodan jaws

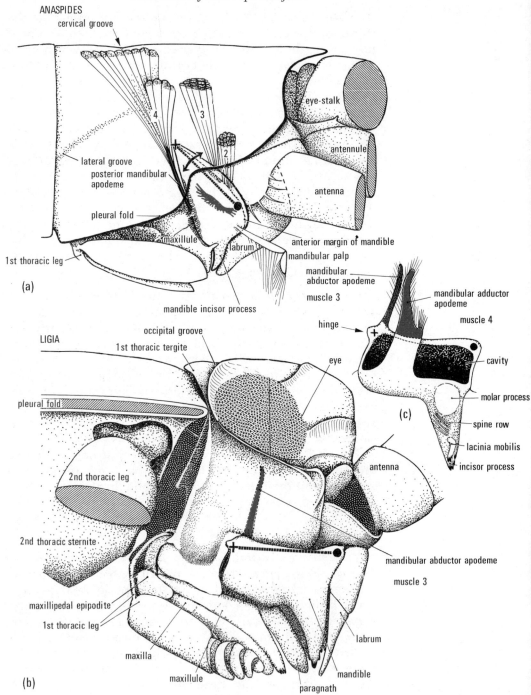

abductor–adductor movements of the mandible taking place in the transverse plane. The original preaxial part of the mandible is much reduced, forming a small expanse just dorsal to the heavy line marking the axis of movement in Fig. 3.5(b) and from it arises a long apodeme bearing abductor muscles 3 which slope upwards and outwards into the bulge of the cheek. The muscles work at poor mechanical advantage, a very common condition for muscles causing recovery movements in contrast to those doing outside work. In *Ligia* muscles 3 part the mandibles and stretch out the relaxed adductor muscles 4. A huge mandibular adductor apodeme extends from the mesial margin of the mandible deep into the head, as shown in Fig. 3.5(c). From it adductor muscles 4 insert on the upper internal surface of the head and on to apodemes from it; the head cuticle is strengthened by a variety of folds, described in the literature as 'lines', but again these provide anti-buckling measures and are not segmental boundaries (cf. *Chirocephalus* and *Anaspides*).

The mandibles of *Ligia* cut strongly by their incisor processes, assisted by the paired lacinia mobilis, a moveable tooth. The molar processes remain, but form flat crushing faces which can grind only a little at their proximal corners. The interior of the mandible is almost devoid of muscles; there is no transverse tendon and most of the associated muscles 5, concerned with the rolling mandible of *Chirocephalus* and *Anaspides*, have become redundant and have disappeared. The labrum in front and the paired paragnaths and maxillae 1 behind the mandibles prevent food from escaping.

Thus the form of the transversely-biting mandibles and their musculature in an isopod such as *Ligia* are derivable

FIG. 3.5 (see opposite). Lateral views of the head of *Anaspides* and of *Ligia* to show one of the methods of obtaining strong biting in the transverse plane from primitive dual-purpose mandibles with strong grinding molar processes and weak biting incisor processes. (a) *Anaspides tasmaniae* Thompson. The axis of swing is oblique and shown by a heavy line ending in a cross, which marks the dorsal articulation, and a spot marks the ventral freer end of the axis. Promotor muscles 2 and 3 roll the molar processes away from each other and part the incisor processes, while remotor muscle 4 causes grinding by the molar processes moving across each other and cutting by the incisor processes (see Fig. 3.4). (b) *Ligia oceanica* Roux. The dorsal end of the axis of swing has become displaced so that the axis is now horizontal. The mandible has very little preaxial region left, but it bears a large abductor apodeme which carries muscle 3. As the mandible swings on its hinge-like articulation it bites with its fellow in the transverse plane. (c) View from the middle line of the left mandible, devoid of muscles, to show the shape of the abductor and adductor apodemes. (After Manton 1964.)

Evolution of arthropodan jaws

from the rolling type of mandible with an incisor process as is possessed by mysids and *Anaspides*.

(iv) *Transverse biting by the mandibles of crayfish and crab.* These Decapoda (Pls. 3(h), (i)) cut their food differently from the Isopoda. Mandibles receive pieces of food from the maxillipeds and chelipeds (first four pairs of trunk limbs) and do not bite directly into the food, as does *Ligia*. The mandibles are relatively smaller and lie well within the contours of the head. The whole of the mandibular gnathal margin is now a stout cutting blade. The labrum does not overhang the mandibles, as in the species described above, but lies within the cutting blades anteriorly, the mandibular palps controlling antero-lateral escape of food (Figs. 3.6(a), (b), (e)).

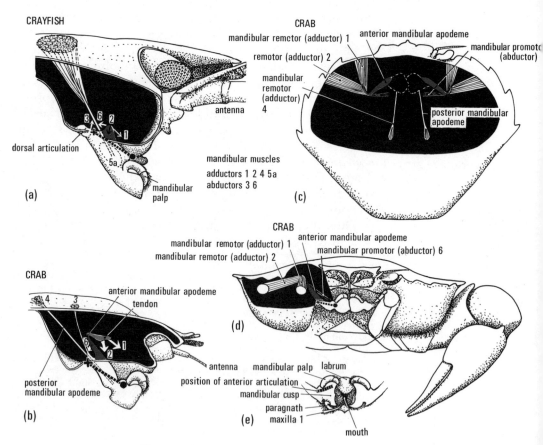

Evolution of arthropodan jaws

The axis of the movement of the mandible on the head is marked by the interrupted line (Figs. 3.6(a), (b)). It is oblique, as in *Anaspides*, but there is no strong dorsal articulation near the cross and no freedom exists at the black spot as in the more primitive mandibles. A main hinge providing the mandibular movement in the crayfish is an elongation of the dorsal articulation along the anterior mandibular margin, while in the crab the dorsal articulation is very weak and a strong anterior articulation lies near the black spot between the epistome and the mandible at the base of the palp. These hinges are derivable from the simple malacostracan type of mandible but the ensuing modes of action of the mandibles of the crayfish and the crab are different.

Besides the posterior apodeme carrying adductor muscles 4, there is a preaxial apodeme, small in crayfish but very large in crab, where this apodeme, with the preaxial part of the mandible, forms a massive cuticular extension into the head which acts as a lever on the mandibular articulation near the black spot (Fig. 3.6(d)). The anterior apodeme of these decapods carries strong adductor muscles, unlike the anterior mandibular apodeme of *Ligia* which is abductor in function. In crayfish and crab adductor muscle 2 arises from the apodeme and passes laterally to the carapace, aided by adductor muscle 1 which passes antero-laterally (Figs. 3.6(a)–(d)). A tendinous flap lies between the tip of the lever-like apodeme of the crab and the origin of muscles 1 and 2. These muscles in the crab become the principal adductors; they are more bulky and stronger than the original adductor muscle 4, which is still present, and their importance in providing very strong cutting depends on the short, strong mandibular articulation with the epistome and the long lever

---

FIG. 3.6 (see opposite). Mandibles and their movements of the crayfish and crab suiting large-food feeding for comparison with *Chirocephalus, Anaspides,* and *Ligia,* Figs. 3.1–5. Direction of pull of some of the muscles is shown by arrows instead of the muscle being drawn. The axis of swing of the mandible on the head is indicated by the dotted line ending in a cross and a spot, but the limited part of this axis which is formed by a hinge-like articulation is not shown, see text. (a), (b) Lateral view of the right mandible of crayfish (*Astacus* (*Potamobius astacus* L.)) and crab (*Carcinus maenas* (L.)) respectively with the head cut back. (c) Dorsal view of the crab to show anterior and posterior mandibular apodemes and their muscles, the position of the gnathal lobes being indicated by white dotted lines. (d) Anterior view of the crab with the head cut back to show the anterior mandibular apodeme and its muscles. The mandible and its muscles are foreshortened in (b) and (d). (e) Ventral view of the mouth region of the crab with the mandibles fully parted disclosing the labrum between the mandibular palps. (After Manton 1964.)

Evolution of arthropodan jaws

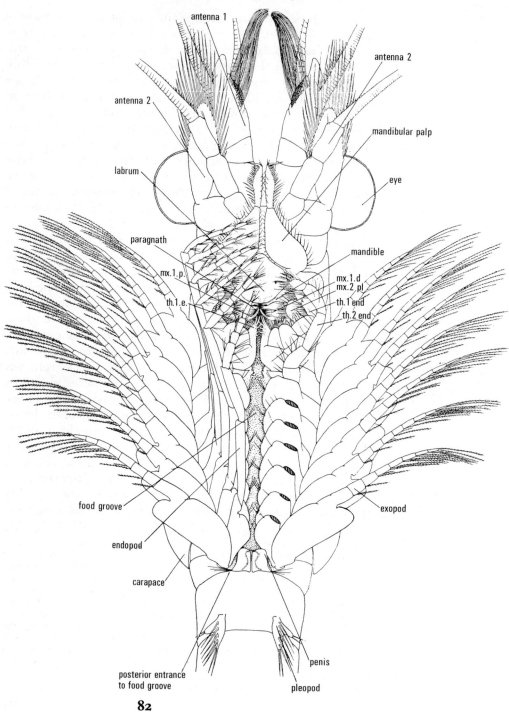

Evolution of arthropodan jaws

produced as described. Adductor muscle 2 represents the functionally transformed mandibular abductor muscle 2 of *Anaspides* (Fig. 3.5(a)). An abductor muscle 6 is marked on Fig. 3.2(a) in *Chirocephalus*; a comparable abductor muscle 6 is also present in *Anaspides* arising from the anterior mandibular margin and inserting on tendinous endoskeleton; in the crayfish and crab abductor muscle 6 is also present, pulling as indicated on Fig. 3.6. The primitive adductor muscles 3 remain in both decapods, as shown. The transverse mandibular tendon is absent and muscle 5 (Figs. 3.2(c), 3.6(a)) becomes progressively more redundant in the two decapods.

Thus the crab shows an ultimate perfection of strong cutting by a mandible using well developed levers, but based upon the skeleto-musculature of more primitive malacostracan mandibles.

(v) *Mandibles used for large-food feeding in certain Peracarida and Syncarida.* In the pelagic large-food feeding peracaridan, *Lophogaster typicus* and in the bottom-living syncaridan *Koonunga cursor*, the route for the food to the mouth is from below upwards, as shown by the heavy arrow in Fig. 3.8(b), in contrast to a route from behind forwards as in the filter-feeding members of these taxa, Figs. 3.7, 3.8(a). The mandibles of the large-food feeders possess massive cusped incisor processes while the molar ridges are less extensive than on the mandibles of filter feeders in the same groups. Diagrams and actual preparations from the same aspect are shown in Figs. 3.8(a), (b) and 3.9(a), (b) of the filter-feeding *Hemimysis* and of the large-food feeding *Lophogaster*. The food which can be negotiated by *Lophogaster* cannot be as hard as some of the food of crabs.

Although a habitual pelagic filter feeder, *Hemimysis lamornae* (Fig. 3.7), when plankton is scarce, can lift food off the bottom, passing it from below upwards to the mouth by the remarkable armature of the mandibular margins and edges of the labrum. These are shown in posterior view from the transverse plane in Fig. 3.10. The right incisor process shears against the left and bites against the movable tooth (the lacinia mobilis). The right lacinia mobilis is more slender and pushes

FIG. 3.7 (see opposite). Ventral view of the filter feeding malacostracan *Hemimysis lamornae*. The thoracic endopods 3–8 are cut short on the right in order to disclose the mouth parts. A food-bearing current passes forwards in the middle line between the thoracic legs, through the maxillary filter plates and out laterally. (From Cannon and Manton 1927.)

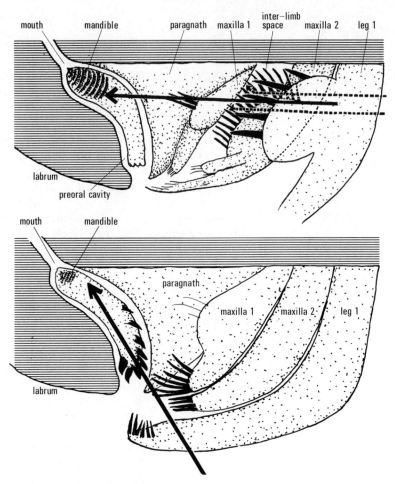

Fig. 3.8. Diagrams to show the routes taken by the food to the mouth in (a) a typical filter-feeding malacostracan, such as *Hemimysis lamornae* (Peracarida, Mysidacea) and *Paranaspides* (Syncarida), and (b) a non-filtering large-food feeder such as *Lophogaster* (Peracarida, Mysidacea) and *Koonynga* (Syncarida), etc. The right half of the head and mouth parts are viewed from the median plane, the body being cut vertically through the median labrum (upper lip) shown by ruling. The paired paragnaths (lower lips) lie on either side of the middle line. The median aspect of the gnathal face of the mandible and of the three following limbs is shown. The diagrammatic spines in heavy black denote those concerned with feeding and bear no relationship to actual size, see Fig. 3.9. The route taken by the food to the mouth is shown by the heavy arrowed line. The interrupted arrowed lines in (a) denote the water current passing from the middle line out through the maxillary filter plate spanning the suction chamber (inter-limb space) between maxilla 1 and maxilla 2.

food onto the left spine row where each large spine is separately inserted on the mandible. On the right the spine row has but one insertion. The molar processes themselves are asymmetrical. Such a pair of mandibles can lift food

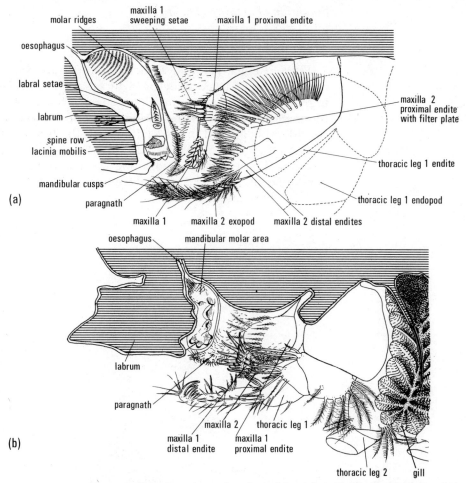

FIG. 3.9. Views from the median plane of the right half of the mouth region of (a) *Hemimysis lamornae* and (b) *Lophogaster typicus* to show the differences between the mandible and mouth parts suited to filter feeding in (a) and large-food feeding in (b). Note the large filter chamber (inter-limb space) between maxilla 1 and maxilla 2 in (a) covered on the median side by the maxilla 2 filter plate. There is no filter plate or a large space between these limbs in (b). The large first trunk-limb endite in (a), which scrapes food off the filter plate, is indicated by dotted lines only. There is no corresponding endite in (b). A diagrammatic representation of the routes of the food to the mouth and of the organs involved in these Crustacea is given in Fig. 3.8. ((a) From Cannon and Manton 1927; (b) from Manton 1928b.)

material up to the molar processes, triturating as it goes; then, after being ground, the material is sucked into the mouth by muscular movement. The mandibular margin of *Lophogaster* (Fig. 3.9(b)) and of the large-food feeding syncarid, *Koonunga cursor* form strong predominantly distal biting cusps without the elaboration seen in *Hemimysis*. The whole

Evolution of arthropodan jaws

feeding mechanisms, serving predatory habits, in selected examples, are given in Chapter 4, §3D.

These Peracarida and Syncarida illustrate yet another modification of basic structure which suits large-food feeding, but here it does not involve a change in axis of swing of the mandible. Instead the biting edges of the mandible are considerably altered in association with the modifications of distal rather than proximal parts of the postoral limbs for feeding. These changes go along with parallel and independent evolution of large-food feeding in two taxa which are not closely related.

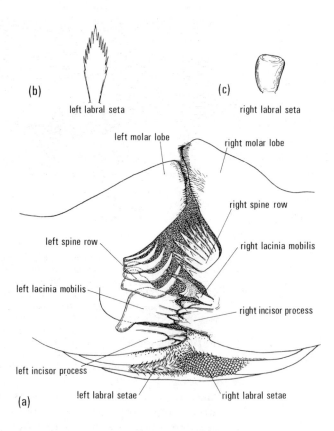

FIG. 3.10. (a) Posterior view of the biting edges of the mandibles and the posterior edge of the labrum in *Hemimysis lamornae*. The mouth lies above and in front of the molar processes. (b) Spine from the left edge of the labrum. (c) Spine from the right edge of the labrum. (From Manton 1928*b*.)

Evolution of arthropodan jaws

D. Conclusions on the jaw mechanisms of Crustacea

Thus amid the great superficial variety of mandibles in Crustacea, much of their evolution is now understandable on a functional and phylogenetic basis. The primitive type of rolling mandibular gnathobase underlies most of them and contrasts not only with the mode of action of the chelicerate gnathobase but with mandibles of all Uniramia. The secondary and varied means whereby the primitive crustacean mandible has become converted to serve strong transverse biting shows only partial and convergent similarities with mandibles of an entirely different type in the Uniramia (see below). There are also many examples of filter feeding having been derived secondarily from Crustacean with primitive raptatory habits and from specialized large-food feeders, with corresponding changes in the mandibles, but these will not be considered here.

E. Endoskeleton supporting mandibular movements in Crustacea

Whether a crustacean mandible performs primitive rolling movements which provide grinding, or whether the food is strongly cut into by the several independently-acquired means of biting in the transverse plane of the body, the mandibles need stronger exoskeletal or endoskeletal support than a normal walking leg. Three types of endoskeletal support are used by crustacean mandibles and by other arthropods, shown in Fig. 3.11. Although internal, all are derived from the ectoderm. Throughout this chapter red on the figures denotes apodemes formed by cuticular intuckings and pink denotes endoskeleton formed by basement membranes or by connective tissue, remaining attached to, or separated from, the ectoderm. In other chapters all types of endoskeleton are usually shown in black, except tentorial apodemes.

(i) *Basement membranes*. The simplest support is provided by basement membranes below the ectodermal cells, often strengthened by tonofibrils from the membrane passing through the ectoderm to the cuticle (Fig. 3.11(a)). This membrane may be thin or thick, or composed of a thicker layer of connective tissue fibres and on to it are inserted mandibular muscles, as in *Chirocephalus*, and tendinous supports from the transverse mandibular tendon (Figs. 3.2(a), 3.11(a))

(ii) *Transverse mandibular tendons*. A massive transverse tendon, carrying muscles from the mandibular concavity, characterizes primitive rolling mandibles. The tendon is

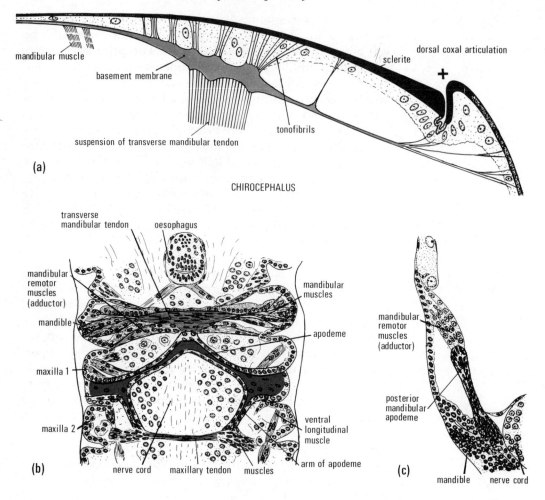

formed by a compact mass of connective tissue fibres. In the adult a few nuclei from the formative ectodermal cells of the tendon may be visible in sections (Fig. 3.2(a)), but in the late embryo abundant nuclei are seen in the horizontal section (Fig. 3.11(b)). The tendon system in sections shows up brilliantly blue in colour after Mallory's triple stain, thus differentiating the tendons from other structures.

The mandibular tendon is an enlarged member of a series of transverse intersegmental tendons which develop from the ectoderm (Manton 1928, p. 934). A very simple adult series is shown in the crustacean *Hutchinsoniella* (Hessler 1964) and by the trilobite *Triarthrus* (Fig. 1.4(a)). A maxilla 2 (maxil-

Evolution of arthropodan jaws

FIG. 3.11 (see opposite). To show three types of arthropodan endoskeleton: subcutaneous connective tissue (pink); internal connective tissue tendons (pink); and cuticular apodeme (red). (a) Transverse section through the dorsal part of the head of *Chirocephalus*, the point marked by a cross being the dorsal mandibular articulation (cf. Fig. 3.2(a)). Black indicates heaviest sclerotization of the cuticle. The subcutaneous connective tissue forms a substantial basement membrane, attached to the cuticle by tonofibrils traversing the ectodermal cells and internally bearing insertions of muscle and tendon. Connective tissue fibrils pass laterally to support the upper part of the mandible, as shown. (After Manton 1964.) (b), (c) Late embryo of *Hemimysis lamornae*. (a) Horizontal section at the upper level of the nerve cord and through the mouth parts. The transverse mandibular tendon is full of cells secreting the tendinous fibrils which form a massive tendon in the adult, such as that shown in Fig. 3.2(a), (c). The horizontal hollow head apodeme is formed by cuticular secretion from the ectodermal intucking shown. (d) Transverse section of the posterior junction of the mandible with the head; from this junction the hollow ectodermal intucking (shown) later secretes cuticle forming the blind stiff posterior mandibular apodeme, whose muscles give strong remotor movement, causing adduction of the incisor processes of the mandibles. The corresponding posterior mandibular apodeme for the adult crayfish and crab are shown in Fig. 3.6(a), (c). (After Manton 1928a.)

lary) tendon is seen in Fig. 3.11(b). Embryonic muscles become attached to the tendons, sometimes covering them completely, and the tendons may sag into adjacent segments, so appearing to be segmental in position, as in the mandibular segment. In most Crustacea the lateral ectodermal connections of the tendons break down in the adult and thus the tendinous endoskeleton becomes entirely internal. It grows with the body and is not shed at ecdysis.

The transverse mandibular tendons become redundant when rolling mandibular movements are replaced by strong transverse adduction. The wide gape of such mandibles is incompatible with the presence of a tendon and muscles linking their cavities.

Similar tendons are formed in myriapods and hexapods and subsequently greatly elaborated. They characterize series of segments moveable on one another (Chapter 8, §2A).

(iii) *Apodemes*. Hollow intuckings of the surface cuticle of the body form apodemes which often support mandibular tendons and also provide sites for the origin of mandibular muscles. A section of a late mysid embryo (Fig. 3.11(b)) shows a pair of horizontal ectodermal intuckings from the maxilla 1 segment which have penetrated into the body, fused with one another in the middle line, where the future apodeme is also united with the mandibular tendon. A pair of posteriorly-directed arms from the apodeme pass backwards to meet the developing maxilla 2 muscles. Later thick cuticle is secreted at the ectodermal surfaces (marked apodeme), so

Evolution of arthropodan jaws

forming a stiff hollow apodemal head endoskeleton which remains open to the sea water. At ecdysis the intucked cuticle is shed and a new and larger apodeme is formed.

A very simple apodeme, formed as a blind intucking like the finger of a glove, develops at the posterior proximal margin of the mandible (Fig. 3.11(c)), lying roughly in the vertical plane. Its inner surface secretes a stiff cuticle, which may or may not be linked by a short section of flexible cuticle with the junction of the mandible and the head. Embryonic remotor muscles 4 (Figs. 3.5, 3.6) are seen attaching to the inner end of the developing apodeme in Fig. 3.11(c).

There are enormous variations in detail and in degree of elaboration of crustacean endoskeleton, associated with the size of the animal and with the strength of mandibular movements, but all are dependent upon elaboration of these three basic types of endoskeleton. There is little uniformity in the apodemes present in various Crustacea, in contrast to the constant form and function of anterior and posterior tentorial apodemes in the Myriapoda and Hexapoda (see Chapters 8–9).

3. Gnathobases of the Chelicerata and Trilobita

In feeding the Chelicerata use paired gnathobases situated on five prosomal limbs behind the mouth in the Xiphosura and on the pedipalps alone in the Arachnida. Apart from the location of the gnathobase on the coxa, there is no re-

FIG. 3.12 (see opposite). The gnathobasic jaws of Chelicerata, Merostomata, Xiphosura, *Tachypleus tridentatus* Leach, viewed transversely to the body. Endoskeleton of connective tissue shown in pink. (a) Leg 5 with the gnathobasic part of the coxa ruled and the rest of the coxa marked by interrupted lines. The paired gnathobases are a little apart. The axis of the promotor–remotor swing, used in walking and burrowing, is marked by a heavy line, a close articulation being present only at the cross where the coxa articulates with a small pleurite. There is no articulation at the black triangle. The anterior proximal margin of the coxa is cut back a little to show the internal cavity; plenty of flexible membrane unites these anterior and posterior proximal margins with the body. Arthrodial membrane is stippled in white at the two pivot joints shown distal to the coxa. (b) The coxa alone, showing the positions of adduction and abduction superimposed. The abductor position, drawn uppermost in surface view and marked as on (a), parts the paired gnathobases; the position of the coxa in (a) is shown behind. Adduction is caused by many muscles pulling on the proximal anterior and posterior margins of the coxa. Abduction is caused by muscles pulling in the direction marked from a small coxal flange or apodeme, working at poor mechanical advantage because the apodeme is situated so near to the fulcrum of movement provided by the coxo-pleural articulation. This adductor–abductor movement is at right angles to the promotor–remotor swing in (a). See also Fig. 2.6(d), (e). (c) Transverse view across the prosoma in front of leg 6. The proximal anterior coxal margin is shown on the left giving origin to the marked muscles and this margin is cut away on the right to show the posterior proximal margin and its muscles. (After Manton 1964.)

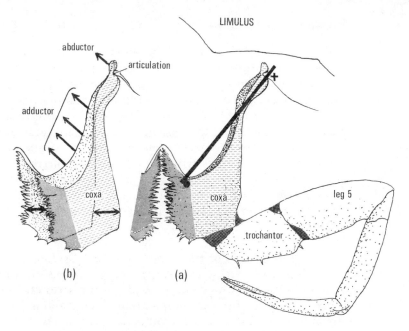
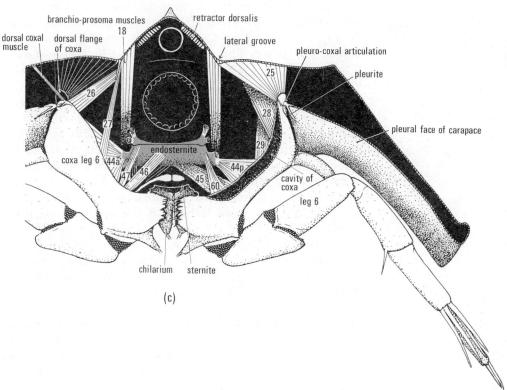

Evolution of arthropodan jaws

semblance to the Crustacea. The movements of chelicerate gnathobases in feeding are adduction and abduction in the transverse plane of the body, a chewing action by each pair, taking place at right angles to the promotor–remotor swing of the coxa on the body. It is the latter which is the basis of crustacean mandibular movements. Strong transverse biting is achieved in some groups of Crustacea and by parallel series of evolutionary changes which independently arrive at the same end result (see *Ligia*, crab, above). The transverse chewing used by all living chelicerates, other than suctorial parasites, appears to be the primitive feeding movement of the group.

A. *Xiphosura*

The living members of the Xiphosura, such as *Limulus* and *Tachypleus*, Figs. 1.9, 3.12, 3.13, probably show some of the basic movements which were used by the extinct Merostomata, but they have developed refinements and a robustness of limb which have contributed to their survival. A pair of preoral chelicerae ending in chelae is followed by five pairs of postoral limbs, each of which carries a gnathobase, marked in Fig. 3.12(a) as for Crustacea, the gnathal lobe bearing horizontal ruling. The positions of these limbs are shown in Figs. 2.6, 3.12(c), 13. The promotor–remotor swing of the coxa on the body takes place about the axis shown in black in Fig. 3.12(a) and is the basis of the walking and burrowing movements (see also Fig. 2.6(a), (b)). The coxa is well articulated to the body at the black cross, but elsewhere is united to the stiff body cuticle by flexible membrane. A chewing movement takes place as shown in Figs. 2.6(d), 3.12(b), the abducted coxa being marked as in Fig. 3.12(a) and behind it is shown the outline of an adducted coxa, the double-headed arrows indicating the directions of movement.

There are plenty of adductor muscles from the proximal margins of the coxa, inserting on to the endosternite (pink) and carapace. The endosternite is formed of compacted connective tissue fibres, as is the transverse mandibular tendon of *Chirocephalus* (Fig. 3.2(a)). Such horizontal and transverse tendons develop from the embryonic intersegments all along the body and they remain so in the adult trilobite (Fig. 2.4(a)) and in many living arthropods. Fusion together of these tendons above the nerve cord in the Xiphosura result in the large endosternite (Fig. 3.12, 3.13).

Evolution of arthropodan jaws

Coxal abduction is caused by the pull of muscle 25 (Fig. 3.12(c)) on a short apodeme or lobe from the coxa projecting into the body dorsal to the fulcrum of movement provided by the coxa–pleural articulation.

Successive pairs of coxae work in opposite phase from one another in chewing, as shown by Fig. 2.6(d), only three pairs of gnathobases at that moment being active. At other moments all may be in action and the large sixth pair may

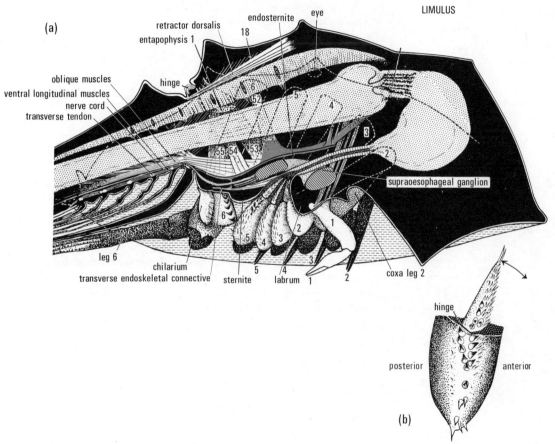

FIG. 3.13. View from the middle line of the left half of the body of *Tachypleus tridentatus* Leach to show the legs and the gnathobases in particular. Endoskeleton of connective tissue shown in pink. There is no gnathobase on the chelate chelicera, leg 1 (marked). Gnathobases 2–5 converge towards the mouth, each bearing a small articulated endite, as shown, and in (b). The 6th gnathobases form strong 'nutcrackers'. The positions of insertion on the body of the wide coxae are shown by numbered dotted lines (see also Fig. 2.6(b)). (b) Similar view of the gnathobase of leg 5 to show the movable endite which can swing forwards and backwards to the vertical position. (For detailed description see Manton 1964, legend to Fig. 16.)

Evolution of arthropodan jaws

heave food material forwards, as shown in Fig. 2.6(e). The longitudinal view in Fig. 3.13 shows how the coxae of legs 2–5 converge towards the mouth. Soft food material, such as worms, can be caught by the terminal chelae of prosomal legs 2–5 and the very flexible, but strong, leg joints permit the chelae to place food in the mid-ventral line for chewing.

The sixth pair of prosomal legs carry the 'nutcrackers', stout cusped gnathobases capable of breaking the shells of bivalve molluscs of considerable size, such as clams, which are seized during burrowing in the muddy or sandy substratum (Figs. 3.12(c), 3.13). This strongly cutting pair of gnathobases lies at the posterior end of the series of feeding limbs, unlike crustacean mandibles situated immediately behind the mouth.

B. *Arachnida*

In the Arachnida the pedipalpal gnathobases are the only pair used for chewing (Fig. 3.14). Arachnids are usually fluid feeders, dependent upon gnathobasic chewing and external digestion of the chewed prey, the resultant fluids being sucked into the mouth. The indigestible parts of the prey are reduced to a pellet which is ejected from the preoral space. In the scorpion the gnathobases never meet; they lie on either side of a preoral space which is closed anteriorly by the bases of the chelate chelicerae and floored by coxal lobes from walking legs I and II. In the spider the gnathobasic lobes are further differentiated from the coxa of the pedipalps and can close together in the middle line, bearing a dense and long armature of setae. The fangs of the chelicerae are here single and highly sclerotized, as are the cheliceral chelae of the scorpion; they fold as shown (Fig. 3.14(b)).

FIG. 3.14 (see opposite). Ventral view of a scorpion (a), (b), and a spider (c) to show the mouth region with the chewing gnathobasic lobes (ruled) formed by the median part of the coxa of the pedipalp (interrupted lines). A preoral space is flanked by the chewing gnathobases and closed in front by the pair of chelicerae, set close together anterior to the mouth. Each chelicera ends in a stout chela in the scorpion and in a curved fang in the spider; these terminal parts are shown in black. The chewing movement of the gnathobases, which never meet in the middle line, is indicated by the double-arrowed line. (a) Scorpion *Parabuthus villosus* Peters. Coxal lobes from walking legs I and II extend forwards to the preoral space and are slightly mobile. The coxae of legs III and IV are fixed on the body and do not move; they are directed towards the small sternite. The pectines are omitted. (b) Dorsal view of *Parabuthus villosus* to show the whole of the chelicerae, the tips only being visible in ventral view (a). (c) Spider *Xenesthis immanis* Auss. The dense armature of setae on the median edge of the gnathobases is indicated. (From Manton 1964.)

Evolution of arthropodan jaws

Thus the feeding gnathobases on the pedipalps of the Arachnida are a modification of the longer series of gnathobases shown by the Xiphosura and work in an essentially similar manner, but modified slightly for fluid feeding. There is no similarity whatever between the modes of operation of feeding gnathobases of Crustacea and Chelicerata.

C. *Trilobita*

Consideration was given in Chapter 2 of the ways in which food material could have been transported forwards, midventrally, towards the mouth in the early armoured arthropods, both with and without gnathobases. The fossil trilobites (Figs. 1.1–5) possessed a series of gnathobasic limbs, more primitive in their arrangement than any chelicerate or crustacean, the series extending backwards from behind the antennae (Chapter 4). The trilobite limbs possess distinguishing features but none show even the beginning of

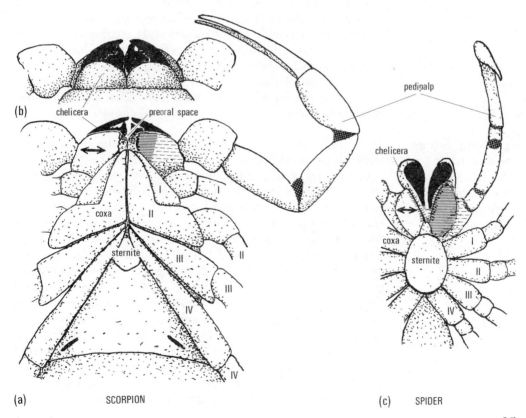

differentiation of a mandible. The trilobite leg base is narrow and simple, contrasting with the Crustacea and Chelicerata.

The serial uniformity of trilobite limbs and their probably primitive coxa–body union, emphasizes the great advances which have been made in feeding habits and associated morphology by even the simplest Crustacea and Chelicerata. The rolling, grinding, mandibles in the former and the transversely chewing gnathobases in the latter are each supported by appropriate endoskeletal structures (pink) bearing strong extrinsic limb muscles (see §2E and 3A above) in contrast to the serially simple endoskeletal bars of trilobites present all along the body (Fig. 1.4 white). The two living groups of marine animals have increased their powers of preparation of food for swallowing and this must promote a more rapid metabolic turnover and therefore a more active life. We do not know for certain how a trilobite used its enormous gnathobases (see Chapter 2, §3B), but the feeble extrinsic limb muscles and the weak coxa–body junction are enough to indicate the level of efficiency of trilobite feeding compared with modern Chelicerata and Crustacea. We do not know why the trilobites became extinct after so long a reign, but an inability to advance their feeding mechanisms on other lines, in face of the active and evolving Chelicerata, Crustacea, and maybe Chordata, may have been a prime factor leading to their end.

4. Conclusions concerning the jaws of the primarily aquatic arthropods

In the early armoured arthropods there was no single pair of appendages modified as jaws; the metachronal rhythm of the ordinary leg movements, together with the very simple coxa–body junctions, clearly seen in trilobites and in *Marrella*, could have resulted in forward transmission of food to the mouth along the mid-ventral line, whether or not a series of limbs bore gnathobases. Trilobites possess very large gnathobases all along the body, *Cheloniellon* on the four postoral segments only, but no strong movements could have been performed by these gnathobases. The Crustacea and Chelicerata both employ gnathobases for food preparation before swallowing, but with much greater efficiency than could have been done by any trilobite. The evolution of feeding habits and structure leading to the grinding or biting pair of crustacean mandibles and to the transversely chewing gnathobases of the Chelicerata are entirely opposed, the one to the other, and represent independent evolutionary trends.

Evolution of arthropodan jaws

5. Jaws and mandibles of the Uniramia (Onychophora, Myriapoda, Hexapoda)

Whole-limb jaws which bite with the tip and not with the base, as in the primarily aquatic arthropods just considered, are associated with a habit of raising food directly from below upwards as considered in Chapter 2, Fig. 2.1(b).

A. *Basic form of uniramian mandibles and the jaws of the Onychophora*

That the jaws or mandibles in the uniramian groups correspond with a uniramous, shortened, usually thickened, ambulatory limb is shown embryologically and plainly by the Onychophora, Fig. 3.15. The paired terminal claws on the walking legs seen in (a) correspond with the stout, highly sclerotized jaw blades seen in (c) and (d).

Each onychophoran jaw is short, both are surrounded by a circular lip and operated by a mass of extrinsic muscles, passing backwards and upwards through several trunk segments, and arising from the jaw base and from a long apodeme arising at the posterior proximal margin of the jaw. The circular lip can be distended and by oesophageal suction clamped down on to the prey, such as a termite or a woodlouse. The jaws slice antero-posteriorly and in alternation, as seen in the trunk limb movements when walking (Fig. 2.4, legs 14–16), but operating within a preoral space (see below) encompassed by the lip. Severed pieces of the prey are then swallowed.

The Myriapoda and Hexapoda possess heavier surface sclerotization. Their jaws or mandibles are segmented in the former and unsegmented in the latter. The massive three-segmented mandible of a iuliform diplopod denuded of muscles is drawn in Fig. 3.16(a) while (b) and (c) show a different kind of segmentation in the mandible of a scolopendromorph centipede. In both, the free end of the mandible (segment 3 in the diplopod and segment 5 in the chilopod), bears cutting teeth and other armature. The proximal segment of the diplopod mandible is strongly articulated on the head along the line between the two crosses (see also Fig. 3.18(a)), but the chilopod mandible is weakly attached to the head by a suspensory sclerite, shown in part in Fig. 3.16(b) and entire in Fig. 3.22(a). The symphylan mandible is two-segmented (Fig. 3.21(b)) and articulated to the head at one point only, marked by a cross.

Unsegmented mandibles characterize the Hexapoda. One of the simplest and most basic in its movements is shown in

Evolution of arthropodan jaws

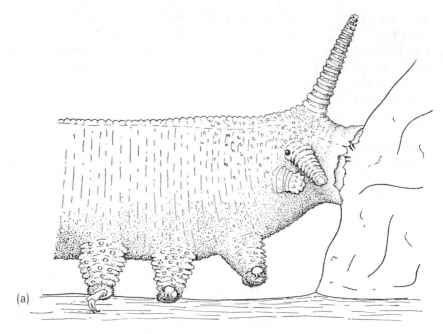

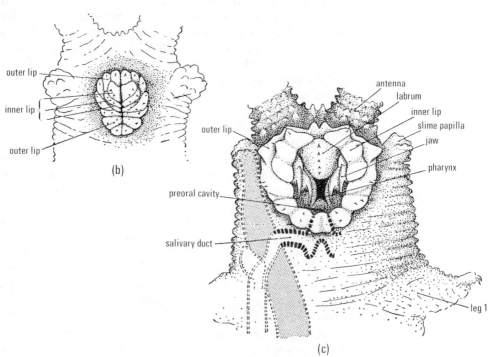

Evolution of arthropodan jaws

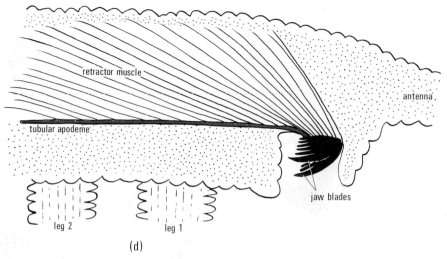

FIG. 3.15 (see opposite and above). Jaws of the Onychophora; the oral region of *Peripatopsis sedgwicki*. (a) Side view of an animal feeding on meat; the inner and outer lips are distended, forming a sucking tube, the tips of the jaw blades being visible in their anterior position. Support of the body on the substratum is secured by walking legs 3 and those behind it. Leg 1 is not used for either prehension or support, because of the position of the jaw apodemes. (b) Ventral view of the mouth with the lips closing the buccal or preoral cavity. (c) Ventral view of the mouth region of a larger specimen showing extended outer and inner lips and the form of the preoral cavity and associated structures. The positions of the salivary and slime glands are shown by dotted lines. The heavy lines on the inner parts of the salivary ducts represent the region lined by thick, smooth cuticle similar to that of the preoral cavity. (From Manton 1937.) (d) Diagrammatic longitudinal view from the middle line of one of the pair of jaws surrounded by the circular lip, dilated as in (c). The two jaw blades on each jaw correspond with the two terminal claws on each walking leg seen in (a). A huge retractor muscle extends from the very short base of the jaw and from a long tubular apodeme extending from the highly sclerotized jaw blades posteriorly through several segments. The musculature operating the jaw is more complex than shown.

Fig. 3.16(d) and its natural position on the head in Fig. 3.23(a). This mandible of the thysanuran *Petrobius* is strongly articulated at the cross, and ample arthrodial membrane links it with the head along its margins, dorsal to the molar lobe or area, so that a mandibular promotor-remotor roll takes place about the axis shown by the heavy line between the cross and the black spot, just as occurs in the gnathobasic mandibles of *Chirocephalus*, also with a vertical axis of movement (Figs. 3.1(c), (d), (e), 3.3(a), (b)). The molar lobes roll across one another and the distal end of the mandibles move together and apart, scraping up unicellular algae. In the locust the mandible is much shorter and the armature on the medial

99

Evolution of arthropodan jaws

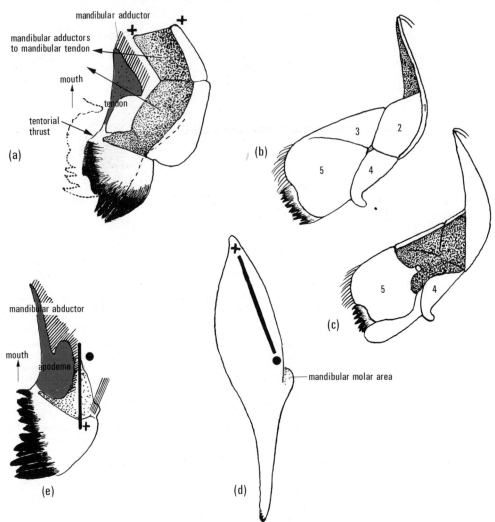

border extends along the whole length of the mandible. The changes in orientation of the axis of movement between the cross and spot is shown by Fig. 3.23(a)–(c) and see below.

The following functional account of the mandibles of myriapod and hexapod classes is minimal for an understanding of the evidence concerning phylogeny. For a full appreciation of the mandibular mechanism in any one group reference can be made to the original account (Manton 1964).

The basic movement of present-day myriapodan mandibles is one of biting together in the transverse plane, while

FIG. 3.16 (see opposite). Mandibles of myriapods and hexapods which bite with the tip of the whole-limb, as in Onychophora (Fig. 3.15), in contrast to the gnathobasic jaws of the Crustacea, Chelicerata, and Trilobita (Figs. 3.1–7, 3.12, 3.14, 1.1, 1.2(b), 1.3(b)). The black cross and spot mark opposite ends of an axis of movement on the head, for further description see text. Heavy stippling indicates the internal cavity of the mandible, all muscles being removed, (a)–(c) shows jointed mandibles in contrast to the unjointed mandibles in (d)–(e). (a) Anterior view of the three segmented mandible of a iuliform diplopod in the abducted position, swinging on a hinge present along the line between the two crosses at the junction of the proximal mandibular segment and the head. The dotted outline indicates the adducted position of the mandible. (b)–(c) Opposite faces of the mandible of a scolopendromorph centipede. The mandible is weakly attached to the head by the suspensory sclerite indicated above, cf. the crosses in (a); the mandible is jointed into five more or less separate sclerites which permit bending of the whole mandible. This mandible lies in a gnathal pouch (see Fig. 3.22(a)). (d) Lateral view of the unjointed mandible of the hexapod *Petrobius*. (e) Posterior view of the unjointed, short, mandible of a locust. The changes whereby the end of the axis of movement marked by a cross comes to lie more ventrally than the spot are shown in Fig. 3.23. (After Manton 1964.)

---

that of the unjointed hexapod mandible is a rolling promotor–remotor movement, much as in the gnathobasic mandible of *Chirocephalus*. From the basic type of hexapod mandibles have developed, by recognizable stages, mandibles capable of strong biting in the transverse plane on the one hand, and protrusible mandibles on the other. The former need a strong articulation with the head while the latter need flexible arrangements which will enable a protrusion of the mandible away from its base. The transitional stages towards these ends are mutually exclusive. The myriapod classes each show their own special proficiencies and they too show strongly biting mandibles and much weaker ones which can be protruded. On the figures in this chapter illustrating myriapods and hexapods, pink, as before denotes endoskeletal tendons; red denotes the anterior and posterior tentorial apodemes; while other head apodemes, notably the posterior mandibular apodeme is shown in white.

(i) *Preoral cavity*. It is useful to define this space so that the comparative aspects of the jaw mechanisms in the various groups can more easily be appreciated.

In the less specialized Crustacea the incisor and molar processes of the mandibles work in a space below and behind the mouth which is closed by the labrum antero-ventrally and by the paired paragnaths postero-laterally, the mandibles themselves flank the preoral space which leads to the mouth (Figs. 3.3, 3.8). A preoral cavity is useful for preventing the

Evolution of Arthropodan jaws

escape of fine particles of food but such a space is less closely confined in large-food feeders such as a crab (Fig. 3.6).

A corresponding preoral cavity houses the business ends of the mandibles of the Uniramia (Fig. 3.18(b)). The labrum, sometimes hinged horizontally into two parts, hangs down between and in front of the mandibles. The lateral edges of the labrum overlap the mandibles. When the mandibular gape is wide, so is the labrum (Fig. 3.19), and the labrum is narrow when the mandibles do not part widely (Fig. 3.23(d)). The lining of the labrum, called the epipharynx, forms the anterior wall of the preoral cavity while its floor is called the hypopharynx. The latter can bear many structures useful in feeding, such as the thick hypopharyngeal cuticular sclerites of diplopods (Figs. 3.18(b), 3.19), or tongue-like extensions or pads (Figs. 3.21(a), 3.22, 3.25(a), (b), 3.26).

A tongue-like hypopharynx is not formed by Crustacea nor are paragnaths present in the land Uniramia because the basic

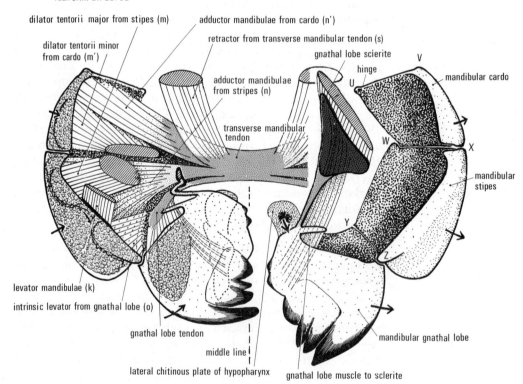

Evolution of arthropodan jaws

routes for the food to the mouth are different (see Chapter 2, §3, Fig. 2.1). The possession of a preoral cavity is a simple functional need which is differently elaborated in the two phyla.

B. *Mandibles of the Myriapoda*

The Myriapoda comprise the Diplopoda (millipedes), the Symphyla, the Chilopoda (centipedes), and the minute Pauropoda (Pls. 4(h)–(i), 5(f), 7(c)–(g); 4(d)–(e); 1(f), 4(j), (k), 5(a)–(e), (g), (h); 4(l), (m) respectively).

The segmentation of myriapodan mandibles appears to be associated with a primitive biting in the transverse plane, but the mechanism is quite different from that of transverse biting in the Chelicerata. Only the Myriapoda among Arthropoda have movably segmented mandibles. A large mandible forming much of the side wall of the head (Figs. 3.1(c), (d), 3.2(a), 3.24) can easily perform rolling grinding movements, as in *Chirocephalus*, derived from the promotor–remotor swing of the leg, but direct adduction and abduction are not easily achieved. Adductor muscles can readily pull a pair of mandibles together (Fig. 3.17), but there is no method whereby mandibular extrinsic muscles can cause abductor movements of mandibles such as those of a iuliform diplopod. All Myriapoda employ the surprising device of using a pair of swinging anterior tentorial apodemes within the head to push the mandibles apart, a method of mandibular abduction not found in any other group.

(i) *Diplopoda.* A iuliform diplopod may be taken as an example of the class. It feeds upon decayed wood and other vegetation for which strong biting is required. The wood is eaten in quantity and bitten into by the stout cusps on the gnathal lobe, or third segment, of the mandible. Crushing and grinding can be performed by proximal armature on the flattened median face of the gnathal lobe (Fig. 3.18(b)), particularly when the mandibles move a little out of phase with one another. The three mandibular segments are stoutly

FIG. 3.17 (see opposite). Anterior view of the isolated mandibles of the diplopod *Poratophilus punctatus* Attems. On the left side the musculature is entire and the gnathal lobe and whole mandible is in a position of maximum adduction. On the right the muscles have been removed from the mandibular cuticle, with the exception of one intrinsic muscle, and the mandible and gnathal lobe are in a position of extreme abduction. The contrasting positions of the lateral hypopharyngeal sclerite is also shown; the heavy arrow on the right indicates the direction of thrust of the anterior tentorial apodeme against the gnathal lobe. Transverse mandibular tendon shown in pink. (After Manton 1964.)

Evolution of Arthropodan jaws

IULIFORM DIPLOPOD

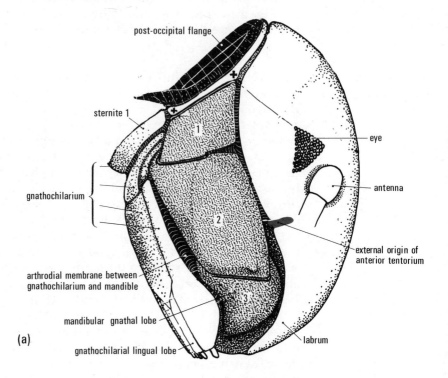

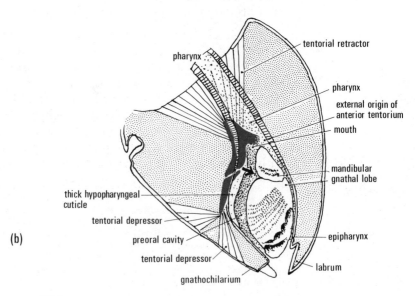

Evolution of arthropodan jaws

hinged to each other along the lines marked W–X and Y–Z on Fig. 3.17. Strong adductor muscles from the two proximal mandibular segments converge onto a transverse mandibular tendon (pink). The range of movement of these two segments, swinging inwards and outwards on the hinge-like articulation of the basal segment with the head, is small. The much greater range of movement of the gnathal lobe could not be implemented by massive muscles passing to the transverse tendon because such muscles, if they existed, would have to shorten and extend more than is possible to normal striated muscles. Instead a tendon from the gnathal lobe bears a very large adductor muscle inserting on the cranium, see right side of Figs. 3.17, 3.20.

The surface cuticle tucks inwards from a notch, shown in Figs. 3.18(a), 3.19, 3.20 to form the *anterior tentorial apodeme* (red). This hollow cuticular bar is flexibly attached at the notch and passes inwards across the wall of the preoral cavity (Figs. 3.18(b), 3.19). The cutting and grinding margins of the gnathal lobes lie in this preoral cavity. A posterior extension from the apodeme carries retractor muscles; a jointed hypopharyngeal process bears depressor muscles (Figs. 3.18(b), 3.19); and very stout protractor muscles arise from the transverse part of the apodeme, Fig. 3.19. These three muscles swing each tentorial apodeme about its external origin where the cuticle is flexible. The protractor and depressor muscles cause the apodeme to exert a thrust in the direction of the arrow marked on Figs. 3.17, 3.18(b). Here a chitinous plate in the wall of the preoral space is pressed against the gnathal lobe, pushing it to the abducted postion (Fig. 3.17, right side) and stretching the relaxed adductor muscles inserting both on the transverse tendon and on the gnathal lobe tendon.

FIG. 3.18 (see opposite). (a) Lateral view of the head of the diplopod *Poratophilus punctatus* Attems, with the overhanging cuticle of the collum segment removed. The post-occipital flange, shown in black, lies at the back of the head, projecting inwards at right angles to the head surface. The three segmented mandible is marked by mechanical tint, the two crosses indicating the ends of the hinge-like articulation of the basal mandibular segments of the head. The external origin of the anterior tentorial apodeme is shown in red; it passes inwards from this notch. (b) Median view of the left half of the head of the same diplopod. The wide face of the gnathal lobe (segment 3) of the mandible fills the preoral cavity. The tentorial apodeme is red; its retractor and depressor muscles are shown here and the huge protractor in Fig. 3.19. When the tentorium swings forwards and outwards about its external origin its thrust is delivered as marked by the arrow, causing mandibular abduction. (After Manton 1964.)

Evolution of Arthropodan jaws

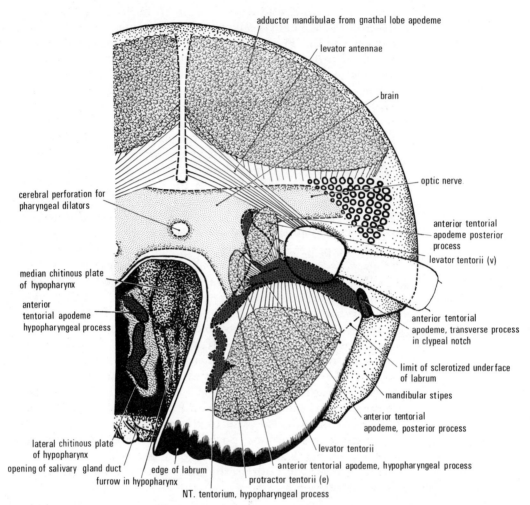

FIG. 3.19. Anterior view of the head of the iuliform diplopod *Poratophilus punctatus* Attems drawn as if the cuticle was transparent, with part of the labrum cut away to expose the hypopharynx, the gnathal lobe of the mandible being omitted. The position of the anterior tentorial apodeme and its processes is shown in red and the heavy marginal sclerotization of the labrum is shown in black. The supraoesophageal ganglion is shown by fine stipple and a convention indicates muscle insertions on the cuticle. (After Manton 1964.)

It is clear that the segmentation of this type of mandible is correlated with the limited but strong proximal adduction by transverse movement and with the wider very strong movement of the gnathal lobes caused by independent muscles, the

Evolution of arthropodan jaws

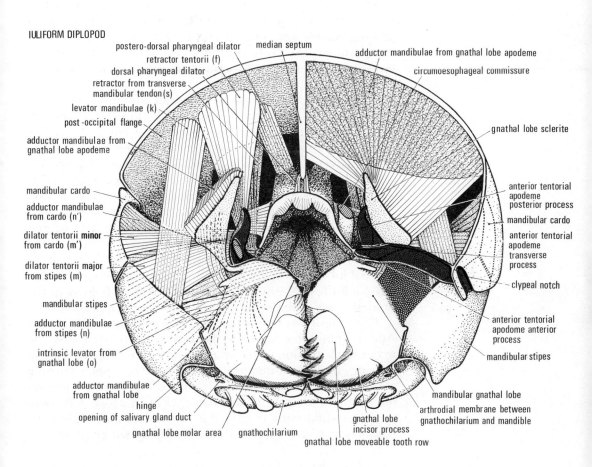

FIG. 3.20. Anterior view of the head of the diplopod *Poratophilus punctatus* Attems with the labrum cut away. The mandibular gnathal lobes are fully adducted. On the right: the preoral cavity is intact showing the transverse processes of the anterior tentorial apodeme (red) lying in the arthrodial membrane (white stipple); the cranial roof has been cut away, leaving the clypeal notch; and the supraoesophageal ganglion has been removed. On the left: the transverse process of the anterior tentorial apodeme has been removed, its cut stump being hatched; the cranial wall is cut away entirely so as to expose the whole of the mandible; the gnathal lobe adductor muscle has been removed in order to show the deeper muscles and the post-occipital flange; and the circumoesophageal commissure is removed. (After Manton 1964.)

whole is correlated with the existence of a paired swinging apodeme which parts the mandibles. It may be noted in passing that the gnathochilarium behind the mouth is indistinguishable anteriorly from the floor of the preoral space or hypopharynx. Only the very tip of the gnathochilarial lingual lobe is free, possessing both anterior and posterior faces.

(ii) *Symphyla*. The Symphyla are small, not more than 4 mm in length, and delicately constructed. They feed on soft decaying vegetable material which is simply cut up by narrow cusped mandibular margins. Symphyla cannot feed on the hard material which is cut and crushed by diplopodan mandibles. The symphylan mandible is two-segmented (Fig. 3.21(b)) and has a twist along its axis so that exact portrayal on paper is not easy. The basal segment articulates with the head at one point only, marked by a cross on Fig. 3.21(b) and permits a little more variation in movement than does the corresponding diplopod articulation. The distal segment, or gnathal lobe, is strongly hinged to the proximal segment and bears a knife-like row of cusps meeting their fellows in the preoral space with no flat crushing faces as in Diplopoda.

The mandibles bite in the transverse plane of the body. Adductor muscles pass from the basal segment to a transverse mandibular tendon (pink, muscles 4, 6, 7 on Fig. 3.21(a)) so causing limited adduction, as in the Diplopoda. Strong adduction of the gnathal lobe is effected by muscle 8 pulling on a stiff gnathal lobe apodeme (Fig. 3.21(a), left side) corresponding with the flexible tendon of the Diplopoda. Abduction is accomplished mainly by a swinging anterior

---

FIG. 3.21 (see opposite). Dorsal reconstruction of the head of *Scutigerella immaculata* Newport to show the musculature, the mandibles, the maxillae and the anterior tentorial apodemes (red), transverse endoskeletal tendons in pink. (a) (Level A) upper level showing the most superficial of the antennal muscles, the adductor muscle from the gnathal lobe, the insertion of muscle 1 and the supraoesophageal bridge with the tentorial suspension. (Level B) a lower level, the postero-dorsal overhanging part of the cranium is sliced off leaving the lateral lobe above the maxilla 1. The junction of head and trunk is shown by the arrow. The antenna together with the gnathal lobe sclerite and muscle 8 is sliced away, so displaying the whole of the mandible. An oesophageal dilator muscle inserting on the anterior tentorial apodeme has been cut, so displaying the transverse mandibular tendon (pink). (b) Lateral view of the head of the symphylan *Scutigerella immaculata* Newport drawn as a transparent object to show the position and component parts of the mandible, its two segments are marked by mechanical tint. The lateral knob-like process from the distal part of the basal mandibular segment is shown lying outside the galea in a strengthened concavity of the latter, the combine forming part of the mandibular abductor mechanism. (After Manton 1964.)

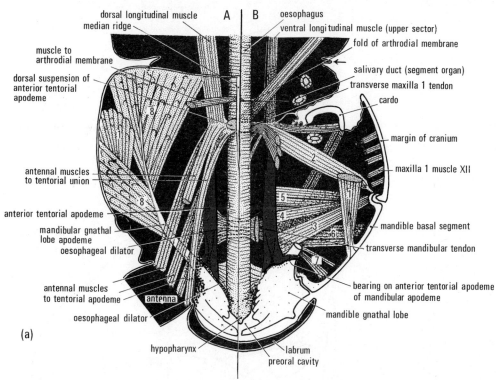

(a)

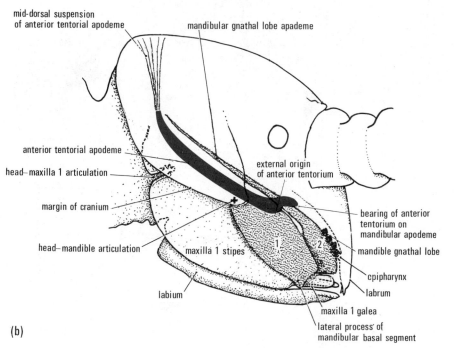

(b)

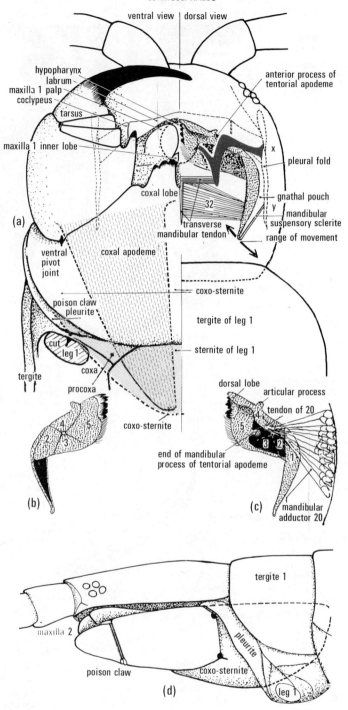

tentorial apodeme, which here pushes on the stiff base of the gnathal lobe apodeme, as shown on the right of Fig. 3.21(b). The anterior tentorial apodeme is fully musculated; it arises from the body cuticle essentially as in a diplopod, but swings mainly from a mid-dorsal suspension on the cranium. A little additional assistance in abduction is provided by the maxilla 1 and a partial use of a promotor–remotor roll (Manton 1964).

The symphylan mandible is typically myriapodan in morphology and in mode of operation and entirely unlike any hexapod mandible.

(iii) *Chilopoda*. A third type of myriapodan mandible is seen in the Chilopoda, carnivorous centipedes of various habits. They all use a large poison claw, which is a modified first trunk leg, used for seizing and breaking into the bodies of armoured arthropods; small arthropods can be swallowed whole. The poison claw, with its strong coxal apodeme projecting deeply into the body, is seen on Fig. 3.22(a), left side. The cusps of the coxal lobes on the median sides of the claw bases (drawn on Fig. 3.22(a), left side) can be used as a 'tin-opener' by a sudden jab, after a lesion has been formed by the claw tips. When a sizeable tear has been made the more delicate mouth parts, maxillae 2, maxillae 1, and mandibles, are used to pull out the soft parts, a process assisted to some extent by digestive juices being poured from the mouth into the wound. If the prey is large the whole head of the centipede may be inserted into the lesion.

The mandibles of chilopods are not visible in a side view of the head as in Diplopoda and Symphyla. The chilopod head is much flattened (Fig. 3.22(a), (d), cf. 3.18(a), 3.21, Pls. 1(f), 5(g), 7(i)). The hexapod *Petrobius* shows a small pleural fold, marked on Fig. 3.23(a), which covers the dorsal end of the mandible. During chilopod ontogeny a similar fold is

FIG. 3.22 (see opposite). (a) Head of the centipede *Cormocephalus calcaratus*. On the left a ventral aspect in surface view shows the poison claw from trunk segment 1, its apodeme projecting internally into the body posteriorly is marked by an interrupted line. The maxilla 1 palp obscures the mandible and between this palp and the labrum lies the entrance to the preoral cavity. On the right the head is drawn in dorsal view, as if transparent, to display the mandible within the gnathal pouch and the tentorial apodeme (red) and transverse mandibular tendon (pink). (b) and (c) Isolated mandible drawn to the same scale as in (a); ventral view in (b); and dorsal view in (c) as on the right side of (a). The gnathal lobe adductor muscle 20 is shown diagrammatically, arising from its tendon marked 'tendon of 20' and inserting on to the cranium. (d) Lateral view of the head showing great flattening and large size of the poison claw. Trunk segment 1 is much thicker. (After Manton 1965.)

formed, but it grows downwards over the sides of the head and unites with the lateral edges of the labrum on either side, as shown in Fig. 3.22(a), right side. The preoral space, which in Diplopoda and Symphyla is flanked by the mandible, is thereby closed laterally so that the mandible at rest lies within a gnathal pouch, marked on Fig. 3.22(a), right side, the pouch opening to the exterior along the free transverse edge of the labrum and covered ventrally by the maxilla 1 palp and poison claw. The mandible is said to be entognathous because of its enclosed position. It is attached to the head not by a stout articulation as in Diplopoda and Symphyla (Figs. 3.18(a), 3.21) but by a suspensory sclerite attached to the point marked Y on the right of Fig. 3.22(a). As in the Diplopoda and Symphyla, muscles pass from the proximal segments of the mandible to a transverse tendon (pink), see muscle 32 on the right of Fig. 3.22(a), other muscles being omitted here. The transverse tendon is very narrow leaving maximum length for muscle 32, and the tendon is united also with the posterior processes of the anterior tentorial apodemes and swings with them.

The mandibular movements of centipedes are unique and consist of a see-saw oscillation about a line passing transversely through the point Y, where the mandibular suspensory rod articulates with the pleurite, and through muscle 32. The proximal end of the mandible, deep in the gnathal pouch, performs wide excursions as indicated by the double arrow (Fig. 3.22(a), right side). A strong outward and downward sweep of the proximal end of the basal sclerite of the mandible causes opposite movements at the gnathal edge, adduction and levation. At the same time a backward movement of the whole mandible, bringing the basal sclerite close to the posterior wall of the cranium, retracts the gnathal margin. Protraction and abduction of the mandible enables it to bite into a cone of soft flesh pulled up by the maxillae 1 and retraction and adduction cuts and pulls the food into the preoral cavity. The degree of curvature of the mandible is controlled by its segmentation and by intrinsic and extrinsic muscles, so enabling the mandible to make the best use of the working space. A gnathal lobe apodeme from the mandible carries a stout tendon giving origin to a huge mandibular adductor muscle which inserts on the cranium, as in Diplopoda (Fig. 3.20, 'adductor mandibulae' on the right) and Symphyla (Fig. 3.21(a), muscle 8 on the left). An indication of this adductor muscle 20 with its tendon of origin is shown here for the centipede, Fig. 3.22(c).

As in other myriapods, the paired anterior tentorial apodemes (red) control and cause mandibular movements. Each apodeme is strongly musculated and swings about its origin at X (Fig. 3.22(a), right). One apodemal arm unites with the upper face of the mandible, as marked on Fig. 3.22(a), (c), and an apodemal anterior process hooks round an articular process on the mandible, so transmitting a thrust from one to the other (Fig. 3.22(a), right side, (b), (c)). A corresponding process exists in Diplopoda (Fig. 3.19). Both the transverse mandibular tendon and muscle 32 of Chilopoda swing back and forth with the mandible because the tendon is linked to the posterior process of the tentorial apodeme (Fig. 3.22(a)). The musculature of the tentorial apodemes and mandibles are complex and are not figured here (see the many drawings in Manton 1964).

(iv) *Conclusions on the jaw mechanisms of the Myriapoda.* The jaw mechanisms of the three myriapod classes (here described in outline) show basic similarities in: (1) their jointed mandibles; (2) their powerful gnathal lobe muscles arising from a tendon or apodeme and inserting onto the cranium; (3) muscles from the proximal segments of the mandible inserting on the transverse mandibular tendon cause limited but strong adduction; (4) the pair of anterior tentorial apodemes swinging within the head, with branches which are to some extent comparable from class to class, cause indirect mandibular abduction by pushing on the mandibles in all classes and additional movements associated with entognathy in the Chilopoda; and (5) the absence of posterior tentorial apodemes (see below). The mandibular mechanism of centipedes is the most complicated among myriapods, but it is based upon the same common attributes, but with entognathy superimposed, in association with the preliminary poison claw contribution to the feeding mechanism.

The jaw mechanisms of myriapods are unique and nothing like them exists in any other class. That of the Pauropoda has not been considered above because details have not been worked out but the mechanism has been much affected by the minute size of these animals and consequent absence of structures meeting the needs of larger animals; for example, the absence of segmentation in the minute rod-like mandible

Evolution of arthropodan jaws

may be correlated with cuticular flexibility alone being sufficient. Pauropodan mandibles are entognathous, but the details differ from those of the Chilopoda. Neither the centipedes nor any other class of myriapod can be considered to represent a primitive condition for the group. Each class has developed its own type of jaw mechanism, based upon the common features noted above, in correlation with differing feeding habits which characterize each class. There is no fundamental resemblance between the mandibular mechanisms of myriapods and hexapods (see further discussion of entognathy in Chapter 6, §3B).

C. *Mandibles of the Hexapoda*

The winged insects (the Pterygota) and the four classes of wingless insects, or Apterygota (the Protura, Diplura, Collembola, and Thysanura) basically possess the same type of mandible, which is a whole limb, unjointed (cf. Myriapoda) and biting with the tip.

(i) *The tentorial apodemes.* The hexapods have in common the possession of two pairs of tentorial apodemes within the head (red on the Figs.), neither pair swings nor has anything directly to do with mandibular abduction (cf. the Myriapoda) other than providing sites for muscle insertion. The point of origin of the anterior tentorial apodeme in three hexapods is marked on Fig. 3.23 and is similar to the origin of the single pair of apodemes in myriapods (Figs. 3.18(a), 3.19, 3.20, 3.21(b), 3.22(a), X on right). The posterior tentorial apodemes, absent in myriapods, arise just anterior to the labium, in *Petrobius* from a very deep and narrow external opening, its dorsal and ventral limits being marked on the left of Fig. 3.23(a). Within the head the apodemes bear many muscles and end blindly. Both apodemes are complex in shape. The posterior pair unite with one another in the Thysanura and Pterygota (Figs. 3.25(b), 3.26) and join, by a different type of union, antero-ventrally in support of the hypopharyngeal region in the entognathous Diplura and Collembola (Figs. 3.29, 3.31).

In association with the ability to bite strongly in the transverse plane of the body, a secondary achievement within the Thysanura and Pterygota (see below), the anterior pair of apodemes fuse together in the middle line of the head (Fig. 3.25(a)) and become very closely associated with the posterior pair (Fig. 3.25(b)) and finally fuse with them also. The tentorium, or head endoskeleton, of a pterygote, such as a locust, spans the head and braces the head capsule against the tension exerted on it by very strong mandibular muscles. Support is gained from the cuticular surface of the head by the four points of origin of the original two pairs of tentorial apodemes. The anterior and posterior tentorial apodemes in the Collembola (Pl. 6(a), (d)) remain separate and are associated with quite different mandibular movements, including the protrusibility of mandibles and maxillae 1 (Fig. 3.29). In the Diplura (Pl. 6(e), (h)) the preoral cavity is very small and no anterior tentorial apodemes are formed from the margin of the diminutive labrum (Fig. 3.30). Crustacea possess no corresponding apodemes in the head.

(ii) *The basic movements and musculature of hexapod mandibles.* Probably the nearest approach to the primitive hexapod mandible is seen in the thysanuran *Petrobius* (Pl. 6(m)), but allowance must be made for the special achievements of this animal. The mandible, in its natural position, but separated from the head in Fig. 3.16(d), is seen partially overlapped by maxilla 1 in Fig. 3.23(a), where the maxilla 1 palp is cut away. The distal end of the mandible is enclosed by the labrum in front, by maxilla 1 galea laterally, and by the rest of maxilla 1 and the labium (maxillae 2) posteriorly. The mandibular movement is a rolling one, as for the gnathobasic mandible of *Chirocephalus*, about an axis situated between the cross, marking the dorsal articulation with the head, and the black spot where the association with the head is not firm. A transverse mandibular tendon (pink) is present (Fig. 3.24) and muscles 8–10 (marked on the right side) converge on the tendon. In the horizontal plane these muscles fan out onto the mandible as shown for *Chirocephalus* (Fig. 3.2(c), muscles 5a and 5b) and contribute to the rolling movement.

Marginal muscles from each mandible, 4 and 5 from the anterior margin and 6 and 7 from the posterior margin, insert on the anterior tentorial apodeme as shown. Other marginal muscles, 1–3 anteriorly and muscle 11 from a slender mandibular apodeme arising at the posterior margin, insert on the dorsal part of the head cuticle. The anterior tentorial apodemes of the Myriapoda are not used in this way. All these muscles, and the anterior tentorial apodemes providing a base for many of them, contribute to the strong rolling movement of the mandibles about the transverse tendon.

The finely cusped tips of the mandibles scrape unicellular algae from surfaces such as those on stones, leaves, seaweed, decaying wood, etc., as the result of the rolling movement. Copious fluids are poured into the enclosed space surrounding the mandibular tips from the enormous salivary glands (Figs. 3.23(a), top left, 3.24, top right). The freed algal cells are sucked upwards in this enclosed space by oesophageal peristalsis. This food material is ground by the pair of molar areas on the mandibles which rub across one another as the result of the rolling movement and the mush is sucked into the mouth, which is situated just anterior to and above the molar areas.

The hydraulic method of collecting the fine food scraped up by the tips of the mandibles represents a specialized achievement by *Petrobius* and contrasts with the methods used by Crustacea in collecting their finely divided food and passing it to the mouth. The similarity between the mandibles of *Petrobius* and simple Crustacea resides only in the rolling movement and presence and similar use of the transverse mandibular tendon (Figs. 3.2(a), 3.24). The former is a derivative of the promotor–remotor swing of a walking leg and can be used equally well on a gnathobasic or on a whole-limb mandible; and the transverse tendon is a potential attribute of all body segments (see this chapter, §2B(ii), D). Thus the differences between the crustacean and the hexapod mandibles are basic and their resemblances are convergent (see further in Chapter 6).

(iii) *Mandibular skeleto-musculature in Thysanura and Pterygota*. As in the Crustacea, the ability of hexapods to feed on large and hard food material opens up the possibilities of greater diversity of habits. The means adopted by Thysanura (Pl. 6(m)) and Pterygota (Pl. 6(i)) in gaining this end are parallel to those employed by certain groups of Crustacea. The details differ and the resemblances are convergent.

A swinging backwards on the head of the dorsal end of the mandible, so that its axis of swing becomes horizontal rather than vertical (Figs. 3.3, 3.5, 3.23, the cross being level with the black spot or even more ventral to the latter in the locust), converts the rolling movement into direct adduction–abduction, the original preaxial part of the mandible being much reduced, as in *Ligia* (Figs. 3.5(b), 3.23).

Correlated changes occur in the endoskeleton. The trans-

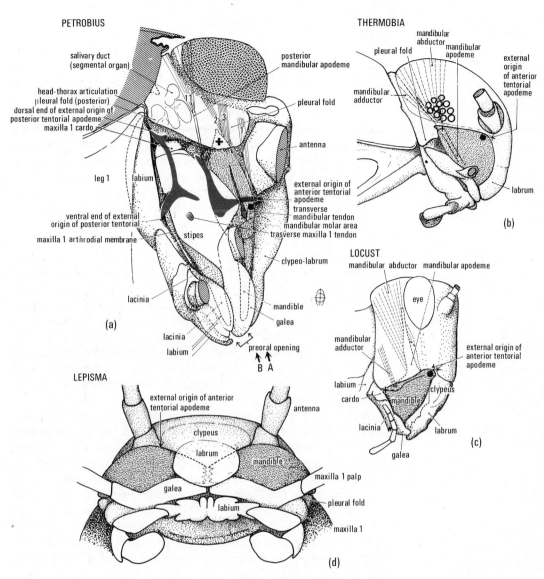

FIG. 3.23. Hexapod heads showing the mandible. Side view of the head of (a) *Petrobius* with the prothoracic tergite cut back; (b) *Lepisma* and (c) locust, drawn as semi-transparent objects to show the change in slope of the head–mandible axis of swing, situated between the cross and the black spot. The end of this axis, marked by a cross, becomes progressively more posterior and less dorsal in position. The positions of the hollow anterior and posterior tentorial apodemes are indicated in red, the arrows A and B showing the planes of the transverse sections through the head in Fig. 3.24. The mandibular apodeme carrying adductor muscles is present in all animals in increasing size. The mandibular muscles are marked by arabic numerals and muscles of the maxilla are marked by roman numerals (for further description see Manton 1964). (d) Ventral view of the head of *Ctenolepisma ciliata* Dufour.

verse mandibular tendon becomes redundant, as in Crustacea, because a wide gape would require striated muscles to change length more than is practicable, if they remained attached to a transverse tendon. *Lepisma* shows muscles 8–10 to be longer than in *Petrobius*; they are adductor in function and have shifted their insertions to the anterior tentorial apodemes (red) which are stabilized by fusion with one another, Figs. 3.24, 3.25(a). The greatest force for the strong adductor movement, capable of cutting into hard grain, comes from adductor muscle 11 arising from the large posterior mandibular apodeme which now arises near the distal mandibular margin (cf. Figs. 3.24, 3.25(a), right sides). Weak abductor muscles, corresponding with muscles 2, 3, 4, and 5 of *Petrobius*, arise near the mandibular hinge (Fig. 3.25(a), left side). These muscles work at poor mechanical advantage, as is frequent in musculature causing recovery movements which elongate relaxed muscles but do no outside work. This thysanuran arrangement for strong biting, seen in *Lepisma* and *Thermobia*, does not permit a very wide gape; see the narrow labrum in Fig. 3.23(d), beyond the margins of which the mandibles cannot abduct. In the Pterygota, with the basic type of biting mandible seen in the locust, the labrum is almost as wide as the head.

The locust type of mandible has for long been considered to be basic for the Pterygota and from it more specialized mandibles have been derived which serve many and different purposes in various orders. The locust mandibles have a large gape and can part to the limit of the wide labrum. This extensive gape has been made possible by the cavity of the mandibles being almost devoid of muscles (cf. muscles 8–10 of *Lepisma*, Fig. 3.25(a)), and by the mandibular adductor apodeme arising proximally on the margin of the mandible and not near the cutting edge, as in *Lepisma* (cf. Figs. 3.25(a), right side, 3.26). A somewhat similar solution to the gape problem has been arrived at by *Ligia*, but other large-food

---

FIG. 3.24 (see opposite). Transverse view from in front of the head of *Petrobius brevistylis* cut at the levels A and B indicated in Fig. 3.23(a) to show the mandible, endoskeleton, muscles, and general anatomy. The anterior face of the mandibular cuticle is seen in level A and a section through the mandible in level B. Anterior tentorial apodemes in red, transverse mandibular tendon in pink. The molar lobes of the mandible are in apposition while the distal mandibular tips are apart. Mandibular muscles are marked by arabic numerals and maxilla 1 muscles by roman numerals. For further description see text. (After Manton 1964.)

Evolution of arthropodan jaws

feeding Crustacea have solved the problem differently.

The locust mandibular adductor musculature is very massive, arising from a large and complex apodeme, and inserting over much of the cranium (Figs. 3.23(c), 3.26). The

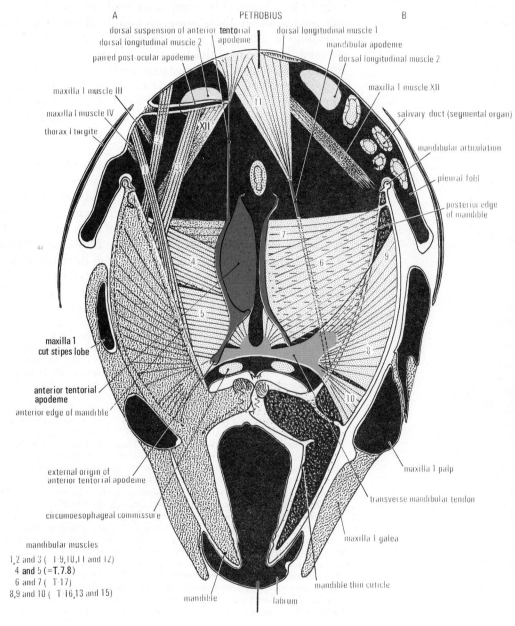

Evolution of arthropodan jaws

abductor muscles are much smaller and, as in *Lepisma*, work at poor mechanical advantage. The mandibular adductor muscles are strong enough to enable the cutting of food by mandibular cusps spread over the whole medial face of the mandible. These achievements are probably dependent upon the evolution of the strong endoskeletal tentorium (red), spanning and bracing the head capsule. This endoskeleton, derived by fusion of the two pairs of tentorial apodemes, is incipient in *Lepisma* (Fig. 3.25(b)) and complete in the locust. Only part of the tentorial surface is used in the locust for muscle insertion. Maxilla 1 muscles (Roman numerals) insert mainly upon the anterior pair of apodemes and the labial muscles on the posterior pair (Figs. 3.25(b), 3.26).

(iv) *Protrusible mandibles suiting fine-food feeding in hexapods.* Feeding on large and hard food in Thysanura and Pterygota requires a mandible which is very strongly articulated with the head in support of adductor and abductor movements in the transverse plane. Quite opposite lines of mandibular evolution are shown by the Collembola, Diplura, and Protura in which mandibular protrusion has been developed, besides other mandibular movements. Mandibular protrusion can only take place if the mandible is loosely attached to the head and this is only practicable if a pleural fold grows down from

FIG. 3.25 (see opposite). Thick transverse sections of the head of *Ctenolepisma longicaudata* passing through the mandible and viewed from in front. Level A is anterior to B. The external origin of the anterior tentorial apodeme (red) is shown in A but the anterior face of the mandible has been removed to display the muscles. The transverse bridge formed by the union of the anterior tentorial apodemes is shown in both levels. Muscle numbering corresponds with that of *Petrobius*. Mandibular abductor 5 is shown and abductor 4 lies just behind it. The apodeme bearing mandibular abductor 2 plus 3 is entire in A and cut short in B, the long cranial insertion of these muscles is also seen in (b) level C. Muscle 2 plus 3, 4, and 5 pull on the very short preaxial border of the mandible. The fan of adductor fibres 8, 9, and 10 from the inner face of most of the mandible to the anterior tentorial apodeme is entire in A and cut away ventrally in B to display the more posterior adductor muscle corresponding with 6 and 7 of *Petrobius*. Adductor muscle 11 is more complex than shown, one large and some small sectors passing behind to the post-occipital flange. (b) Transverse sections, as in (a), passing through the maxilla 1, level C being anterior to level D. Muscle numbering corresponds with that of *Petrobius*. Muscle XI, the protractor–depressor to the cardo–stipes hinge, and the stipes muscles VI and X are shown inserting on to the anterior tentorial plate (red) in C, muscles VI and X are cut in D so displaying the more posterior stipes muscles VII and VIII which insert on the posterior tentorial plate (red). Muscle X runs in several sectors. The cranial retractors of the lacinia, muscles III and IV, are shown in C and the promotor of the cardo, muscle XII in two sectors, lies immediately behind muscles III and IV. The labial segmental organ is cut showing the end sac (marked) from which a duct passes up to the pleural fold and down to the median exit duct. (After Manton 1964.)

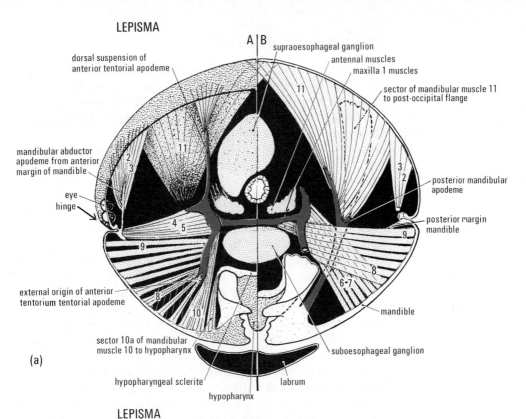
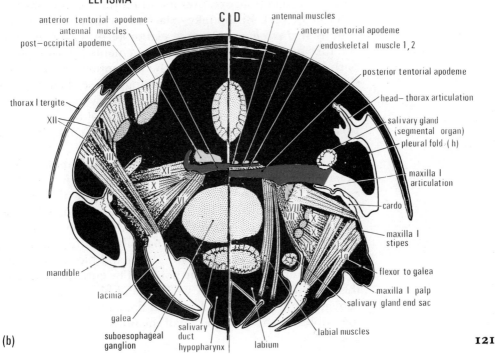

I2I

the side of the head, as described above for Chilopoda (§5B(iii)), so enclosing the mandibles in a gnathal pouch. In hexapods maxillae 1 are also enclosed along with the mandibles in this pouch. Only examples from the Collembola and Diplura will be considered here.

(v) *Collembola.* The mandibles and maxillae 1 are entognathous, that is to say they are contained in a pouch on either side of the head, which is formed ontogenetically as in Chilopoda, see above, by downgrowth of a pleural fold on the head which fuses on either side with the labrum in front and the labium behind. The gnathal pouch is deep and the mandibles within it are long and narrow. Each pouch communicates with the much smaller median preoral cavity which at rest houses the biting tips of the mandibles and maxillae 1. There is no strong articulation between the mandible and the head, only a very flexible junction formed by a suspensory ligament which permits the mandible to be protruded and retracted. No precision of cutting by the mandibular gnathal margins could be achieved without some further mandibular support and this is gained, like fingers controlling a pencil, by distal cuticular structures against which the mandible executes its promotor–remotor swing or roll (see below). Biting takes place at the end of the remotor roll, as in *Petrobius*. There is elaborate provision for these rotator and counter-rotator movements, terms which perhaps describe more readily the promotor–remotor roll of the mandibular mechanism. The 'fingers' (see below), holding the distal end of the mandible allow no gape, as in transversely biting mandibles; the parting of the mandibular tips results only from the promotor swing (counter-rotator movement).

The complexity of structure within the collembolan head is great. Only some salient features which refer to the mandibles can be considered here. A dorsal view of the head of *Tomocerus* is shown in Fig. 3.27 with the superficial surface structures removed on the right and a deeper level displayed on the left, showing the entire mandible within the gnathal pouch. Maxilla 1, underneath the mandible in this view, projects more deeply into the pouch. The cuticular surface of the mandible is mottled. A whole mandible in lateral view is shown in Fig. 3.28(a) and in (b) a dorsal view of the pair of mandibles, with an indication of the movements and muscles

Evolution of arthropodan jaws

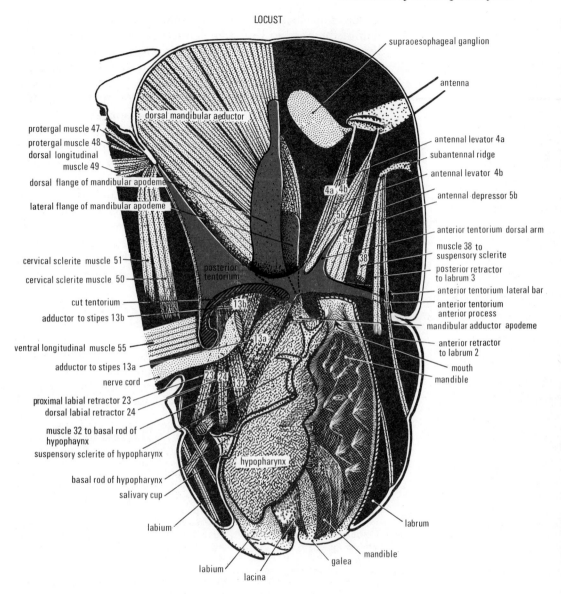

FIG. 3.26. Sagittal half of the head of the pterygote insect *Locusta migratoria* L., the hypopharynx being left intact. The circumoesophageal commissures have been removed in order to display the antennal and mandibular muscles and the tentorium (red). The upper sector of the suspensory sclerite of the hypopharynx sends one branch to the oesophagus and the other unites with the mandible close to the origin of the mandibular (adductor) apodeme. The mandibular muscles are removed on the left to show the posterior part of the cranium and tentorium. The muscle numbers are those of Albrecht (1953). (After Manton 1964.)

which cause them. The longitudinal reconstruction of the left half of the head in Fig. 3.29 shows the preoral cavity limited by the epipharynx above; a very substantial hypopharynx, quite separate from the labium below (cf. the gnathochilarium of a diplopod where the hypopharynx forms its anterior face, Fig. 3.18(b)); and the tips of the left mandible and maxilla 1 projecting into the preoral space.

The suspensory ligament attaching each mandible to the head permits mandibular protrusion and other movements (Figs. 3.27, left side, 3.28(b)). Muscles 1 and 2 pass from a dorso-lateral mandibular ridge to the cranium above it. The rest of the mandibular muscles arise from the internal mandibular cavity and from the margins of this cavity: muscles 3, 6, 9, 11, 12 insert on the cranium, muscles 13, 15, 16, 17 insert on a substantial, but narrow, transverse mandibular tendon (pink), and muscles 7 and 8 insert onto the anterior tentorial apodeme (red). (For further details see Manton 1964).

The *anterior tentorial apodemes* arise, as in myriapods and other hexapods, from the cuticle at the labral margins. The pair end blindly and separately ventral to the transverse mandibular tendon (Figs. 3.27, 3.29). The apodemes bear several branches and carry the important cuticular boss against which the mandible rotates (Figs. 3.27–29). A crescentic frontal sclerite on the labrum holds the mandible

FIG. 3.27 (see opposite). Antero-dorsal (frontal) reconstructuction of the head of the collembolan *Tomocerus longicornis* (Müller) viewed from above, the anterior surface cut away to show the endoskeleton, gnathal pouch, and the form and musculature of the mandible and maxilla 1. Tentorial apodemes are shown in red. Transverse segmental tendons and their connectives are shown in pink; the posterior endoskeletal plate is lightly stippled, the gnathal pouch cavity is white. Uncut surfaces of mandibular and maxilla 1 cuticle are mechanically mottled and muscle insertions on the dorsal cuticle are shown by a convention. Cut muscle ends are finely stippled. Mandibular muscles are marked by arabic numerals, the numbering of Hoffmann (1908), and the maxilla 1 muscles I–XII are marked by roman numerals. The oesophagus is cut. Level T shows a view of the mandibular muscles, which are further cut away in level U to show the transverse mandibular muscles 15 and 16 which unite with their fellows below the transverse manibular tendon (pink). The counter-rotator (remotor) muscles from the anterior margin of the mandible are removed. In level U muscle 2 and part of muscle 1 are cut away from their origins on the dorsal–lateral mandibular ridge; the antennal muscles, the dorsal suspensory muscle, and the posterior dorsal–ventral suspensory muscle are all removed (the latter is entire in level T). The proximal parts of maxillae 1 are shown, muscles III and IV being entire in level T and cut short in level U. The anterior tentorial apodeme (red) is almost entirely shown in level T, branch ii to the superlingua and branch iii to the preoral cuticle above (anterior to) the mandible being marked. The edge of the suboesophageal ganglion is indicated by a dotted line. (For further description see text and Manton (1964) Figs. 32–3.)

Evolution of arthropodan jaws

on its outer face (see the same figures). These apodemes carry mandibular muscles and are rigid, as in other hexapods. They serve mandibular movements by their rigidity, in contrast to the swinging tentorial apodemes of the myriapods which provide mandibular abductor mechanisms.

The distal fine biting cusps from the mandible lie beyond the more proximal molar grinding ridges, shown diagram-

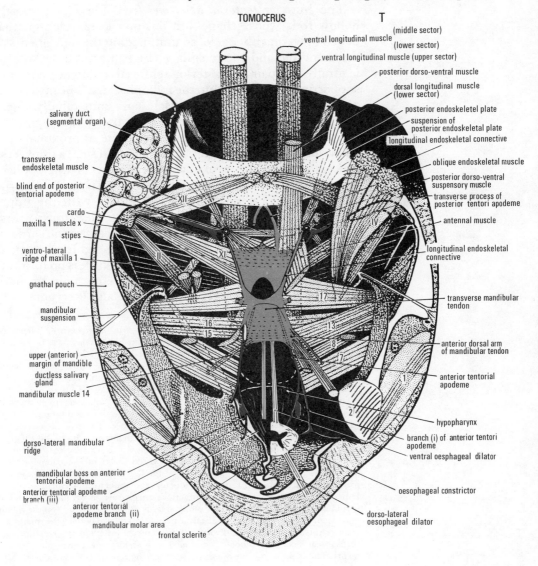

matically in Fig. 3.28. The molar grinding takes place, as in other rolling mandibles, on the remotor or rotator phase. The mandibular muscles causing these movements and those of protrusion and retraction, are listed on Fig. 3.28. By fine and precise movements collembolan mandibles cut into vegetable and animal food, fungal spores, etc., followed by grinding as in all animals using the rolling movement.

The *posterior tentorial apodemes* (red) form long, mostly hollow rods whose shape and position keeps each gnathal pouch open on either side, so unimpeding the movements of the mandible and the maxilla 1 (Figs. 3.27, 3.29). Antero-ventrally the pair of posterior tentorial apodemes end in a supporting complex within the hypopharynx. The main shaft carries labial and maxilla 1 muscles and ends blindly where the pair are united by a transverse muscle; each apodeme carries many other muscles passing in a proximal and posterior direction. Laterally from the main shaft a transverse process crosses the proximal wall of the gnathal pouch, ending in fibrils passing through the gnathal pouch wall to the outer cuticle in a position corresponding with the upper lateral intucking forming this apodeme in *Petrobius*, Fig. 3.23(a), left side. The transverse process carries the cardo of maxilla 1. Protrusion of the maxilla is effected by alterations in the flexures between cardo and stipes. There is no fusion of the anterior and posterior tentorial apodemes as in the hard and large food feeders (cf. Lepismatidae and Pterygota).

FIG. 3.28 (see opposite). The cóllembolan *Tomocerus longicornis* (Müller). (a) Shows the mandible in lateral view; the distal serrated lobe is downwardly directed and the grinding molar area projecting from the medial face gives an upward process extending to the mouth. This process is foreshortened in (b). The stout cusps on the posterior edge of the molar area are indicated as if the mandible was transparent. (b) A diagram of the mandibles in their natural position in the head in antero-dorsal view to summarize the functions of the muscles. In using the diagram it must be remembered that no allowance can be made for foreshortening and for muscle pulls being exercised at an angle to the plane of the paper. The frontal sclerite and mandibular boss on the anterior tentorial apodeme, which limit the range of movement of the tips of the mandibles, are shown in black. Rotator movements (curved arrow) and rotator muscles are shown on the right and counter-rotator movements (curved arrow) and muscles on the left. The muscles which can give predominantly protractor, retractor, adductor, and abductor movement are listed. Different combinations of muscle contractions can give very many different kinds of small movements. Grinding by the molar areas is accomplished only on the rotator movement but scratching by the distal cusps may be effected both by the rotator and counter-rotator movements. Protractor movements which may puncture food surfaces can take place alone or combined with abductor, counter-rotator movements. The effect of any one muscle depends on the action of other muscles. (After Manton 1964.)

Evolution of arthropodan jaws

(vi) *Diplura*. These hexapods also have entognathous mandibles and maxillae 1. The margins of the dipluran mandibles, as in Collembola, carry the primitive rotator, counter-rotator muscles, but dipluran mandibular protrusibility is differently contrived and the transverse mandibular tendon (pink) is used differently from that of the Collembola. Most of

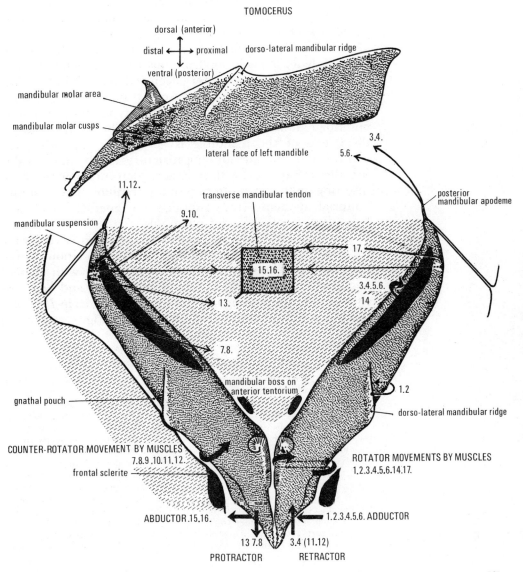

the muscles of the mandibles and maxilla 1 of *Petrobius*, Collembola, and Diplura can be homologized with one another. The same complement of muscles is present in each group, a very general phenomenon in Uniramia, but apparently homologous muscles do not always perform similar functions in different classes. For example, mandibular protrusion in Collembola is implemented by muscles 13, 7, 8 inserting on the anterior tentorial apodeme, but in Diplura by muscles g, and those alongside it, inserting on large transverse mandibular tendons (cf. Figs. 3.27 with 3.31).

All features appertaining to dipluran entognathy suggest independence in evolution from that of the Collembola. The dipluran gnathal pouch is very deep and the mandible is only attached to the head by muscles, there is not even a suspensory ligament. The preoral cavity is so small that an anterior tentorial apodeme could hardly be formed and the distal end of the mandible is adequately supported and controlled by the surrounding structures and by the movable prostheca (Fig. 3.31) with its long tendon ending in muscle b inserting on the posterior face of the cranium. The posterior tentorial apodeme (red) supports the gnathal pouch and hypopharynx much as in Collembola.

The dipluran transverse mandibular tendon is very wide (cf. Figs. 3.27 with 3.31) and more distal in position than in Collembola. About this wide tendon, fused with that of maxillae 1, the mandible swings by rotator and counter-rotator muscles fanning out from the tendon, as in *Petrobius* in a plane at right angles from the paper of Fig. 3.31; cutting

---

FIG. 3.29 (see opposite). Reconstruction of a sagittal half of the head of the collembolan *Tomocerus longicornis* viewed from the middle line. Mandibular muscles are marked by arabic numerals, maxilla 1 muscles by roman numerals. The transverse maxilla 1 and mandibular tendons (pink) and the posterior endoskeletal plate are cut, as are the mandibular muscles 4–6 which cross to the other side of the head, and mandibular muscles 15 and 16 and maxillary muscle X which are transverse in position. Almost the whole of the anterior tentorial apodeme (red) is shown, branch (i) being indicated by dotted lines. The internal part of the posterior tentorial apodeme (red) is shown with its superficial arm alongside the maxilla 1, but the details of its fade-out into the elaborate cuticular thickenings of the hypopharynx are not shown. The more important branches of the tendinous endoskeleton visible in this aspect are shown. The median opening of the paired (salivary) ducts of the labial segmental organs is more complex than shown, the ducts open into a median diverticulum of the gnathal pouch. The latter is freely open to the exterior between the paired parts at the marked median gap, but distal to this there is a close (possibly fused) union between the two halves of the labium over the stippled area. (After Manton 1964.)

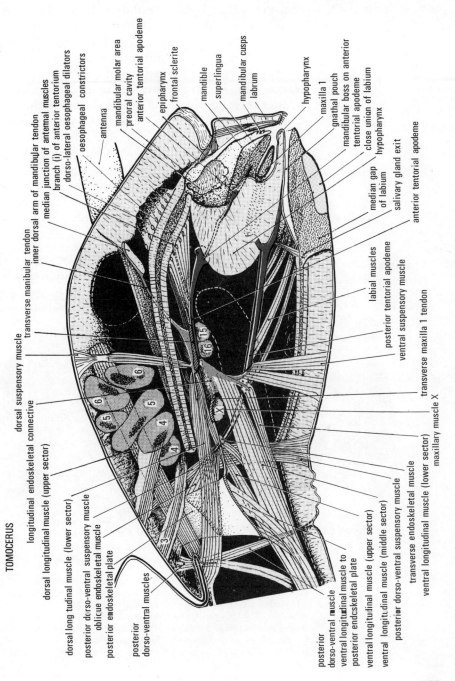

Evolution of arthropodan jaws

by the tips of the mandibles is thereby effected. The transverse mandibular tendon in Collembola does not form an axis of mandibular movement, nor does it carry rotator and counter-rotator muscles. Muscles from the cavities of the mandibles diverge towards their insertions on the narrow transverse mandibular tendon of Collembola but such muscles converge in the Diplura where their functions differ. (For a fuller account of the differences between the entognathy of Collembola and Diplura see Manton 1964, p. 65).

It is considered that the differences in the details of entognathy in Collembola and Diplura, including differences in use and form of the transverse head tendons and in the muscular mechanisms of mandibular protrusion, indicate independent acquisition of protrusible mandibles in the two classes. Consideration of much more detail is required to

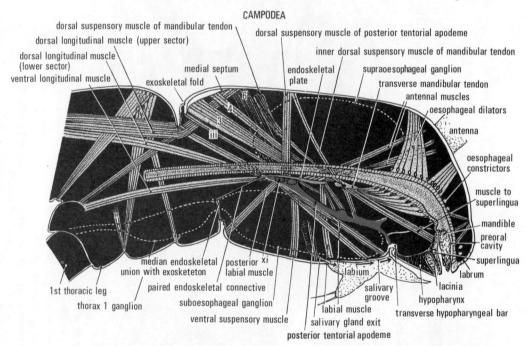

FIG. 3.30. Reconstruction of a sagittal half of the head of the dipluran *Campodea staphylinus* to show the relationship of the transverse tendons (pink) of the posterior tentorial apodeme (red) of the left side and the muscles. There is no anterior tentorial apodeme. The opening of the salivary duct into the posterior end of the median labial groove is indicated and the direction of flow towards the mouth is shown by arrows. On the left the section is sub-median, passing through the coxa of the first thoracic leg. The positions of the ganglia of the nervous system are indicated by dotted lines. (After Manton 1964.)

Evolution of arthropodan jaws

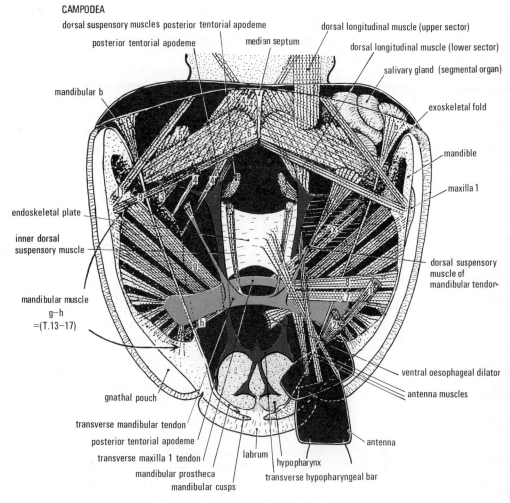

FIG. 3.31. The dipluran *Campodea staphylinus* West. Antero-dorsal view of a reconstruction of the head, drawn as a transparent object. The antennal base, the antennal and superficial mandibular muscles, the unobscured muscles of maxilla 1, and some coils of the labial gland (segmental organ) are shown on the right, these structures being omitted on the left in order to display the mandible. Superficial mandibular muscles 7–8 are omitted on the left. Transverse segmental tendons are pink and the paired posterior tentorial apodeme and its processes are shown in red. Mandibular muscles bear arabic numerals and maxilla 1 muscles bear roman numerals. The numbering suggests the homologies believed to exist between the muscles of *Tomocerus* and *Campodea* (Figs. 3.27 and 3.31), where similar numbering is employed. Mandibular muscles 1 and 2 are present in Collembola and Diplura alone; muscles 3–11 here can be recognized also in *Petrobius*. The muscles between the arrows on the left, between muscle fibres g and h, represent muscles 13 to 17 of *Tomocerus* but do not correspond in detail; they correspond more closely with *Petrobius* muscles 8–10, Fig. 3.24. For further consideration see Manton (1964).

Evolution of arthropodan jaws

substantiate this conclusion (see Manton 1964). The redundance of the anterior tentorial apodeme in Diplura is related to the minute size of the preoral cavity.

(vii) *Conclusions on the jaw mechanisms of Hexapoda.* The mandibles of the hexapods considered above appear to be derivatives of: (1) unjointed, whole-limb mandibles which bite with the tips; (2) mandibles basically using the promotor–remotor movement of a walking leg. This is seen, with refinements, in *Petrobius* today. Supporting these movements are; (3) paired anterior tentorial apodemes principally serving the mandibles and (4) paired posterior tentorial apodemes whose origin is associated with the labium; (5) these four apodemes fuse together in association with large or hard-food feeding and secondary strong biting in the transverse plane. In entognathous classes the tips of the posterior tentorial apodemes form elaborate supporting units, sometimes joined, within the hypopharynx but the rest of these apodemes are free. In all these five features the Hexapoda contrast with the Myriapoda and indicate independence of evolution of hexapod mandibles from those of myriapods. The Symphyla in no way exhibit a mandible intermediate between these two subphyla.

The head endoskeleton supporting mandibular movements of hexapods is constructed from the same three systems utilized by Crustacea (see §2E above) and most other Arthropoda. Sub-ectodermal basement membrane or thickened connective tissue are present. Horizontal ectodermal connective tissue tendons are large in the mandibular segment, as in the more primitive Crustacea and in Myriapoda (Figs. 3.2a, 3.17, 3.22, 3.24). Unlike the Chilopoda and Diplopoda, a maxilla 1 tendon in hexapods is also present (cf. Figs. 3.17, 3.22 with 3.23(a)) and is sometimes fused with that of the mandible (Figs. 3.27, 3.31), as frequently occurs in Crustacea and in the corresponding endosternite of the Chelicerata (Fig. 3.13). These transverse mandibular tendons disappear along with the acquisition of strong transverse biting in both Crustacea and Hexapoda for similar functional reasons. It is the head apodemal system consisting of fixed anterior and posterior pairs of tentorial apodemes, the former associated with the mandibles and the latter with the labium, which is characteristic of the Hexapoda and differentiates this subphylum from all other arthropods.

Evolution of arthropodan jaws

Two types of evolution have been exploited by the several classes of hexapods: (1) biting in the transverse plane permitting ingestion of hard or large food has been acquired secondarily by the Lepismatidae among the Thysanura and by the Pterygota; and (2) entognathy, whereby both mandibles and maxillae 1 become covered by a pouch and movements of protrusion as well as basic promotor–remotor roll take place. The differences in mechanism of the entognathous mandibles and maxillae in the Collembola, Diplura, and Protura (the latter not considered here) indicate parallel evolution of this condition in these three classes.

Thus the jaw mechanisms of the Onychophora, Myriapoda, and Hexapoda are each distinct, but all have in common the use of the tip of the whole leg for biting in contrast to the gnathobase of the Crustacea, Chelicerata, and Trilobita (Table 3.1, p. 65). The primitive biting in the transverse plane by segmented myriapodan mandibles contrasts absolutely with the secondary transverse biting acquired by the Lepismatidae and the Pterygota. The many specialized uses of the mouth parts of the Pterygota are beyond the present scope, but the biting of the locust type of mandible appears to be basic to them all.

A study of the functional morphology of the jaw mechanisms of extant arthropods demonstrates without doubt the independence of the evolution of the Chelicerata, the Crustacea, and the Uniramia. An understanding of how jaw mechanisms are constructed and how they work is perhaps the most difficult to achieve but it is the most rewarding in providing evidence of phylogeny (Chapter 6).

4 *The evolution of arthropodan types of limbs and the effects of habits on the limbs of* Limulus *and Crustacea*

1. Introduction

IN the whole of the animal kingdom there are only a limited number of ways in which legs are formed. In annelids and arthropods a pair of limbs is borne by some or most segments of the body. The vertebrates use only two pairs of limbs but many trunk segments contribute to the ontogenetic development of each limb. The advantages accruing from the use of two rather than a much larger number of leg pairs is shown in Chapter 7, §6C(vi), Fig. 7.11. The five-rayed echinoderms use a multitude of tube feet as well as movable spines for locomotion, usually segmentally arranged along each of the five arms, their positions corresponding with the series of skeletal ambulacral ossicles.

The flatworm *Rhynchodemus* shows a remarkable method of walking by non-permanent limb-like protrusions of the body. Most flatworms progress upon a continuous carpet of mucus which is secreted from the anterior end of the body. *Rhynchodemus* economizes in the laying down of mucus and leaves a track of mucus 'footprints', separated from one another, as it 'walks'. Figure 4.1(a) is taken from a photograph of this worm in side view. The body is thrown into leg-like projections, each of which stands on a mucus pad. The animal is moved forwards by the body flowing into and out of the leg-like projections. In the figure the anterior end is raised ready for the next 'footfall' and secretion of mucus, while the posterior end has just been lifted from a mucus pad. Figure 4.1(b) shows a view from above of a track of mucus 'footprints' left by the two sides of the body.

The more primitive arthropods possess permanent legs on most of their body segments and these paired structures are fairly similar to one another serially along the body (e.g. Figs.

Evolution of arthropodan types of limbs

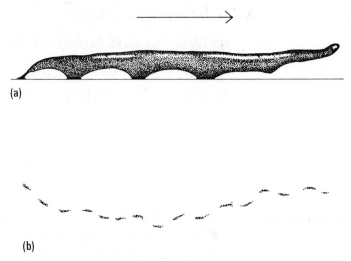

Fig. 4.1. (a) Side view of the flatworm *Rhyncodemus* sp. walking over a damp surface, the body flowing in and out of temporary leg-like lobes, each of which secretes a mucous pad. (b) Shows a dorsal view, on a smaller scale, of a track of mucous pads left by the animal. (After Pantin 1950.)

1.1, 1.2, 1.6, 2.4). Divergencies from this condition are of three kinds and all are associated with function. The limbs on the head and prosoma and different parts of the trunk become differentiated one from another in shape and serve special purposes, such as jaws, and walking and swimming limbs. Secondly, the limbs may disappear entirely from certain segments and become enlarged on others. Thirdly, very rarely the dorsal and ventral elements composing some segments become dissociated or more than one pair of limbs are carried by one segment, as in the posterior thoracic region of the branchiopod *Triops* (Pl. 2(a)).

It has been shown in Chapter 2 (§3 and 4) how the primarily aquatic armoured arthropods employ a basically different method of feeding from the terrestrial Uniramia (Onychophora, Myriapoda, Hexapoda). The legs in the two groups are also different. Those of the Uniramia are unbranched while branched limbs, usually biramous, characterize the former. The contrast is shown in Fig. 4.5. An embryonic limb is usually a simple finger-like projection which gradually increases in length. The rudiment may remain simple or it may become more complex by outgrowths developing from a limited number of sites (Fig. 3.1(a)). A small number of exites and endites can grow from the outer

Evolution of arthropodan types of limbs

and mesial sides respectively of the leg base (protopod) which remains uniramous. An exopod or outer ramus can arise further from the body near the distal end of the protopod and the rest of the main axis forms the endopod, used for walking in those animals which can walk. These parts of crustacean limbs are marked in Figs. 4.5(a), 4.6; the protopod in Crustacea is often divisible into two segments known as the coxa and basis. In Chapter 3 the protopod of the mandible has been referred to simply as the coxa. The limbs of only a few of the fossil armoured arthropods are known in detail and these are illustrated in Chapter 1 and Fig. 4.2. The legs of the fossil Merostomata are not well known, though we have the living representative *Limulus* and allied genera. In the trilobites, in *Marrella*, and in *Limulus* a type of biramous limb is present which differs from those of Crustacea. It has already been shown in Chapter 3 that the use and movements of the chewing gnathobase of the Chelicerata differs from the rolling grinding gnathobase of Crustacea. The similarities between the biramous limbs of Crustacea and the other armoured arthropods appears to be superficial and the differences marked and important. If this is so then the limb structure groups the arthropods at least into the same three divisions as does a consideration of their jaws (Chapters 3 and 6).

2. The biramous limbs of trilobites, of other fossil arthropods, and of *Limulus*

A uniramous flagelliform antenna is the simplest kind of sensory limb and is widely present among many arthropodan groups, but not in the Chelicerata.

The post-antennal limbs of trilobites form an even series on the cephalon and trunk, differing only from one another in size (Figs. 1.1, 1.3, 4.2). The protopod, or basal uniramous part of the leg, appears to be unsegmented. The grooves lying across the lateral part of this leg in *Triarthrus* are doubtless strengthening devices, as are so many comparable grooves in extant arthropods on the head, body, and limbs. The tendency to suppose that all such grooves are evidences of evanescent intersegments is to be deplored. The proximal coxal endite or gnathobase, on the mesial side of the leg base, has been considered in Chapter 3. The main axis of the limb forms a walking endopod in trilobites, in *Marrella, Burgessia, Cheloniellon,* and on the prosoma of chelicerates.

The nature of the outer branch present on the limbs of many of the early arthropods is a matter of particular interest

Evolution of arthropodan types of limbs

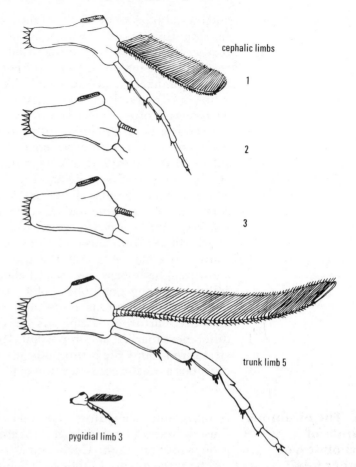

FIG. 4.2. Limbs of the trilobite *Triarthrus eatoni* (Hall); see whole animal in Fig. 1.3. The limbs behind the antenna are essentially similar all along the body, the three head limbs (cephalic limbs 1–3) are followed by larger trunk limbs and small limbs on the pygidium (pygidial limb 3 being shown). (After Cisne 1975.)

because we need to know whether there is a basic similarity to or a difference from crustacean limbs. Only recently have the limbs of certain trilobites and other extinct arthropods been sufficiently elucidated to make this determination possible. As will be shown below, the crustacean protopod carries proximally one or two respiratory exites or epipods and a single, more distal, exopod (Figs. 4.5(a), 4.8(a)–(c)). The shapes of both may differ considerably.

Only one outer branch exists on the limbs of trilobites and *Marrella*. It arises distally on the protopod and thus appears

Evolution of arthropodan types of limbs

to correspond with the crustacean exopod and not with an epipod. In *Triarthrus* (Fig. 4.2) the outer branch arises much as does the exopod of the crustaceans (cf. Figs. 4.2 with 4.5(a)), but the structure of the branch differs in *Triarthrus* and *Anaspides*. The outer branch of *Triarthrus* bears very many flat, probably respiratory, lamellae while the exopods of the thoracic limbs of *Anaspides*, of *Hemimysis,* and of the larval lobster (Figs. 3.7, 4.5(a), 4.10(a)) bear fine setae used in swimming, each seta being much too small to accommodate an internal respiratory flow of blood. In the trilobite *Olen-*

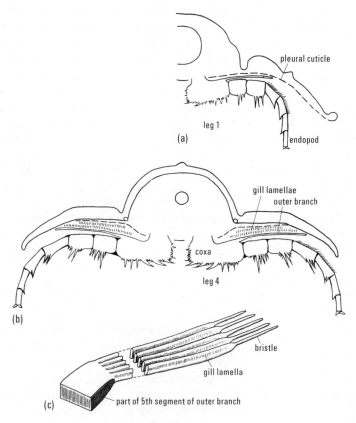

FIG. 4.3. Limbs of trilobites. (a, b) Transverse section through the body of *Olenoides* to show (a) the 1st postantennal limb and (b) the 4th limb (1st trunk limb behind the head). The gill lamellae seen in (b) project backwards from the limbs in front. (After Whittington 1975a.) (c) Reconstruction from sections of the lamellae of the outer branch of *Ceraurus pleurexanthemus* Green, Ordovician, passing through part of the fifth segment of the outer branch of the limb. (After Størmer 1939, from Tiegs and Manton 1958.)

oides the outer branch of the limb is much shorter and wider than in *Triarthrus*. It arises from the distal dorsal face of the protopod, marked by the right-hand hatched areas on legs 1–7 in Fig. 1.1. The outer branch is divided into a proximal lobe bearing lamellae, which are probably respiratory, and project backwards over the limbs behind, and a small distal lobe which bears no lamellae. The origin of this outer branch is seen in transverse section in Fig. 4.3(a), leg 1, and the lamellae of preceding legs are cut over leg 4, Fig. 4.3(b). A reconstruction of the lamellae from the trilobite *Ceraurus*, made from sections, shows the shape of the lamellae (Fig. 4.3(c)). Their size is such that they could have housed a respiratory flow of blood, as exists in the exites of *Anaspides* (Fig. 4.5(a), Pl. 3(f) and branchiopod crustaceans. In *Marrella* the outer branch of post-antennal limb 2 and posteriorly are much the same as in *Olenoides* (Fig. 1.6).

Burgessia shows more differentiation among its limbs. As in trilobites, three post-antennal pairs were borne by the cephalon and these carried a flagelliform outer ramus (Fig. 1.7). The following seven pairs of trunk limbs each bore a single flat, wide, plate-like outer ramus, overlapping those behind. The exact origin of this outer ramus from the protopod of the leg (coxa) has not been ascertainable from the fossils. Thus we are left with the clear evidence of the two trilobites *Olenoides* and *Triarthrus* showing that the outer rami of their legs correspond in position of origin fairly well with the exopod of Crustacea, but that the trilobite outer ramus differed in carrying lamellae which were probably respiratory and that no proximal exites or epipods were present.

The Merostomata, as far as we know, possess no antennae. One pair of preoral chelicerae, ending in a pincer, formed by the penultimate segment being elongated against the terminal segment, is situated in front of the mouth, thus corresponding in position with the trilobite antenna. The Aglaspida may have possessed an elongated chelicera (*Aglaspella*, Fig. 1.14(c)), while the Xiphosura and Arachnida both possess a short chelicera (Figs. 1.9, 3.14).

Prosomal limbs in the Xiphosura are mostly uniramous, the endopod being used for walking and burrowing. The sixth prosomal limb carries an apparent exite (or flabellum) proximally on the coxa (Fig. 4.4(a)) which is not respiratory. The seemingly biramous opisthosomal limbs of *Limulus* form respiratory and swimming organs, five pairs being sunk in a depression of the body covered by the genital operculum (Fig. 1.9). Each branchial limb (Fig. 4.4(b), (c)) is flattened and the endopod very short (4b). From the lateral expanded part of the limb arise some 150 large, flat, wide gill plates, shown more plainly in the longitudinal slice (Fig. 4.4(c)). Metachronal beating of the five pairs of limbs, each limb fused along its median margin with its fellow, results in water being taken in and squeezed out from the interlimb spaces, as in *Chirocephalus* and *Branchinella* (Figs. 2.8–10), the bare distal part of the limb marked 'outer lobe' in Fig. 4.4(b), (c) acting as a valve enhancing control of the water circulation, much as is done by the distal parts of the limbs of the branchiopods (Pl. 2(b), Figs. 2.9–10). The respiratory plates of *Limulus* bear some resemblance to the much narrower lamellae of *Olenoides* and *Marrella*, but are present in such large numbers as to arise from the whole length of the limb proximal to the valve-like outer lobe. The compact form of the respiratory limbs is an advantage in burrowing and the strong beatings of these limbs also provides a swimming thrust. The five highly efficient branchial limbs serve the needs of a large, strong, and active animal.

The five pairs of prosomal limbs behind the mouth of *Limulus* show little differentiation in structure one from another. The fan of large digging terminal spines on the sixth leg is suited to shovelling the substrate during burrowing (Fig. 2.6), the other four pairs of limbs end in small pincers used in feeding.

The huge eurypterid *Mixopterus* (Fig. 1.12) displays a general scorpion-like appearance, carrying short, strong chelicerae in front of the mouth and postoral pedipalps followed by four pairs of legs corresponding with the walking legs of arachnids (cf. Fig. 1.13). But these four leg pairs are all very different from one another, more so than in any scorpion or other arachnid, none of which show such advanced serial differentiation. These legs are used in opposite phase, as in scorpions, as shown by the fossilized tracks which also bear deep indentations caused by the strong, blade-like sixth prosomal limb (cf. Pl. 7(a), (b)). The rest of the body is limbless, as in arachnids, although the mesosoma may have borne reduced limbs which lay flat against the sternum and covered gill-like structures (Fig. 1.10(d)).

Thus some of the fossil Merostomata show a more

Evolution of arthropodan types of limbs

advanced condition of serial differentiation of their limbs than the modern chelicerates. The elongation and enlargement of the prosomal limbs of eurypterids brings them far beyond the lateral contours of the body and, with no further limbs projecting behind them, the set can be well fanned out, which is a locomotory advantage (Chapter 5, §3, Fig. 5.2). But even so, the stride length taken by *Mixopterus* was much shorter than that of the scorpions. The wide, flattened terminal podomeres on the most posterior prosomal limb of *Mixopterus* suggests as good a swimming ability as is provided by the last thoracic leg pair of the swimming crab *Macropipus* (Pl. 3(h)). Here there are similar blade-like terminal podomeres on the most posterior pair of thoracic legs (§3A(iv)). These crabs can occur very far from shore, in deep water, and are good swimmers.

3. The biramous limbs of Crustacea

Crustacean limbs exhibit bewildering complexity and variety in their adaptations to special needs, such as walking, burrowing, swimming, jumping, filter feeding, detritus feeding, large-food feeding, and special modes of prehension. The most detailed correlations exist between limb structure and the exact environmental circumstances in which the animals live. Every limb can differ in structure from every other limb on the same animal and the whole of the often remarkable morphology is related to habits of life (e.g. chydorid and macrothricid Branchiopoda, Fryer 1968, 1974). The basic form of crustacean limbs, but not the complexities, and some of the major modifications existing in the various groups, will be described here.

A. *Basic plan*

As already noted, the common plan of crustacean limbs consists of a protopod, often segmented into coxa and basis

FIG. 4.4 (see opposite). Limbs of *Limulus* (Xiphosura): (a) 6th prosomal limb (see also Fig. 3.12) showing walking and digging endopod, the coxa (interrupted lines) with biting gnathobase (ruled) and lobe (flabellum) from the coxa with which may represent an epipod or an exopod. (b) The contrasting opisthosomal limb used for swimming and respiration; the position on the animal is shown in Fig. 1.9(b). Each limb is flattened and possibly biramous, a left limb is cut away from its fellow along their median union (dotted line) and viewed from behind. Up to 150 thin, overlapping branchial lamellae arise from the unjointed basal part of the limb, which terminates in an outer lobe, free from gill lamellae, and a small jointed endopod. (c) Opisthosoma cut longitudinally through the gill lamellae. The external surface of the opisthosoma is seen above. The outer lobes shown in (b, c) form valves which both protect the gills and assist inter-limb intake and expulsion of water. Up to 150 gill lamellae are borne by each limb, many more than drawn. A copious blood supply passes to the gills from the branchial arteries shown. The remotor muscle causing the limb beat lies in the plane of the drawing but the promotor muscle lies to one side.

Evolution of arthropodan types of limbs

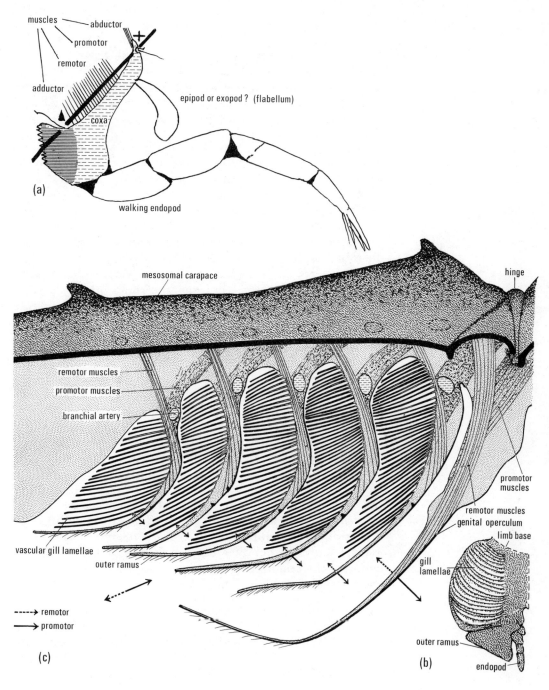

Evolution of arthropodan types of limbs

(Figs. 4.5(a), 4.6, 4.8). The outer or lateral margin of the protopod may or may not bear: exites proximally; an exopod distally; one or more endites from the mesial margin; the main axis forming an endopod (Figs. 4.5(a), 4.8(a)–(c)). The relative sizes of the parts vary greatly and the limbs of minute Crustacea such as *Bathynella* and *Hutchinsoniella* are expected to be simpler, with relatively smaller limb lobes and larger setae than in larger crustaceans.

(i) *Exites*. These are usually concerned with respiration. They may be the major site of gaseous exchange between the blood and the outside water, as in *Anaspides* and *Chirocephalus* (Figs. 4.5(a), 4.8(c), 2.7, 2.8); they may bear gills of complex form, as in crabs and lobsters (Fig. 4.9(b), (c), 4.11(c), (d)); or, as long epipods, they may control the intake of water for respiration and direction through the gill chamber under the carapace in these same animals (Fig. 4.9(a), (d), ep. 1). Exites can also control water currents, such as the outgoing flow from the interlimb filter chamber between maxilla 1 and maxilla 2 of a filter-feeding mysid (see Chapter 3, Figs. 3.8, 3.9).

(ii) *Endites*. These are usually concerned with feeding and are best developed on the head and close behind it on maxillipeds (trunk limbs 1–3, Figs. 4.8, 4.9(b)–(d)). A maxilla 2 endite may form a single filter plate in a mysid (Figs. 3.8(a), 3.9(a)) or every trunk limb can carry less efficient filtratory endites as in a typical anostracan branchiopod (Figs. 2.9, 2.10). Four masticatory endites are seen on the maxilla 2 of the lobster larva (Fig. 4.8(d)); the exite and exopod on the outer side of this limb are fused to form a large vibratory plate (scaphognathite), which draws water out of the gill chamber under the carapace fold and is the major cause of the respiratory flow (Figs. 4.8(d), 4.9(a)). The first maxilliped, or first trunk limb behind the head, of the lobster or crab (Fig. 4.9(d)) shows two endites used in feeding, a slightly larger endopod than on maxilla 2 (Fig. 4.8(d)), and a separate exite and exopod on the outer side of the limb.

(iii) *Exopods*. These are usually concerned with swimming. On the thorax of a mysid, a lobster larva, and of the adult *Anaspides* the beating of these organs, either by a backward and forward movement, or by an elliptical swirl, results in swimming (Figs. 3.7, 4.10(a)). Sometimes the endopod also

Evolution of arthropodan types of limbs

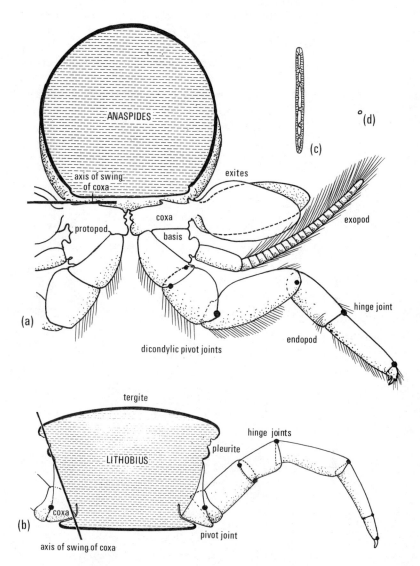

FIG. 4.5. Biramous and uniramous limbs. The black spots indicate the positions of the distal hinge joints and more proximal pivot joints, the latter being paired articulations on the lateral faces of the legs (see Figs. 4.27 and 4.28). (a) Posterior view from the transverse plane of the 6th thoracic leg of the crustacean *Anaspides tasmaniae*. The axis of swing of the coxa on the body is shown on the left. (b) Anterior view, from the transverse plane of a middle thoracic leg of the centipede *Lithobius forficatus*, the axis of swing on the body, marked on the left, approaching the vertical. (c) Transverse section through an exite of (a) showing its flat shape and internal vascular spaces. (d) Transverse section through a seta from the exopod of (a), cf. filaments of the exopod of a trilobite in Fig. 4.3(c). (After Manton 1969.)

145

Evolution of arthropodan types of limbs

participates in swimming, as in branchiopods. Particular swimming organs are formed by the endopod and exopod being equal in size, flattened and well furnished with setae, as in the swimming thoracic limbs of copepods and the abdominal swimming pleopods of shrimps and lobsters (Figs. 4.18, 4.10(b)). In the minute *Hutchinsoniella* the whole of the trunk limb endopod, exopod, and exite together are used in swimming (Figs. 4.6, 4.7).

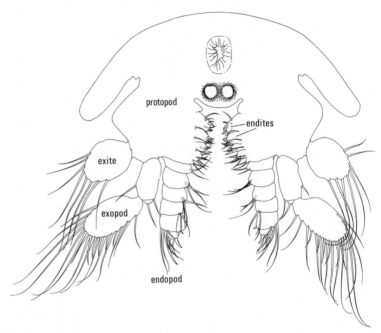

FIG. 4.6. A pair of thoracic limbs of *Hutchinsoniella macracantha*. (After Sanders 1963.)

(iv) *Endopods*. These are used for walking and burrowing, when these habits are practised, and can form the most substantial and conspicuous part of the leg of the larger Crustacea, situated, for example, on the thorax of the Malacostraca (Fig. 4.10(b)). Such simple uniramous limbs are clearly secondarily so because biramous limbs frequently precede them during ontogeny (see below) or exist in the more primitive members of the same taxon. The simple uniramous endopod of the adult crab or woodlouse is a convergent, parallel evolution to the primitively uniramous leg of a myriapod or an insect (Fig. 5.1). The distal end of crustacean endopods may show modifications for gripping food, as in the maxilliped of copepods (Fig. 4.12(b)) and the crushing chela on the fourth thoracic segment of the lobster (Fig. 4.10(b)). The terminal segments of the eighth thoracic leg is flattened and used for swimming by the swimming crab *Portunus*, cf. the eurypterid *Mixopterus* (Fig. 1.12). Other uses of the endopods in conjunction with exopods and exites have been mentioned above.

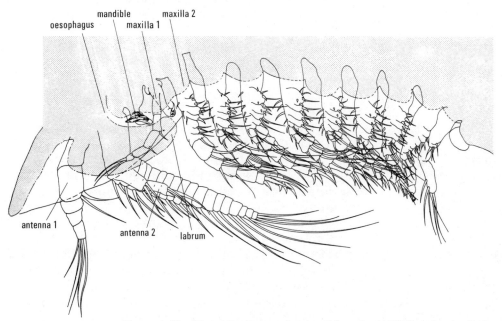

FIG. 4.7. *Hutchinsoniella macracantha*, viewed from the middle line, showing limbs on the right side of the body. (After Sanders 1963.)

(v) *Stenopodia and Phyllopodia*. Crustacean limbs are seen to be of two extreme types, but with many intermediates. Sometimes these types have been considered to be of phylogenetic significance, a point of view which can hardly be maintained today. At one extreme are limbs with rami rounded in transverse section, such as the thoracic limbs of *Anaspides* (Fig. 4.5(a)), seen also in the trilobite *Triarthrus*. Such a limb has been called a stenopodium. Then there is the contrasting bath-shaped, leaflike limb or phyllopodium, as seen in *Chirocephalus* and *Branchinella* (Figs. 2.9, 2.10, 4.8(c) and in *Sida* (Pl. 2(b)), where exites and endites bend backwards from the main axis of the limb and the distal part

147

of the leg can also bend in a posterior direction. The legs shown in Fig. 4.8(a), (b) are more flattened in one plane, as are the mouth parts of the lobster in Fig. 4.8(d), (e) which bend in an anterior direction. Function determines whether a simple biramous leg, without exites or endites, is rounded as in the barnacle thoracic limbs (Fig. 4.13) or flattened as on the 2nd–5th swimming thoracic limbs of a copepod (Fig. 4.18). Phyllopodia with exites and endites have probably been evolved more than once in association with known functions. A limb such as the trunk limb of *Hutchinsoniella* (Figs. 1.15(a), 4.6) might have had the potentiality for modification into the various types of limbs of the larger Crustacea, but one suspects that a simpler limb gave rise to those of both the Branchiopoda and the Cephalocarida (*Hutchinsoniella*).

B. *Serial differentiation of crustacean limbs*

A far higher degree of differentiation between limbs situated on the various segments is shown by the Crustacea than by most Trilobita, Merostomata, and other Chelicerata.

The antenna of trilobites, *Burgessia*, and *Marrella* correspond, at least functionally, with antenna 1 of Crustacea. The former are uniramous and flagelliform, while some swimming Malacostraca bear two or three rami on antenna 1. A serial uniformity of head and trunk limbs, as in trilobites, is never present in Crustacea. The three post-antennal limbs on the head of *Burgessia* differ considerably from those behind and there is greater serial differentiation in *Cheloniellon*, where an apparent antenna 1 and antenna 2 are borne on the cephalon in a preoral position, followed by biramous limbs on the succeeding segments, the anterior four pairs of which appear to carry gnathobases (Fig. 1.8). The Crustacea contrast in showing much greater serial differentiation of their limbs, which, to some extent, is bound up with the differentiation of a head, thorax, and abdomen.

The head typically bears paired antenna 1, antenna 2, mandible, maxilla 1, and maxilla 2, but the hinder limit of the

FIG. 4.8 (see opposite). Crustacean limbs possessing most of the basic parts: protopod, endopod, exopod, exites, and endites. Contrast the flattened phyllopodial limbs (b–e) with the rounded stenopodial limbs of *Anaspides*, Fig. 4.5(a). (a) *Paranebalia longipes*, 5th thoracic leg. (After Calman 1909.) (b) *Hutchinsoniella macracantha*, 5th thoracic leg. (After Sanders 1963.) (c) *Chirocephalus diaphanus*, median view of a trunk limb in its natural position with backwardly directed exites, endites, and the distal endopod and exopod. (After Cannon 1928.) (d) *Homerus americanus*. 2nd maxilla of first larva of the lobster. (After Herrick 1865.) (e) *Homerus americanus*. 1st maxilliped of first larva of the lobster. (After Herrick 1895.)

Evolution of arthropodan types of limbs

head may not be clearly defined, especially in *Hutchinsoniella* (Fig. 1.15), where the maxilla 2 closely resembles the succeeding trunk limbs (Fig. 4.7) as in no other crustacean. The maxilla 2, or last head limb, is usually strongly demarcated from the trunk limbs. Antenna 1 and 2 are usually sensory but antenna 2 forms the principal swimming organs

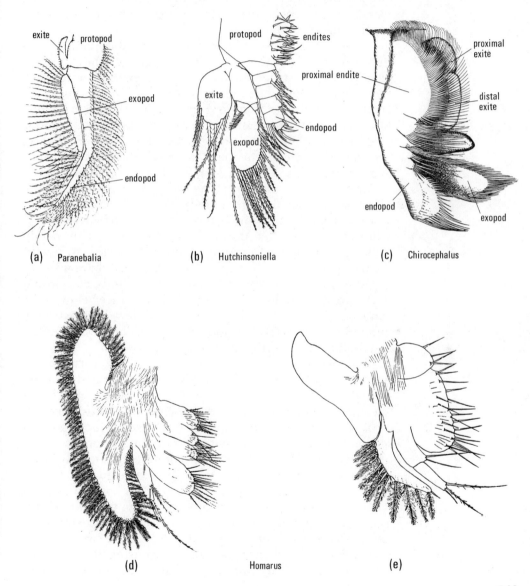

of Cladocera and can bear copulatory arrangements in males used for holding the females. The clasping organs can be highly elaborate.

The paired crustacean mandible, situated just beside the mouth, differs in its structure and movements from the feeding organs of trilobites and chelicerates (Chapter 3). The two further pairs of head limbs, maxilla 1 and maxilla 2, are usually small, concerned with feeding, and differ from one another in structure.

The trunk is divisible into thorax and abdomen. Limbs may be restricted to the thorax and if present on the abdomen, usually differ in structure and function from thoracic legs. All thoracic limbs are much alike in *Chirocephalus* and used for swimming and filter feeding (Figs. 2.9, 2.10). The first trunk limb is a maxilliped dealing with food by its tip in Copepoda (Fig. 4.12(b)). The single maxilliped of a mysid or syncarid malacostracan cleans the maxillary filter plate with its proximal endite (Figs. 3.8(a), 3.9(a)). Three pairs of maxillipeds are present in decapod Crustacea where both endites and the distal parts of the limb are used particularly in feeding (Fig. 4.9(b), (d)). The remaining thoracic limbs are locomotory; swimming in function in the Copepoda (Fig. 4.18); mainly walking in many Malacostraca, but some species swim with the thoracic exopods (Figs. 4.10(a), 3.7). Adult barnacles are sessile and use their long biramous thoracic limbs as a casting net for food collection (Fig. 4.13, Pl. 2(c) and see below). The ostracods have a very short body and few limbs, widely different from each other and used for walking, swimming, and food collection (Fig. 4.17 and see below).

Limbless segments forming an abdomen serve many purposes, often balancing the body in the water and carrying genital and faecal products well clear of feeding limbs (copepods, cephalocarids, and branchiopods—Fig. 1.15, Pl. 3(b), (c)). The body terminates in a telson which is not a segment, but the unsegmented embryonic remains of the posterior end which has never become segmented. It bears the anus (Figs. 6.3, 6.4). A caudal furca on the telson, however large (Figs. 1.15, 4.12, Pl. 3(b)), is not homologous with a pair of legs. A furca is present also in trilobites (Figs. 1.1, 1.2). A special provision in swimming Malacostraca is a 'tail fan'. It consists of a short wide telson to the side of which lie much expanded endopods and exopods of the 6th abdominal limbs, together forming a wide fan used in swimming

Evolution of arthropodan types of limbs

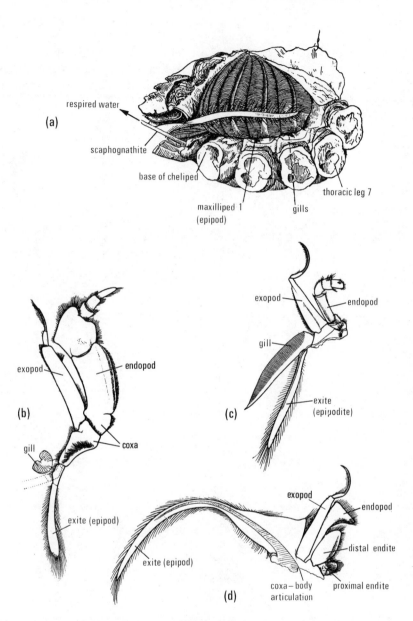

FIG. 4.9. Mouth parts and gills of a crab, *Carcinus maenas*. (a) Lateral view of the organs in the left branchial chamber, exposed by removal of the overlying carapace. The scaphognathite (exite and exopod of maxilla 2 fused) vibrates and bales water out of the gill chamber, water entering at the bases of the thoracic legs. (b) 3rd maxilliped of crab. (c) 2nd maxilliped of crab. (d) 1st maxilliped of crab. (After Borradaile 1922.)

151

Evolution of arthropodan types of limbs

(Fig. 4.10(b) and see below). The rest of the abdominal limbs form swimming or copulatory organs and they can carry the eggs, as in female crabs and lobsters. Other single highly differentiated limbs are the crushing great claws of the lobster, one of which may weigh almost 2 lb in an adult

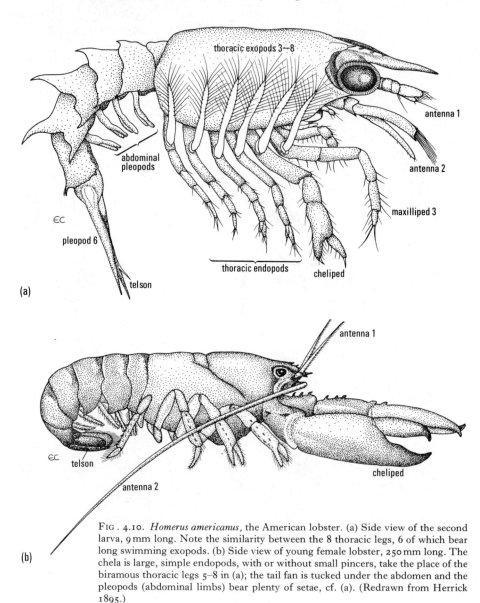

FIG. 4.10. *Homerus americanus*, the American lobster. (a) Side view of the second larva, 9 mm long. Note the similarity between the 8 thoracic legs, 6 of which bear long swimming exopods. (b) Side view of young female lobster, 250 mm long. The chela is large, simple endopods, with or without small pincers, take the place of the biramous thoracic legs 5–8 in (a); the tail fan is tucked under the abdomen and the pleopods (abdominal limbs) bear plenty of setae, cf. (a). (Redrawn from Herrick 1895.)

Evolution of arthropodan types of limbs

weighing a little over 11 lb and proportionately much more in a 25-lb lobster.

C. Changes in crustacean limbs during ontogeny

Some crustaceans hatch as nauplius larvae with few segments (e.g. most Branchiopoda, Cephalocarida, Copepoda, Cirrepedia, penaeid prawns, etc.). As segment number increases from a preanal growth zone, limbs progressively develop and may gradually assume their adult features (e.g. *Chirocephalus* —Fig. 1.15(f), (g)). But in other crustaceans there may be dramatic changes in the limbs. The creeping nauplius larva of *Amphiascus* (Copepoda) has a clasping organ on the antenna 2, used for holding food, and a gnathal lobe (Fig. 4.12(a)). At one moult this larva changes its shape and segment numbers and now similar feeding claspers are situated on the maxilla 2 and maxilliped and not on the antenna 2 (Fig. 4.12(b)). The larval stages of *Squilla* remould the form of their limbs several times during larval life, absorbing some and growing them again in different shapes.

Even when there is a long embryonic development (e.g. Cladocera, Leptostraca, Peracarida, Syncarida, etc.) and a crustacean hatches with all its segments formed, the limbs they bear may not be as in the adult. The 9-mm second larva of a lobster (Fig. 4.10(a)) possesses swimming exopods on the 3rd–8th thoracic segments. The 2nd maxilliped (2nd thoracic limb of the first larva—Fig. 4.11(a)) has a very small exopod. In the fourth larval stage the 3rd maxilliped still has a conspicuous exopod (Fig. 4.11(b)), but the exopod is very small on the 5th thoracic segment (Fig. 4.11(c)) and it is gone from the last thoracic legs by the fifth larval stage (cf. Figs. 4.10(a), 4.11(b)). The growth of a chela is seen also in the 9-mm larva and adult lobster in Figs. 4.10(b), 4.11(c). Many other examples of this phenomenon could be given. Many Decapoda hatch at post-naupliar stages and their larval limbs undergo changes before becoming adult.

D. Special habits of Crustacea, the correlated limb structure, and crustacean evolution

Three crustacean habits of great interest will be considered briefly. The first is the raptatory habit leading sometimes to specialized large-food feeding; the second is filter feeding; and the third is rapid escape jumping through the water. All except primitive raptatory feeding have been independently acquired many times and contribute to the success of the groups concerned. There are many other habits worthy of consideration with which structural peculiarities have evolved, but they cannot be dealt with here.

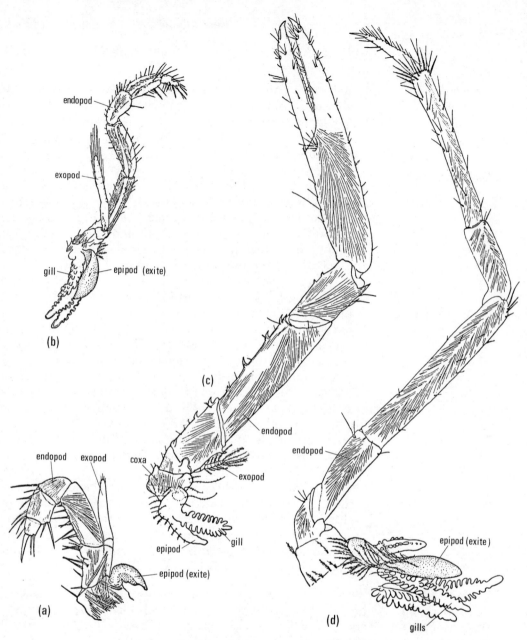

FIG. 4.11. Limbs of lobster larvae to show the position of the gills on the exite (epipod) and the progressive disappearance of the exopod which results in the simple uniramous endopod of the adult thoracic limbs 5–8. (a) Left 2nd maxilliped of first larva, anterior view. (b) Right 3rd maxilliped of fourth larva, dorsal view. (c) Left 5th thoracic limb (1st periopod) of fourth larva, ventral view, the exopod now being very small. (d) Left 8th thoracic leg (4th periopod, or walking leg, Fig. 4.10) of 5th larva, dorsal view; the exopod has now disappeared. (After Herrick 1895.)

Evolution of arthropodan types of limbs

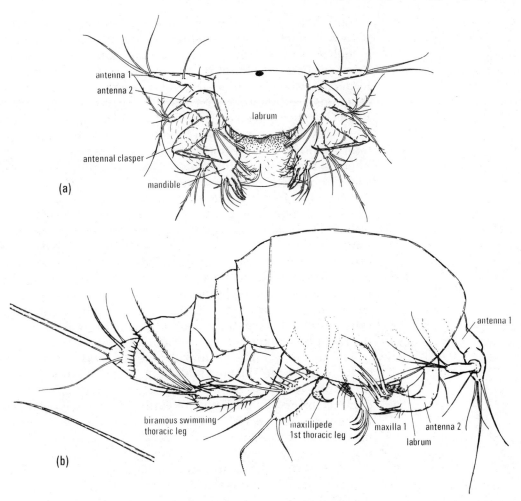

FIG. 4.12. Harpacticid copepod *Amphiascus* sp. (a) Nauplius larva 0·096 mm long. (b) First copepodite larva, showing great change in body form at one moult, including the loss of the clasping hook on the naupliar antenna 2 and its replacement by a similar hook on the maxilliped (1st trunk limb). The caudal setae are cut.

(i) *The raptatory habit and its derivatives.* Particle feeding by *Hutchinsoniella* is restricted to zones rich in flocculent detritus on the bottom (see below). A sparser supply of edible particles might have encouraged simple raptatory feeding, both on the substratum and in the water above it. The most advanced raptatory feeders, capable of eating the hardest or largest food, appear to be end terms of this evolutionary trend (e.g. *Squilla,* crab, Pl. 3(i)). Bottom-living copepods and ostracods (Pl. 3(a)) are largely raptatory. This habit has given place many times to filter feeding, an activity dependent upon a higher order of limb differentiation and co-ordination between successive limbs than possessed by simple raptatory feeders. *Calanus* is a filter-feeding copepod, but most of the calanoids are raptatory. *Cyclops* (Pl. 3(b), (c)) is raptatory. Even *Calanus* can feed on large live or dead animals in winter. The early Malacostraca were probably bottom-living raptatory animals. Many have developed complex modes of filtering the water as they adopted a swimming habit, but some remained raptatory, such as the primitive mysid *Lophogaster* (Fig. 3.9(b) and Manton 1927), the majority of Mysidacea now being filter feeders. Predacious feeding utilizes the head appendages and from 1–3 anterior thoracic limbs. The mandibles bear strong distal cusps; few or no grinding surfaces; good articulations with the head; and powerful musculature (Ch. 3, §2A, C). The distal parts of the more posterior feeding limbs are well provided with biting or shredding setae or cusps, see the setae shown diagrammatically in Fig. 3.8(b) on the mandible, maxilla 1, maxilla 2, and the tip of the maxilliped or 1st trunk limb.

The tips of one pair of maxillipeds are used to grip prey or food masses in the copepods (Fig. 4.12(b)). In the large predatory malacostracan *Squilla* a very strong prehensile organ is formed by the 1st trunk limb, where the large distal podomere clasps back on the penultimate one. One pair of maxillipeds is present in all Peracarida, the base of this limb often being elaborated for assistance in filter feeding (Figs. 3.8(a), 3.9(a)) and the limb-tip for large-food feeding (Fig. 3.5(b)). The lobsters, crabs, and shrimps possess three pairs of maxillipeds. In the two former the 4th thoracic limb forms a grasping or crushing claw or chela, developing initially as shown in Figs. 4.10(a), 4.11(c), where the penultimate podomere grows forwards as a finger opposing the terminal podomere. Smaller chelae may be present on several thoracic legs in other malacostracan species such as crawfish and prawns. Typically, crabs run sideways, e.g. shore crab (Pl. 3(i)), but the portunid swimming crabs swim by the flattened and modified last pair of thoracic legs (see Pl. 3(h) and legend).

(ii) *Filter and suspension feeding in: barnacles, copepods, branchiopods, cephalocarids, malacostracans, leptostracans, euphausids, and ostracods*. This is a widespread method of food collection within the Crustacea, but no one of the many methods employed today can be regarded as primitive or ancestral to the others. Minute species can collect finely divided particles by the metachronal movements of their simple limbs without much further elaboration (e.g. *Hutchinsoniella* (Figs. 4.6, 4.7)). Some small Cladocera possess very elaborate filter-feeding mechanisms (Fryer 1968, 1974). Larger crustaceans, if they filter feed, need efficient systems for this purpose.

The barnacles (Cirripedia) form a casting net from their long setose thoracic endo- and exopods (cirri) and the net cleans itself as it curls up after each cast, passing food material to the mouth (Pl. 2(c)). The mechanism, however, is not at all simple. From a position with curled up or retracted limbs, the body and limbs are projected outwards through the opened valves of the carapace (shell). The force causing extrusion of the body and limbs is largely hydrostatic in origin, a special blood pump forcing blood into the cirri, so extending them. There are no extensor muscles to the hinge joint on each cirrus, but a long flexor muscle passes along the whole length of each exopod and endopod (Fig. 4.13). The vascular system is canalized, enabling the pump to force blood into the cirri and elsewhere; there is no heart as in a normal arthropod (for details see Cannon 1947).

Some copepods (Pl. 3(c)), such as *Calanus*, use one stationary pair of filter plates on maxillae 2. The gentle swimming movements of the head endopods from antenna 2, mandible, and maxilla 1 create lateral swirls which suck water from the mid-ventral region through the filter plates. The particles so retained are scraped off the plates by the tips of the maxillipeds and then pushed forwards to the proximal parts of maxillae 1 and to the mandibles. Each filter plate is constructed as described below for the Branchiopoda (for further details of *Calanus* see Cannon 1928).

Filter plates are similarly constructed in copepods, in branchiopods, and in filter-feeding Malacostraca. Filter plates borne on limb endites are illustrated in Figs. 4.8(c), 4.14(c), (d), and more diagrammatically in Fig. 4.16 for all types of branchiopod legs shown, except the ancestral and the *Triops*-types. Filter-plate setae are long, closely spaced in the same plane, and bear setules on either side which span the gap between the main setae. The whole forms a net so fine that

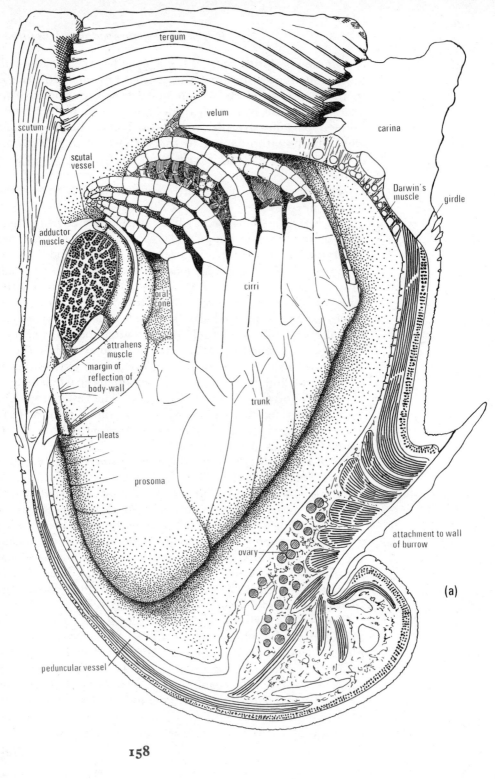

Evolution of arthropodan types of limbs

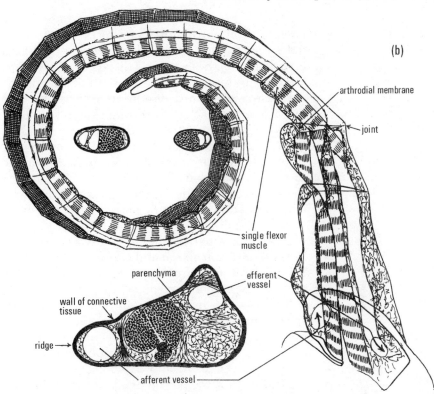

FIG. 4.13. The pedunculate barnacle *Lithotrya*. (a) Reconstruction of the whole animal in which the right half of the capitulum and peduncle have been omitted. The ventral side is uppermost while the dorsal parts of the body are below. (b) Reconstruction of a pair of cirri (exopod and endopod) borne by the two segmented protopods, to show the musculature and blood vessels, together with sections through a cirrus at different levels. The whole limb × 42; sections × 80. (After Cannon 1947.)

high magnifications of a microscope may be needed to reveal the setules.

The more primitive Branchiopoda use a long series of similar limbs, each bearing a filter plate. The limbs beat in metachronal rhythm, as shown in Fig. 2.10, the backstroke starting when the limb is roughly in the transverse plane of the body (see Chapter 2, p. 54 and Fig. 2.8(c)). During the forward stroke (Fig. 2.10, legs 6–10) the inter-limb space enlarges. The filter plates cover the gap between one leg and the next on the medium side; distally the legs overlap by their terminal backward curvature; and laterally the inter-limb space is closed by the backwardly directed exites. Thus water

159

can only enter an inter-limb space of increasing size by passing through the filter plates (Pl. 2(b)). Suspended particles are retained on the plate. During the backstroke (Fig. 2.10, legs 1–5) the exites and distal parts of the leg flap so as to leave an unimpeded outflow from the inter-limb space, now decreasing in volume. A swimming current is thus produced, but also some backwash through the filter plate, enough to dislodge the filtered food with help from minute comb-setae situated at the base of each filter plate which comb the one in front. The forward flow of water in the median food groove, produced as explained in Chapter 2, §4, carries the filtered food particles to the labral region where entanglement in labral secretion enables the particles to be pushed by maxillae 1 to the grinding mandibles (see Chapter 3, §2B).

The Cladocera (Branchiopoda) use a shorter series of trunk limbs than *Branchinella* and *Chirocephalus*, only 5 or 6 pairs. Filter plates are borne by 5 leg pairs in *Sida* but only two, more efficient, pairs are present in *Daphnia*. In *Sida* Pl. 2(b) shows legs 3–6 in the suction phase, with large inter-limb spaces into which water is being sucked through the filter plates, as in *Branchinella* (Fig. 2.10). Legs 1 and 2 on the contrary are approximating, so obliterating the inter-limb space between them and expelling the water laterally and ventrally. In *Daphnia* the five pairs of trunk limbs are very different from one another (Fig. 4.14); their movements are highly co-ordinated, but the metachronal rhythm along the trunk is distorted a little so that legs 3 and 4, bearing the filters, beat back almost simultaneously. Outlines are given in Fig. 4.15 of the limbs devoid of setae, the animal lying ventral side upwards. In (a) the 3rd and 4th trunk limbs are approximately at the end of the forward stroke, causing maximum suction through the filter plates and in (b) these limbs are at the end of the backstroke. Currents (1) and (4) are sucked into the inter-limb space through the filter plates. On the backstroke the volume of the inter-limb space is reduced, the exopods flap away from the body proximally and away from the limb behind distally, and currents (2) and (2) pass out backwards. Water in the median space between the limbs is squeezed forwards and currents (3) and (5) enter the median food groove. These currents unite to form an orally directed food current in a mid-ventral food groove, bearing particles filtered off from the water by the filter plates. The 1st trunk limb is moving forwards while the 3rd and 4th are

Evolution of arthropodan types of limbs

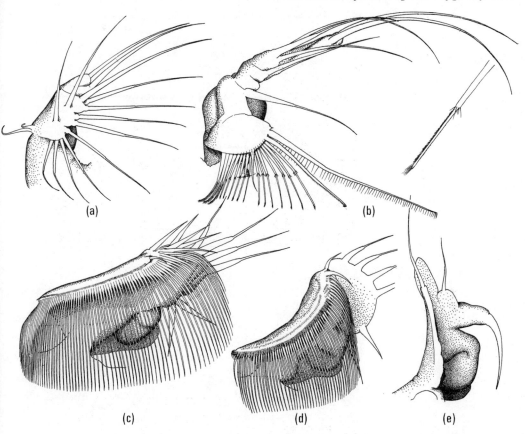

Fig. 4.14. *Daphnia magna*. (a)–(e) Median views of right trunk limbs 1–5 respectively, in their natural positions, but separated from one another (their origins from the body lie towards the bottom of the page). A seta from the middle of the gnathobase of limb 2 is shown on the right. (After Cannon 1933a.)

moving backwards; on the backstroke of trunk limb 1 current (6) is sucked in, as shown, and on the reverse movement current (7) passes to the inter-limb space 1–2. The parts are so arranged that current (8) must come from the food groove. Thus the 3rd and 4th limbs produce intermittent spurts of water into the hinder end of the food groove while the 1st trunk limb is sucking water out of the food groove anteriorly. An oral current is thus produced by pressure from behind and suction from in front. The filter plates are scoured by the gnathobasic lobe of trunk limb 2 which combs filtered residue from the anterior part of the 3rd gnathobasic filter while the

Evolution of arthropodan types of limbs

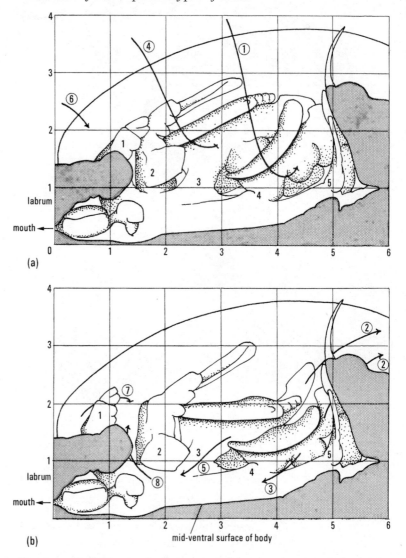

FIG. 4.15. *Daphnia magna*. Outlines to show the movements of the limbs and the feeding currents. For description see text. (After Cannon 1933a.)

posterior part is scoured by comb setules on the 4th gnathobase at the base of the filter setae. Just behind maxillae 1 secretions from the labral glands pass into a region of comparative quiet and into this space the gnathobasic lobes of the 2nd trunk limb pass the filtered particles. These particles, entangled in labral gland secretions, are pushed forward to maxillae 1 and on to the mandibles at irregular intervals.

Complex as is the feeding mechanism of *Daphnia*, far greater complexity exists in the Chydoridae and Macrothricidae (Fryer 1968, 1974) where very diverse filtering mechanisms exist with great serial differentiation between the trunk limbs. These crustaceans show how far they have advanced in complexity and diversification from the more primitive conditions seen in the unrelated merostomes, trilobites, and other known early armoured arthropods.

The possible methods of evolution of the many kinds of branchiopod filter plates are shown in Fig. 4.16. A primitive hypothetical limb, top right, possesses no well-formed endites. Two different types of gnathobase, the *Triops*-type and the *Lepidocaris*-type, could have come from this. A filtratory gnathobase, derivable from the same *Lepidocaris*-type could have given rise to the *Estheria*-type and the *Sida*-type. A long seta in *Sida* projects from the distal margin of the gnathobase. This seta is still seen in *Moina* and *Diaphanosoma* and suggests that the large filter plates of *Moina* and *Daphnia* have been derived from a gnathobase, the distal parts of the limb having been reduced, while in *Diaphanasoma* the gnathobasic filter has come into line with the filter plate borne on the more distal endite and in *Chirocephalus* the fusion of the two endites is complete. The Notostraca, alone among the Branchiopoda, excepting the few predators, lack filtering endites. They spend much of their time on the bottom, their legs beating in metachronal rhythm and they can swim slowly (Pl. 2(a)). Their setose, knobbed endites retain particles in the middle line which are then passed backwards and up towards the ventral body surface. From this position they are pushed oralwards by forwardly directed gnathobases, a unique slope for a branchiopod gnathobase, see Fig. 4.16. This brief inspection indicates the existence of divergent lines of evolution of filter feeding within the Branchiopoda (further details will be found in Cannon 1933 and Fryer 1968, 1974).

The feeding mechanism of the cephalocarid Crustacea, e.g. *Hutchinsoniella macracantha*, 3 mm long (Fig. 1.15(a)), is remarkable for its simplicity, in contrast to the elaborate filter feeding of branchiopods and particularly of Cladocera of much smaller size. Small Crustacea may be expected to be simpler in construction than larger ones which are similar in other respects. The simplicity of *Hutchinsoniella* is probably primitive and has been preserved because of the particular

Evolution of arthropodan types of limbs

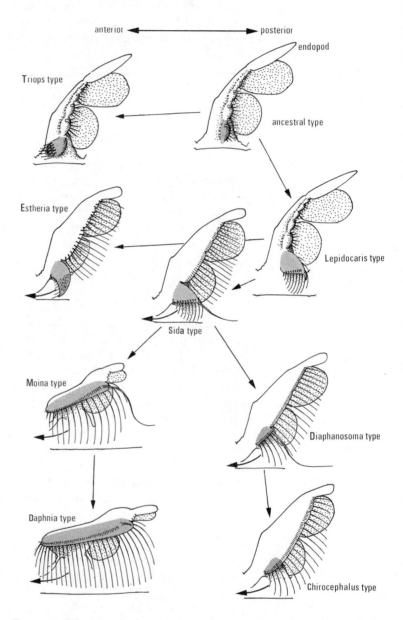

FIG. 4.16. Diagrammatic representation of branchiopod trunk limbs to show their possible evolutionary derivations, based upon a functional analysis of the morphology of the animals. The gnathobase is indicated by mechanical stippling. Curved arrows show direction of flow from the inter-limb space to the food groove. (After Cannon 1933a.)

habitat in which the animal is found. *Hutchinsoniella* is almost wholly confined to living in and on the 2-cm thick flocculent unconsolidated zone situated over soft sub-tidal sediment. The animal is a non-selective deposit-feeder, taking in large masses of the flocculent material. The trunk limbs are transversely flattened, the space between successive limbs being shorter, antero-posteriorly, than the thickness of the limbs (Fig. 4.7). The large protopod bears a series of small endites armed with short, curved spines and more slender, longer setae (Fig. 4.6). The limbs beat in metachronal rhythm and the exites, exopods, and endopods behave much as in a branchiopod. Water is sucked into the inter-limb space on the forward stroke and then expelled from this space on the backstroke as a swimming current. Suspended particles are held up on the endites, their longer setae overlapping the leg in front when the legs are close together, so dislodging particles which are forced back to the median space by the curved spines. These spines on each leg pair come close together at the end of the backstroke because the protopod moves in an ellipse, as shown in Fig. 2.7. Detritus is shifted forwards by the endites moving forwards and apart. Material is passed from limb to limb and also to a slightly more dorsal position on the preceding limb. The endites possess their own muscles and actively flick forwards during the forward stroke. Thus food material is passed forwards towards the mouth, close to the body mid-ventrally, but without the elaborations in filter plates and collecting currents possessed by the Branchiopoda living in much less concentrated suspensions of food material. The filter-plate feeders can collect minute food such as protozoa, single algal cells, etc.

The evolution of some crustacean feeding mechanisms and general habits. Two most important aspects of the feeding of *Hutchinsoniella* are the benthic habit and the morphological simplicity of the food-collecting mechanism. From such simplicity the much more elaborate filter mechanisms of swimming branchiopods might have evolved and with the advancement of filtering, replacing the simple detritus feeding, immense possibilities became opened up for inhabiting diverse pelagic environments. The Cladocera, for example, show an enormous adaptive radiation in which the structure of the body, carapace, and limbs suits a multitude of quite distinct micro-environments within even one lake (Fryer 1968, 1974).

It is possible that the Copepoda also had a benthic origin, crawling detritus feeders preceding copepods with raptatory and swimming habits (Fryer 1957). The bottom-living harpacticid copepod shown in Fig. 4.12 partly grasps its food with the large clasper on the antennae 2 of the nauplius and on the maxilliped of the copepodite larvae and adult and partly scrapes up food from the bottom. The parasitic mode of life could have been derived from raptatory habits. Pelagic filter feeding, dependent upon so much limb specialization, can hardly be other than an end term in copepod evolution.

Malacostracan filter feeding. In the branchiopod examples mentioned above and in the Cephalocarida the curvature of the limb is directed posteriorly and without curvature there would be no hydraulic efficiency. Filter-feeding *Malacostraca,* such as *Nebalia* and *Euphausia* use many pairs of filter plates but a mysid or *Paranaspides* uses a single, more efficient, pair of filter plates borne on maxillae 2 through which water passes from the middle line. Metachronal rhythm of each plate and adjacent limbs in mysids and *Paranaspides* is similar in principle to that of the Branchiopoda, but the limbs curve forwards, not backwards (cf. Figs. 2.9, 2.10 with 3.8(a), 3.9(a)). The one type of filter could not have given rise to the other without passing through a non-functional intermediate stage when the limb did not curve in either direction. Therefore the malacostracan filtering maxillae 2 must have evolved independently from the filtering limbs of the Branchiopoda.

A forwardly directed mid-ventral current along the thorax is produced in many, but not all, filtering mysids and Syncarida and supplies water to the maxillae 2 filter plates. In *Hemimysis* (Fig. 3.7) each swirling thoracic exopod causes an axial flow of water towards the base of the leg. Here the set of the limbs on the body directs the flow of water forwards. The postoral limbs are shown diagrammatically in median view on a right half of the body in Fig. 3.8(a) and an actual preparation, with leg 1 removed, is seen in Fig. 3.9(a). On either side between maxilla 1 and maxilla 2 there is a large inter-limb space, or suction chamber. This chamber is intermittently closed laterally by an exite, ventrally by the distal parts of maxilla 2 being held against the paragnath and distal parts of maxilla 1 and medially by a proximal filter plate on maxilla 2. A large endite on the 1st thoracic leg covers the maxilla 2 filter plate on the median side (Fig. 3.7). Metach-

Evolution of arthropodan types of limbs

ronal movement of the limbs alternatively draw water into each suction chamber through the filter plate from the midventral space (interrupted arrows on Fig. 3.8(a)), the water passing out laterally. Finely divided material is left on the surface of the filter plates. An abundance of scraping setae, with setules in many planes, are situated on a basal endite of each trunk limb 1; metachronal rhythm causes them to scrape the plate and pass the material on to the proximal brush setae of maxillae 1, which then push food on to the molar processes of the mandibles. The parts of the limbs concerned with the manipulation of food are indicated diagrammatically by coarse black setae on Fig. 3.8(a), while an actual preparation is shown in Fig. 3.9(a).

The requirements for filter feeding from a simple set of limbs such as maxilla 1, maxilla 2, and trunk limb 1 are very precise (Fig. 3.8(a), 3.9(a)) and account for the general similarities between the filter-feeding mechanisms of some mysids, *Paranaspides, Anaspides,* and others; but the differences between the taxonomic groups suggests that the filtering mechanisms are parallel evolutions and do not represent a basic mode of malacostracan feeding which existed before the various taxa separated. Indeed it seems probable that ancestral Malacostraca were bottom living in habit and may have been detritus feeders or scavengers (Chapter 6, §2 E(vi)).

Filter feeding, originally practised in the open water, in some groups has given place to a modified filter-feeding mechanism being used in the protection of tubes or algal growths, as in the Tanaidacea (Dennell 1934, 1937).

A different modification of malacostracan filter feeding in *Nebalia* (Leptostraca) has already been mentioned in Chapter 2 (Fig. 2.11). The full series of thoracic legs here form paired para-median filter walls within a carapace. Each wall is formed by two rows of hooked setae per limb, the hooks interlocking with those of adjacent limbs. Metachronal beating of the flattened thoracic limbs within the carapace draws a current of water from in front into the median space between the filter walls. Water enters the enlarging inter-limb spaces, passing through the filter walls and flowing out laterally and distally as a swimming current. The filtered particles are cleaned off the filter walls by the spiked setae shown (Fig. 2.11); two rows on each leg project through the walls when adjacent limbs come close together. The forwardly directed brush setae from the basal parts of the thoracic limbs then take over the transport of food to the mouth, in this case pushing the particles against the incoming current. The mouth parts, as seen in a mysid, are modified in *Nebalia* to meet the flow of filtered food coming from behind (Cannon 1927).

The euphausids also make a filter wall from overlapping, forwardly directed setae from the thoracic coxae and endopods. They swim through the water with an open basket slung below the thorax which catches plankton organisms under 40 μm in size (Barkley 1940); but this is a very different structural arrangement from the thoracic filter walls of *Nebalia* within the carapace. The euphausids live in open water while *Nebalia* usually lives in sheltered places, under stones and in weeds.

The Ostracoda (Pl. 3(a)) also demonstrate the indisputable independence of evolution of filter feeders within the same order. Most species live on the bottom, but they can swim. These small crustaceans are covered by a close-fitting bivalve shell. The genera *Pionocypris* and *Doloria* possess a vibratory plate spanning the space within the shell, situated between the soft parts attached to the shell and the free margin of the shell (the carapace). A current of water is thereby drawn through the shell cavity from in front and small particles of food are grabbed by the anterior limbs (Fig. 4.17). But the

FIG. 4.17 (see opposite). Diagrammatic side views of ostracods (a) *Doloria* and (b) *Pionocypris* with the left valve of the shell removed. Organs of similar function are provided by different limbs in the two genera; the table shows the comparisons.

	Pionocypris vidua	*Doloria levis*
Food stream through the shell produced by	Vibratory plate of maxilla 1	Vibratory plate of maxilla 2
Food collected by	Mandibular palps	Mandibles
Food entangled by	Labral secretion	Labral secretion
Food mass passed by mandible on to	Maxilla 1	Endopod of maxilla 1
Food mass passed by maxilla 1 on to biting mouth parts which are	Gnathobase of mandibles	Three endites of maxilla 1, two endites of maxilla 2, and part of exopod of maxilla 2
Food prevented from escaping and pushed back on to mouth parts by	First trunk limbs	First trunk limbs
During bottom feeding food may be scraped up by	Antenna 2	Mandibles

(After Cannon 1931.)

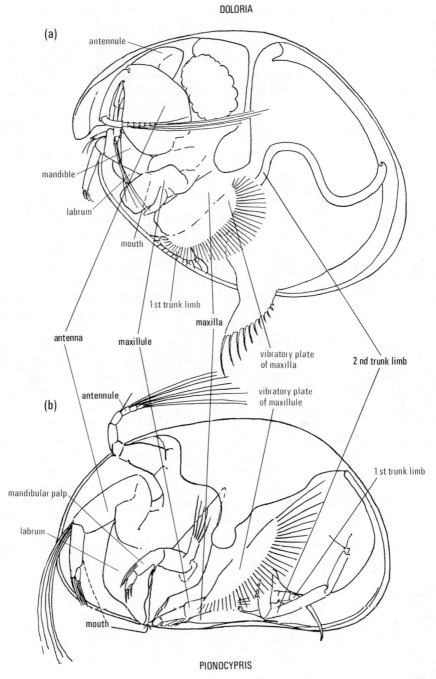

parts involved in this feeding are different in the two genera, as shown by the table given in the legend (for further information see Cannon 1931).

There are many other ingenious ways of collecting suspended food used by crustaceans. The little prawn *Atya* lives in flowing water. It holds fans or rosettes of setae on the chelipeds facing the current and then 'licks' off the collected particles. Some decapods have elaborated ways of sweeping up detritus of food value from the surrounding surfaces, as in *Galathea* (Nicol 1932).

Summary. The solutions to the food problems within the Crustacea are almost limitless. It is probable that a benthic habit, using simple limbs, in metachronal rhythm, for food collecting and for locomotion characterized ancestral Crustacea. The next step may have been feeding on larger or more sparsely distributed particles and so to raptatory habits. The end terms of such trends are the strong predators of today such as lobsters (Malacostraca), *Euchaeta* (Copepoda), and *Leptodora* and *Polyphemus* (Branchiopoda). Filter feeding has been evolved very many times, usually from simple raptatory feeding but sometimes from advanced large-food feeders such as the amphipods, where *Haustorius* possesses an entirely secondary filter-feeding device. Even *Anaspides* uses its filter mechanism in a remarkable way, to sieve off particles it scrapes from stones in flowing water, the particles being first confined so that they are not swept away (Manton 1930).

The possible mechanical principles involved in filter feeding are, however, limited and thus the same ones have been exploited independently by different groups of Crustacea during their evolution.

a) A food-laden current is necessary, either existing outside the animal, as in *Atya,* or, more usually, produced by the metachronal rhythm of the limbs. The current can result from the actions of a long series of limbs, such as in *Chirocephalus*, or from a short series of more efficient ones, such as in *Paranspides*, where the maxilla 2 filter is unaided by the thoracic limbs (cf. *Hemimysis*).

b) Filter plates are necessary for abstracting particles from a current. A long series of filter plates exists in the Branchiopoda such as *Chirocephalus* and *Sida* and in Malacostraca such as *Nebalia* and euphausids. Two filter plates are present in many Cladocera such as *Daphnia* and one pair only in

Diaptomus (Copepoda), in many mysids (Peracarida), and in *Paranaspides* and *Anaspides* (Syncarida).

c) Filter plates must be cleaned and food from them passed forwards to the mouth. This is usually done by metachronal rhythm of limbs producing either an orally directed current, or mechanical means, or both. The filtered water must also be disposed of.

d) An entirely different method is less frequently used, that of sweeping up surface particles in still water (*Galathea*), or using a casting net (barnacles).

e) Small particles in fairly concentrated suspension in the water can be abstracted without filter plates, as in *Hutchinsoniella*, and larger particles are so collected by *Triops*.

(iii) *Escape jumping through the water*. Two quite different examples are shown by copepods and prawns.

In the copepods (Pl. 3(b), (c)) four pairs of thoracic limbs lack exites and endites and the endopods and exopods together form flattened plates. Each pair of limbs is united by a complex sclerotization called a coupler (Fig. 4.18). This structure is not present outside the Copepoda and until recently its significance has been unknown (Perryman 1961). The coupler is the key to the achievement of rapid jumping.

Gentle continuous swimming is accomplished by anterior endopods on the head. Sudden rapid jumps are effected by almost simultaneous, but metachronal, movement of the four pairs of swimming thoracic limbs, moving through a large angle of swing (Fig. 4.19(c), (d)). The large angle is made possible by the complicated anatomy associated with the coupler, strength to the movement being ensured by complex skeletal arrangements.

The proximal anterior rim of the coxa extends laterally as a strongly sclerotized bar which articulates with the middle of the lateral edge of the tergite marked h in Fig. 4.19(a), (b). The pair of limbs swing forward and backward about the axis $h-h$ but only to a small extent.

The coupler consists of two half-cylinders of cuticle, one within the other; they are fused distally and open anteriorly. The outer half-cylinder is united at its base with the proximal medial margin of the coxa on either side (Fig. 4.18). The coupler also lies against the ventral body surface and its proximal end is firmly articulated with the posterior end of a median prong arising from a large intersegmental sclerite in

Evolution of arthropodan types of limbs

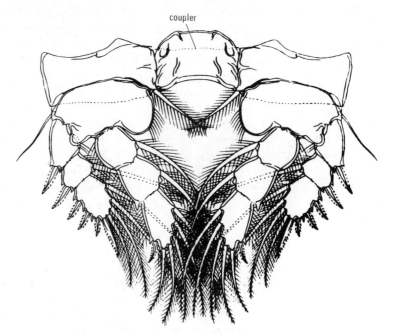

FIG. 4.18. Pair of thoracic swimming limbs of a copepod united by the coupler. The protopod consists of two segments, coxa and basis, and the exopod and endopod are each formed by three segments.

front, x in Fig. 4.19(b). This sclerite, formed from the sternal cuticle, projects anteriorly and posteriorly from a transverse strongly sclerotized intersegmental vertical flange, the upper edge of the flange forming a stout horizontal bar attached laterally to the tergite at either side at p, Fig. 4.19(a), (b). Flexible cuticle lies ventrally around the leg base. The promotor–remotor swing of the leg and coupler, taking place largely about the mid-ventral sclerotized prong at x, approaches 110° in contrast to the range of movement of 50–60° or less used by the legs of crustaceans and other arthropods with simple coxa–body junctions. Simultaneously with the remotor swing, the ventral sclerotizations swing about the transverse intersegmental bars p–p by approximately 10°, so contributing to the angle of swing of the legs. Thus the limbs swing indirectly about three axes, h–h, p–p, and the median point x.

Moreover, there is a skeletal hoop marked in Fig. 4.19(b) which supports an area of flexible cuticle behind the limb from which arise most of the remotor muscles. That the

Evolution of arthropodan types of limbs

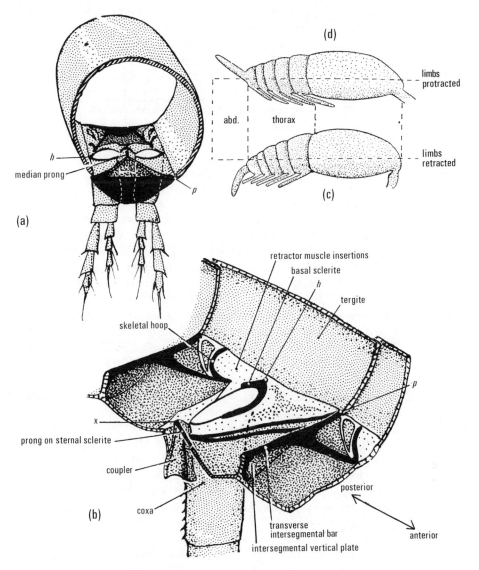

FIG. 4.19. The coupler mechanism of the swimming thoracic limbs of *Calanus finmarchicus*. (a) Anterior view of cuticle of one segment to show the coxal cavities opening into the body cavity; the highly sclerotized anterior rims of each coxa extend to and articulate with the tergo-pleural arch at h. Two intersegmental transverse bars unite and articulate with the tergo-pleural arch as shown, and they form the inner edges of the intersegmental vertical plates. (b) Internal view of the cuticle of one segment. The median prong on the posterior end of the sternal sclerotization, extending posteriorly from the intersegmental plates, firmly articulates with the coupler and forms the fulcrum of movement of the coupler and legs. (After Perryman 1961.)

173

origin of these muscles is on soft cuticle at a little distance from the coxal rim and not from the coxa itself, is a device reducing the range of swing by the muscle fibrils on their tonofibrils attached to the cuticle. Similar devices are provided by the arcuate sclerite at the spider's femur–patella joint and by the origin of the principal retractor muscle 1 of the collembolan springing organ where similar needs exist (Chapter 10, §5D(ii), Fig. 10.5, and Manton 1972). The whole skeletal system concerned with the copepod's jump is of such strength as to prevent unwanted flexures which might detract from the force of the jump. No such system of moving exoskeletal parts has been evolved by any other Crustacea (for further details see Perryman 1961).

The prawns and shrimps comprise another group of good jumpers. The abdomen from an outstretched position is strongly flapped downwards and forwards under the thorax; the tail fan (Fig. 4.10(b)), formed by the telson and large flat 6th abdominal legs, provides plenty of resistance to the water, as does the tail of a fish or a whale using quite different movements. The crustacean's balance during the tail flap is also ensured by horizontally expanded flat exopods projecting forwards from the base of antenna 2. There is usually a dorsal hump on the posterior part of the thorax, associated with the intersegmental flexures and internal musculature. The whole shrimp-like shape, or caridoid facies, as it is called, is associated with this jumping habit. Many essentially bottom-living crustaceans, such as lobsters, still retain the jumping ability and by flapping the abdomen, shoot backwards through the water at speed or into cover.

The musculature concerned with jumping by an abdominal flap is unique among Crustacea. In most arthropods muscle fibres, or sectors of muscles, extend directly from an origin to an insertion, crossing some movable joint and so causing displacement of the one part on the other. The deep abdominal and thoracic muscles of *Paranaspides* (Fig. 4.20) are typical in general form of the many crustaceans which employ rapid backward jumps through the water. The dorsal muscles are twisted like ropes. The deep ventral muscles are substantial and elaborate and, incidentally, form the most edible part of a lobster. The sectors of these muscles do not lie in one plane or in a straight line. They pass through several segments and loop about one another. The exact way in which these muscles work has not been ascertained. A parallel

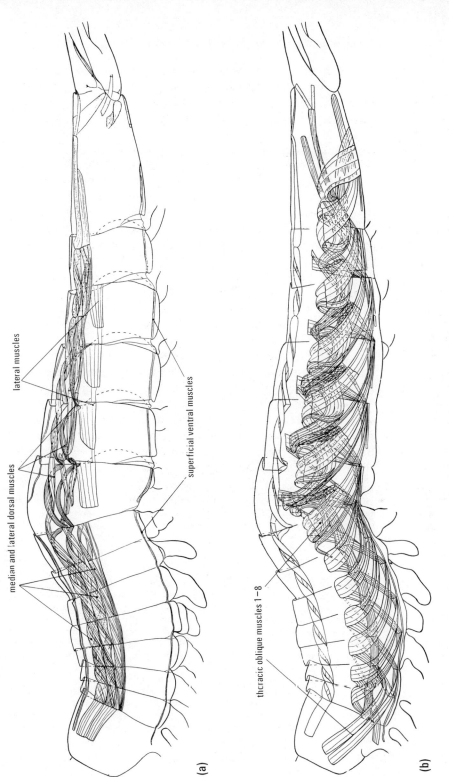

FIG. 4.20. Lateral views of the right side of *Paranaspides lacustris*. (a) Showing the dorsal and lateral muscles and the superficial ventral muscles. (b) Showing the main ventral muscular system. (After Daniel 1932.)

evolution of twisted dorsal muscles and looped complex deep oblique muscles is seen in the terrestrial thysanuran *Petrobius*, which also jumps with the abdomen, using a catapult principle (Chapter 9 and Evans 1975).

(iv) *Conclusions concerning crustacean limbs and habits.* Many other examples could be given of special habits of Crustacea and of the particular morphology associated with them, but enough has been mentioned to indicate the high levels of habit diversification and of associated structure which have made the various habits possible and led to the great success of the group. The plasticity of the biramous limbs of crustaceans is responsible for much of this habit diversification, a plasticity conspicuously absent in the early armoured arthropods. Biramous limbs of stereotyped form existed in the trilobites and many early armoured arthropods; the merostomes have given rise to one very efficient arrangement of limbs, that of *Limulus*, but apart from the latter none of these arthropods have persisted to the present day. The trilobites evidently did achieve a variety of habits but their limited ability to diversify their limbs, in contrast to the Crustacea, may have led to their extinction in competition with Crustacea of more versatile habits, more flexible structure, and speedier movement. The early chordates, the bottom living ostracoderms, may well have gobbled up the slow-moving early arthropods while the nimbler evolving

FIG. 4.21 (see opposite). To show the difference in the method of stepping by a jointed limb and by a parapodium. (a) Dorsal view of the field of leg movement, relative to the body, during the backstroke of a jointed limb. The tarsal claws do not slip on the ground and remain at the same distance from the middle line throughout the backstroke. The beginning of the backstroke is at A–B with the leg outstretched. At A–C the leg is half-way through the backstroke, effective leg length being shorter than in A–B. At A–D the leg has elongated again to the end of the backstroke. (b) The leg of a polydesmoidean diplopod in two positions with different degrees of flexure between the podomeres. Adjustment of inter-podomere flexures during stepping maintains the tarsal claws firmly on the ground without slipping. (c) A parapodium during stepping cannot alter its length by the method shown in (b). The parapodium starts the backstroke in approximately the transverse plane of the body (segemnt 1); the parapodium is then short and the aciculum partly withdrawn. During the backstroke the parapodium elongates and the aciculum is progressively protruded. Thus the parapodial chaetae and acicula do not slip on the ground (segments 1–4). The black spots indicate parapodia in contact with the ground. The diagram is based upon photographs of the polychaete *Sphaerosyllis* walking and the segments shown represent one metachronal wave. (d, e) *Nereis diversicolor,* contracting muscles shaded, as if seen by transparency. (d) Profile of a parapodium retracted during preparatory stroke in crawling (cf. segment 1 in (c)). (e) Profile of a parapodium extended during the propulsive backstroke (cf. segments 2–5 in (c)). (Figs. (d), (e) after Mettam 1967.)

Evolution of arthropodan types of limbs

crustaceans escaped, but this is just a speculation. Only *Mixopterus* (Eurypterida) possessed as striking a serial differentiation of its limbs (Fig. 1.12) as any crustacean. But *Mixopterus*, although in advance of the modern Arachnida in this respect, must have still retained primitive limb joints permitting only limited movement (see Chapter 10, §2, §5D(i), Pl. 7(a), (b)) resulting in small angles of swing of the leg, plainly indicated by the small antero-posterior distance

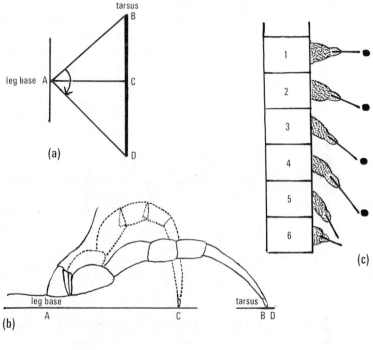

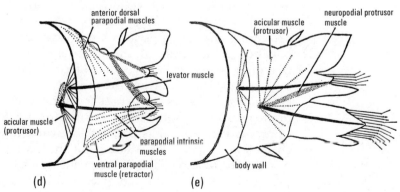

4. The limbs of annelids

(stride lengths) between the grouped footprints forming the track. These features must have resulted in slow progression.

Since the fossil record provides no evidence of the origin of the primarily marine arthropods, it becomes of interest to examine the functional morphology of modern annelid limbs and see if any of them might have had the potentiality of transformation into a jointed arthropod limb.

It has been shown that the post-antennal limbs of the primarily aquatic arthropods, whether they are flattened (phyllopodial) or rounded (stenopodial) in shape, are based upon a biramous plan, with or without proximal exites and endites. A limb secondarily used for walking only in the larger, so-called 'higher', crustaceans (Malacostraca) is formed by the endopod alone, all other rami and lobes having disappeared, often during the early life of the species (e.g. exopods in Fig. 4.11). The limbs are jointed, that is, podomeres are formed by cuticular sclerotization and are separated by more flexible cuticle called arthrodial membrane at each joint (see below). Bending of the limb is effected by flexures at each joint and the combined effect may give shortening of the distance between the limb tip and the base (Fig. 4.21(b), cf. the distances AB and AC). The joints are worked by antagonistic pairs of muscles or by flexor muscles (Fig. 2.2(d), (e) where the muscles are indicated by thick arrowed lines; joint mechanisms are considered below, §6A, B).

Annelid limbs are best developed in many polychaetes as a biramous parapodium, one pair on almost every segment. An upper branch, the notopodium, and a lower branch, the neuropodium, are each supported by a strongly sclerotized rod, the aciculum (Fig. 4.21(c), (e)) and distally many projecting chaetae of various kinds extend from the chaetal pouches at the ends of the noto- and neuropodia. Each aciculum arises in a deep ectodermal pouch. Protrusor muscles (*am*) fan out from the inner end of each pouch and antagonistic muscles (*vpm*) pass from the walls of each pouch to the parapodial walls (Figs. 4.21(d), (e), 4.23(a), (b)).

Parapodia are used as paddles for swimming; most polychaetes are capable of walking on the bottom using their neuropodia as legs (Pl. 1(c)–(e)); while the burrowing movements of others are assisted or caused by parapodial action. Some polychaetes can stand well clear of the substratum on

Evolution of arthropodan types of limbs

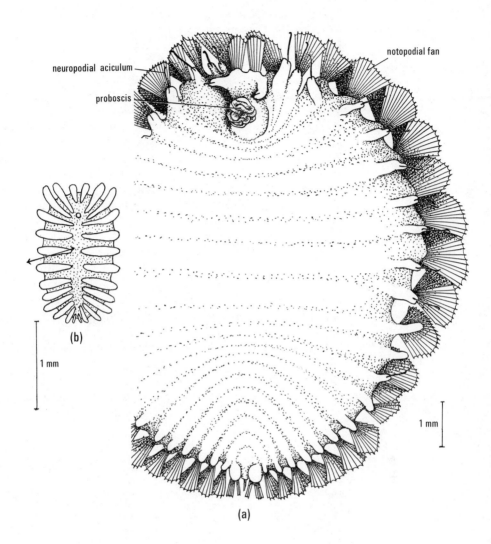

FIG. 4.22. The polychaete *Spinther arcticus* Sars. (a) Length about 9 mm, in ventral view, with the proboscis partly extruded, showing 22 segments bearing neuropodia, on some of which the hooked acicula have fallen out. The lateral end of the notopodial fans of chaetae project dorsal to the neuropodia. (b) Diagram of a young individual 1·4 mm long and 0·76 mm wide, in dorsal view showing the narrow form of the body, approaching that of normal annelids, and the elongated notopodial (elytral) ridges across the dorsal surface. Much less cephalization has taken place here than in the adult, only three pairs of notopodial–elytral ridges have swung forwards laterally to lie anterior to the eye-bearing tentacle of the prostomium. (After Manton 1967.)

their neuropodia as they walk, e.g. *Hermonia hystrix* (Pl. 1(b)), others, such as *Spinther arcticus* possess longish leg-like neuropodia with very long hooked acicula used in anchoring the animal to sponges during feeding (Fig. 4.22). The notopodium of *Spinther* is complex and spread over most of the dorsal surface of each segment of this slow-moving worm, so forming a hedgehog-like protective armour of spines. The notopodium is a wide transverse area, outlined in Fig. 4.22(b), bearing a ridge of spiky chaetae which overhang the body laterally as seen in ventral view in Fig. 4.22(a), so protecting the long neuropodium also. Similarly placed protective spines are found in the totally unrelated diplopod *Polyxenus* (Pl. 6(j), (l)), the functional significance in the two being the same.

The parapodia in most errant polychaetes are moved by their promotor–remotor swing on the body about a vertical axis formed by anterior and posterior hinge lines between each parapodium and the body wall. An almost vertical axis of swing is marked in Fig. 4.5(b) for a centipede. Parapodia move in metachronal rhythm (Fig. 2.4 and Pl. 1(c)–(e)). The phase difference between successive parapodia is small (leg $n+1$ is less than 0.5 of the duration of a pace in advance of leg n), so that the tips diverge during the propulsive phase, while those performing the forward stroke, off the ground, converge (Pl. 1(c)–(e)) and see Chapter 2, §2D and Chapter 7, §4A). Parapodia of a pair are moved in opposite phase in

---

FIG. 4.23 (see opposite). The polychaete *Spinther arcticus* Sars. Slightly oblique transverse sections through the young individual shown in Fig. 4.22(b). (a) At the level of the arrow in Fig. 4.22(b) (the section passes through the 6th neuropodium and the more median end of the left dorsal notopodial–elytral ridge). The position of the middle part of this ridge and of its deeply sunk chaetae is drawn as seen by looking into the body cavity. The bases of the dorsal chaetae are set in the septum situated just anterior to them, the septal muscles forming dorso-ventral strands; distally the chaetae are cut short. The most median of these muscle strands are continuous with the dorsal circular muscle layer. The lateral gut diverticula lie as shown in between the successive segmental septa. (b) Transverse section through the 9th neuropodium of the young individual shown in Fig. 4.22(b). The acicular sac of the neuropodium contains the long, functional, hooked aciculum, here cut short, and three young replacement acicula, the youngest of which has no demarcation between the terminal blade and the shaft. Short, simple chaetae lie round the aciculum. The transverse notopodial–elytral ridge, with its deeply sunk chaetae reaching the septum in front, is indicated beyond the plane of the section. The position of the gut is shown by interrupted lines. The dorsal chaetae are cut short externally. Note the deeply sunk acicular sac reaching to the most median part of the septum in contrast to (a); and in the adult animal in Fig. 4.22(a) the sac lies very far indeed from the middle of the body and in line with the exposed part of the aciculum. (After Manton 1967.)

Evolution of arthropodan types of limbs

contrast to the limbs shown in Fig. 2.5(a), (b). The number of parapodia forming each metachronal wave varies greatly in different polychaetes and at different times in the same animal. Shorter waves are seen when walking and longer waves during swimming.

It is the walking movements of a parapodium, when no slipping on the substratum takes place (Pl. 1(b)), which are of

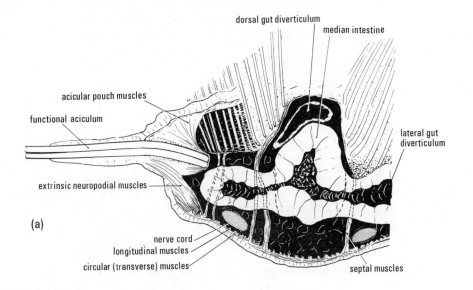

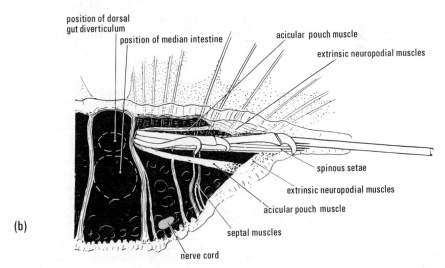

interest for comparison with the movements of a walking arthropodan leg, e.g. a whole leg of a myriapod or hexapod, or the walking endopod of a woodlouse or of a lobster. The movements differ as do the mechanisms. A typical walking leg, as viewed from above (Fig. 4.21(a)), is outstretched at the beginning of the backstroke when it is placed on the ground, line AB and position AB in Fig. 4.21(a), (b). The leg swings backwards from its base, so that the tip, relative to the base, lies along the line BD during the full backstroke. Flexures of the leg joints shorten the leg to position AC when half-way through the backstroke and elongation of the leg follows to the end of the backstroke AD when the limb-tip is raised from the ground. If the leg could not flex to its position half-way through the backstroke, slipping on the ground by the tarsus would reduce the effectiveness of stepping.

The polychaete when crawling moves the parapodium quite differently. The beginning of the backstroke starts from an almost transverse position, Fig. 4.21(c), segment 1 (cf. Fig. 4.21(a) position AB). The neuropodial aciculum is pushing on the ground opposite the black spots in Fig. 4.21(c). As the backstroke progresses (segments 2–4) the aciculum remains on the ground owing to the active elongation of the neuropodium itself together with the progressive protrusion of the aciculum, the parapodium at the same time swinging on vertical anterior and posterior hinge lines between the base of the parapodium and the body. When no further elongation of the parapodium is possible the aciculum is raised from the ground. Parapodial shortening and a pulling in of the aciculum start, but the remotor backstroke may continue for a

FIG. 4.24 (see opposite). The polychaete *Nereis diversicolor* to show the parapodium, some of the parapodial and trunk muscles, and acicula in black. (a) Anterior half of a segment from middle body region, posterior view. Acicula and muscles of the left side removed; oblique muscles numbered. (b) Segments as seen from the mid-sagittal plane looking into the parapodial coelom, anterior end to the right. (c) The adjacent segment has the dorsal longitudinal muscles removed, the oblique muscles are numbered 1–8 and commisural vessels are lettered E, G, J. (After Mettam 1967.) *am*, acicular muscle from neuropodial aciculum to tip of notopodium. *cdm*, dorsal cirrus muscle. *ch*, notopodial chaetal sac. *chs*, chaetal sac suspensor. *ci*, ciliated organ. *cn*, connective tissue cap connecting acicular heads. *dlm*, dorsal longitudinal muscle. *dml*, lateral diagonal muscle. *dmm*, median diagonal muscle. *dp* 1–5, dorsal parapodial muscles. *dt*, transseptal diagonal. *dv*, dorsal vessel. *dvm*, dorso-ventral muscle of body wall. *f*, intersegmental fibres. *in*, intestine. *nc*, nerve cord sheath and supraneural muscle. *no*, head of notopodial aciculum with radiating acicular muscles. *nr*, dorsal fibres of parapodial intrinsic (3) elaborated as neuropodial retractor muscle. *p* 1–3. parapodial intrinsic muscles. *nu*, head of neuropodial aciculum with radiating acicular muscles. *pc*, parapodial coelom. *sp*, septum. *vlm*, ventral longitudinal muscle. *vpm*, ventral parapodial muscles. *vv*, ventral vessel.

Evolution of arthropodan types of limbs

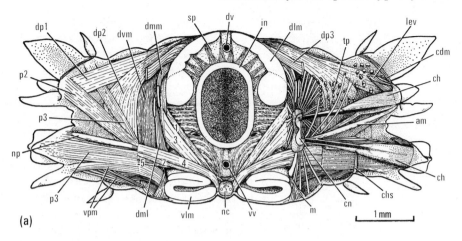
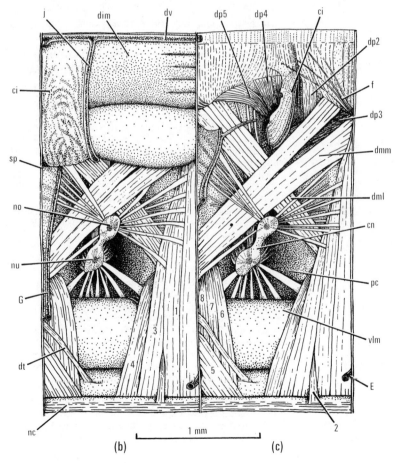

183

Evolution of arthropodan types of limbs

moment, see Fig. 4.21(c), segment 5. Segment 6 shows further shortening of the parapodium and the beginning of the forward swing which will complete the cycle and repeat the position seen on segment 1. Figure 4.21(c) is taken from the minute polychaete *Sphaerosyllis*, while Fig. 4.21(d), (e) show the much larger *Nereis diversicolor*, yet the essential movements are the same (Mettam 1967, Manton 1974). Swimming movements differ in that the range of parapodial action against the water is increased, largely by the occurrence of body undulations, trunk muscles providing extra force for the parapodial movements (Gray 1939), but the longitudinal trunk musculature is not used in this way during crawling.

The mechanisms of movement of a jointed leg and of a parapodium are entirely different. A parapodium cannot exert a propulsive thrust on the ground during shortening as can a jointed leg between positions AB and AC in Fig. 4.21(a), and thus the polychaete when crawling starts its backstroke in a position roughly corresponding to AC. The musculature of *Nereis* and of *Nephthys*, worked out so admirably by Mettam (1967), accounts for all parapodial movements and all the complex musculature shown in Fig.

FIG. 4.25 (see opposite). Diagrams, based on *Peripatopsis* spp., to show features associated with the mode of action of the lobopodial limb. Cut longitudinal muscles are cross-hatched, the nerve cord is stippled, and haemocoelic spaces are black. The superficial ectoderm, subcutaneous connective tissue, and circular muscles are white and the superficial double layer of oblique muscles is marked by a similar convention on body and legs. Muscle numbers 3, 8, 12, 14, 15, and 16 are those of Snodgrass (1938). Muscles 14a and 14b are additional. The segmental organs and leg glands are omitted. They lie in the haemocoelic space dorso-lateral to muscle 16, except for the exit ducts. The pericardial nephrocytic tissue is unlabelled. (a) Transverse section passing through a leg and viewed from in front, showing the posterior inner face of the leg with sectors of the dorsal remotor muscle 8 fanning over it; the antero-posterior fibres of muscle 16 are cut. One of the two pedal nerves passes into the leg. The proximal posterior rim of the leg, where the layer of oblique muscles from the ventral and lateral body wall unite, is indicated. Single-headed arrows show the direction of flow of blood up the subcutaneous lateral channel, into the pericardium dorsally, into the heart through ostia, and from the perivisceral haemocoel through the ostia in the pericardial floor, so into the pericardium. The double-headed arrows indicate a reversible flow through the ostia in deep oblique muscle sheet 3 between the perivisceral and the lateral haemocoels (see Fig. 4.26(a), (b)) and a flow between the haemocoelic cavity of the leg and the lateral haemocoel. (b) Similar section, viewed from in front, but not cutting through a subcutaneous haemocoelic space on the trunk, and showing the muscles which traverse the median haemocoelic space of the leg. The dorsal end of remotor limb muscle 8 fans far behind the leg and attaches widely to the body wall. The long levator from the foot passes internal to the bladder of the segmental organ. The anterior internal face of the leg (not shown) is much as seen in (a) but the promotor muscles 7 and 11 are less bulky than are the remotor muscles 8 and 12 drawn in (a). From Manton 1967.)

Evolution of arthropodan types of limbs

4.24 is employed in maintaining body stability and in causing parapodial movements. The relatively larger neuropodium of *Spinther* has a basically similar musculature (Fig. 4.23).

By no stretch of the imagination could the acicular method of stepping be converted into that of a jointed leg. The far-fetched notion of Sharov (1966) concerning *Spinther* being a radially symmetrical descendant of the coelenterates is belied by the normal polychaete shape of the early stages of *Spinther*, not seen by Sharov (1966) (Fig. 4.22(b)). He did not recognize the enormous notopodium of *Spinther* or the absence of true radial symmetry. The neuropodium is no lobopodial limb basic to that of all Arthropoda as he supposed (see further Anderson 1966, Manton 1967).

5. The uniramous lobopodial limbs of the Onychophora and associated respiratory system

It is important to study the movements and functional morphology of the only lobopodial limb extant today and found throughout the Onychophora. Here we do not draw a

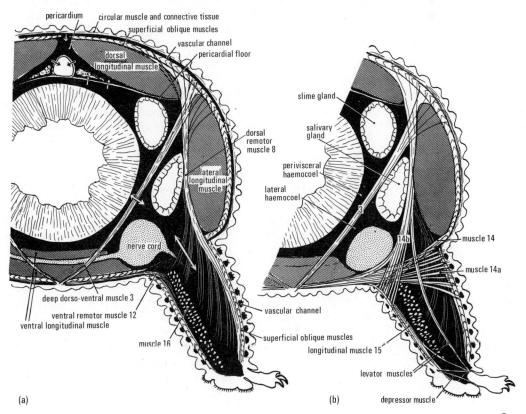

blank, as with the parapodium, but find a type of limb which could never have arisen from a parapodium but which could have given rise to the jointed legs of the myriapods and hexapods by easy transitional stages, all of which would be workable, with no postulation of functionally impossible intermediate conditions.

The Onychophora stand clear of the ground on their short ventro-laterally directed limbs, one pair to each trunk segment (Pl. 4(a)–(c)). The lobopodia move in slightly untidy metachronal rhythm (Fig. 2.4), each pair of legs being either in similar or in opposite or in any other phase relationship. The length of each wave is also variable. The movements of the right legs only in Fig. 2.4 are indicated by the wavy lines, the thicker parts representing the backstroke of the legs relative to the head end. Each leg is put on the ground well after the leg in front is raised because the blunt, wide limb-tip allows no other method of stepping. During brisk walking the forward and backward strokes are of almost equal duration but this is not always so (see Chapter 7).

The body wall of the onychophoran trunk and limbs is similarly constructed. Within the cuticle and ectoderm superficial layers of oblique muscles, set roughly at right angles to one another, and circular muscles lie within a substantial layer of subcutaneous connective tissue fibres. There is no other skeleton and this connective tissue carries the insertions of all the muscles and is thickest on the trunk (Fig. 4.25). This body-wall musculature passes insensibly into that of the limb wall. A thick cylinder of longitudinally arranged muscle fibres lies within the superficial muscles along the trunk, divisible into dorsal, lateral, and ventral longitudinal sectors. The pericardial floor spans the haemocoel above the gut and carries the heart mid-dorsally. An oblique muscle sheet 3 lies continuously along the body on either side and separates a perivisceral haemocoel from lateral haemocoelic spaces opposite the entrances to the leg cavities (Fig. 4.25). Promotor and remotor muscles leave the anterior and posterior proximal margins of the lobopodium and insert on the body wall ventrally and dorso-laterally (see dorsal and ventral remotor muscles 8 and 11 marked in Fig. 4.25(a)). Muscles 14, 14a, and 14b control the size of the leg base and a long levator muscle from the foot passes dorsalwards as shown in Fig. 4.25(b). A stout transverse muscle 16 passes antero-posteriorly across the lumen of the leg; its contraction contributes to limb elongation (Fig. 4.25(a), (b)).

Muscle sheet 3 is perforated by several ostia per segment which can presumably be opened or closed according to the tension on the surrounding muscles (Figs. 4.25(a), 4.26(a), (b)). Blood circulation passes from the pericardium into the heart via ostia, permitting flow into the heart but preventing regurgitation back into the pericardium, and then to the various organs of the body. The flow reaches the perivisceral haemocoel and can either pass into the limb on its mesial side and out on the lateral side, or pass directly to the lateral circular channels which lie internal to each transverse ridge of the body wall, shown in black in Fig. 4.25(a); the section in 25(b) shows no vascular channel because its plane lies between the furrows. In the leg the vascular channels lie roughly in the circular position and so are cut transversely in Fig. 4.25, cf. the longitudinal section of the body wall in Fig. 4.26(c). These lateral channels of the trunk empty into the pericardium dorsally.

Two types of ostia, besides those in the heart wall, influence blood circulation. Some pass through the pericardial floor and appear to transmit blood from below upwards to the filtering nephrocytic tissue on the dorsal face of the pericardial floor (Manton 1937). The ostia in muscle sheet 3 could transmit blood in either direction between the perivisceral and the lateral haemocoelic spaces.

Limb movements are very various. During the promotor–remotor swing, used in walking, a shortening takes place when halfway through the backstroke, accomplished by the limb muscles but with or without a change in limb diameter or in internal limb volume. At times the diameter of the leg increases and so internal volume on shortening remains the same. But when the animal squeezes through narrow places, each limb can completely shrink against the body, appearing like a wart. In this case the haemolymph must pass from the limb into the trunk when all the limb muscles contract at once. Moreover, one limb at a time can be flattened out in this manner (Pl. 4(c)).

It seems probable that the lateral haemocoelic space above the entrance to each leg forms an isolating mechanism, controlled by the ostia in muscle sheet 3. The flow from the leg and through the ostia (see arrow in Fig. 4.25(a)) to the perivisceral haemocoel would permit speedy contraction of one leg only by its intrinsic muscles, and a flow in the opposite

Evolution of arthropodan types of limbs

direction could inflate a limb and assist its elongation caused by appropriate muscular contraction. It appears that leg movements in Onychophora are partly controlled by direct action of intrinsic and extrinsic limb muscles and partly by haemocoelic pressure, in turn controlled by tension of many trunk muscles. Muscles 14, 14a, and 14b could stabilize the leg and prevent overinflation by haemocoelic pressure generated in the lateral haemocoel of the trunk. Pressure exerted by the heart musculature maintains the basic blood circulation, but it is the trunk muscles which control flow of haemolymph which at times is concerned in leg movement.

This isolating mechanism of each lobopodium is unique and accounts for the great flexibility in the uses to which a series of legs can be put by the Onychophora (see Chapters 6 and 7). It is quite wrong to claim that extension of a limb is done by increase of fluid pressure from the trunk, because such extension can be effected entirely by intrinsic limb muscles with no change in limb volume.

The contrasts between the mechanisms of stepping by a polychaete parapodium and an onychophoran lobopodium are extreme. These mechanisms bear not the slightest resemblance to one another. The parapodium of *Nereis* and of *Nephthys* possesses extrinsic muscles causing the promotor–remotor swing, superficially suggestive of muscles of similar function in Onychophora; but Mettam (1967) suggests that primitively the parapodium was worked by intrinsic muscles alone and that the extrinsic muscles are

FIG. 4.26 (see opposite). Onychophoran vascular system. Haemocoelic spaces are black and cut muscle fibres are cross-hatched. (a) *Macroperipatus* sp. Whole mount of part of piece of the deep oblique muscle sheet 3, seen in Fig. 4.25, to show the component series of very large fibres united by connective tissue (white) and one ostium, lined by peritoneum. (b) Horizontal longitudinal section of the same, passing perpendicular to the surfaces of muscle sheet 3 at the level shown by the arrow in (a) and cutting through the middle of one ostium. No distinction is made in the drawing between the connective tissue fibres which invest each muscle fibre and unite them, staining bright blue with Mallory's triple stain, and the very thin surface peritoneum which does not stain blue. The white double-headed arrow indicates the probable flow of blood through the ostium in either direction, according to the momentary requirements of the limb mechanism which determines the amount of fluid needed in the lateral haemocoel. (c) Horizontal section through the mid-lateral body wall of *Peripatopsis moseleyi* to show the vascular channels which lie internal to each annular ridge on the body wall. The ridges bear papillae surmounted by sensory spines. A shortening of the body approximates the ridges by a deepening of the gully between them, and an extension of the body causes the reverse, but neither movement influences the integrity of the lateral vascular channels which bring blood from the limb and lateral haemocoelic spaces to the pericardium. (After Manton 1967.)

Evolution of arthropodan types of limbs

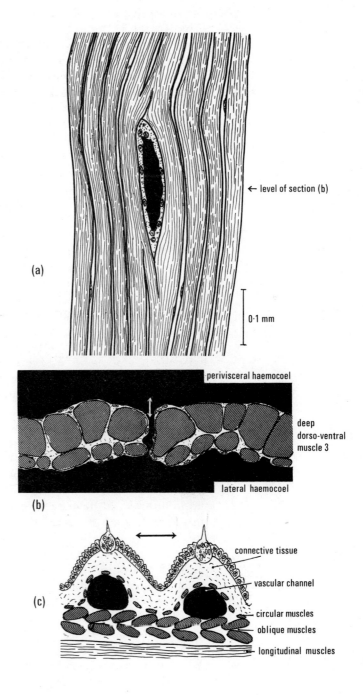

derivatives of part of the oblique musculature which braces the body wall. If this is so, there is no homology between the extrinsic muscles of these polychaetes and of the limbs of the Uniramia. The Onychophora possess the deep oblique muscle sheet 3, which braces the body, as well as possessing extrinsic lobopodial muscles; and the jointed legs of myriapods and hexapods show a subdivision of muscle sheet 3 into the various deep dorso-ventral and deep oblique muscles (see Chapter 8, §2C(iii), (iv)), all additional to the extrinsic limb muscles.

The parapodium depends for support on coelomic pressure which provides, with the trunk musculature, rigidity against which the parapodial muscles can work; but the mechanism of a lobopodium has nothing to do with the coelom, reduced in the adult onychophoran to the end sacs of segmental organs and gonoducts. The lobopodial mechanism is bound up with a haemocoelic trunk. No such haemocoelic annelids exist today, but one suspects that the lobopodium and the trunk haemocoel evolved together in some line of metamerically segmented marine worms.

An analogy to an incipient lobopodium can be seen in the flatworms, the fluid content of their parenchyma being essentially haemocoelic. The temporary limb-like extensions of the body of *Rhynchodemus* (Fig. 4.1(a)) can be considered as analogous to incipient lobopodia. Such temporary limbs could have become more permanent. But this is a purely speculative analogy because there is no evidence of relationship between these flatworms and annelids, but it suggests how a haemocoelic, lobopodial segmented worm might have existed at the same time as the ancestors of parapodial coelomate worms. It will be shown below how a lobopodium could have been transformed into a uniramous jointed limb such as present in myriapods and hexapods (§6E).

The tracheal system of the Onychophora has hitherto appeared to be unique in that the minute tracheae do not branch and arise from numerous ectodermal depressions. It is usual in myriapods and hexapods for fewer, larger tracheae to arise from a limited number of spiracles, the tracheae dividing internally to supply the organs. But it is probable that the tracheae of Onychophora, 2 µm in diameter, are related to hydrostatic pressure changes which occur locally within the body and accompany shape changes and limb

Evolution of arthropodan types of limbs

movements. Such small tubes are not easily deformable by outside pressures. Exactly the same type of tracheae exist in the centipede *Craterostigmus* (Chapter 8, §5D and Manton 1965, Pl. 6), where hydrostatic changes must be considerable during the head movements. Thus the peculiar tracheal system of the Onychophora cannot be considered to be a primitive feature, but on the contrary is associated with habits and locomotory mechanisms. Tracheal peculiarities of other centipedes are also associated with particular habits (Manton 1965, §8(iii)).

6. The basic construction of the uniramous trunk limbs of myriapods and hexapods

There is reason to suppose that the Uniramia are a homogeneous group (Chapter 6; Manton 1972) and that the jointed uniramous legs of myriapods and hexapods evolved from a lobopodium, not exactly resembling the onychophoran lobopodium, but similar in principle. With the advent of cuticular rigidity caused by sclerotization and/or calcification, annular zones of stiff cuticle must have given rise to podomeres separated by zones of lesser sclerotization which are more flexible and formed the arthrodial membrane at each joint. These changes can be seen during ontogeny.

The walking legs of trilobites, merostomes, and crustaceans, on the contrary, represent the endopod of a biramous limb. In many Malacostraca the exopod, endites, and exites have disappeared, leaving a simple walking leg (see above §3A(iv)). This endopod is an independent and parallel evolution to the basically uniramous limbs. Simple similarities between jointed limbs of two independent origins may be expected to be convergent, but the many dissimilarities can be shown to be related to the independent evolution of the several taxa and to the differentiation of their habits and correlated morphology.

The coxa–body joint within the Myriapoda and Hexapoda, although providing the promotor–remotor swing, is so fundamentally diverse in different taxa as to supply many phylogenetic implications (see Chapters 5, 6 and 10). Between the coxa and the more distal part of the leg (called the telopod) lies a pivot joint which provides the principal levator movement of the telopod and gives also the antagonistic depressor movement (Figs. 4.28(a), 5.3, 5.4, 5.6, 5.8, 5.9, 5.12, 5.14, 5.15). There may be one more distal pivot joint, but sometimes there are several. Usually a series of hinge

Evolution of arthropodan types of limbs

joints, whose muscles cause flexures only, lies along the rest of the telepod which ends in a tarsal claw or pair of claws. There are examples of joints in process of disappearance and others in partial formation in myriapods in particular. The essential features of the hinge and pivot joints of the telepod, together with their musculature, will now be considered, leaving the specializations and the coxa–body and other joints until Chapter 5.

A. *Hinge joints* The simplest hinge joint shows a short arthrodial membrane dorsally, where it unites strongly sclerotized parts of adjacent podomeres which butt up (Figs. 2.2(e), 4.27(a)). Thereby an articulation, or hinge, is formed at one point between the two successive podomeres, but laterally and ventrally the arthrodial membrane progressively increases in expanse, thus permitting the distal podomere to be pulled a little way into the cavity of the proximal podomere by the ventrally arising flexor muscle shown, the proximal margin of the distal podomere moving as indicated by the arrows. Little flexure at the joint is possible unless there is marked difference in diameter of the margins of the podomeres. An incipient hinge joint (Fig. 4.27(b)) shows the arthrodial membrane to be more limited in extent, the podomere diameters more nearly alike and the dorsal sclerotization is undivided. Such an arrangement allows some flexibility but much less than in Fig. 4.27(a).

B. *Pivot joints* The simplest pivot joint possesses two articulations between adjacent podomeres, usually situated in the middle of the anterior and posterior faces of the leg. Movement of the distal podomere at the joint is up and down about an axis passing between the anterior and posterior pivot articulations (Fig. 2.2(d)). A pivot articulation may be a simple cuticular projection, large or small, from the proximal podomere which

---

FIG . 4.27 (see opposite). To show the construction of simple hinge and pivot joints between limb podomeres. (a) Hinge joints between femur and postfemur of the diplopod *Schizophyllum sabulosum*. (b) Imperfectly formed hinge joint between tarsus 1 and tarsus 2 of the diplopod *Callipus longobardius*. (c) Oblique view of the diplopod *Blaniulus guttulatus* showing pivot joint from the trochanter fitting into the concavity in the margin of the prefemur into which the pivot articulates; ample arthrodial membrane lies between the podomeres dorsally (coxa and trochanter are fused). (d) Transverse section through the trochanter–prefemur pivot joint of the diplopod *Tachypodoiulus niger*. (e) Anterior view of the pivot joint between prefemur and femur of a large spirobolid diplopod drawn as if transparent. (After Manton 1958a.)

Evolution of arthropodan types of limbs

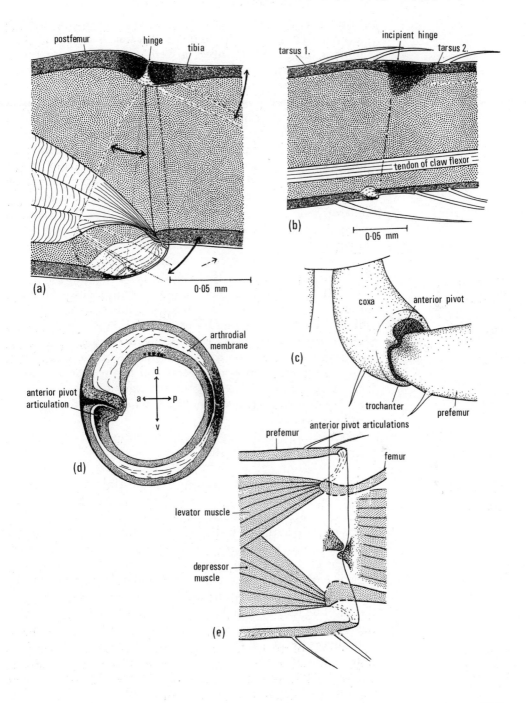

Evolution of arthropodan types of limbs

fits into a depression on the anterior margin of the more distal podomere (Fig. 4.27(c)). The transverse section through a joint in Fig. 4.27(d) shows a well-formed anterior pivot, but posteriorly there are only heavy local sclerotizations at the overlapping podomere margins, but with very short arthrodial membrane between them. Dorsally and ventrally at the joint there is considerable difference in diameter of the overlapping edges of the podomeres and extensive arthrodial membrane is here present, so that antagonistic muscles, pulling on the dorsal and ventral proximal margin of the distal podomere, can flex it up and down about an equatorial axis passing through the articulating pivots. One of these well-formed pivots is shown in Fig. 4.27(e), drawn as if transparent; the pivot on the posterior face of the leg is not shown.

C. *Extrinsic leg muscles*

These muscles usually arise from the proximal margin of the coxa and cause the promotor–remotor stepping movement of the legs. They insert within the trunk on any convenient sites, which may be apodemes (Fig. 3.11(b), (c)); endoskeletal plates formed of connective tissue, as in *Limulus* (Fig. 3.13) and many hexapods or the thick subcutaneous connective tissue of Onychophora; or directly on to the sternal and/or tergal faces of the cuticle (Figs. 3.12, 4.29, 5.13, 9.4(c)). The muscle fibres pull against tonofibrils which pass through the ectodermal cells.

Extrinsic muscles arising deeper in the leg serve special purposes in particular taxa.

D. *Intrinsic leg muscles*

These confer rigidity and mobility at the joints, as is required. Where there are pivot joints all along the leg, or almost so, then there are antagonistic pairs of short muscles working every joint (usually so in Pterygota). Those muscles causing the levator movement of the distal podomere, raising the leg from the ground, are more slender than the massive depressor muscles which depress the distal podomere firmly in the direction of the substratum and contribute to the locomotory thrust. The levator and depressor muscles are short, traversing only one joint; see the muscles in the two left-hand podomeres of the leg in Fig. 4.28(a). Since there is no animal today without its own particular accomplishments, an ideally simple limb musculature cannot be shown. But the intrinsic muscles of the leg of a iuliform diplopod in Fig. 4.28 can be

Evolution of arthropodan types of limbs

taken as fairly typical, even if very substantial, in correlation with burrowing by the motive force of its legs. The dorsal view in (b) shows how the flexor muscles from hinge joints and depressor muscles from pivot joints use both faces of the podomere for insertion of their fibres. Some sectors of the flexor muscles cross more than one joint on the posterior face of the leg.

Where there are two successive pivot joints beyond the coxa–body junction and the following joints are hinge joints,

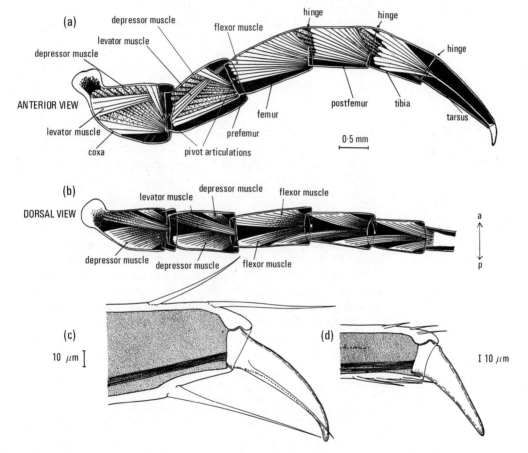

FIG. 4.28. (a) An anterior view of the leg of a spirobolid diplopod to show the positions of the hinge and pivot joints and the musculature. The pivot joint between coxa and trochanter is vanishing while that between trochanter and prefemur is functional. (b) Dorsal view of the same. (c, d) Tarsal claws of the diplopods *Tachypodoiulus niger* and *Lophoproctus lucidus* respectively showing dorsal articulation with the tarsus and the ventral tendon of the flexor muscles which traverse the tarsus as shown in (a). (After Manton 1958a.)

Evolution of arthropodan types of limbs

the flexor muscles of the hinge joints depress the distal podomeres and hold the leg tip firmly on the ground, but with the usual absence of extensor muscles at hinge joints (Fig. 4.28), propulsive leg extension is effected by the more proximal depressor muscles, combined with the remotor muscles working the whole leg, which together press the limb-tip against the ground throughout the latter part of the backstroke (Fig. 4.21(a), position AC to AD). Haemocoelic pressure can extend hinge joints when off the ground, but provides no locomotory thrust during walking.

The only long muscle within a simple leg, traversing more than one joint, is the depressor muscle from the tarsal claws which pulls, usually on a long tendon, and inserts on the penultimate, or on several more proximal podomeres also. Long intrinsic muscles occur in legs adapted for speedy movements (see Chapter 5, §4B).

E. *Derivation of a simple walking leg*

The limb features just described could readily have been derived from a lobopodium with the advent of increase in surface sclerotization. The Onychophora possess a thin sclerotized surface layer of cuticle but they do not lack the ability to form heavy sclerotization when required, as seen in the paired claws on their legs, in the heavy paired jaw blades, which correspond with claws, and in the long jaw apodemes (Fig. 3.15(d)). Joints could have been formed in the manner described above, becoming more elaborate in larger species and in taxa with more specialized habits (see Chapters 5 and 10). The superficial oblique musculature of the onychophoran leg wall might readily have divided up, with re-orientation of the fibres, to form the short intrinsic muscles crossing one leg joint only (Fig. 4.28). Muscle 16 would become redundant and disappear, and the long claw levator

FIG. 4.29 (see opposite). The scolopendromorph centipede *Cormocephalus calcaratus*. (a) Ventral view of the cuticle of the leg base, the muscles removed, and drawn as a transparent object to show the sclerites, costa coxalis (black), coxal apodemes, etc. and the anterior coxa–trochanter articulation between the end of the costa coxalis and the proximal rim of the trochanter. The aspect is directly ventral to the sternite and therefore shows an oblique anterior view of the coxa, trochanter, and prefemur. Arthrodial membrane is stippled. (b) Ventral view of one segment to show the principal extrinsic muscles inserting on the sternite; dorsally inserting extrinsic limb muscles cannot be seen in this aspect. The costa coxalis is shown in white. The extrinsic limb muscles are: *pr.co.t.*, *pr.co.a.*, *pr.co.b.*, *lev.tr.*, *ret.tr.s.*, *ret.co.s.*, *ret.tr.t.*, *dvc.v*. The meanings of the abbreviations are: *pr.* promotor, *lev.* levator, *ret.* retractor, *co.* coxa, *tr.* trochanter, *s.* sternite, *t.* endoskeletal tendon of sternal longitudinal muscle, *dvc.v*. ventral sectors of dorso-ventral muscle complex (see Manton 1965).

Evolution of arthropodan types of limbs

from the foot might have been retained in part as the ubiquitous tarsal claw flexor of jointed limbs. Some examples will be given below showing how plastic muscles are and how they can alter their origin or insertion or both according to functional needs (e.g. Fig. 8.14). The extrinsic limb muscles of Onychophora, Myriapoda, and Hexapoda correspond in

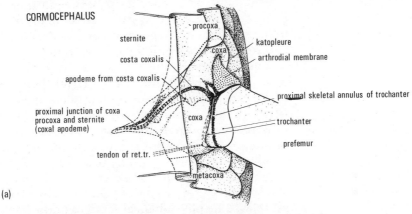

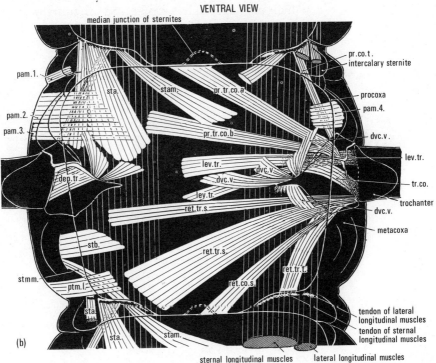

essentials. Some more detailed similarities between the muscles of the Onychophora and various myriapod and hexapod groups will be given in Chapter 8, §2.

With increase in surface sclerotization, which led to the formation of segmental sclerites on body and legs, the musculature must have become increasingly dependent upon attachments to the hard parts, although the sub-ectodermal connective tissue layer is still used for many muscle insertions in the trunks of myriapods and hexapods (see Chapter 8, §2A) and some muscles pull on arthrodial membranes rather than on sclerites (see the site or origin of the remotor muscles of the copepod swimming limb, Fig. 4.19(b)). But with the progressive dependence upon joint articulations and on antagonistic muscles supported by them, there followed a reduction in dependence upon the haemocoel for causing trunk and limb movements, compared with the onychophoran lobopodium. Primitive mechanical uses of the haemocoel became superseded, but they persist in part, for example in causing the extension of hinge joints lacking extensor muscles. The physiological services of the haemocoel remain in all myriapods, hexapods, and other arthropods.

5 Special requirements of land arthropods: stepping; leg rocking; and joints in the Uniramia

1. **Introduction** In the previous chapter the many shapes exhibited by biramous legs have been dealt with, together with the uses to which they are put among modern arthropods. In the early fossil species there is limited variety in limb morphology in the best known trilobites and in the early Merostomata, including the Eurypterida and Xiphosura.

The Crustacea present an entirely different picture with their vast variety in limb morphology and in the functions fulfilled by these limbs. There could hardly be a greater contrast than this.

In all these aquatic groups there are many species possessing a walking endopod and these endopods appear to be homologous from group to group. When the walking habit predominates then the other parts of the biramous limb, the exopod together with exites and endites, disappear or become reduced, leaving often the endopod alone. But it must be remembered that this endopod is simply part of the biramous limb. It does not represent the whole uniramous limb of the Uniramia (Onychophora, Myriapoda, Hexapoda). All walking limbs, whether they be whole limbs of the Uniramia or the endopods of the primarily aquatic arthropods, are dependent for movement upon sclerotized cuticle forming podomeres alternating with arthrodial membranes at the joints. Such jointing is essential to permit a shortening of the leg half-way through the backstroke, as shown in Fig. 4.21(a), or at some other part of the backstroke according to the field of movement of that particular limb (Fig. 5.2).

No such jointed leg could have evolved from any limb resembling that of modern polychaetes where the mechanism of movement is entirely different (Chapter 4, §4). We have no

Special requirements of land arthropods

information whatever as to the origin of the biramous limb of the primarily aquatic arthropods but, as shown in Chapter 4, §4–6, a lobopodium essentially as in the Onychophora has the potentiality of transformation into the jointed whole-limb walking leg of the Myriapoda and Hexapoda.

2. Hanging stance and leg length

Many arthropods possess very short legs when they live in confined spaces such as within decaying logs and similar environments. Here no particular speed of locomotion is needed and none can be provided by animals with short legs such as the crevice-living Collembola and similar animals. Longer legs enable the execution of longer strides and greater speeds of movement when the pace duration and angle of swing of the legs remains the same. This may lead to more open living and a greater range of food material. Arthropods differ from larger animals, such as ungulates, in presenting a relatively larger surface area to air currents, although their actual weight is very small, and if they stood up on their legs as do the ungulates, they would blow over with the greatest ease. Short-legged species such as *Peripatus* and *Pauropus* (Fig. 5.1) do stand up on their legs but most arthropods hang down from them.* It is desirable for most arthropods, terrestrial and aquatic, to keep the ventral surface of the body as close to the ground as possible and a hanging stance has therefore been developed as shown for Crustacea, Arachnida, Myriapoda, and insects in Fig. 5.1 and Pls. 4(k), 6(m), 8(h). This means that the leg from its origin on the body rises laterally to form a 'knee' and then descends again to the ground. The crustacean *Ligia* has an exceptional method of folding its legs so that quite long legs obtain maximum cover from the flanks. Unless the legs are short, or not very flexible, there is a knee podomere which is marked by interrupted lines in Fig. 5.3. Particular mobility may exist at either end of this podomere or at the distal end only.

There is no reason to suppose that the leg is composed of a similar number of podomeres in all arthropods. There are probably three comparable points on the leg; one is its origin; the second is the tarsal end impinging on the ground; and the third is the knee, a podomere marked by interrupted lines in

---

*It is the lack of appreciation that arthropods cannot stand up on long legs, and indeed the non-recognition of what are the more primitive types of arthropod at the present day, which has led to the caricatures of *Scutigera* and others in some modern text-books.

Special requirements of land arthropods

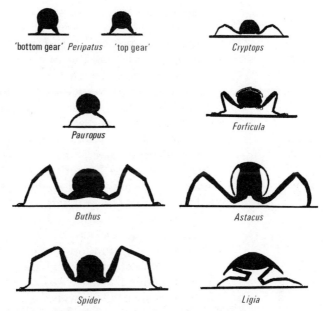

FIG. 5.1. Single segments of a series of arthropods in transverse section, with their limbs drawn relative to the ground as seen in living animals. The two figures of *Peripatus* represent the same segment employing different gaits. The scorpion and spider show walking legs 3, *Astacus* walking leg 2, and *Ligia* walking leg 6. (After Manton 1952a.)

Fig. 5.3. These three points were probably the first to be differentiated on an evolving jointed leg; but there is every reason to suppose that different numbers of joints and podomeres were formed proximal and distal to the knee in different taxa. The opposite view, expressed *ad nauseam* in the literature, claiming that all arthropods have a basically similar number of podomeres is here rejected (Manton 1966), and a monophylectic derivation of the Arthropoda is also rejected. The actual number of podomeres is simply a correlate of the functions the limb is called upon to perform (Manton 1966, 1972).

3. Fields of movement, leg numbers, and the evolution of hexapody

Where legs are short, such as those of *Peripatus* shown in Fig. 5.1, Pl. 4(a)–(c) the fields of movement of one leg and the next do not overlap one another; see the heavy vertical lines depicting the position of the limb-tip relative to the head end throughout the backstroke in Fig. 5.2. By contrast, the centipede *Scutigera*, with very long legs, shows an enormous amount of overlap of the fields of movement, the heavy vertical lines again depicting the backstroke. Not only do the fields of movement overlap greatly, but the fanning out of the fields of movement, although reducing the overlap considerably, results in the anterior legs shortening throughout the backstroke while the posterior legs extend throughout the backstroke, two very different movements, and both are propulsive. The effect of the great overlap in fields of movement is a necessity for very precise stepping or the animal would 'tread on its own toes' and in any case only a small range in gait is possible with such restrictions. The crustacean *Ligia* has only seven pairs of walking legs but again their fields of movement overlap very considerably and because of this the animal is restricted to one gait and changes speed mainly by changing the pace duration. The diagrams of a scorpion, of *Galeodes*, and of an earwig in Fig. 5.2 show the advantage of reduction in leg number in spreading out the fields of movement so that they scarcely overlap. And this spreading out also allows a longer leg. The scorpion has only four leg pairs used for walking and the stepping of the 1st and 4th legs are extremely different just as is the stepping of the 1st and 14th legs of the centipede *Scutigera*. *Galeodes* is one of the fastest running arachnids and it runs on only three and not four prosomal legs, the first pair of prosomal legs being sometimes used as antennae. In the pterygote insect *Forficula* (earwig) there is no overlap in the fields of movement and consequently great freedom in choice of gaits by this animal.

Two examples of hexapody, the use of only three pairs of legs, have been shown in Fig. 5.2, one an unusual arachnid and the other a winged insect. The mechanical advantages of using only three pairs of legs are so great that it is not surprising to find that hexapody has been evolved very many times. Even the prawns frequently walk on the bottom of the sea on only three pairs of legs. A spider crab walks forwards on three pairs of legs whereas most crabs walk sideways on four pairs. The terrestrial hexapods comprise five groups, the Pterygota, with wings, and the wingless Diplura, Protura, Collembola, and Thysanura (Pl. 6). There is every reason to suppose, as will be shown later, that hexapody has been

FIG. 5.2. (see opposite). The fields of movement of the legs of certain Arthropoda. The heavy vertical lines represent the movements of the tip of the leg relative to the body during the propulsive backstroke. The positions of the legs at the beginning and end of the backstroke are shown by thin lines. For further description see text. (After Manton 1952*a*.)

Special requirements of land arthropods

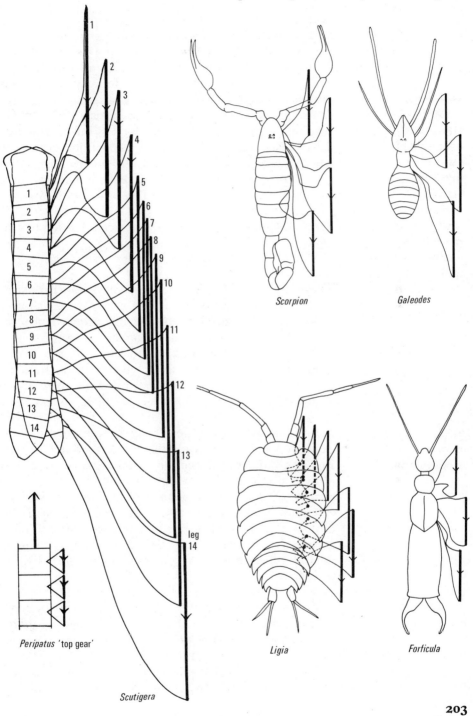

Peripatus 'top gear'

Scorpion

Galeodes

Scutigera

Ligia

Forficula

Special requirements of land arthropods

evolved in parallel by these five groups and that there has never been any such animal as an ancestral six-legged insect. As far as terminology goes it is better to restrict the word insect to the Pterygota because it is unlikely that they are closely related to any one of the four apterygote hexapod groups.

4. Leg structure and stepping suited to produce (i) strong movements and (ii) the mutually exclusive fast movements

A strongly constructed leg suitable for a iuliform diplopod employing the motive force of the legs for burrowing, or for the anterior third of the body of a geophilomorph centipede, where very strong holding onto the substratum is desirable, needs a strongly constructed leg skeleton with strong extrinsic and intrinsic leg muscles. The power of a muscle can be expressed as the force it puts out multiplied by the distance through which hard parts are displaced divided by time, i.e.

$$power = \frac{force \times distance}{time}$$

Therefore if the force is to be great, the distance must be small and the time must be long, in other words slow, strong movements are needed. If, on the other hand, speed is desirable then the force must be small, the time must be small and the distance as long as possible. These two sets of requirements are mutually exclusive and therefore legs constructed for speed are very different from those constructed for strength.

A. *Legs suiting strong, slow movements*

Let us consider first the strong movements and take the example which has already been described in Chapter 4; see Fig. 4.28(a), (b). Maximum transverse sectional area of muscle gives maximum tension and therefore a plentiful supply of short, wide muscles are present in legs conspicuously constructed for strength, as in this leg of a iuliform diplopod. The extrinsic muscles also are wide and short. The podomeres themselves are strongly constructed with limited flexibility between them; contrast Figs. 4.28(a), (b) and 5.3(a) with 5.3(c).

B. *Legs suiting fast movements*

In Fig. 5.3(c) the diameters of the joints at the intersegments are very different from those just described, so permitting easy and wide flexures of one podomere upon another, but such freely moving joints are also weak and can dislocate very easily, see the greatly emarginated femur–tibia joint of the fast-running centipede *Lithobius* (Fig. 5.6, Pl. 5(e), (g)). Fast

Special requirements of land arthropods

leg movements, on the contrary, usually provide both long strides and rapid stepping. A large alteration in the positions of the hard parts at a joint, relative one to another, can cause a large change in the flexure or extension at the joint, as is needed in causing maximum effective changes in leg length at different moments (Figs. 4.21(a), 5.2). Long muscles are suitable to serve such joints because the same proportionate shortening of a long and a short muscle will give a greater absolute shortening of the longer muscle and therefore cause the greater change in the position of the hard parts at a joint. Greatest physiological efficiency of a muscle, however, results from contractions which approach the isometric, that is a condition in which muscle tension greatly increases but changes in muscle length are slight. So again, long muscles are suitable for causing fast and wide movements at joints and leg bases.

Long intrinsic muscles traverse several leg joints, as seen in Fig. 5.6, in contrast to the strong, slow-moving legs of a diplopod, Fig. 4.28(a). The extrinsic leg muscles of fast-running centipedes fan out in separate sectors over both sides of the sternites (Fig. 4.29) and extend upward to the tergites (Fig. 5.6), a great contrast to the short, wide, ventrally situated extrinsic diplopod muscles attached to an endoskeleton (Fig. 5.3(a)). A remarkable diplopod which has secondarily developed carnivorous habits and fast running shows secondary long intrinsic leg muscles (Fig. 5.4), with superficial resemblances to centipedes and marked contrasts to the normal type of diplopod intrinsic musculature (Fig. 4.28(a)). The extrinsic muscles of the lysiopetaloid diplopods are also longer and extend forward to the segment in front, thereby gaining length. The contrast between these two types of millipede limbs is very striking and see further below.

C. *Muscle numbers*

Compensation for the small force put out by the long fast-moving muscles on the legs of fleet runners is an increase in muscle number. There is limited space within the leg, but all manner of contrivances enable an increase in the number of extrinsic muscles. Extra coxal apodemes provide sites of origin for such extrinsic muscles and in the fleetest myriapod with the longest legs, *Scutigera*, fusion of the coxa with the pleurite above it serves the same purpose. By such means the two pairs of extrinsic muscles present in most diplopods is increased to four pairs in the fleet lysiopetaloid diplopods

Special requirements of land arthropods

(Figs. 5.3(a), 5.4) and the increase in number of extrinsic muscles in centipedes is even more spectacular; 13 in the Geophilomorpha (Pl. 5(a), (b)), 19 in the Scolopendromorpha (Pl. 1(f)), 20 in the Lithobiomorpha (Pl. 5(g)), and 34 extrinsic muscles per leg in the long-legged Scutigeromorpha (Pl. 4(j), (k)), the legs being shown in Fig. 5.14(b) but not the muscles (see Manton 1965, Figs. 66–72), a truly remarkable achievement for an arthropod because it is not easy to house such a large number of extrinsic muscles working each leg. Further, the very long, but strong legs of *Scutigera* need greater strength than those of all the shorter legged myriapods and in the Scutigeromorpha alone the pleurite just above the limb, the katapleure, has become folded giving an external knife-like projection providing strength to the fused katapleure and coxa.

D. *Sliding sternites*

There is another curious way in which extra muscles can be brought in to add remotor force to leg movement where it is advantageous for the legs to be short and well covered by the flanks. Figure 5.5 shows a stereogram of a millipede in which the transverse section is roughly half cylindrical in shape, promoting easy rolling up in the dorso-ventral plane. But this little creature is capable also of curling up laterally at the same time as hanging on to a ceiling and again in that position no legs project or can easily be harmed. The extra force to the limbs is provided by the freedom of the sternal plates from the rest of the skeleton. The coxae are well articulated on the sternites near the middle line and these sclerites themselves slide backwards and forwards relative to the rest of the skeleton by their own muscles and thereby they enhance the range of movements of the legs. Several groups of diplopods, including the pill millipedes, have invented this remarkable way of enhancing leg movement both in range and in strength, trunk musculature being drawn upon for additional

---

FIG. 5.3 (see opposite). Views from the transverse plane of legs of (a) a iuliform diplopod; (b) *Lithobius* (Chilopoda); (c) *Polydesmus* (Diplopoda). The legs are in their natural positions and the axes of swing of the leg on the body are marked. The knee podomere is marked by interrupted lines; the coxa–body joint articulation is marked by open rings; pivot-joint articulations are marked by black spots within a ring; and hinge-joint articulations are marked by simple dorsally situated black spots. The principal muscles working the joints, an antagonistic pair at pivot joints and flexor muscles only at hinge joints, are marked by the arrowed lines. Only in (a) are extrinsic muscles shown attached to the tracheal pouch apodemes. (After Manton 1966.)

Special requirements of land arthropods

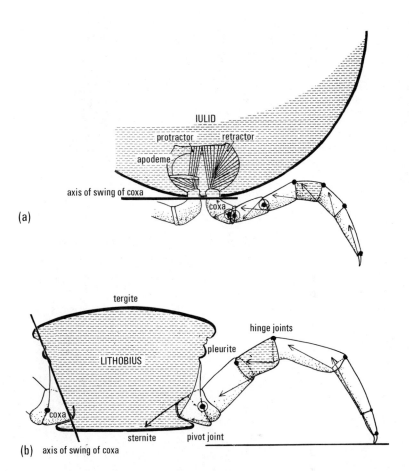

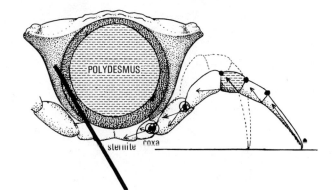

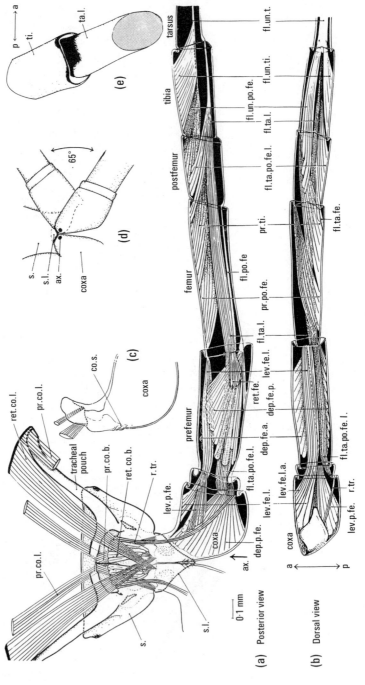

Fig. 5.4. Limb of the fast running, scavenging nematophoran diplopod *Callipus longobardius*. (a) Posterior view; (b) dorsal view; (c) base of coxa showing articulating folds, *co.s.*; (d) diagram showing a 65° rotation of the coxa about its almost vertical axis *ax.*; (e) postero-dorsal view of hinge joint between tibia and tarsus 1, heavy sclerotization being shown in black. (After Manton 1958b.)

Special requirements of land arthropods

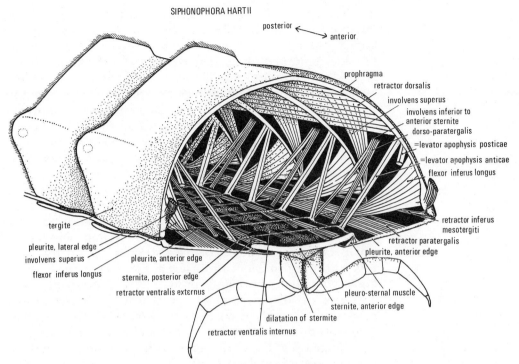

FIG. 5.5. Oblique anterior stereogram of trunk segments of the diplopod *Siphonophera hartii* showing the manner of overlap of the sclerites and the principal muscles; for further description see text. (After Manton 1961.)

remotor leg force (see Manton 1954, 1958a, 1961b). A ventral view of sliding sternites in action is shown in Pl. 4(f).

E. *Muscle leverage* The leverage of a leg muscle is of importance in providing a maximal propulsive force. The origins of muscles contributing to propulsive forces are usually so placed as to give maximal leverage, while muscles doing little or no outside work, but causing recovery movements and extension of contracted propulsive muscles, frequently function at very poor mechanical advantage, as was shown for the jaw muscles in Chapter 3; compare the adductor and abductor muscles in Fig. 3.5(b), (c).

F. *Position of axis of swing* The position of the axis of swing of the leg on the body has a great effect on the provisions for strong, slow movements such as required for burrowing diplopods, and fast move-

209

Special requirements of land arthropods

ments, suitable for pursuing prey or for escape. These differences are apparent even when the leg lengths are the same. The horizontal axis of swing shown in Fig. 5.3(a) provides a limited range of movement and the closer to the ground the ventral surface of the body is held the shorter will be the field of movement of the leg tip on the ground, but the stronger will be the backstroke. By contrast, an approximately vertical axis of swing of the leg on the body, such as seen in Figs. 4.5(b) and 5.3(c), allows the ventral surface of the body to be held very close to the ground and a maximum range of movement of the leg tip on the ground is then possible. All the fast-moving myriapods possess a more or less vertical axis of swing to the leg.

There is a very interesting secondary adaptation to swift habits to be found among the fast-running scavenging lysiopetaloid diplopods. Their leg structure is illustrated in Fig. 5.4(a), (b), and a ventral view of the coxa on the body in Fig. 5.4(d). Here a horizontal promotor–remotor swing of the coxa on the body moves about a vertical axis, seen end-on as a black spot. There is no coxal swing about a horizontal axis as in normal diplopods (cf. Fig. 5.3(a)) and instead the parts are so arranged that the coxa works, as it were, on a hinge in a vertical position on the sternite. This is a very weak mechanical contrivance, quite unsuitable for burrowing, but it provides the centipede-like wide field of movement of the tip of the leg and is doubtless the explanation for this surprising modification found throughout a large group of relatively fast-moving diplopods (Nematophora, Manton 1958*b*, 1961*b*).

FIG. 5.6 (see opposite). Leg and trunk musculature of the centipede *Lithobius forficatus*. (a) Anterior view of trochanter, prefemur, femur, and tibia to show the superficial muscles on the anterior face of the leg. (b) Anterior view of the leg with the superficial muscles shown in (a) omitted. The trunk is represented by a thick transverse piece of a segment bearing a long tergite, viewed from in front at the level of the tendon of the lateral longitudinal muscles; these muscles are not seen because they lie fore and aft of their tendon. The costa coxalis, the proximal skeletal annulus of the trochanter, and the anterior pivot articulation between coxa and trochanter are shown in black. (c) Posterior view of the leg and of a thick section of the trunk passing near a ventral intersegment and through a typical long tergite and observed from behind the level of the pleural furrow. The tendons of the lateral and sternal longitudinal muscles are confluent, as are these muscle masses. The ample arthrodial membrane between coxa and trochanter is stippled in white on both figures. (For further description see Manton (1965), Fig. 63.)

Special requirements of land arthropods

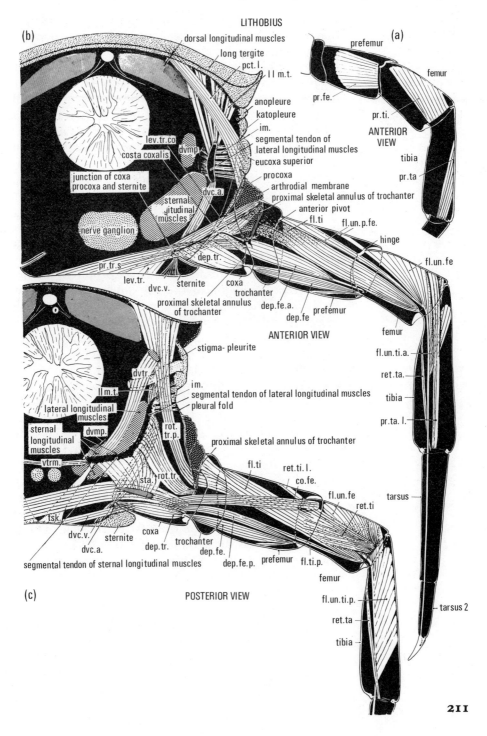

Special requirements of land arthropods

5. Leg rocking in myriapods and in hexapods

In an animal with a plantigrade stance, such as present in the majority of Pterygota, the tarsus remains squarely on the ground throughout the backstroke and most of the distal joints of the legs are provided with antagonistic muscles (Fig. 5.14(a)). On the other hand, the majority of myriapods and the four apterygote groups of hexapods are unguligrade, possessing sometimes one but usually two terminal tarsal claws. The distal podomeres of these legs usually are worked by hinge joints possessing only flexor muscles, as shown in Figs. 4.28(a), 5.3, 5.6, 5.10(b), 5.11(c), 5.13. It is obvious that propulsive extension, as is needed in the positions already noted in Fig. 5.2, cannot always be effected by flexor muscles of hinge joints. Propulsive extension of these hinge joints is done by the proximal depressor muscles, both intrinsic, near the base of the leg, and extrinsic between the leg and the body, which press the limb-tip firmly on the ground, and, with the force from the remotor swing, effect propulsive extension.

Indirect extension of the leg can be accomplished more easily if the incompressible dorsal face on the leg, which lies in between the series of hinge joints, is rotated a little forward during the backstroke, here called leg-rocking.

A. *Myriapod methods of leg-rocking*

Reference to Fig. 2.2(a), (b) shows that the swing of the coxa on the body in a iuliform diplopod automatically brings the dorsal surface of the leg slightly forward during the backstroke and vice versa. The axis of the normal promotor–remotor swing alone of a centipede leg is shown in Fig. 5.7 and in Figs. 5.8 and 5.9, the independent leg-rocking mechanism can be seen. The heavy line in Fig. 5.7 shows the normal axis of swing, the legs on segment S show the remotor position; note the large dorsal expansion of the coxa and the large expanse of arthrodial membrane between this part of the coxa and the telopod or the rest of the leg. The segment marked L shows the promotor position of the limb. The two roughly vertical lines in Fig. 5.8 indicate the backwards and forwards rock of the dorsal part of the coxa on its basal union with the sternite.

The muscles which cause this rock are many and strong and a few of them are indicated in Fig. 5.9(a): *pct.1., tcx., rot. tr.p.* Here leg 10 is in the promotor position and two muscles passing forwards and backwards from the dorsal part of the coxa to the tergite above clearly pull in opposite directions and contribute to the rock. There is a very large muscle

Special requirements of land arthropods

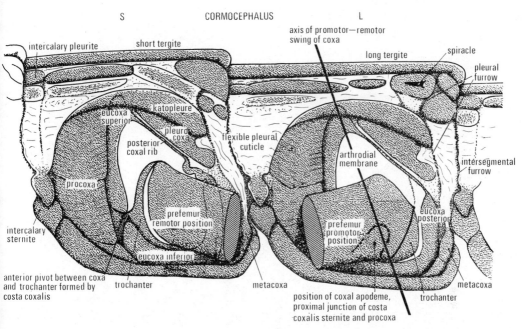

Fig. 5.7. Diagrammatic side view of the scolopendromorph centipede *Cormocephalus nitidus* to show the general morphology of trunk segments and the basis of leg movements. Major sclerites, which are considerably sclerotized, are mottled, more flexible pleural cuticle and arthrodial membrane at the coxa–trochanter joint are white, and pleurites possessing little sclerotization are not demarcated. The stigmaless S-segment, with the shorter tergite, shows the remotor position of the telopod; the stigma-bearing L-segment, with the longer tergite, shows the promotor position of the telopod. The coxa is similarly drawn on both segments although it swings about the axis indicated by the heavy line. The labelled posterior coxal rib forms the postero-dorsal sector of the coxal cylinder. For further description see Manton (1965).

leaving the posterior corner of the trochanter and passing almost directly upwards to the posterior part of the tergite above (Fig. 5.9(a) *rot.tr.p.*). Contraction of this muscle cannot do other than swing the coxa forwards about its ventral articulation. In this animal, *Lithobius*, the coxa itself forms an incomplete ring, being broken posteriorly as shown (cf. Figs. 5.7 and 5.8, Pl. 5(g)), and this allows the trochanter to be pulled well into the soft flanks of the body, giving a maximum remotor movement to the leg. There are many more rocking muscles than shown (see Manton 1965, Table 2, p. 364, Figs. 3, 48, 49, 52–4, etc.). In Fig. 5.9(b) the rocking muscles of the very long legs of *Scutigera* (Pl. 4(j)) are omitted because there are so many muscles packed so close

213

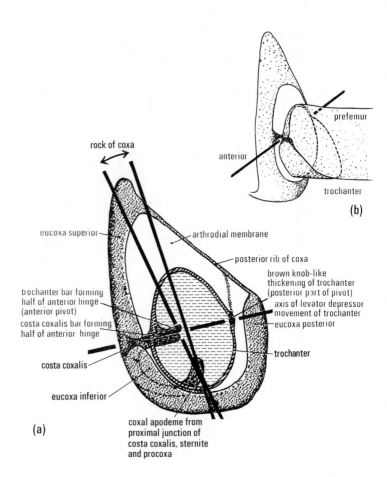

FIG. 5.8. (a) Diagrammatic lateral view of the coxa of the centipede *Cormocephalus nitidus* illustrating the two leg movements, based upon the coxa, which are additional to the promotor–remotor swing of the coxa on the body shown in Fig. 5.7. The telopod is cut off just distal to its pivot joint with the coxa. The costa coxalis of this large species is less localized than it is in Fig. 5.9(a). From the costa coxalis a stout bar projects, in the parasagittal plane, backwards into the trochanter, forming the ventral half of the anterior hinge-like part of the coxa–trochanter pivot joint. The posterior part of the coxa–trochanter pivot is very loose, a sclerotized knob on the rim of the trochanter being supported by a weak sclerite traversing the expanse of arthrodial membrane between the coxa and the trochanter. A strong coxal apodeme provides the fulcrum for the parasagittal rock of the coxa, indicated by the two heavy lines. (b) Oblique anterior view of the coxa and trochanter of *Cryptops anomalans*. The axis of the levator–depressor movement taking place between the coxa and the trochanter is shown, the anterior pivot articulation in this smaller species being simpler than in (a). (After Manton 1965.)

Special requirements of land arthropods

together, but a fusion between the coxa and the katapleure is depicted here, a device which increases the strength of the coxa and also increases the number of muscles responsible for the rocking movement.

In all the myriapods the coxal rock takes place about the ventral sternal articulation of the coxa and the rocking muscles may be quite simple where the species are small in size. In the Symphyla (Fig. 5.10(a), (b), Pl. 4(e)) the pleurite situated above the coxa is shown in both diagrams. In (a) the principal rocking muscles *tep.* and *tcx.* swing the coxa in opposite directions, just as do the corresponding muscles in the centipede shown in the previous Fig. 5.9(a). In the minute Pauropoda (Fig. 5.11(b), Pl. 4(m)) there are relatively fewer muscles and no pleurite, but the arrangement of the insertions of the promotor and remotor muscles on the coxa are such that the remotor muscles *ret.rot.co.b.* can do no other than swing the dorsal surface of the leg forwards during the remotor backstroke, the antagonist being *pr.rot.co.a.* These two muscles represent several separate muscles in other groups. Muscles *tep.pct.* in centipedes both enhance the forward rock during the backstroke.

Thus in all four myriapod groups there is a comparable rock, swinging the dorsal surface of the leg forwards during the backstroke and vice versa, and in all the swing takes place on the coxa–sternite articulation. The arrangement is a little different in the Iuliformia but the principle is the same in all the Myriapoda.

B. *Hexapod methods of leg-rocking*

The apterygote groups of hexapods also profit by a comparable rock to that of the Myriapoda but they achieve it in different ways.

(i) *The Diplura.* Figure 5.12(a) shows a side view of the leg base and sclerites of the dipluran *Japyx* (Pl. 6(h)). The pleurites fit tightly round the coxa and the heavy vertical line indicates the axis of the promotor–remotor swing of the leg. There is no rocking here between the coxa and the sternite but a most remarkable rock takes place within the leg itself. The leg of *Japyx* is seen in Fig. 5.12(a) and that of *Campodea* in Fig. 5.13(a). In both there are unique muscles, labelled *rot.fe.* in Fig. 5.13(a), arising from the proximal posterior part of the trochanter and extending obliquely forwards and downwards to the ventral face of the femur. The

Special requirements of land arthropods

femur–trochanter pivot joint is quite abnormal; it possesses a strong well formed anterior articulation and no posterior articulation. The result of the contraction of these oblique muscles is to rotate the femur slightly within the rim of the trochanter, there being sufficient play posteriorly for this movement to take place about the strong anterior pivot articulation (Fig. 13(b), (c)).

This rotation of the femur is transmitted to the distal part of the leg in a remarkable manner. A unique pivot lies at the femur–tibia joint. Its anterior component, marked W in Fig. 5.13(b), (c) control the movement of the tibia at this joint. The contraction of muscles *rot.fe.* will have the effect of twisting the femur in the direction of the heavy arrow in Fig. 5.13(c) and this twist is transmitted to the tibia by the nature of the remarkable femur–tibia joint. The next joint of the leg is perfectly normal and is shown in Fig. 5.13(e). But the oblique muscles *rot.fe.*, originating in the trochanter, are responsible, in effect, for twisting the dorsal face of the leg forwards during the backstroke, the same advantage as has been obtained by the myriapods in other ways.

(ii) *The Protura.* In spite of their minute size, the Protura (Pl. 6(k)) also have a leg-rocking device and it differs from those of all other groups. Figure 5.12(c) shows a lateral view of the base of the leg on the body of *Eosentomon*. The two circular black spots on the diagram indicate the articulation of the coxa with the sternite ventrally and with pleurite 1 dorsally, which forms an arc above the coxa. The heavy vertical line shows the promotor–remotor swing of the leg, but this is not fixed dorsally. Pleurite 1 slides against a facet on pleurite 2 so that movement here will alter the position of the dorsal edge of the coxa; by this pleural articulation the dorsal face of the coxa swings forwards during the backstroke and vice versa, as is indicated by the double arrowed line.

FIG. 5.9 (see opposite). Lateral views of the centipedes (a) *Lithobius forficatus* and (b) *Scutigera coleoptrata*. The diagrams show the disposition of the sclerites; flexible pleuron (white); arthrodial membrane; and skeletal ribs. The legs in *Lithobius* are cut off through the prefemur; leg 10 is turned forwards, leg 11 backwards, and leg 12 is turned upwards at the coxa–trochanter joint. A few of the leg-rocking muscles are shown. In (b) legs 6, 7, and 9 are cut short, leaving a little more of leg 8. The coxa, katapleure, and pleuro-coxa form one fairly rigid functional unit. Legs 6, 7, and 9 are detached at the distal breaking zone on the trochanter. Leg 8 shows two successive dorsal hinge articulations divided into strong dorso-lateral components. The tergites over legs 7 and 8 are fused and spread forwards and backwards covering the next adjacent legs. For further description see Manton (1965).

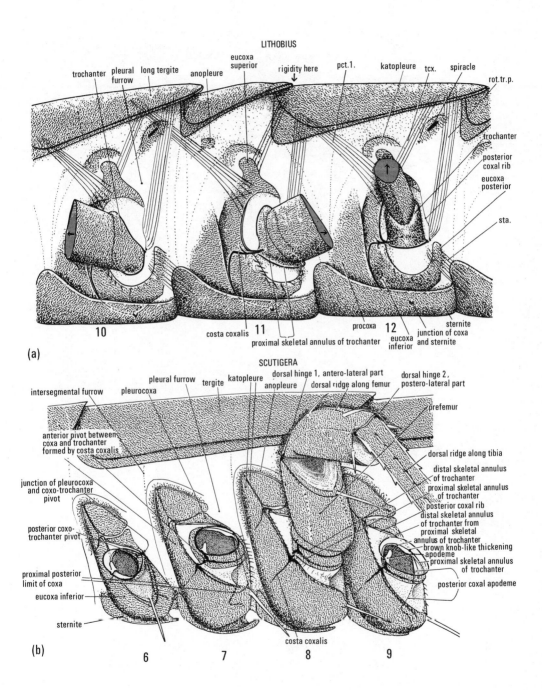

Special requirements of land arthropods

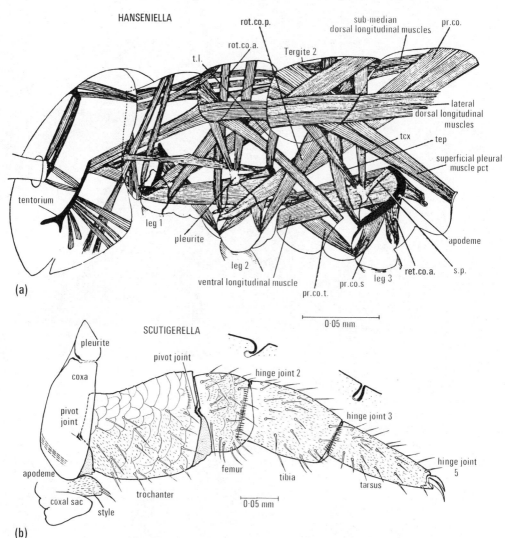

Fig. 5.10. The symphylan *Hanseniella agilis* (Pl. 4(e)). (a) Lateral view of the head and anterior leg bases and trunk muscles. (Redrawn from Tiegs 1940; from Manton 1966.) (b) Posterior view of leg 9 to show the interpodomere articulations.

Three types of coxa–body junction in hexapods remain to be considered and all are extremely different one from another.

(iii) *The Collembola*. These hexapods (Pl. 6(a), (d)) are perhaps the most surprising of all because there is no coxa–body articulation at all, the whole of the leg and trunk

Special requirements of land arthropods

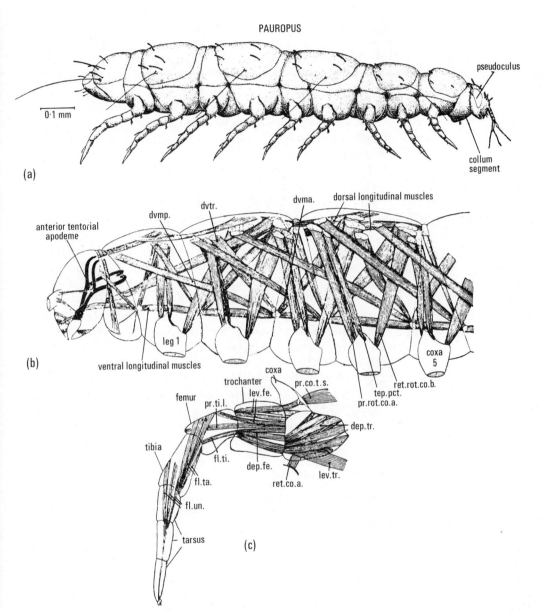

FIG. 5.11. The pauropod *Pauropus sylvaticus* (Pl. 4(m)). (a) Lateral view of an adult showing the nine pairs of legs and the large tergites on every other segment. (b) Lateral view of the head and anterior muscles of the trunk and legs. (c) Posterior view of the leg. For further description see text. (After Tiegs (1947), with replacement of muscle labelling; from Manton 1966.)

Special requirements of land arthropods

morphology being modified in correlation with hydrostatic jumping by the terminal springing organ (Chapter 9). The coxa is suspended from the trunk in a very elaborate manner (Fig. 9.5). The more distal leg joints are also stabilized so that no dislocation or extension of the leg can be produced by sudden increase in internal hydrostatic pressure. A coxa freely suspended as shown in Fig. 9.5 can of course rock in any direction as well as undergo movements of many kinds.

We have seen examples of hexapods and myriapods in which the principal coxal articulation lies with the sternite, although there may be very many other differences. In the Thysanura and Pterygota the articulation of the coxa is with the pleurite above and not with the sternite below. Plenty of mobility exists ventrally between the coxa and the body in contrast to the other examples which have been given above.

(iv) *The Pterygota*. Here the whole organization of the leg base is dependent upon a fixed pleurite (Fig. 5.14(a)) which is rigidly united with the tergite above and by a skeletal strut with the sternite below. With this fixed pleurite articulates the upper outer margin of the coxa. There is ample arthrodial membrane between the coxa and the sternite antero-ventrally and there is usually a separate sclerotization, the trochantin, situated in this membrane. In many Pterygota there is extremely free movement ventrally between the coxa and the body including the movements shown by the arrows in Fig. 5.14(a); adductor and abductor movements of the coxa on the body such as these are impossible to most of the groups so far considered. But in very many Pterygota the parts are so

---

FIG. 5.12 (see opposite). Diplura and Protura. (a) Lateral view of the thorax of the dipluran *Heterojapyx novaezealandiae* (Pl. 6(h)), to show the trunk and limb sclerites, the limb base, and the axis of swing of the leg on the body; the head end is to the left. The edges of most sclerites are not sharply defined but merge or fade into the more flexible surrounding cuticle. Leg 1 is turned forwards and leg 2 backwards at the coxa–body joint, the axis of swing being shown by the heavy vertical line. (b), (c) The proturan *Eosentomon* (Pl. 6(k)). (b) Lateral view of the mesothoracic segment, the coxa being seen end on. The head end is to the right. The positions of the two coxal articulations are marked by black spots. The axis of the promotor–remotor swing of the coxa is marked and the coxal rock, caused by a sliding of pleurite 1 against pleurite 2 at the sliding facets, is shown by the arrows within the coxal cavity. The short vertical arrows, three above and three below, mark the intersegmental and intrasegmental levels of flexibility. (c) Ventral view of mesothoracic segment to show the subdivision of the sternite into the zones marked, the position of the coxa–sternite articulation, and the median sternal apodeme. The horizontal arrows mark the levels of intersegmental and intrasegmental flexibility. (In part after Prell 1913; from Manton 1972.)

Special requirements of land arthropods

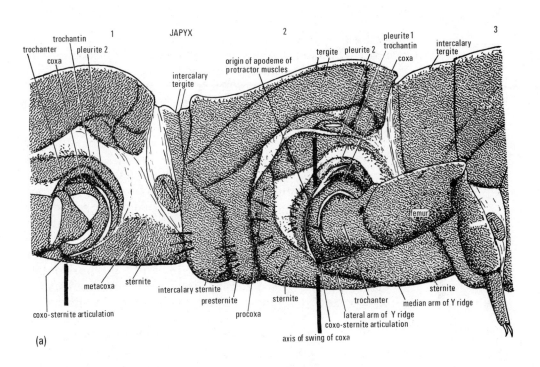

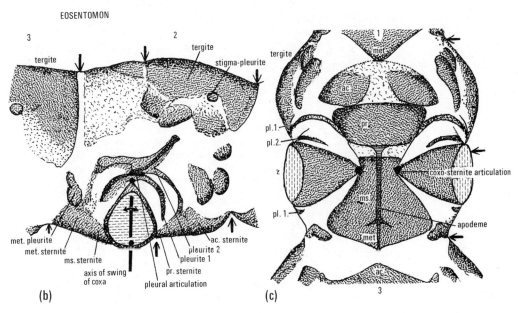

arranged that the coxal movement on the body of some if not all three legs is largely restricted to a promotor–remotor swing about a slightly oblique transverse axis, a second coxal articulation often being present between the trochantin and the coxa. Without the fixed pleurite the evolution of flight would not have been possible because it depends on the rigidity of these hard parts and the muscles associated with them.

(v) *The Thysanura*. Here again we have a principal articulation of the dorsal part of the coxa with a pleurite, but the arrangement could not be more different from that of the Pterygota. The pleurite is a mobile structure, not a rigid one, and the whole working of the leg depends on the mobility of the pleurite. In the Machilidae (Pl. 6(m)) the arrangement on the first leg is not quite the same as on the second and third legs, but on the latter there is a strong articulation anteriorly between the coxa and the anterior end of the horizontally placed pleurite (Fig. 9.8(b)). The coxa can execute substantial levator and depressor movements against this horizontal pleurite, but the promotor–remotor swing of the leg on the body is contrived quite differently from that in any other hexapod or myriapod. The movement is carried out by the pleurite on the body and on the remotor swing the posterior end of the pleurite deforms the flexible pleural membrane of the body so providing a remotor movement to the coxa via the very substantial coxa–pleural articulation at the anterior end of the pleurite.

In the Lepismatidae (Pl. 6(g)) the pleurite lies anteriorly against the coxa, with the basic articulation at its dorsal end and here the promotor–remotor swing does take place

---

FIG. 5.13 (see opposite). The dipluran *Campodea* sp. (Pl. 6(e)). (a) Diagrammatic transverse section to show the skeleto-musculature of thorax and legs. Level F on the right lies posterior to level E on the left. (b)–(d) The femur–tibia joint of the Diplura. Distal views of the end of the femur of *Heterojapyx* (Pl. 6(h)). (a) with the tibia downwardly flexed on the femur and (b) with the tibia much more depressed, as indicated by the solid arrow on (d), so exposing the dorsal process Z, bearing the tendon of the tibial extensor muscle, *lev.ti.*, see (d); the points X and Y marked on both (c) and (d) show the movement. A very strong femoral anterior component of the pivot joint is marked W; there is no posterior component. The right hand arrow on (c) shows the direction of rotation of the femur caused by contraction of the femoral rotator muscles *rot.fe*. A and B, see (a); this rotation is transmitted to the tibia (lower arrows) by the articulation between processes W and Z at the femur–tibia joint. (d) Posterior view of the femur–tibia joint of *Campodea*, its structure being essentially as in *Heterojapyx*. (e) Posterior view of the tibia–tarsus hinge joint of *Campodea*. (From Manton 1972.)

Special requirements of land arthropods

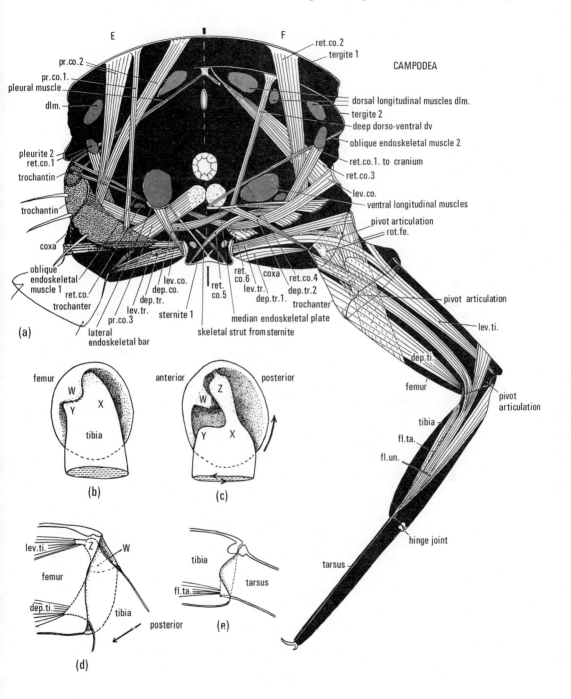

Special requirements of land arthropods

between the coxa and the pleurite. But there are very many unique features concerning the leg base of the Lepismatidae (see Chapter 9). The many peculiarities of the coxa–body joint and of its muscles in the Thysanura are associated with jumping rather than with walking or running gaits which are executed by the animals. These gaits will be considered later on together with the facilitating morphology.

C. *Conclusions concerning the coxa–body joint and the leg-rocking mechanism of myriapods and hexapods*

An attempt has been made to summarize the contrasts in the coxa–body articulation of the various groups of myriapods and hexapods in Fig. 5.15, first column of diagrams. The differences between the myriapods and the five hexapod groups are so great as to make it inconceivable that one type of coxa–body junction could ever have given rise to another. The myriapod groups are essentially similar to one another but the five apterygote groups are not. It is clear that the contrasting junctions of the leg with the body must have evolved independently in many-legged ancestors and at the same time there must have been parallel evolution of the hexapod state.

Besides the actual coxa–body junction we have the great variety in which a rocking mechanism has been established

Fig. 5.14 (see opposite). Transverse views of the body and an appended prothoracic leg of a locust and of a plantigrade chilopod, *Scutigera* (Pl. 4(j)), for comparison between the modes of articulation and consequent movements of the coxa on the body and of the jointing of the legs. Tergites and sternites where cut in section are shown in black. Pleurites, cut in (b) but entire in (a) are white or lightly stippled. The coxa is mechanically shaded; arthrodial membrane is shown by white stipple; the coxa–body articulation or point of closest union between the two is ringed; pivot joints and hinge joints are marked by ringed and unringed black spots respectively. The direction of pull of the muscles controlling the position of the podomeres are marked by thin arrowed lines. (a) Prothoracic leg of the locust viewed from in front. The antero-lateral part of the sternite is cut away on the right-hand side of the diagram in order to expose the whole of the pleurite, which carries the coxa–body articulation. Elsewhere the coxa is linked to the sternite and pleurite by arthrodial membrane, in which is located the trochantin. The coxa is free to swing in any plane about the pleural articulation. The heavy arrowed lines lie on arcs drawn from this articulation and show the abductor and adductor positions of the coxa, clearly seen by comparison of the positions of the coxal lobe marked X on either side. This coxal movement is impossible to a myriapod. (b) A leg of *Scutigera coleoptrata* from the middle part of the body. The pleurite (katapleure) is united to the tergite by flexible pleuron only and moves with the coxa. The close union between coxa and body lies with the sternite at the ventral margin of the coxa (cf. the dorsal pleural articulation in the locust). The movement of the coxa on the body is a promotor–remotor swing and on an axis lying in the transverse plane, and there can be a parasagittal rock as shown in Figs. 5.8 and 5.9. The extra hinge joint marked 4 on the diagrams of certain diplopods and chilopods corresponds here with the many tarsal joints. (After Manton 1966.)

Special requirements of land arthropods

and again we cannot conceive of one such rocking mechanism ever having given rise to another. A further consideration of the phylogenetic implications of present knowledge of functional morphology of the leg base in myriapods and hexapods will be given in Chapter 6 and arachnid leg-rocking is done in different ways (Chapter 10, §5C).

6. Particular leg joints of certain myriapods and hexapods

A comparative survey of the leg joints of myriapods and of hexapods is shown in Fig. 5.15, the conventions being explained in the legend. The promotor–remotor swing usually takes place between the coxa and the body, but there are

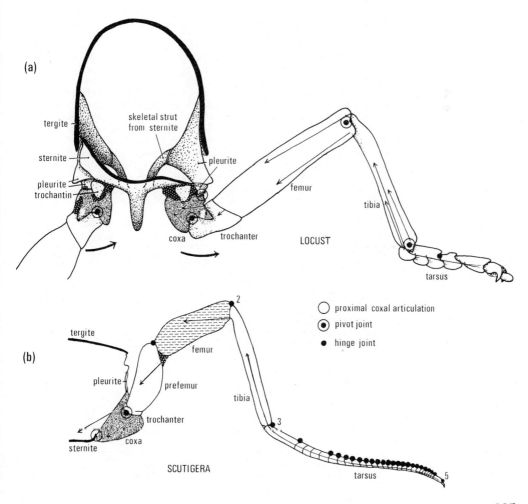

Special requirements of land arthropods

modifications in the Thysanura. The great variety of coxa–body junctions has already been stressed. There are also various ways in which leg-rocking is accomplished, frequently at the coxa–body joint but not invariably so.

FIG. 5.15 (see opposite). A summary of joint construction in the legs of Myriapoda and Hexapoda shown in a diagrammatic and comparative manner. (From Manton 1973b.) Each leg joint is viewed proximally and end-on, with the anterior face of the leg to the left and the posterior face to the right, the left leg being extended from the body in the transverse plane. The concentric circles represent the overlapping podomeres at a joint, antero-posterior leg-flattening being omitted. Articulations are shown by black spots, heavy lines passing through them showing the axis of movement at each joint. Where different classes or orders share the same morphology at one particular joint, only one diagram is given (see the arrows). The leg joints are not all comparable from taxon to taxon because the leg appears to have been divided up differently both proximal and distal to the knee (Manton 1966). Whether the coxa is ventrally directed from the body, as in most Diplopoda, Collembola, Pterygota, and Machilidae, or laterally directed, as in most Chilopoda and Lepismatidae, or obliquely in between these positions, as in Symphyla, Pauropoda, Diplura, and Protura, the axis of the promotor–remotor swing of the coxa on the body lies in the transverse plane of the body. If the left side of the body between the mid-dorsal and mid-ventral lines be opened out so that it lies flat, the axis of swing of the coxa on the body can be drawn as a vertical line, as shown on the body–coxa diagrams for all Myriapoda and Hexapoda. This axis is at right angles to that of the levator–depressor movement at the coxa–trochanter joint. (a)–(c) show the Myriapoda. All joints of the Diplopoda are given and the resemblances and differences between the Symphyla, Pauropoda, and Chilopoda are illustrated by two rows of diagrams. The diplopod rock depends on the ventral position of the axis of swing of the coxa on the body and is shown in Fig. 2.2(b). (d)–(h) show the hexapod classes, a common type of coxa–trochanter joint being present in Collembola, Diplura, and Pterygota.

The *coxa–body* junction in all provides the promotor–remotor swing of the leg-base on the body, but the support for this movement is obtained in fundamentally different ways in the several classes, that of the Thysanura being most divergent from the rest and the Lepismatidae and Machilidae differing considerably from each other.

The *coxa–trochanter* joint in all is basically a dicondylic pivot, usually with an equatorial axis permitting levator–depressor movements. The posterior articulation is sometimes reduced or absent while the anterior articulation is enlarged (Chilopoda, Protura), and only in the Thysanura is the axis markedly displaced towards the ventral face, so enhancing the range of movement above it.

Trochanter–femur joints usually permit levator–depressor movements, by either equatorial pivot or dorsal hinge-joint articulations. A rocking joint here in the Diplura is a derivation of an equatorial dicondylic pivot joint. There is very little movement at the trochanter–femur joint in Protura and in Pterygota and Thysanura it permits slight retractor movements about a dorso-ventral axis.

Femur–tibia joints usually possess a dorsal hinge articulation, except in Pterygota and Thysanura where dicondylic pivot articulations with horizontal axes of movement are present, associated with the presence of antagonistic muscles and absence of leg rocking.

Tibia–tarsus joints are all similar with dorsal hinge articulations.

The large number of leg joints in the Diplopoda and the small number in Collembola are associated with functional needs. For further description see text and Manton (1958a, 1965, 1968, 1972).

Special requirements of land arthropods

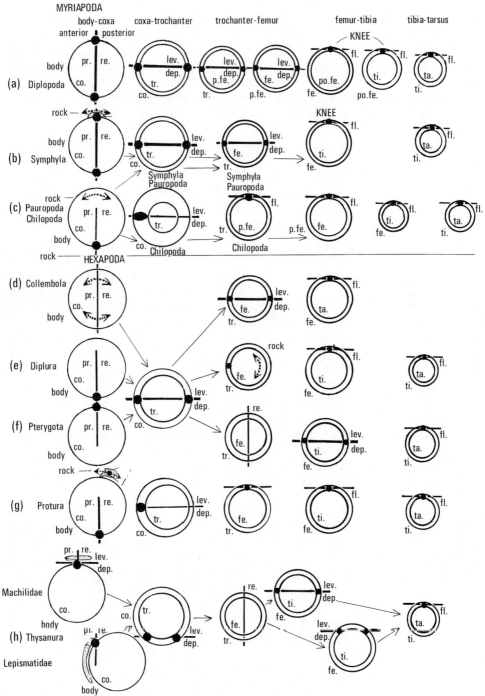

227

Special requirements of land arthropods

A. *Joints giving levator–depressor movements*

The next main joint on the legs is extraordinarily uniform whether it lies between coxa and trochanter, as in Chilopoda, Diplura, Thysanura, Pterygota, and Collembola (Figs. 5.3, 5.14(a), 9.4, 9.5, 9.8, 9.11, 9.12), or between trochanter and prefemur as in the Diplopoda (Figs. 4.28(a), 5.3(a)). The axis of movement at this joint is horizontal, usually taking place about two pivot articulations situated on the anterior and posterior faces of the leg and giving the principal levator–depressor movements to the legs. Where leg strength rather than speed of action is important, the anterior and posterior pivots are both well developed, as shown by the Diplopoda, Symphyla, and Diplura in Fig. 5.15, but where great speed of movement is desirable as in Chilopoda, the most remarkable joint to be found in the animal kingdom lies between the coxa and the trochanter. Figures 5.7–9 show an enormously elaborate anterior pivot articulation which traverses the arthrodial membrane between the coxa and trochanter, lying at right angles to the coxa, these podomeres being of very different diameters. There is often practically no posterior articulation at this pivot joint. The anterior articulation is so strong that it forms a more or less horizontal hinge between the coxal and trochanterial parts; and this narrow bar carries the whole of the distal part of the telopod and is shown particularly clearly on segments 11 and 12 of *Lithobius* in Figs. 5.8 and 5.9.

The Thysanura show an unusual displacement of the coxa–trochanter axis of swing towards the ventral face of the leg, giving greatest movement dorsal to this axis (Fig. 5.15, transverse level h, column 2). This and other features of the joint serve the thysanuran jumping gait (see Chapter 9, §5).

B. *Joints providing active flexor movements of the leg*

The more distal limb joints mostly call for little comment. The majority are simple hinge joints, with a single dorsal point of close articulation. When strength requirements demand it, double articulations are formed dorsally, as seen in Figs. 5.4(e), 5.9(b).

The extra number of pivot joints in the Diplopoda is associated with the need for turning the leg through an S-shaped bend in order that the knee may rise close in to the sides of the body although the leg originates mid-ventrally (Fig. 5.3(a)). Entire coverage of the legs by the flanks is useful in burrowing. There are two other most surprising joint

Special requirements of land arthropods

specializations in myriapods which are worthy of consideration.

C. *Skeleto-musculature associated with pushing by the dorsal surface in the Diplopoda*

Although most diplopods burrow by head-on shoving (see Chapter 8), the Polydesmoidea possess strongly developed keels projecting over the legs dorsally. These animals insinuate themselves between layers of decaying leaves by pushing upwards with the dorsal surface, once the head end has made the initial entry. The very strong leg can be flexed closely under the keel (Fig. 5.3(b)). Stong straightening of all the legs can push the dorsal surface upwards, so widening the split in the decaying leaves or wood. The large polydesmoidean diplopod *Platyrhacus* (Fig. 5.16) possesses two hinge joints modified to carry extensor muscles as well as the normal flexor muscles, these extensors assisting in strong leg extension and dorsal surface pushing.

A cuticular apodeme *kl*, drawn and labelled in Fig. 5.16(b) but drawn only at the prefemur–postfemur and at the femur–tibia joints on Fig. 5.16(a), project into the proximal podomere from the strong dorsal hinge articulation. The length of this stout cuticular apodeme is sufficient to provide

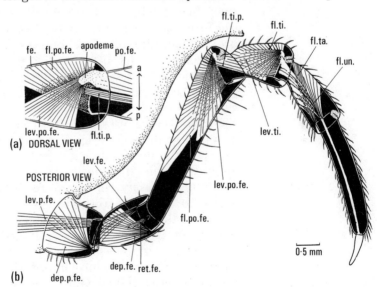

FIG. 5.16. The leg of a large and powerful polydesmoidean diplopod, *Platyrhacus* sp. (a) Anterior view of the leg flexed below the keel projecting from the tergite. (b) shows in clearer detail the apodeme *k.l.* bearing the additional extensor muscle to the hinge joint *lev.po.fe.* Two such modified hinge joints are shown on the leg in (a) at the femur–postfemur and at the postfemur–tibia joint. (After Manton 1958*b*.)

Special requirements of land arthropods

considerable leverage for the extensor muscles *lev.po.fe.* and *lev.ti.* which antagonize the normal flexor muscles at these joints *fl.po.fe.* and *fl.ti.*

This example shows not only extreme adaptation to the immediate needs of the animal but also that the absence of extensor muscles at a hinge joint is not due to any inability to develop them. One must accept that the presence of flexor muscles at hinge joints simply suits the animal best and that leg extension by proximal depressor muscles is usually the more appropriate device.

D. *Synovial cavities in myriapodan joints*

The fluid contained within synovial cavities of vertebrate endoskeletal joints facilitates smooth movements at these positions. Such a device has hitherto been unknown outside the vertebrates. However, the diplopods surprise us by developing synovial cavities within the cuticle at certain hinge and pivot joints where both strength and smoothness of movement is facilitated. Figure 5.17(a) shows a normal hinge joint with a small amount of arthrodial membrane dorsally and ventrally. Figure 5.17(c) shows a split in this arthrodial membrane dividing it into an inner and an outer layer, the space between being filled with fluid. Ventrally this permits the formation of sliding facets between the cuticles of the proximal and the distal podomeres; and dorsally the highly sclerotized cuticles of the adjacent podomeres are rolled round each other to some extent, giving great strength in this position and also there is the small upper extension of the synovial cavity, so plainly visible ventrally, *s.fe.po.fe.* No such synovial cavities have been described from any other group of arthropods. Figure 5.17(b) shows both a synovial cavity at the postfemur–tibia joint and the additional apodeme $k1$ bearing the unusual *lev.ti.* muscle. This muscle and the synovial cavity contribute to the strong tergal pushing exerted by the force from the legs.

Within the polydesmoidean species, both large and small, the use of synovial cavities is exploited to the full in the proximal part of the leg. It has been shown above how strong leg podomeres usually have limited differences in diameter at the joints between them (Fig. 4.28(a)), while legs constructed for speed gain wide flexures at the joint by the formation of proximal and distal emarginations of the podomeres and much greater expanses of arthrodial membrane are thus exposed when the leg is extended (Fig. 5.6). Such emar-

Special requirements of land arthropods

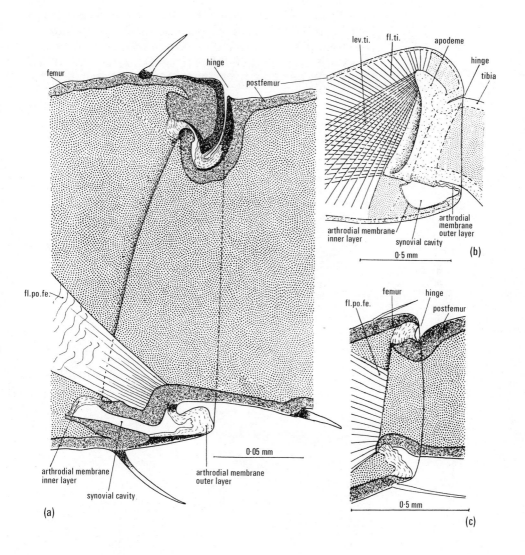

FIG. 5.17. Hinge joints. (a) Femur–postfemur joint of *Polydesmus angustus* showing synovial cavity dividing the cuticle (arthrodial membrane) at the joint into an inner layer and an outer layer within which the sliding facets of hard cuticle work. Only one muscle is shown, the flexor to the postfemur. The dorsal elaborations forming the hinge are very strong and the darker tone indicates the greater sclerotization. (b) Postfemur–tibia joint of *Platyrhacus* sp. showing both synovial cavity and the additional and unusual levator *lev.ti.* to the tibia. (c) The femur–postfemur joint of a spirobolid diplopod. For further description see text. (After Manton 1958*b*.)

Special requirements of land arthropods

ginated joints are weak, as shown by photography, and readily give unwanted flexures. Figure 5.18(a) and (b) show a polydesmoidean coxa in exactly the same position. The levator and depressor positions of the prefemur relative to the coxa in Fig. 5.18(a) and (b) are very different, yet this feat is done with no external exposure of arthrodial membrane.

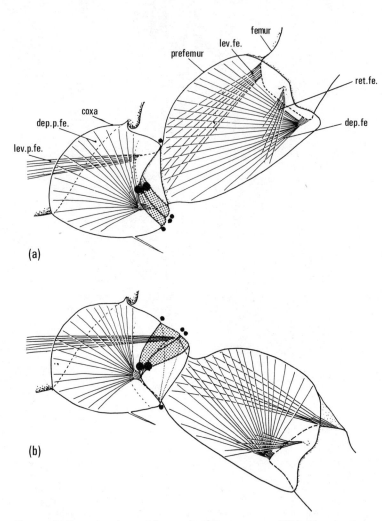

FIG. 5.18. Posterior views of the proximal leg segments of *Polydesmus angustus* to show the wide range of intersegmental movements. In (a) the prefemur is elevated to a maximum and in (b) it is depressed to a maximum. The trochanter is marked by mechanical stipple and bears a line round the middle. For further description see text. (After Manton 1958*b*.)

Special requirements of land arthropods

The mechanism of this movement is one of the most remarkable in the whole of the Arthropoda and involves an elaborate use of synovial cavities. The distal margins of the coxa, dorsal and ventral, are marked by the most left-hand of the small black spots. The stippled zone is the trochanter bearing a central furrow, marked by two small black spots at the outer margin. The large black spots show the position of the pivot articulations between coxa and trochanter and between trochanter and prefemur (Fig. 5.18 *p.fe.*). The highly truncated proximal end of the prefemur sinks through the trochanter and into the coxa and is worked by the antagonistic muscle shown, *lev.p.fe.* and *dep.p.fe.*, the depressor muscles, as would be expected, being much the stronger with large transverse sectional areas. There are ample synovial cavities between coxa and trochanter and between trochanter and prefemur and this results in the dorsal half of the trochanter being entirely covered by the coxa in position (a) and the ventral half being entirely covered by the coxa in position (b) in Fig. 5.18. A very extensive swing takes place about the two closely placed pivot articulations (heavy black spots). There is no exposure of arthrodial membrane anywhere at this elaborate double joint and no weakness as in other widely flexing joints of arthropods. This whole wonderful polydesmoidean joint enables the leg when flexed as in Fig. 5.3(b) to be strongly depressed proximally, so raising the ventral face of the body away from the substratum and delivering a dorsal thrust from the tergites and keels. This diplopodan joint complex and the centipede coxa–trochanter anterior pivot are perhaps the most remarkable joints to be found within the Myriapoda. Two other remarkable joints will be mentioned later in the Arachnida (Chapter 10) and again the functions are known and the details are unique.

7. The legs of Carboniferous and modern Myriapoda

Having surveyed the basic form and the functions served by the legs of the Uniramia, together with a sample of the highlights of specializations for particular purposes, it is of interest to look at one of the best preserved of the early fossil uniramians, *Arthropleura armata* (Fig. 1.16). This animal reached sizes of over 5 feet but why the Carboniferous arthropods went in for gigantism we have no idea. We also have little information as to how these animals lived. Leg structure, recorded in some detail by Rolfe and Ingham

Special requirements of land arthropods

(1966) is reproduced here in Fig. 5.19. The differences between this leg and those of modern myriapods shown in other figures here is very great. We can now make some functional suggestions concerning *Arthropleura*. The small difference in diameter between the overlapping margins of

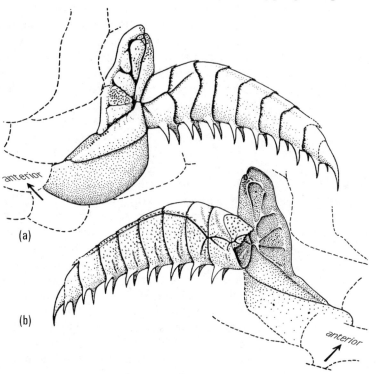

FIG. 5.19. Carboniferous myriapod *Arthropleura armata*. Reconstructions of left leg of a medium-sized adult in oblique foreshortened view. (a) Antero-ventral, (b) postero-ventral view. For further description see text. After Rolfe and Ingham (1966) and Fig. 1.17.

one podomere and the next means that flexures between successive podomeres could only have been slight, a feature accounting for the large number of podomeres and the absence of a knee. The heavy spines, terminally and ventrally along the leg, suggests a backstroke which scraped the leg on or in a rough, possibly softish, substratum rather than a leg standing cleanly on the ground as in modern myriapods where flexures are so much easier. Some strengthening ridges lay along the legs of *Arthropleura*. A leg of a large arthropod, possessing the flexibility of modern forms, would be expected to show clear pivot and hinge articulations indicated exter-

Special requirements of land arthropods

nally by local heavy sclerotizations, such as those shown by *Limulus* and *Scutigera* (Figs. 3.12, 5.9(b)), but *Arthropleura* does not. Together these limb features of *Arthropleura* suggest a slowness and weakness of movements compared with modern forms.

The attachment of the legs to the body of *Arthropleura* was supported by an elaborate anterior plate or pleurite on to which the coxa was articulated. Similar support is provided in modern myriapod legs by the procoxa in chilopods (Fig. 5.9(a); see also Manton 1965, Figs. 73–5) where the form and mobility or otherwise of this pleurite is correlated in detail with the needs of the animal. The two large paired plates overlapping each sternite and the leg bases may have covered respiratory devices (cf. Eurypterida, Fig. 1.10(d)). There were no spiracles.

Comparison between the legs of *Arthropleura* and of all the myriapodan legs considered above emphasizes the great advances that have been achieved in leg construction and adaptive radiation by the modern myriapods.

The legs of the Pterygota have not been considered in detail here because many are so well known. Larval stages are often uriguligrade but the adults gain security on inclined and vertical surfaces and even upon ceilings by virtue of the plantigrade tarsus usually bearing a few articulated joints, and terminal elaborations such as an adhesive pulvinus, or by sticky glandular secretions from the tarsal claw tendon. The podomeres of pterygotes exhibit endless adaptations to habits, such as the prehensile first leg of the praying mantis; the strong burrowing limbs of the mole crickets; the jumping 3rd limbs of locusts and grasshoppers; and the stridulating apparatus formed in various ways by a number of species in which a limb often participates. On the whole it may perhaps be said that the Pterygota modify their limbs in association with special requirements of a variety of kinds, but, except in a general way, not for speed, which, when required, is obtained by flight. The apterygote groups and the myriapods contrast by exhibiting so many examples of the attainment of fast-running, or of jumping escape reactions in Thysanura and Collembola, all involving joint specializations rather than much elaboration of the podomeres themselves. The short-legged crevice-livers do not need speed and are probably secondarily short-legged and this applies to most of those with burrowing habits.

6 *The broad subdivisions of the Arthropoda*

1. Introduction

WE have surveyed in outline a sufficiency of accurately described arthropod morphology, as a result of recent work, together with the largely new understanding of the functions carried out by these structures. We have not yet covered the whole anatomy of any one arthropod; this will be done for certain groups in later chapters and as far as space permits. But the functional morphology so far considered allows a start to be made on a new appraisal of the evolution of the Arthropoda as a whole and of its subdivisions, an appraisal now based upon substantial evidence.

It has been a zoological pastime for a century or more to speculate about the origin, evolution, and relationships of the Arthropoda, both living and fossil. There has been plenty of scope, since these animals exceed in numbers of species the whole of the rest of the animal kingdom put together, and, in the absence of direct fossil evidence, discussions have, of necessity, been largely speculative.

Apart from the general thesis that the arthropod phylum has sprung from metamerically segmented coelomates, there are few points relating to its ancestry on which any real agreement has, until recently, been reached; indeed, there has been uncertainty whether the vast arthropod assemblage is a 'natural' group or whether it comprises more than one line of descent derived independently from segmented coelomates. In the Arthropoda the basic features of annelid morphology are evident; Cuvier recognized this by including both within the major *embranchement* Articulata as did Lankester when he ranked them, with annelids and rotifers, as a sub-phylum of the Appendiculata. The new features that appear in the Arthropoda: the firm exoskeleton, coupled with growth by ecdysis; jointed appendages; striated muscle; haemocoel and ostiate heart; are advances in the sense already considered in Chapter 1, and not specializations suiting particular circumstances. They are all important, for on them depends the immeasurable superiority of the arthropod over the annelid. This high measure of success enjoins caution in rejecting the belief that such advances may have been repeatedly called forth from distant ancestors, and indeed we have to recognize that any theory of arthropod phylogeny, monophyletic or otherwise, commits us to a good deal of convergence, the independent evolution of fairly similar structures such as tracheae, compound eyes, plastron respiratory structures, the shapes and enroling ability of pill-millipedes and of certain isopods (Pl. 3(d), (d1), (e), (e1), and legends), and many other examples.

The phylum was named by von Siebold in the Siebold–Stannius 'Lehrbuch' in 1848 and defined in the following terms: 'Animals of completely symmetrical form, and with jointed locomotory organs. The central nervous system consists of a ganglionic ring surrounding the oesophagus, and of a ventral ganglionic chain growing out from the latter.' His further subdivisions of the Arthropoda do not accord with modern knowledge.

A review of the theories of arthropod evolution was given by Tiegs and Manton (1958). This review is now in many ways out of date, because of the abundance of new correlated evidence of arthropod phylogeny which has accumulated during the last thirty-five years from the field of functional morphology. The accumulation of evidence over a large field was necessary before the deductions could be made. This evidence is not speculative, it is sound; and the broad conclusions which have been reached tally with and are substantiated by the recent work on functional comparative embryology of annelids and arthropods, the latter using the illuminating concepts of 'fate-maps' of presumptive areas of the blastula, or of its equivalent, as employed in vertebrate studies (Anderson 1973). This is the stage at which the fundamental framework of the body is established. The study of functional morphology of arthropods over certain fields has yielded more detailed evidence than any embryology can be expected to provide, but in other fields the embryological evidence is the more compelling. It is most significant and satisfactory that the two lines of modern work are in such close agreement.

The broad conclusions (Fig. 6.1) emerging from these studies are very different from the speculations about ar-

Broad subdivisions of the Arthropoda

thropod relationships existing in the literature, most of which lack any functional considerations. Unless the functions of the various parts of an animal are understood in some detail, and that means gaining an understanding as to how the whole animal works, no sound conclusions can be reached. Many existing theories depend upon the assumption of functionally impossible ancestral stages. It is obvious that the past history of animals must comprise organisms which are workable at all stages of their evolution. Yet this simple truth has not taken its place in the existing speculations concerning ancestors of the arthropods or of the supposed phylogenies within the group.

2. Similarities and differences of phylogenetic significance between arthropodan taxa

A fresh assessment can now be made of (a) the significance of the similarities shown within each of the major groups indicated in Fig. 6.1 and (b) the differences between these groups. Firstly we will consider the tagmata, or differentiated regions, of the body; secondly the form of the limbs; thirdly the composition of the head where one exists; and fourthly the structure and mode of action of the jaws or mandibles. The reasons for uniting the Onychophora, Myriapoda, and Hexapoda into one phylogenetic group, the Uniramia (Manton 1972) will be considered below. The evidence to date suggests that the status of a separate phylum should be given at least to each of the Uniramia, Crustacea, and Chelicerata and possibly also to the Trilobita.

A. *Body regions, or tagmata, of arthropods*

Although the body of some annelids, especially the tubicolous polychaetes, is divided into two or more regions which differ in their anatomy and function, this tagmosis (tagmatization) is much more marked in the Arthropoda. The tagmata of the body are largely, but not entirely, composed of segments. A body segment is defined in §2C(iii) below. There is much variation between the tagmata of the several arthropodan groups and similar tagmosis is indicative of affinity.

(i) *The head, cephalon, and prosoma.* Anteriorly there is frequently a head, carrying sense organs, the mouth, and feeding organs. In trilobites a cephalon, and in myriapods and hexapods a head, is well differentiated from the trunk. All three groups have one pair of sensory antennae. Hexapods often possess a particularly mobile neck. The Crustacea possess a head of constant anterior composition but posteriorly the head is usually not strongly demarcated from the

Broad subdivisions of the Arthropoda

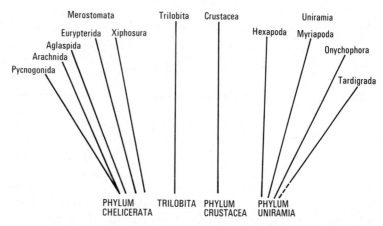

FIG. 6.1. The grouping of arthropodan taxa on the basis of comparative jaws and of trunk limbs. On the right the phylum Uniramia bite with the tip of a whole-limb and the trunk lacks biramous limbs. The three subphyla, Onychophora, Myriapoda, and Hexapoda have probably evolved independently from a primitive stock of early terrestrial Uniramia. In the middle the phylum Crustacea uses mandibular gnathobases, primitively by a rolling movement giving grinding of food. Biramous trunk limbs are present and the exopod may carry swimming setae but never a close series of flat lamellae. On the left the phylum Chelicerata also uses gnathobases but bites by primitive adduction in the transverse plane, not by rolling. The opisthosomal limbs of *Limulus* appear to be basically biramous with many flattened gill lamellae carried by the outer branch. The phylum Chelicerata and the phylum Crustacea appear to be entirely distinct. The Trilobita are also entirely distinct from Crustacea, the outer rami of trilobite biramous limbs carrying flattened, probably respiratory, lamellae. The Trilobita show many primitive features and perhaps stand nearest to the Chelicerata and early merostomes but far removed from the phylum Crustacea. The Pycnogonida and Tardigrada are also tentatively placed where their relationships appear to lie.

trunk. Two pairs of sensory antennae are present followed by feeding limbs. The onychophoran head bears one pair of antennae, feeding and defensive organs (Figs. 1.1, 1.2, 1.3, 1.7, 1.15(a), 3.1, 3.5, 3.7, 3.15, see further below §2C, D and Table 2, p. 493).

In the merostomes and arachnids (Chelicerata) an anterior unit, called the prosoma, carries feeding and sense organs as well as a series of walking legs, but there are no antennae (Figs. 1.9–1.12, 1.14(a)–(c), Pl. 8). The prosoma normally carries six pairs of limbs. The first pair, the chelicerae, are preoral and usually end in chelae held in front of the mouth, hence the name Chelicerata. In the Arachnida the next pair of limbs, the pedipalps, are usually concerned in feeding, their gnathobases lying in a para-oral position (Fig. 3.14). Behind them lie four pairs of walking legs.

The anterior end of the body of a pycnogonid (Fig. 1.16)

239

terminates in a large or small suctorial proboscis. There is no head or differentiated prosoma or cephalon closely resembling that of any other arthropod; the anterior chelate limbs, the palpi, and the dorsal eyes are reminiscent of the Arachnida, but the ovigers in the males and some females are unique (see §2E(v)).

(ii) *The trunk: including thorax; abdomen; pygidium; opisthosoma; meso- and metasoma*. Behind the head or the cephalon the trunk may consist of many, usually similar, segments, as in Onychophora and most Myriapoda (Fig. 2.4, Pls. 4, 5, 7(c)–(i)). In Trilobita, Crustacea, and Hexapoda a series of limb-bearing segments, usually forms a unit called the thorax; it consists of three segments in hexapods (Pl. 6(a)–(m)), up to about 30 in Crustacea and more or fewer in the Trilobita, but this unit is only very roughly comparable from group to group. In trilobites a pygidium of unseparated limb-bearing segments forms a posterior growth zone from which free segments are detached anteriorly and added to the thorax during life (Figs. 1.1, 1.3). In the Crustacea an abdomen, formed by a series of segments with or without limbs, lies posterior to the thorax and the body then terminates in a conspicuous telson (Figs. 1.15(a), 4.10, §C(iii)). A well developed abdomen lies behind the thorax of hexapods.

In the Merostomata the posterior part of the body is articulated with the prosoma forming an opisthosoma composed of a graded series of free segments, with some differentiation between them in Eurypterida (Fig. 1.10). This region may be divisible into two tagmata, loosely called meso- and metasoma, showing greater differences between their segments, as in *Mixopterus* and scorpions (Figs. 1.12–14). In *Limulus* the mesosomal segments are welded together dorsally, their limbs being free ventrally and used for swimming and respiration (Fig. 1.9). The metasoma is represented here by a long, strong spine articulated with the mesosoma and used in burrowing. In many arachnids, such as araneomorph spiders, all traces of adult segmentation within the opisthosoma have disappeared, and in mites, harvesters and others, even the distinction between prosoma and opisthosoma has been lost. The embryonic opisthosoma possesses up to eleven segmental rudiments and a telson.

The earliest representatives of the Chelicerata, such as the

Broad subdivisions of the Arthropóda

Aglaspida (Fig. 1.14(c)) show tagmosis comparable with the better known merostomes. The cephalon of *Euproops* (Fig. 1.14(a)), a xiphosurian, with long genal processes, is reminiscent of the Trilobita (Figs. 1.1, 1.2) but the limbs may not have been biramous as in the trilobites.

There are many other partially known arthropods from the Burgess Shale. Some possessed long and elaborate anterior limbs different from the rest, e.g. *Leanchoilia superlata* Walcott, and *Emeraldella brocki* Walcott. These and other early fossil arthropods are presently under reinvestigation. *Naraoia compacta* Walcott (Fig. 6.2) appears to be a trilobite in which the dorsal sclerites over the whole trunk are united (Whittington, in press).

Thus the types of tagmata of the body are in general characteristic of each of the major groups of arthropods listed in Fig. 6.1 (see further below).

B. *Types of arthropodan limbs*

The limbs within each of the main groups of arthropods in Fig. 6.1 are constant in basic form, but we still need to know more about the limbs of the early armoured arthropods, although some are well preserved and well described. Biramous limbs existed in trilobites and some other early arthropods as in *Marrella* (Fig. 1.6) and *Naraoia* (Fig. 6.2). The limbs of the later Merostomata are probably derived from a biramous type. The Crustacea possess biramous limbs which differ decisively from those of trilobites. The Uniramia all possess unbranched uniramous limbs.

(i) *The limbs of Trilobita and of Crustacea*. There is a fundamental distinction between the limbs of trilobites and of crustaceans. The outer rami of the former, arising distally on the coxa, carry many flat lamellae (Figs. 1.1, 1.3(b), 4.2, 4.3(c)) which are of a size and shape capable of housing a respiratory circulation (Tiegs and Manton 1958). The exopods of Crustacea, when of a comparable length, arise similarly placed, that is distally on the coxa (Fig. 4.5(a)) but carry swimming setae far too small to have any respiratory function, other than the creation of water currents for swimming or the aeration of gills or other respiratory structures (Figs. 4.8(c), 4.9, 4.11, 4.16). Crustacean limbs are far more complex than those of trilobites, sometimes bearing one or two exites (epipods) proximally on the coxa; these are not represented in trilobites, and many Crustacea also possess

Broad subdivisions of the Arthropoda

a series of coxal endites (Figs. 4.6, 4.7, 4.8(b)–(e), 4.9(d), 4.16). A single large coxal endite or gnathobase is present in trilobites all along the body in the two species with best known limbs (Figs. 1.2(b), 1.3(b), 1.4(b), 1.5).

The simplest of crustacean antennae 1 consist of a long simple ramus as seen in the trilobites (Figs. 1.1–3, 1.15(a), 3.1). Such a limb could not be more simply contrived and thus the resemblance means little beyond the existence of a simple sensory limb in both groups. But antenna 1 in many Crustacea is complex (Figs. 3.7, 4.10, 4.12(b)).

The limbs of Crustacea show a multitude of structural variations associated with habits and needs, while the post-antennal trilobite limbs are essentially similar all along the body in known species. The superficial similarity between trilobite outer rami and crustacean exopods, such as those of the thoracic legs of *Anaspides*, may be due to their use in swimming, but apart from this there is fundamental dissimilarity between a trilobite and a crustacean limb, the exopods of the former bearing many flat gill lamellae and in the latter simple setae. In trilobites and in bottom-living crustaceans the somewhat similar endopod is used for walking (Figs. 4.3, 4.5(a), Pl. 3(f), (g), (i)). When this function predominates, the crustacean limb is reduced to a stout endopod alone (Fig. 4.10(b)) as occurs in no known trilobite. Finally the union of the trilobite coxa with the body is small, lacking an articulation, and cannot allow restricted, but strong movements, promoted in Crustacea either by articulations, or by the great width of the coxa–body junction, as in *Hutchinsoniella* and the flattened branchiopod limbs. So in all, the trilobite limbs are differently constructed from those of Crustacea and the latter show a far greater capacity for functional variation.

(ii) *The limbs of Chelicerata: Merostomata; Arachnida; and early merostome-like arthropods*. The limbs of extant Chelicerata appear to have a biramous basis but are modified for special needs. The mesosomal limbs of *Limulus*, fused in pairs, appear to be biramous, with up to 150 flat, thin gill lamellae appended to the outer ramus (Fig. 4.4(b), (c)), the endopod being short. On the prosoma the postoral limbs are formed by stout endopods, each with a strong gnathobase; the sixth pair possesses an outer lobe which may represent an exopod. Little is known about the limbs of early merostomes beyond those shown in Figs. 1.14(a)–(c), 6.2. The eurypterids

Broad subdivisions of the Arthropoda

possessed very well-developed endopods, often highly differentiated one from another on the prosoma, with a chelate preoral chelicera (Fig. 1.11(a) as in 1.14(c)). No antenna is present.

The Middle Cambrian arthropods from the Burgess Shale (Fig. 6.2) show an anterior, possibly preoral antenniform limb, followed in *Leanchoilia superlata* by an elaborate, possibly first postoral limb. Outer rami bearing many flat plates appear to characterize the trunk limbs in at least *Leanchoilia* and *Naraoia*. Such limbs suggest that the absence of lamellae-bearing outer rami on the postoral prosomal limbs of eurypterids and *Limulus* may be secondary, and that lamellae-bearing outer rami may have been characteristic of the evolving merostomes and have been retained, almost in typical form, on the opisthosomal limbs of *Limulus*. The tendency to form lamellae or slat-like respiratory organs persists in the gill books of terrestrial scorpions, although the early aquatic scorpions appear to have had externally project-

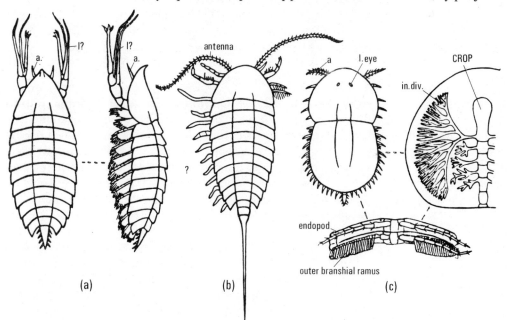

Fig. 6.2. Middle Cambrian Arthropoda from Burgess Shale. (a) *Leanchoilia superlata* Walcott, 80 mm; (b) *Emeraldella brocki* Walcott, 4 mm; (c) *Naraoia compacta* Walcott, 30 mm, after Størmer (1944). *a.*, preoral antennule; *in.div.*, intestinal diverticula; *l.eye*, lateral eye; *I*, first (preoral) limb. The lower drawing lacks the large gnathobases and proximal endites which have been found (Whittington, in preparation). (After Størmer, from Tiegs and Manton 1958.)

Broad subdivisions of the Arthropoda

ing gills (Fig. 1.14(d)). Whether chelicerae developed from preoral antenniform limbs we do not know, but the long shaft of the chelicerae of *Aglaspella* and of the eurypterid *Pterygotus* (Figs. 1.14(c), 1.11(a)) suggests that this may be so; and modern Crustacea show how easily a chela can arise on different limbs along the thorax of the Decapoda. The initial development of the great chela of a lobster is shown in Figs. 4.10(a), (b), 4.11(c).

Thus, on the basis of limb structure, it may be suggested that the Chelicerata are fundamentally distinct from the Crustacea, but exhibit strong resemblances to the Trilobita. However, the limbs of the Chelicerata are serially differentiated and those of the trilobites are not. If the Chelicerata have any distant relationships with other large taxa it could be with the trilobites but not with the Crustacea.

(iii) *The limbs of the Pycnogonida* differ from those of all other arthropodan groups. They are jointed and uniramous with no indication that they have ever been biramous (Fig. 1.16), in contrast to the other primarily aquatic arthropods. The union of the four pairs of walking legs with the body is almost unique in that lateral projections from the trunk carry the basal leg-segment. The coxa–body articulation is also unique (see Chapter 10, §9).

(iv) *The limbs of the Uniramia*. In the Uniramia a quite different type of limb is present which is never branched. In the Onychophora a soft-walled lobopodium shows a construction and mechanism which, by easy functional transitional stages, could have given rise to the uniramous jointed limbs of Myriapoda and Hexapoda (Chapter 4, §6E). This lobopodium could not have been derived from a polychaete parapodium (Chapter 4, §§4, 5) or from either of the two biramous types of limb present in the Crustacea on the one hand or in the Chelicerata and Trilobita on the other. The limbs of the Uniramia, when jointed, show a variety of modifications correlated with function, particularly on the head, but such flexibility in limb structure does not indicate relationships either with the Chelicerata or with the Crustacea.

C. *The cephalon, prosoma, and the heads of arthropods*

In many arthropods the cephalon or head, as the case may be, forms a well-defined anterior tagma bearing optic and tactile organs; feeding limbs associated with the mouth; and a concentration of nervous tissue, roughly called a brain, composed of supra- and suboesophageal components united by circumoesophageal commisures. The composition of the head end is quite distinct in each of the larger taxa (see Table 2, Chapter 11); in fact the Arthropoda could be classified, or divided into major groups, on their head end structure alone.

(i) *Types of anterior end of the body* . The Trilobita possessed a cephalon which formed a dorsal shield articulated with the most anterior segment of the trunk. The cephalon bore paired compound eyes dorsally and ventrally three pairs of limbs in *Olenoides* and *Triarthrus*; the paired preoral antennae are simple long rami and the three postoral pairs of limbs closely resemble the biramous limbs of the trunk (Figs. 1.1, 1.3). No other known arthropod possesses such a four-segmented cephalon except *Burgessia* (Fig. 1.7, Hughes 1975), whose affinities probably do not lie near the trilobites.

In *Marrella splendens* there was a wedge-shaped dorsal cephalic shield bearing paired lateral and postero-lateral projections, their tips curving posteriorly, but there were no eyes (Fig. 1.6). Paired antennae arose below the head shield, as shown, and behind them a pair of long setose limbs which may perhaps be called second antennae. Segments of the post-cephalic body followed. Again this head is unique.

Cheloniellon (Fig. 1.8, Pl. 1(a)) possessed a small cephalic shield bearing eyes dorsally and two pairs of preoral limbs ventrally; possibly both were sensory and perhaps corresponded with antenna 1 and antenna 2 of Crustacea. The next free segment, appertaining to the trunk, carried two pairs of biramous legs and the segments behind each carried one pair. All tergites were articulated, but laterally expanded, so combining to give much the same external flat but circular form to the main body as in *Burgessia*, where the head shield spread backwards and outwards as a carapace covering the leg-bearing trunk segments and the legs (see dotted outline in Fig. 1.7). Thus these early arthropods show no common plan in their cephalic organization.

The Merostomata are more uniform (Figs. 1.10–12, 1.14(a)–(c)) showing the anterior end of the body or prosoma to have been covered by a dorsal shield or carapace which bore the eyes. Below this shield in an aglaspid *Khankaspis bazhanovi* (Repina and Okuneva 1969) there appear to have been biramous limbs, the outer ramus carrying gill lamellae. In *Aglaspis spinifer* (Raasch 1939) were six pairs of limbs,

Broad subdivisions of the Arthropoda

chelicerae, and five pairs of simple walking legs; one specimen shows indications of limbs on the following free segments bearing expanded pleural lobes (Fig. 1.14(c)). Essentially the same construction of the anterior end of the body is present in *Limulus* (Figs. 1.9, 3.12, 3.13). Here, as in the extinct Merostomata and in modern Arachnida, the cephalon is not demarcated from the prosoma bearing the walking legs. One pair of chelate preoral limbs, the chelicerae, used in feeding (Figs. 3.13, 3.14), is present except in some early merostome-like arthropods.

In the Crustacea the head usually is not clearly defined from the trunk, but has a quite distinctive composition. Anteriorly it bears paired preoral sensory antennae 1 and antennae 2 and an embryonic pair of pre-antennal somites, without limbs, in some classes, which form much adult tissue. The ganglia of these segments contribute to the brain. The ganglia of the antenna 1 and antenna 2 segments form respectively the deuto- and tritocerebrum, adding themselves on behind the protocerebrum, so contributing to the supraoesophageal ganglion or brain. A pair of mandibles, of the composition considered in Chapter 3, lies in a para-oral position behind antenna 2. The postoral paired maxillae 1 follow, whose ganglia, with those of the mandibular segment, contribute to the suboesophageal ganglion. Such a head is present in the Cephalocarida where limbs of the segments behind the maxilla 1 resemble one another (Figs. 1.15(a), 4.7). In the malacostracan embryo, described below, the maxilla 1 mesodermal somites behave differently from the rest (black in Figs. 6.3, 6.4), perhaps indicating a parallel to the Cephalocarida in showing a primitive posterior limit to the crustacean head. The antennae 1, antennae 2, and mandibles are the three limb pairs possessed by the nauplius larva characteristic of Crustacea (Fig. 1.15(a)). This larva is sometimes suppressed as the egg becomes more yolk-laden. These three limbs are usually precociously developed on embryos lacking a naupliar stage (Figs. 6.3, 6.4).

Posteriorly the limit of the crustacean head is variable. Maxillae 2 usually follow maxillae 1 as a pair of feeding limbs; the head may then merge into the trunk, as in the Branchiopoda and Leptostraca (Malacostraca). Frequently the next posterior pair of limbs forms maxillipedes used for feeding, one pair in Copepoda and Syncarida, three pairs in Decapoda. Where the thoracic tergites are not free, a cephalothorax results with no clear demarcation showing where the head ends. Certain slight folds or grooves in the cuticle used to be considered to mark segmental boundaries, but most of these are mechanical strengthening devices preventing buckling of the cuticle under the strain of the jaw and other muscles and do not mark intersegmental planes.

The Onychophora have a quite different and distinct head, bearing preoral antennae and para-oral jaws and a pair of postoral limbs converted into defensive slime papillae. There is no embryonic pre-antennal segment and the trunk with its walking limbs follows without external demarcation (Figs. 2.4, 3.15(a), (b)).

The hexapod classes have a common type of head which bears eyes dorsally. A pre-antennal segment is usually embryonic and is followed by the preoral antennal segment and the embryonic premandibular segment. The para-oral mandibles and postoral maxilla 1 segment follow, with their paired limbs, and maxilla 2 limb pair is fused into a labium.

The preoral part of myriapod heads is similar to that of the hexapods, but the postoral myriapod head appears to be basically composed of two segments to which a third segment has been added in whole or in part in three of the four classes (see below, §5A).

(ii) *Ontogenetic development of a head.* In some recent books figures are shown of an imaginary basic arthropodan head consisting of a preoral unsegmented acron (see §C(iii)) bearing the eyes and followed by a series of similar cylindrical segments which are supposed to have fused together to form the head. The mouth, originally behind the prostomium or acron, is supposed to have moved posteriorly, so leaving preoral limbs. This concept is far removed from the facts of development of the head and is an unlikely picture of the phylogenetic origin of the arthropodan head.

Every segmenting egg produces an embryo which at first lacks metameric segmentation such as seen in the adult animal. As growth and differentiation proceed, segmental rudiments appear, usually in a ventrally situated germ band. The most anterior of these segmental units, with their embryonic mesoderm, ectoderm, rudimentary nervous components, and developing limb buds, shift forwards lateral to the rudiment of the stomodoeum and mouth. In this manner some limbs and ganglia become preoral, but the mouth does

Broad subdivisions of the Arthropoda

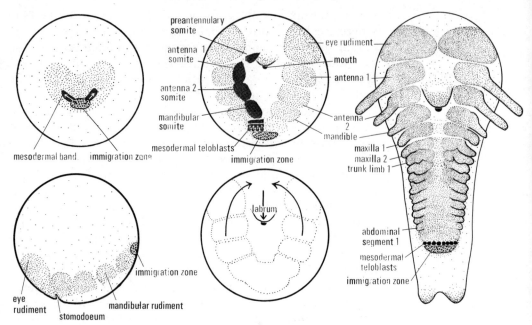

FIG. 6.3 (see above). Diagrammatic views of the developing germinal disc of a malacostracan crustacean to illustrate head formation, the migration of the parts, and the growth of the trunk. Concentrations of ectodermal cells are indicated by stippling. The black areas bearing white stipple show mesoderm lying internal to the ectoderm. The blastoporal area is marked by the same convention in all diagrams. The white arrows show the direction of either migration of a mass of mesodermal cells or the formation of rows of descendants from mesodermal teloblasts. (a) Ventral view of early germinal disc in which the ectodermal thickening is V-shaped and mesodermal immigration is taking place from the blastoporal area. (b) Lateral view of the same. (c) A later stage in which the V-shaped thickening of the germinal disc has divided into optic and segmental rudiments as marked. The stomodoeum and labrum are forming and three mesodermal somites have separated from the head mesodermal band. The preantennulary mesoderm has an independent origin. Posteriorly a row of four mesodermal teloblasts on either side is giving off rows of descendants, the most anterior already forming the mesodermal somites of the maxilla 1 and maxilla 2 segments. (d) Diagrammatic representation of the movements taking place in the embryo shown in (c). The labrum and mouth are migrating posteriorly while the head bands are shifting anteriorly. (e) Later stage in which the optic rudiment developed from non-segmental embryonic tissue is well established. The labrum and mouth are now much more posterior in position with the rudiment of antenna 1 and antenna 2 lying anterior to them. Rudiments of many trunk segments are already established from the rows of teloblasts lying posteriorly. Endodermal immigration from the blastoporal area is still continuing as it was in stages (a)–(c). The unsegmented posterior remains of the embryonic tissue forms the telson.

FIG. 6.4 (see opposite). Lateral reconstructions of the same malacostracan embryo (a mysid) to show the growth of the mesodermal somites. The somites of teloblastic origin are indicated in black, as is the preantennulary mesoderm, while the head-band mesoderm forming the three nauplial segments is stippled. At first the tail end of the embryo is folded forwards at the caudal flexure and as development proceeds the embryo straightens. (a)–(d) Show embryos of advancing ages. In (c) all the somites are established, including 6th and 7th abdominal somites, and their cavities are starting to form. In (d) the preantennulary somite and those from the maxilla 2 segment and posteriorly have grown up to the dorsal surface. The rudiment of the dorsal longitudinal muscle is forming from the outer part of the mesodermal somites. The preantennulary somite provides musculature over the ectodermal fore-gut and forms ultimately an anterior aorta which joins on to the heart formed dorsally from the maxilla 2 and more posterior segments. Maxilla 1 somite remains ventral in position and thus there is no segmental tissue dorsally here in the head. (After Manton 1928a.)

Broad subdivisions of the Arthropoda

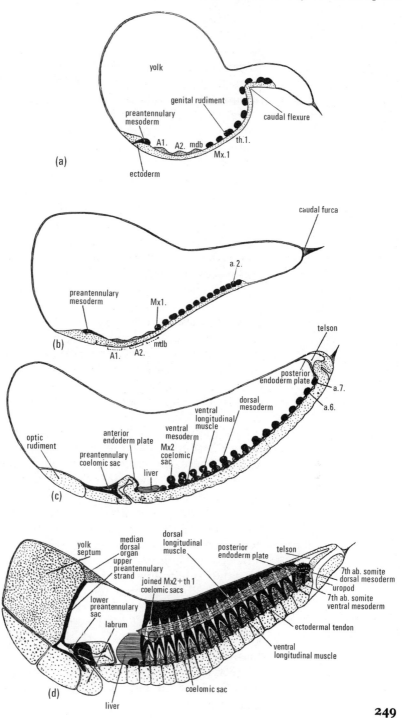

249

not shift out of one segment and into the next. The stomodoeal intucking from the ectoderm, whose external opening gives the mouth, becomes covered ventrally by the downward and backward outbulging of the body wall which forms the labrum. A preoral cavity is partially enclosed by the labrum and flanked by the tissue of the preorally shifting lateral parts of the anterior segments. The dorsal part of the head is usually formed from non-segmental tissue derived from the embryo. Thus the anterior segments are never cylindrical, but merely arise from initially ventral segmental structures and are roofed by unsegmented tissue derived from the early embryo lacking metameric segmentation. The eyes and protocerebrum develop from the anterior non-segmental embryonic tissue, the acron (Fig. 6.3 and see §C(iii)). More posteriorly the segmenting ventral band does give rise to cylindrical trunk segments from the maxilla 2 backwards. It is thus fruitless to try to decide, as has often been attempted, where the dorsal segmental limits lie on the head because there are no such limits in the non-segmental dorsal tissues.

This summary of how a head is formed in arthropods can perhaps be more readily understood from Figs. 6.3, 6.4 which represent diagrammatic reconstructions of actual embryos of a malacostracan crustacean; the general principles of growth are common to most arthropodan heads. Figures 6.3(a) and (b) show ventral and lateral views of early differentiation of a yolky egg. Fine stippling shows the concentration of the superficial embryonic blastoderm into a V-shaped area and at the base of the V an immigration zone, representing the blastopore (heavy dotted area), gives rise to the mesodermal and endodermal cells which sink inwards, thus establishing mesodermal bands (white stipple on black), which grow forwards below the concentrating ectodermal cells on either side. Figures 6.3(d) and (e) show later stages. The ectoderm has formed the beginning of the eye rudiment on either side followed by those of the antenna 1, antenna 2, and mandibular segments. The mesodermal bands have spread forwards also and differentiated into the somites of the antenna 1, antenna 2, and mandibular segments. The preantennulary somite has a separate origin direct from the blastoderm and lies in the position shown. The labral projection forms and grows backwards, see arrow in Fig. 6.3(d), where the paired arrows indicate the lateral forward shift of the ventral segmental units of the germ bands. The mouth in Fig. 6.3(c) is level with the anterior end of antenna 1 segment, but in (e) the mouth lies at the anterior end of the mandibular segment owing to the shift shown by the three arrows in Fig. 6.3(d). In this crustacean three pairs of head limbs are formed from the embryonic germ bands established on the blastoderm, the ganglia of the antenna 1 and antenna 2 segments becoming preoral, while the mandibular ganglia remain behind the mouth, contributing to the sub-oesophageal ganglion. The dorsal parts of these segments are formed by the embryonic blastoderm which is not segmental.

After mesodermal immigration has established the head band mesoderm, two transverse series of large cells called ectodermal and mesodermal teloblasts, one row internal to the other, give off rows of descendants which establish mesoderm and ectoderm of the segments of the maxilla 1, maxilla 2, and the trunk region which grows out from the posterior end of the embryo. Such teloblasts are not present in all Crustacea and there are various ways in which the mesoderm is established in other Arthropoda (see Anderson 1973). In the malacostracan Mysidacea the mesodermal somites of maxillae 1 remain ventral in position, but posterior to this level the somites spread up the sides of the body to the dorsal face, and, with the ectoderm, establish the trunk segments which are thus cylindrical (Fig. 6.4(a)–(c)).

(iii) *Definitions of a segment; the acron; and the telson*. A segment of the body of an arthropod may be defined as a unit, repeated numerically, whose initial differentiation starts in the mesoderm and is followed by the formation of paired mesodermal somites, solid or hollow, paired ganglia of ectodermal origin and paired limb rudiments.

Segments are laid down along most of the developing embryo or larva, but anteriorly and posteriorly blastodermal tissue derived from the outer layer of cells of the early cleaving, but non-segmental embryo, persists in the adult to varying extents. Anteriorly this unsegmented tissue forms the acron, giving the front and dorso-lateral parts of the head which bear the eyes, forms the protocerebral ganglionic rudiments and establishes the surface of the head dorsal to the metamerically segmental ventral head rudiments. As in vertebrates, the ganglia in the non-segmental parts of the brain serve as correlation centres and supply sense organs. To claim that such anterior nervous tissue and other derivatives of the acron are segmental is to disregard all normally accepted criteria of a segment (Manton 1960).

Posteriorly the tissue from the embryo lacking metameric segmentation persists as a telson, situated behind the last trunk segment (Figs. 6.3(e), 6.4). The telson may be large or small in the adult and it may or may not bear a large or small caudal fork, paired processes deriving their mesoderm, if they have any, from somites anterior to the telson. The telson itself carries the anus and has no true pair of legs, no pair of mesodermal somites, nor a pair of ganglia, and, like the acron, represents the remains of the cleaved but non-segmented early embryo.

(iv) *Further composition of arthropodan heads*. The Uniramia have differently shaped embryos and many fundamental differences from crustacean ontogeny (Anderson 1973) but heads are formed in an essentially similar manner. The head of the Onychophora (see §2C above) possesses limbs on all segments behind the acron; only one segment, the antennal, becoming preoral during ontogeny. The formation of mandibles on the second body segment instead of on the fourth as in the Myriapoda and Hexapoda may be related to a very early adoption of carnivorous feeding by the simplest of jaw movements not far removed from those of the walking legs.

In all Uniramia the posterior limit of the head is clearly defined, cf. Crustacea, although the postoral segmental components are not exactly the same in the four myriapod classes. A similarity in preoral composition of the heads in Myriapoda and Hexapoda is confirmed by the embryonic and evanescent pre-antennal segment, carrying limb rudiments in the myriapod *Scolopendra* (Heymons 1901) and the insect *Carausius* (Leuzinger, Wiesmann, and Lehmann 1926) and the embryonic limbs on the evanescent premandibular segment in the myriapod *Hanseniella* (Tiegs 1940) and the pterygote insect *Pieris* (Eastham 1930). Thus only the antennal segment is obviously preoral in the adult. A possible functional explanation of the convergent similarity in preoral segment number between Crustacea and the myriapod–hexapod groups may lie in a need for space for the development of a good-sized labrum and preoral cavity. In all other respects the heads of Crustacea differ from those of the Uniramia. A head capsule of sclerotized cuticle is present in

Broad subdivisions of the Arthropoda

the Myriapoda and Hexapoda. The four-segmented trilobite cephalon, as in *Olenoides* and *Triarthrus*, with one pair of preoral antennae followed by limbs similar to those of the trunk (§2 above), is quite distinct from other arthropodan heads.

Thus, a consideration of the anterior structure of arthropods, the prosoma of the Chelicerata and the quite distinct cephalon of the Trilobita and heads of Crustacea and Uniramia places them in the same categories as does a consideration of the limbs (Fig. 6.1). A further assessment of the Uniramia is given below in §5.

D. *The jaws of arthropods*

The jaws and mandibles of arthropods have been considered in sufficient detail in Chapter 3 to show the basic similarities and differences between these organs in the various groups, together with their derivations. A summary of the conclusions is given in Fig. 6.5. There is a fundamental

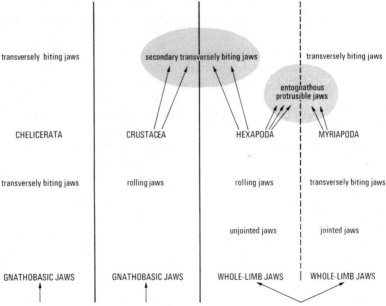

FIG. 6.5. Diagram showing the conclusions reached concerning the distribution of the principal types of mandibles or jaws (below) and the derivation of the jaw mechanisms (above). The heavy vertical lines indicate an entire absence of common ancestry between the jaws referred to on either side; an interrupted vertical line indicates separate evolutions of the jaw mechanisms of Hexapoda and Myriapoda which probably had a common origin; and the shaded areas indicate mandibular mechanisms showing convergent similarities derived from unlike origins. (From Manton 1964.)

253

distinction between the whole-limb jaws of the Uniramia which bite with the limb-tip and the gnathobasic jaws of the primarally aquatic arthropods which bite with the limb-base. There are no jaws in Pycnogonida.

(i) *Types of jaws.* Evolution of the two types of jaw was probably associated with persistent habit differences (Chapter 2), but the gnathobases of Chelicerata and Crustacea are entirely different. In the living arthropods we have three mutually exclusive types of jaw:

(a) the chelicerate gnathobase whose movement is derived from the adductor–abductor movement of a walking leg;

(b) the crustacean gnathobase, very differently constructed, with a movement derived from the promotor–remotor swing of a walking leg; and

(c) the whole-limb jaw or mandible of the Uniramia where biting is done with the tip.

The extant Chelicerata chew or bite their food by movements in the transverse plane of five pairs of postoral prosomal coxae in *Limulus* and by one pair, the postoral pedipalps, in Arachnida. The gnathobasal processes of walking legs I and II of scorpions are not mobile. The Crustacea contrast with a grinding, rolling movement of a single pair of mandibles in the more primitive members of the group. The leg distal to the gnathobase is reduced or absent. Strong secondary transverse biting has been evolved several times within the Crustacea and by different means. Grinding of food becomes incompatible with strong transverse cutting, but these latter achievements are secondary and not primitive or directly comparable to the basic transverse chewing in Chelicerata.

(ii) *Jaws of the Uniramia.* The whole-limb jaws of the Uniramia use no basal lobes for biting, only the tip. This whole-limb mandible differs in the three component subphyla:

(a) it is unjointed in the Onychophora where food is sliced by an antero-posterior movement similar to the stepping of the locomotory legs, the jaws being short and ventrally directed with large blade-like terminal claws (Fig. 3.15).

(b) The Myriapoda have jointed mandibles which bite in the transverse plane. Adduction is done by muscles but abduction of a wide, jointed mandible is impossible by any arrangement of extrinsic or intrinsic muscles. Abduction is

Broad subdivisions of the Arthropoda

accomplished by the surprising action of swinging anterior tentorial apodemes within the head capsule which push the mandibles apart (Figs. 3.17, 3.19–22). As noted above, primitive transverse biting also exists in the Chelicerata, but by the unjointed gnathobase, not by the tips of a whole-limb, all the details being different.

(c) In the five hexapod classes the whole-limb mandible is unjointed and the primitive movement is a rolling, grinding one derived from the promotor–remotor swing of a walking leg, the grinding being done by the surfaces of molar lobes, as in *Petrobius* and Collembola (Figs. 3.24, 3.28, and Manton 1964, Figs. 19–24). There are many cases of secondary strong transversely biting mandibles suitable for dealing with large, hard food which have been obtained by changes in the position of the axis of swing of the mandible, changes in its shape and by the formation of suitable mandibular apodemes. Two pairs of tentorial apodemes are needed within the head capsule and they are fixed, see *Lepisma* and locust, Figs. 3.25, 3.26, not mobile as in the Myriapoda. Their fusion forms the pterygote tentorial endoskeleton which braces the whole head capsule, giving it sufficient rigidity to bear the tension of the enormous mandibular adductor muscles and providing direct sites for the insertion of the extrinsic muscles of the other mouth parts. There is convergence between strongly biting hexapods (Fig. 3.26) and certain crustaceans (Fig. 3.5(b), (c)) in the achievement of transverse biting suiting large, hard food, but the details are quite different.

These three types of whole-limb mandibles ((a)–(c) above) form fundamental distinctions between the three uniramian sub-phyla, Onychophora, Myriapoda, and Hexapoda. Each type of mandible lacks any plausible potentiality for giving rise to any other type, if functional continuity was to have been maintained. The symphylan mandible is decidedly myriapodan in nature and does not bridge the gap between the Myriapoda and Hexapoda. It is probable that these three types of mandible (sharing the one basic similarity of being a whole-limb), were differentiated early and represent one of the many contrasting features between the evolving Onychophora, Myriapoda, and Hexapoda (see §5).

Thus the three types of mandible or jaw in the Chelicerata, Crustacea, and Uniramia substantiate the probable independence in evolution of these three groups on the grounds of functional morphology. One cannot conceive of any one of the three being capable of giving rise to either of the other two with no loss of efficiency.

(iii) *Trilobite gnathobases.* The evolutionary position of the trilobite gnathobase and of the group itself is far less certain.

The long series of large but weak gnathobases are carried by a remarkably small coxal head on the body in the two best-known examples (Figs. 1.1–5). The whole coxa serves the feeding mechanism, not only the gnathobase. It is probably much too facile to suppose that such a coxal head represents that of a primitive ancestral coxa capable of advancement either towards the Crustacea and/or the chelicerate type of gnathobasic set-up. It is more probable that the small non-articulated coxal head is essential to trilobites to permit the variety of coxal movements which must have been used (see Chapter 2). A more elaborate, larger, articulated coxal head would probably have been a disadvantage to them. The long series of similar limbs behind the single pair of antennae appears to be a primitive feature and all the post-antennal limbs presumably contribute to the feeding mechanism. In *Limulus* a postoral series of limbs also combine to form a feeding mechanism but the series is short and therefore the prosomal limbs can be orientated so that their gnathobases converge towards the mouth, a mechanical advantage possessed also by the eurypterids (Figs. 1.9, 1.10, 1.11, 3.13).

It is possible to conceive of an archi-trilobite-like arthropod coming to resemble a merostome-like arthropod (Figs. 1.14(a), 6.2(b), (c)), but it is much more difficult to see how a crustacean could have arisen in this way. A habit of using a series of gnathobases for feeding persisted in trilobites, but a reduction in limb number, permitting an orientation of gnathobases towards the mouth, would pave the way for the evolution of the *Limulus* type of coxa–body junction, gnathobase, and feeding technique. The more progress made along such a supposed evolutionary line the greater would have been the available food range because harder food could have been tackled. A progressive fusion of the trilobite cephalic shield with the tergites of the immediately posterior trunk segments would provide the basal stability needed for the shorter, stronger series of feeding gnathobases seen in the Xiphosura. The trilobite cephalon, which in many species possessed long genal spines, might become very like the prosoma of some aglaspids and Xiph-

Broad subdivisions of the Arthropoda

osura (Fig. 1.14(a)–(e)). However there is no evidence that such changes ever took place. The characteristic chelicerae of the Chelicerata and their absence in the Trilobita is hardly a stumbling block because chelae can be seen to develop so easily on many different legs in the Crustacea, and, as noted above, long-shafted chelicerae existed in some aglaspids and eurypterids, a not impossible change of structure and function from that of an antenna, associated probably with a developing ability to deal with larger, harder food.

Two Cambrian arthropods are shown in Fig. 6.2(a), (b); they possess long antennae but show no obvious resemblances to trilobites or merostomes. *Naraoia*, Fig. 6.2(c) of the same age, is probably a trilobite with fused trunk terga (Whittington, in press). A most striking feature of the many known Cambrian arthropods is their difference from one another, from the trilobites, merostomes, and crustaceans and the absence of intermediates between them. The status of each may be just as independent as those of trilobites, merostomes, and crustaceans.

A supposed contrary evolution of an archi-trilobite towards a crustacean is very difficult to conceive. The simplest crustaceans we know already possess a rolling, grinding gnathobasic mandible, quite unlike the chelicerate gnathobase. They eat fairly fine food particles collected by the trunk limbs and passed forwards, as in *Hutchinsoniella*, but without proximal well developed simple pairs of trunk gnathobases; or food is brought forwards to the mandibles for grinding and for shredding and cutting (Chapter 3), by simple raptatorial collection of larger food by a few pairs of limbs. But the Crustacea never employ any feeding device with a roughly radial convergence of limbs round the mouth as do the Chelicerata (Figs. 1.9, 2.6, 3.13).

E. *Conclusions concerning the relationships of the larger groups of Arthropoda*

(i) *Chelicerata, Crustacea, Uniramia, and their embryology.* The evidence derived from the comparative functional morphology of limbs, jaws, and the organization of the head end of the body (§2A–D) tallies in demonstrating basic similarities within each of the three taxa, the Chelicerata, the Crustacea, and the Uniramia. The same evidence shows the great differences between these three taxa, differences so deep-seated as to indicate that the rank of phylum should be applied to each.

The evidence of comparative embryology of annelids and

of arthropods supports this polyphyletic interpretation of arthropod relationships into at least these three phyla.

The clitellate annelids display a particular modification of the polychaete pattern of development, in which spiral cleavage is retained but the trochophore is eliminated in favour of the direct, lecithotropic development of the metamerically segmented body. The clitellate modification of annelid development provides a functional intermediate to the intralecithal cleavage and blastodermal development of the Onychophora, indicating at the very least that the Onychophora are related to segmental ancestors of the annelids. The onychophoran pattern of development underlies all of the developmental diversity and specialization exhibited within the Myriapoda and Hexapoda, supporting the view that the onychophoran–myriapod–hexapod assemblage is a unitary phylogenetic group, the Uniramia. Within the Uniramia, however, the Onychophora are not ancestral to either the Myriapoda or Hexapoda and the Hexapoda are not descended from any of the sub-groups of the modern Myriapoda. The embryological evidence gives clear indications that the Onychophora, Myriapoda, and Hexapoda have each diverged independently from a common lobopod ancestry.

The Crustacea are a second unitary group of arthropods, with a much more distinct unity in their embryonic development than the Uniramia. The mode of development in Crustacea is based on spiral cleavage and a configuration of presumptive areas whose subsequent development is as a nauplius. The details of this sequence leaves no doubt that the Crustacea are unrelated to the Annelida or to the Uniramia, except in so far as all three groups are members of a larger, spiral cleavage assemblage of invertebrates. The metamerically segmented coelomates which gave rise to the Crustacea cannot be identified.

The Chelicerata comprise a third unitary group of arthropods, again with a distinctive theme of embryonic development. Their specialized total cleavage and configuration of presumptive areas obscure any evidence of past relationships, except that a link with the Uniramia can be ruled out. Possible relationships with either the Annelida or the Crustacea cannot be confirmed or denied embryologically. A phylogenetic origin of Chelicerata independent of those of the annelids and other arthropods is also possible, since there is no certainty that their ancestors even had spiral cleavage (Anderson 1973).

(ii) *Trilobite internal structure.* The Trilobita appear to be perhaps distantly related to the Chelicerata on the evidence already considered, but in addition, the wonderfully preserved internal tissues of *Triarthrus*, claimed by the X-ray work of Cisne (1975), should be considered. We hardly dared to expect such detailed preservation after 300 million years in unsquashed fossils! In particular *Triarthrus* shows a series of intersegmental transverse tendons with muscles attached to either end, just as appear in many crustacean embryos and modified somewhat in most adults (Fig. 3.11(b), Chapter 8, §2A, and Manton 1928*a*, text Fig. 23), but they are not

Broad subdivisions of the Arthropoda

recognizable anterior to the mandibular segment, doubtless owing to the more advanced structural modifications associated with head formation in Crustacea than in trilobites. The tendon series in *Triarthrus* extends uniformly from just in front of the antenna (Fig. 1.4(a)). Similar intersegmental tendons are figured by Hessler (1964) for the adult crustacean *Hutchinsoniella* but the similarity with the trilobite does not indicate trilobite–crustacean affinity because similar tendons are developed in the Myriapoda, but in the adult they often become more complex.

Triarthrus shows muscles diverging anteriorly, posteriorly, and upwards from the lateral ends of the transverse tendons to the tendinous intersegmental junctions of the dorsal longitudinal muscles with the tergites above. These muscles are marked in Fig. 1.4 as *dvp* (*dvmp*), *dva* (*dvma*), and *dvv* (*dvr*). The former lettering in each case represents the labelling used by Cisne for *Triarthrus* and for the similar muscles in the crustacean *Hutchinsoniella* by Hessler (1964) (Fig. 1.15). Similar muscles exist in the myriapods and hexapods and the lettering in brackets gives the labelling used originally by Bekker (1926) for certain chilopods and adopted by Manton (1965, 1966) for the more detailed work on the Chilopoda and Pauropoda (see Figs. 8.9, 8.13, 8.16). Modifications of Bekker's labelling is used for the muscles of somewhat similar positions and functions in the Hexapoda (Manton 1972). These muscles, forming the deep oblique and deep dorso-ventral muscle systems in myriapods and hexapods, probably arose in the Uniramia by subdivision of muscle sheet 3 (Figs. 4.25, 4.26(a), (b)) of an onychophoran-like ancestral stage. These deep oblique muscles characterize limb-bearing segments which are freely moveable on one another. The deep oblique series, *dvmp*, *dvma*, are concerned with providing body rigidity and an anti-undulation mechanism, where one is needed (see Chapter 7). They strongly stiffen and slightly shorten the body. Their antagonists, the deep dorso-ventrals, *dvr*, only have to extend relaxed deep oblique muscles, doing no outside work, and are therefore less bulky than the deep obliques (see Manton 1973, Appendix on uniramian musculature). The deep obliques can become much elaborated in serving special needs in myriapods and hexapods (Manton 1965, 1972, 1973).

Thus the serially simple form of the segmental tendons, the deep oblique and the deep dorso-ventral muscle systems in trilobites shows a primitive state of these systems, but does not indicate particular affinity with either Crustacea or Uniramia. These deep muscles are all small in size in *Triarthrus* and much greater force must have been exerted by the stouter dorsal and ventral longitudinal muscle systems, which are present in all arthropods and annelids. The absence of the lateral longitudinal muscles in trilobites and of recognizable superficial oblique muscles indicates that body flexures were mainly dorso-ventral and not in the horizontal plane, as indeed is indicated by the exoskeleton; the hinge line between one segment and the next made lateral movement impossible.

It should also be remembered that much change in muscles takes place after death, including shrinkage and disintegration. An animal as large as *Triarthrus* would have possessed much stouter muscles than those depicted by Cisne; many would have been composed of multiple sectors; and massive muscles concerned with enrolment, not seen at all by Cisne, would have obscured the slender basic system. The enormous enrollment muscles of a diplopod are shown in Manton 1954, text-figs. 4(d), 8, muscles *fl.in.long.in., fl.in.long.ex., inv.in.in., inv.in.s., ob.in., lev.ap.an., lev.ap.post.*; comparable muscles in text-fig. 3(a); see also Manton 1961*b*, text-fig. 35 and Manton 1973, p. 369.

(iii) *Trilobite and crustacean modes of development and the relationships between these two taxa.* The methods of growth of the body of trilobites and of crustaceans is strikingly different. There are some variations in the modes of laying down of the body regions and segments, but the majority in each group show their own characteristic growth patterns.

The Crustacea either add segments progressively from a pre-telson growth zone during a series of larval stages (Fig. 1.15(b), (c)), so adding to the nauplius segments; or segments are added in the same manner during embryonic life as shown in Figs. 6.3, 6.4. The usual process of segment growth is one of gradual differentiation along the body, the older, more anterior segments being the more advanced. Divergencies from this even differentiation from before backwards serves special needs in some larval developments, such as those exhibiting a sudden metamorphosis, as between naupliar and copepodite stages in copepods, or naupliar and cypris stages in barnacles. Some Malacostraca possess larvae with pre-

Broad subdivisions of the Arthropoda

cociously formed anterior thoracic and abdominal limbs, serving needs of swimming and seizing food, but all these divergencies are secondary to the basic even development from a growth zone just anterior to the telson.

A review of the ontogeny of trilobites is given by Whittington (1959) which includes the basic mode of development and the divergencies shown by various trilobites. The earliest known stage is a protaspis larva 0·25–1 mm in length, subcircular in outline and approaching a subspherical shape (Fig. 1.2(g)). Growth of this larva produced the cephalic furrows and rings and a protopygidium became divided from the cephalon (Fig. 1.2(c), (d)). A number of larval stages followed during which the differentiation of the cephalon proceeded. At the same time the protopygidium developed segmental grooves, spines, etc., the anterior segments being the more fully formed (Fig. 1.2(e), (f)). One by one these separated and formed the post-cephalic thoracic region of the body (Fig. 1.2(e), (f), (h)–(l)). Each species of trilobite had a constant adult number of thoracic segments, additional segments to the pygidium must have been the result of continued activity of a posterior growth zone. In *Ceraurus* new pygideal segments ceased to appear before the thorax was completed, whereas in many other trilobites the reverse was true. No crustacean has anything remotely resembling the development of adult trunk segments from the pygidium of a trilobite, where the pygidial segments with presumably functional limbs separated to join the articulated thorax, cf. segment formation in Crustacea in Figs. 1.15(g), 6.3, 6.4.

(iv) *The status of the Trilobita.* Thus the evidence to date, based upon the differences between: the composition of the trilobite cephalon and crustacean head; the structure of the limbs; and the method of growth of the body in the trilobites and Crustacea, is against the existence of any affinity between these two taxa. We are left also with a possible association between early trilobites and early merostome-like arthropods and thus perhaps to a more distant one with the later Chelicerata. Many other early armoured arthropods may or may not have been similarly related. But the Trilobita and Chelicerata may represent two phyla which have reached the arthropodan grade of organization independently, see also above, §2D(iii).

(v) *The status of the Pycnogonida.* The affinities of this

Broad subdivisions of the Arthropoda

structurally uniform group of arthropods are obscure (Calman and Gordon 1933, Hedgpeth 1947, 1954, 1955, and others). The variation within the Pycnogonida is small. Their probably secondarily uniramous limbs borne on large lateral protuberances from the body contrast with the limbs of most of the primarily aquatic Arthropoda which are differently constructed and are, or probably have been, essentially biramous. The absence of jaws of any kind, the number and form of the limbs anterior to the walking legs, the cheliphores, palpi, and ovigers, contrast with the anterior ends of all other primarily aquatic arthropods.

The protonymphon larva of pycnogonids (Fig. 1.16(c)) does not resemble the larva of any other group. The protonymphon larva possesses three pairs of anteriorly directed legs; the first becomes the cheliphores, the second and third, after regression and regeneration, form the adult palpi and ovigers (when present) while the four pairs of normal walking legs, but sometimes five or six pairs, develop progressively on the trunk.

The fossil record is of interest. There are many specimens of the Lower Devonian *Palaeoisopus* (Fig. 6.6). The chelifores are preserved lying over the proboscis and paired ovigers are present. The four pairs of legs originated from trunk protuberances, strengthened by annuli, as in modern pycnogonids. The trunk terminated in a narrow, jointed opisthosoma. There are other fossil animals seemingly related to the pycnogonids. In all, such evidence suggests great antiquity for the pycnogonids.

There has been hesitancy to accept the pycnogonids as related to the Arachnida, although some zoologists have felt inclined to place them there. The Arachnida, unless parasitic, constantly possess preoral chelicerae, postoral pedipalps followed by four pairs of walking legs and have no proboscis. This same number of anterior limbs in similar positions is possessed by the Eurypterida and by *Limulus*. The ticks and mites (Acari) within the Arachnida possess a complex anterior feeding unit, the gnathosome, consisting of an enlarged labrum, an epipharynx projecting from the pharynx, a hypostome formed by fused pedipalpal coxae, and chelicerae in various shapes. This whole structure is much more complex than is the pycnogonid proboscis (Fig. 1.16(b)), but the latter is relatively very large compared with the small trunk.

Broad subdivisions of the Arthropoda

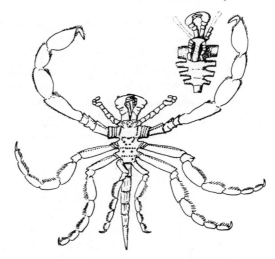

FIG. 6.6. Reconstruction of *Palaeoisopus problematicus* Broili, 130 mm, Devonian. The anterior limbs are superimposed on the proboscis. (After Lehmann, 1959.) Above and to right the ventral aspect. (From Broili, 1933.)

The following points are pertinent to the problematic status of the Pycnogonida. It is shown in Chapter 10, Fig. 10.9, how the limbs of Pycnogonida and Arachnida differ from all other Arthropoda in the absence of a promotor–remotor swing of the leg at the coxa–body junction. In Arachnida the coxa is almost (as in spiders) or entirely fixed, not movable, on the body (Fig. 10.2). The arachnids show how the possible movements at any particular leg-joint have changed, e.g. a secondary morphological provision giving a promotor–remotor swing at the coxa–trochanter in Solifugae and Acari but at the trochanter–femur joint in Opiliones and Amblypygi, and there is no such movement at any joint in scorpions and spiders (see Ch. 10, §5C). It is clear that arachnids have been capable of altering their primitive joint construction as well as dividing their legs into different numbers of podomeres in the several orders. This flexibility in construction of joints is shared by the Pycnogonida. Their principal need in feeding and swimming is a levator–depressor movement of the entire leg which is provided by a horizontal, longitudinal axis of movement at the coxa–body joint, unique among arthropods (Fig. 10.9). It is suggested below that swaying of the pycnogonid body is most important. The principal need for the legs of most arthropods, on the contrary, is a basal promotor–remotor swing, useful for walking and for swimming but by quite different means from those used by pycnogonids (Chapter 10, §9).

In walking and swimming pycnogonids perform their leg movements with the irregularity characteristic of arachnids (Figs. 10.7, 10.8). The metachronal rhythm of pycnogonids shows a phase difference between successive legs of about 0·5 of a pace, so that successive legs are in opposite phase, a feature characteristic of arachnids and also of *Limulus* during gnathobasic chewing (Figs. 2.6(d), 10.7, 10.8). Continuous swimming is thereby obtained. This phase relationship stands in marked contrast to other arthropods; even hexapodous arachnids contrast with uniramian hexapods in this mode of stepping. It is probably significant that this same type of leg rhythm is present in a xiphosuran, the arachnids, and pycnogonids but not in other taxa.

Secondly, serious consideration should be given to the overall shape of pycnogonids. This is indicated by the newer appreciation of the significance of body shapes of many arthropods, e.g. those of the Diplopoda and Chilopoda and of their component orders (see below, §5D, Chapter 8, §§3 and 5). We should take as meaningful the superficial similarity between the shapes of pycnogonids, the crustacean Caprellidae† (Pl. 3(g)) also living and feeding similarly on hydroids, and the Devonian fossils such as *Palaeoisopus*. All have very long flexible claws. We do not know for certain how the large extant deep-sea pycnogonids feed but there are plenty of large sedentary hydroids, pennatulids, holaxonian Alcyonaria with slender branches, much as in the small shallow water *Isis*, and sponges upon which the large deep-sea pycnogonids may feed.

It is possible that an acquired small body and long slender legs may confer a camouflage effect, as does the shape of stick insects, and that a gentle swaying movement, leaving the tarsi in place, is of importance. The proboscis needs to move only a little in order to suck up one polyp after another while the body remains firmly adhering from the claws. The leg-jointing admirably suits the production of a swaying movement (Chapter 10, §9, Fig. 10.9), useful both as a camouflage device, as in many insects, and for facilitating feeding. It is

†A *Caprella* once said to a clam
What a jolly fine fellow I am
I have gills in my middle and legs I can twiddle
And my tail is reduced to a sham.

probable that pycnogonid leg structure is associated more directly with swimming, climbing, and swaying than with walking. Other animals living an exposed life are in need of protection against predators. The polychaete *Spinther* is similarly exposed when adhering by its long, hooked neuropodial acicula to a living sponge, its dorsal surface being in effect a 'hedgehog' of spines (Figs. 4.22, 4.23).

Thirdly, it is possible that an emphasis on suction and the feeding on stationary prey, a most unusual feature, resulted in the anterior morphology of pycnogonids and of the Devonian *Palaeoisopus* and others. A large mobile proboscis, anterior elongation and the modification of the anterior legs in modern pycnogonids, so that they no longer resemble closely those of the marine chelicerates or arachnids, is useful in picking off polyps, the rest of the body swaying and moving but little. The convergent similarity between these animals and Caprellidae is probably more significant than we yet appreciate; *Caprella* has even reduced the abdomen to a vestige (Pl. 3(g)).

Pedipalpal chewing in arachnids is accompanied by external digestion, digestive juice flowing from the mouth over the food being chewed. How the aquatic scorpions fed we do not know, but external digestion is not a suitable mode of feeding in the water where dissipation of digestive juices must take place. When soft food, such as hydroid or other polyps, can be sucked into the mouth by normal oesophageal suction, it is but a small step for more elaborate suctorial morphology to have been evolved. It is possible that the success of this mode of feeding led to changes in the chelicerae and pedipalps of some archi-arachnid ancestors. Since pycnogonids exist with five and six instead of the normal four pairs of walking legs, there would appear to be little difference between this and the presence of additional ovigers between the palps and first walking legs in at least the males of many species.

Reduction in the diameter of the pycnogonid body and shortening of the posterior region behind the legs has been accompanied by unusual outpushing of the viscera into the legs, but this occurs also in the carapace of some crustaceans. Probably the large surface/volume ratio and slow movements account for the absense of both excretory and respiratory organs.

What is known of pycnogonid embryology does not help us. However, that of the arachnids shows that a scorpion-like ontogeny could not have given those of other arachnid orders but all might have come from an ontogeny not unlike that of *Limulus*. How the eurypterid ontogeny compared with that of the xiphosurids we do not know.

It is noteworthy that land arachnids existed in the Lower Devonian, e.g. *Alkenia* (Størmer 1970), at the same time as scorpions were still aquatic and gill-breathing. This implies that several groups of arachnids had already become distinct from one another before climbing on to the land. If this was so, then the derivation of the pycnogonids from one such aquatic group of arachnids might reasonably have occurred.

Thus a consideration of the functional significance of body shape and leg-joint construction in pycnogonids, together with what little is known of arachnid diversification while still in the water, shows that a reasonable case can be made for pycnogonid-arachnid affinity (see also Chapter 10, §10).

(vi) *Crustacean evolution*. Reference to crustacean evolution has already been made in Chapter 1, §1E and Chapter 4, §§3A(v), 3B, and 3D. Within the Crustacea we are gaining considerable and very detailed information concerning the correlations between structure, function, habits, and general ecology in certain of the large crustacean groups, shown, for example, by the work of Fryer on Cladocera (1968, 1974) which is too complex to be summarized here (see Chapter 4, §§3A(v), 3B, and 3D). We have no direct evidence of the relationships between the five major groups, the Branchiopoda, Ostracoda, Copepoda, Cirripedia, Malacostraca, and the smaller taxa, the Mystacocarida and Cephalocarida (see Chapter 1, §1E). The latter in many respects probably is the most primitive group of Crustacea living today, but so many limbless posterior segments can hardly be a primitive feature (Fig. 1.15(a)). The limbless segments are very flexible (Fig. 1.15(b)) and the even width of the whole body is reminiscent of small bottom-living harpacticid copepods. It is possible that living in flocculent deposits presents a physical aspect in common with the interior of cow-pats, where the snake-like larvae of the fly *Anisopus* live (Pl. 8(k)). Here the smooth body, without limbs, progresses by snake-like undulations passing along it. Each backwardly progressing convexity presses against the resistant medium and so forces the body forwards. The even proportions and flexibility of both the cephalocarid and harpacticid crustaceans may be of locomotory significance in respect of the medium

Broad subdivisions of the Arthropoda

and should not be regarded necessarily as primitive crustacean features.

Some of these crustacean groups are represented by fossils in the Lower Cambrian, such as types of ostracods and phyllocarids of uncertain affinity, but, as noted in Chapter 1, some are too advanced or similar to modern groups, in these and in later deposits, to give us direct information (see discussion in Tiegs and Manton 1958). There were no copepods, cirripedes, or branchiopods until much later. The presence of a nauplius larva throughout the modern Crustacea (Fig. 1.15(f), (g)) suggests a fundamental unity within the group.

The naupliar limbs are still used for swimming in the adult Mystacocarida (1 mm long); many adult Branchiopoda, e.g. Cladocera and presumably *Lepidocaris* (Fig. 6.7), swim with the antennae 2; and copepods, when swimming slowly, use the endopods of antennae 2, mandibles, and maxillae 1 (Pl. 3(b), (c)). The nauplius of *Hutchinsoniella*, and its subsequent development by a series of larval stages, makes it not difficult to envisage a differentiation of the Branchiopoda, Copepoda, and Malacostraca from some common stem (Sanders 1963), with some simple grade of organization leading to the more specialized classes, the Ostracoda, Cirripedia, and Branchiura. The wonderful preservation of *Lepidocaris*, 3 mm (Fig. 6.7), from the Devonian Rhynie Chert, in which the fragments of the animal can be studied directly, using immersion oil high-power lenses on the transparent chert, reveals another animal of branchiopodan affinities with swimming antennae 2, simple biramous trunk limbs posteriorly and those in front bearing filtratory gnathobases and smaller distal endites (Fig. 6.7, Cannon 1933, and Scourfield 1926). But apart from the simple thoracic endites, although the gnathobasic setae are complex (see also Fig. 4.16), *Lepidocaris* merely illuminates evolution within the Branchiopoda but not the origin of the group.

There is the Devonian *Cheloniellon* (Fig. 1.8), preserved in considerable detail, with perhaps two pairs of antennae, no mandibles and four succeeding pairs of gnathobases, but the trunk limbs appear to be trilobite-like and this animal does not really help us with the origin of Crustacea. In any case *Cheloniellon* is too late in geological time. There are many other incompletely known fossils such as *Waptia*, but as far as we know lacking convincing crustacean affinities.

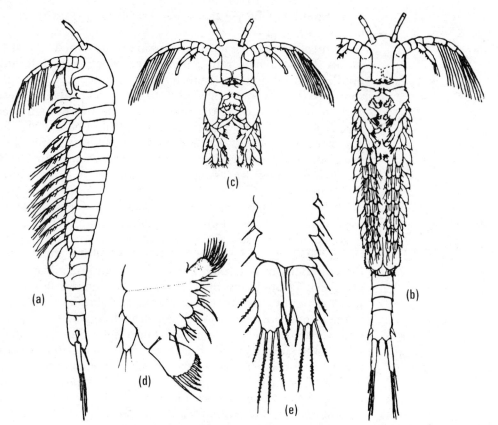

FIG. 6.7. Reconstructions of the branchiopod crustacean *Lepidocaris rhyniensis* Scourfield, 3 mm, Devonian. (a) female, from the side; (b) female, from below; (c) head of male, ventral view; (d) one of the first pair of trunk limbs; (e) one of the posterior (7th to 11th) pairs of limbs. (After Scourfield 1926, from Tiegs and Manton 1958.)

It has been suggested by Cannon, Fryer, and others that ancestral Crustacea must have been small and bottom-living and later gained the ability to swim actively, filter feed, or become powerful predators of large size, such as the largest crab *Macrocheira* with a leg span of 3·5 m. From small animals may have arisen the various adaptive radiations shown by branchiopods, copepods, and malacostracans.

Examples of filter-feeding and of predatory or scavenging malacostracans have been mentioned in previous chapters and it may be suggested now that the differentiation of the diagnostic malacostracan features, an eight-segmented thorax and an abdomen (consisting of seven segments in

Leptostraca, partially fused 6th and 7th segments in *Lophogaster* and *Gnathophausia*, and entirely fused to an apparent 6th segment in most other malacostracans (Manton 1928a, b)) could only have occurred during a bottom-living existence of small archi-malacostraca. This could have resulted in the walking legs (endopods) of a thorax becoming differentiated from more feeble abdominal limbs behind, so permitting a fanning-out of the fields of movement of the thoracic legs as leg length increased, giving longer strides and better walking ability (Chapter 5, §3, Fig. 5.2). Today there is further fanning-out, seven of the eight thoracic legs of an isopod are used for walking, only four of the eight in the crab and lobster while some spider crabs use only three pairs, thereby avoiding mechanical interference of long legs during forward progression. It is probable that malacostracan abdominal segments continued to bear small limbs because the Malacostraca took to a swimming as well as a bottom-living habit fairly early. Then, with the improvements in swimming capacity by the formation of pleopods, tail fan, swimming thoracic exopods, and a large exite on antenna 2, filter feeding in the water above the substratum would have become practicable. The metachronal swimming movements of the abdominal pleopods of *Anaspides* is well seen in Pl. 3(f). Filter feeding, as suggested above, probably evolved independently in the larger subdivisions of the Malacostraca.

The Phyllocarida are an imperfectly known, heterogeneous group of arthropods well represented in the Cambrian. They include the earliest known Malacostraca. Most phyllocarids lacked a crustacean type of head, but a caudal furca and bivalved carapace were usual. *Canadaspis* (Briggs 1976, in press) was a near leptostracan, but much larger than the modern *Nebalia,* possessing a crustacean type of head, 8 biramous flattened thoracic limbs, and a 7-segmented abdomen lacking limbs and furca. The maxillule carried median armature and a maxilla was differentiated from the thoracic limbs by lesser length. This small degree of maxillary differentiation is much as in *Hutchinsoniella* (Fig. 4.7) where, alone among living crustaceans, the maxilla is undifferentiated from the thoracic limbs. Thus crustacean taxa must have been diversified by the Lower Cambrian.

A consideration of the evolution of the Uniramia forms the latter part of this chapter, but it is fitting first to consider various types of parallel evolution within the Arthropoda.

Broad subdivisions of the Arthropoda

3. Convergent similarities among arthropods

Very many examples of convergent evolution of similar structures exist among the arthropods. Some have been mentioned above in paragraphs 1 and 2E(v). Consideration here will be limited to transversely biting mandibles, to entognathy, and to organs of sight.

A. *Transversely biting mandibles*

Examples have been given of striking convergent similarities between advanced types of mandible which can bite strongly in the transverse plane (Chapter 3). The means of obtaining a strongly cutting mandible, capable of dealing with hard food, are various in the Crustacea and Hexapoda and have been independently evolved.

B. *Entognathy*

The converse type of advance also occurs, namely the opposite to very strongly hinged biting organs. A weakly attached mandible capable of protrusion is useful for piercing and other purposes. The molar processes of such mandibles can shred food material very finely, as demonstrated so beautifully by Goto (1972). A very weak, flexible junction between the mandible and the head needs protection and this is done by the overgrowth of the mandible by folds from the head on either side to give the condition known as entognathy (see onychophoran and collembolan ontogeny shown in Fig. 6.8 and Chapter 3).

Entognathy is not a taxonomic character indicating affinity, as has been claimed, but it is a convergent feature turning up in widely different taxa and always associated with protrusible and piercing mandibles capable of varied movements. The many occurrences of entognathous protrusible mandibles within the Uniramia is shown in Figs. 6.5, 6.8, 6.9. The whole-limb mandibles characteristic of the Uniramia

---

FIG. 6.8 (see opposite). The form and development of a gnathal pouch in arthropods. (a) and (b) ventral views of the oral region of *Peripatopsis sedgwicki* Purc. with the round lips closed over the jaws and preoral cavity in (a) and with the lip open, as it is when pressed on to the surface of prey, in (b). The mouth lies behind the labrum, and the jaw blades, two to each jaw, slice from before backwards at the sides of the mouth. (After Manton 1937.) (c) and (d) Oral views of embryos of *Peripatus edwardsii* Blanch., the younger is (c), redrawn from Kennel (1886). Paired oral folds lie at the side of the jaws and then unite with each other behind them (d) enclosing the preoral cavity. Later the folds unite in front of the labrum. (e), (f), and (g) oral views of embryos of the collembolan *Anurida maritima* Laboulbene, the younger in (e) redrawn from Folsom (1900). The oral folds at first (e) lie lateral to both mandibles and maxillae 1 and unite later (f) with the labrum and labium (maxillae 2) to form the gnathal pouch. (g) lateral view of the same stage shown in (f). (From Manton 1964.)

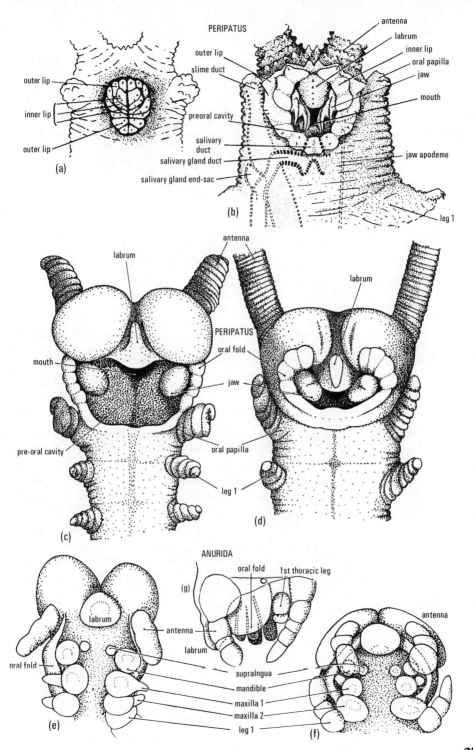

Broad subdivisions of the Arthropoda

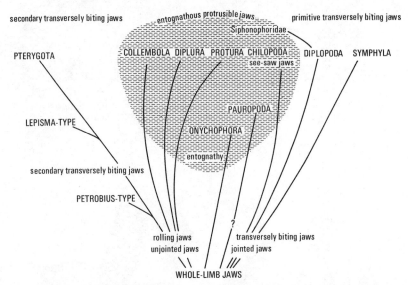

Fig. 6.9. Diagram showing the conclusions reached concerning the interrelationships and evolution of the jaw mechanisms of Onychophora, Myriapoda, the apterygote classes, and the Pterygota. The shaded area indicates convergent evolution of entognathy and protrusible mandibles. The Pauropoda possess an unsegmented mandible, but there is no evidence as to whether this has been derived from a primitively segmented or unsegmented mandible. (From Manton 1964.)

remain unjointed (Fig. 6.5 on the left) and become jointed on the right. Primitive transversely biting mandibles persist in Diplopoda and Symphyla (on the right); secondary transverse biting (on the left) has been evolved from primitive rolling, grinding mandibles in the Lepismatidae and Pterygota. The middle shaded region (Figs. 6.5, 6.9) embraces entognathous protrusible mandibles, independently arising in Collembola, Diplura, Protura, Onychophora, Pauropoda, Chilopoda, and the Siphonophoridae among the Diplopoda. The structural details whereby entognathous mandibles, and sometimes maxillae 1 also, have been achieved are different in every group shown. Protrusible entognathous mandibles have clearly been convergently acquired.

The Crustacea also show many independent evolutions of piercing protrusible mandibles in copepods, isopods, and Branchiura. Some possible intermediate stages towards entognathy are illustrated by the longitudinal halves of two

Fig. 6.10 (see opposite). Longitudinal halves of the head end of two copepods (a) *Amphiascus* and (b) *Idya* viewed from the median plane. These copepods show possible stages in the evolution of crustacean entognathy. For further description see text.

Broad subdivisions of the Arthropoda

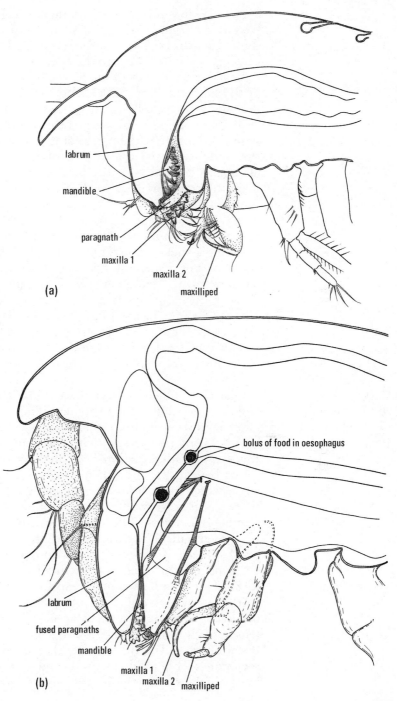

273

copepods viewed from the middle line in Fig. 6.10. Both are free-living feeders from the surfaces of weed, stones, and sometimes the outside of larger animals. In *Amphiascus* (Fig. 6.10(a)) the labrum overhangs the cusped tips of the mandibles; the distal ends of the paired paragnaths meet the edge of the labrum, almost enclosing the cusped tips; and close behind lie the strong maxilla 1 cusps followed by maxilla 2 and the maxilliped with long, strong, scraping distal tips. In *Idya* (Fig. 6.10(b)), the paragnaths have fused together along the middle line so that a sort of trunk is formed by the elongated labrum and fused paragnaths. The cusps of the mandible and maxilla 1 work at the distal end of this trunk; the former inside it and the latter outside; maxillae 2 and maxillipeds both end in single strong scraping tips. The long oesophagus can now exert greater suction, as a bolus of fine material is sucked up and passed to the alimentary canal. It could be but a small step further for a copepod to fuse the labrum and the paragnaths together and enclose the mandibles reduced to stylets, as in many ectoparasitic copepods. The stylets may be elaborate, curved or straight, serrated or otherwise; some even possess transverse jointing. But presumably all are derivatives of the crustacean gnathobase, the mandibular palp having disappeared long ago.

C. *Organs of sight*
(by Professor G. A. Horridge, F.R.S.)

We can distinguish between the action of an eye to detect light and its possession of a field of view. There are only three ways in principle that are used by arthropods to achieve directional vision, (a) by an eye-spot with screening pigment; basically this can be called the pin-hole method; (b) by a lens with a retina beneath it; and (c) by a receptor having a long, thin shape which in the limit acts as a light guide. Therefore where these features are found they cannot be used as arguments of homology between different groups: they are more likely to be convergent on account of the physical limitations (Fig. 6.11). Screened eye-spots occur in many larval or nauplius eyes and in adults, as in Collembola. A small retina in the focal plane of a lens confers excellent vision of the outside world (often over only a small field) in spiders, some insect ocelli, and in some lower crustaceans. A light guide without a lens confers directional vision in some larval and adult (e.g. termite) eyes. In the best-developed compound eyes the receptor is a light guide with its end at or near the focus of a lens, which narrows the field of view and increases sensitivity.

Most crustaceans and insects have the last-mentioned system.

Having ruled out factors where optics are significant, we are left with the arrangements of cells and tissues as the characters of eyes on which to base homologies. At this level we have another caution to enunciate. Histological details are incredibly similar over the whole animal kingdom, suggesting that a limited number of genes, enzymes, and protein-building blocks are available. Comparative histology has never been a subject with an evolutionary aspect because cilia, synapses, myofibrils, and the rest of the components at this level seem to be, by and large, the only organelles available.

The following points bear upon the question of converging evolution of compound eyes in different groups of arthropods.

Given a simple ocellus, there is no structural difficulty in making the field of view narrow. If this process happens there is then a strong functional advantage in bringing ocelli together in large numbers. The process is partially complete in several existing forms, e.g. larvae of ant-lions. Simultaneous evolution of mechanisms of correlation between ocelli would be essential, so the process of evolving a compound eye is a natural progression although at first slow.

The simple eyes of Onychophora, spiders, scorpions, *Limulus* dorsal eye, adult insect ocelli, myriapod eyes, and similar eyes of minor groups in and around the arthropods are not one family of structures. Eyes have apparently evolved many times as they have in annelids, molluscs, and coelenterates: it is not a difficult matter when the sensory cells have peripheral cell bodies anyway. Reviews of this scattered topic will be found in Plate (1924) and Bullock and Horridge (1963).

Concerning possible homologies between ocelli of various kinds and their possible modifications, such as frontal organs, in different classes of arthropods, not sufficient is known about their central connections to make conclusions about their relationships. The whole literature on this difficult field is mentioned by Paulus (1972) who hesitates to make any significant conclusions about the phylogeny of the arthropod groups.

Compound eyes appear with the earliest crustaceans and with the earliest insects; they also occur among the most

Broad subdivisions of the Arthropoda

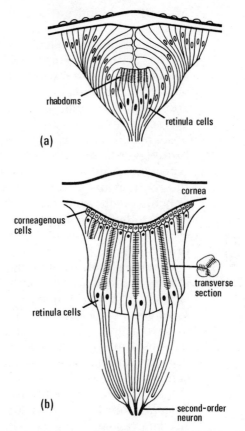

FIG. 6.11. The two fundamental types of simple eye which provide a representation of the outside world. (a) Larval eye (stemma) of the water beetle *Dytiscus*, showing a specialized region of cornea with lens and group of receptor cells. (After Grenacher 1879.) (b) Ocellus of dragonfly with separate layer of clear corneagenous cells and arrangement of receptors in groups like ommatidia, pointing in different directions (after Ruck and Edwards 1964). We can argue as follows from the structure. In (a) the receptors form a shallow layer that could be in the focal plane of the lens. In (b) the long rhabdom columns presumably act as light guides and therefore each has a narrow field of view looking in its own direction. Each type, therefore, is able by a different method to form a composite picture of the outside world. The same two mechanisms are found also in compound eyes of different types. (After Horridge.)

primitive living members of both groups. Compound eyes were well-developed in trilobites, where the lens of calcite crystals in each facet was apparently so aligned that the double refraction was eliminated. Together with the large number of facets, this suggests that fields of view were narrow.

Limulus has a compound eye which resembles that in

crustaceans and insects only in being compound. The units of cells are quite unlike the ommatidia of the other compound eyes. They differ in the components of the optical system, in pattern and number of receptor cells, in having an eccentric cell that has no receptor properties, and in distribution of screening and epithelial cells. Moreover, an efferent nerve supply to the receptor cells has recently been found (Fahrenbach 1974). If *Limulus* is our nearest relative of the trilobites then nothing can be inferred from their compound eyes in relation to the compound eyes of insects and crustaceans.

One simplifying factor is that the compound eyes of insects and crustaceans are composed of units (the ommatidia) which are basically the same in the following ways:

(1) In cell components. Per ommatidium there are four cells forming the crystalline cone; two cells which start life against the cornea and become the two principal pigment cells in insects; about eight primary sensory photoreceptor neurons which are sister cells from three divisions of one founder cell; and finally a group of accessory pigment cells which are related to neuroglia.

(2) In fine structural detail and physiological mechanisms. Although compound eyes of both groups, especially insects, are varied, they are basically organized in similar ways. Even the arrangement of the rhabdom as a stack of crossed plates, characteristic of Crustacea, is found in some insects of widely scattered orders, although perhaps this is also a functional convergence.

(3) In biochemistry of eye screening pigments. Apparently diversity is as great within the separate groups as between them. Pteridines predominate in Crustacea, ommachromes in insects. A great deal more biochemical information is needed among the primitive groups before significant conclusions can be reached.

(4) The basic arrangement of neurons. What little is known of neural interaction behind the compound eye suggests that insects and crustaceans are closely similar. In particular, the receptor axons, synapses of the lamina, neurons of the lamina and medulla, and arrangement of axons in regular chiasmata between neuropiles, is remarkably similar. *Limulus* is clearly rather different, arachnids are very different. One major contrast is that the crustacean compound eye, even in the most primitive living examples, is usually, but not always, stalked. In structure, physiology, and

function, this stalk has no relation at all to a limb. Eye-stalk muscles are innervated from the protocerebrum. If the crustacean eye-stalk had any relation to a limb, antenna 1, antenna 2, or any of the other appendages would surely have proprioceptive sense organs, and it does not in any case known (but only the most advanced have been examined). The function of the eye-stalk is not understood in the lower Crustacea. In advanced Crustacea the eye-stalk movements partially compensate for body tilt, and for displacement of the visual world as the animal moves. The only known function of the movable eye-stalk is to sharpen up vision by eye tremor (again studied only in decapods and *Daphnia*).

Another major difference, physiological this time, is that crustacean eye pigments are hormonally controlled whereas those of insects are controlled by the direct action of light in each ommatidium individually. *Limulus* alone has efferent nerves to the receptors. Only the advanced members of the two groups have been studied.

All of the above, in my opinion, can well be explained by independent evolution of the compound eye in crustaceans and insects. A possible requirement to explain compound eyes by convergence might be that the distant ancestors could have had similar simple eye structure with a similar chain of interneurons behind. Their simple eyes would then have to be basically present-day ommatidia. The convergence theory requires that a common tendency of ommatidia to be multiplied materializes similarly in the separate groups. Similarly the neuron chain beneath the eye would have to make the same changes towards similar optic-lobe structure in their reduplication. We are then left with the problem of explaining the great diversity of simple eyes that are found today, and the fact that modern simple eyes always have a much simpler neuron chain behind them than compound eyes. Moreover, their central connections, where known, are unlike the central connections of compound eyes. Therefore compound eyes in crustaceans and insects must have evolved from simple eyes that are found nowhere today. But the evidence from the fields of functional morphology and embryology are against common ancestry of crustaceans and insects. Or else the common ancestor of these two groups had compound eyes for which there is no evidence. Nor is there evidence supporting a supposed loss of compound eyes in many crustacean orders.

4. Grades of uniramian evolution: Monognatha, Dignatha, and Trignatha, and their phylogenetic groupings

The word *taxon* is used to define, roughly, any unit of classification used by systematists which embraces animals believed to have phylogenetic affinity. But terms such as phylum, sub-phylum, class, family, order, and species (and there are many more), indicate a hierarchy in which the higher category embraces a number of categories of the next group, i.e. one phylum may contain several sub-phyla and one sub-phylum may contain several classes. Phyla may be entirely unrelated and usually carry this significance, while the subdivisions of one phylum, if correctly assessed, should represent the phylogenies of the animals concerned. This is sometimes a counsel of perfection, which increasing knowledge is striving to obtain. The strong evidence based on functional morphology which is being discussed in this book, suggests the recognition of at least three independent phyla, the Chelicerata, the Crustacea, and the Uniramia, instead of the single monophyletic phylum of Arthropoda. Unassessed are many different types of early arthropods besides the trilobites.

The evolution of the phylum Uniramia may now be considered in further detail. Besides the taxa mentioned above which stand on a phylogenetic basis, we also have to recognize unrelated groups showing grades of organization which have been reached independently. The idea was first elaborated for the vertebrates. The mammal-like grade has been attained or approached several times independently by separate groups of reptiles. The question has been considered: are the reptiles themselves a grade and not a phylum? Have they evolved independently more than once from the Amphibia or from the forebears of the Amphibia? Consensus of opinion on the fossil record probably puts them into one phylum, but only probably.

At one time it was suggested that the Monognatha, Dignatha, and Trignatha (Fig. 6.12) represent phylogenetic stages of evolution within the Uniramia based upon the postoral composition of the head. This can no longer be upheld.

One can describe the Onychophora as monognathan, possessing one pair of feeding limbs behind the mouth (Fig. 3.15). The Myriapoda are basically dignathan, with mandibles and maxillae (a gnathochilarium in the Diplopoda), but possessing varied posterior additions to the head. Part of the collum segment is added to the gnathochilarium in the

Broad subdivisions of the Arthropoda

Diplopoda (Dohle 1964); the entire leg-like second maxilla segment is added in the Chilopoda and similarly the maxilla 2 segment of Symphyla, but the appended symphylan limbs are not like the pterygote insect's labium in their detail. The Chilopoda have all but incorporated the poison claw segment into the head. The Pauropoda are simply dignathan. The five hexapod classes (the Pterygota and the four wingless apterygote classes) all have a common trignathan head type with three pairs of feeding limbs behind the mouth, the mandibles, maxillae 1, and the maxillae 2 fuse to form a labium.

It is not possible to imagine one of these three head types ever having given rise to another (see next section); there is no reason to suppose that the trignathan hexapod head ever went through a dignathan state, such as seen in myriapods, and neither the dignathan nor trignathan heads need have gone through a monognathan state. All three probably evolved independently along with different manners of jaw evolution from a whole-limb (Chapter 3, and further in Manton 1964, 1972). The superficial trignathy of the Symphyla does not mask the fundamental difference between this class, with jointed, transversely biting mandibles, and swinging single pair of tentorial apodemes; and the trignathan hexapods, with unjointed, basically rolling mandibles and two pairs of rigid

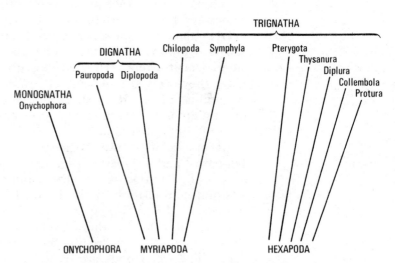

Fig. 6.12. Diagram showing the probable relationships of the component taxa of the Uniramia, as indicated by the converging lines, together with the grades of organization, Monognatha, Dignatha, and Trignatha, which do not indicate taxonomic relationship. (From Manton 1973a.)

tentorial apodemes. The terms Monognatha, Dignatha, and Trignatha indicate no more than grades of organization, not taxa. They may be useful descriptive terms, but do not mask the reality of the three uniramian subphyla, the Onychophora, Myriapoda, and Hexapoda. The Symphyla are clearly myriapodan and in no way represent a type of arthropod leading towards the hexapods.

5. The evolution of the Uniramia: Onychophora, Myriapoda, and Hexapoda

A. *Heads and limbs*

The morphological similarities between all the Uniramia are: the unbranched uniramous limbs (cf. biramous limbs of the primarily aquatic arthropods and their derivatives on land, the Arachnida); and the mandibles, formed from a whole-limb which bites with the tip and not with the base. The Myriapoda all possess a sclerotized cuticular head capsule, absent in Onychophora, and of similar segmental preoral composition to that of the Hexapoda. The myriapodan head is basically dignathan, with various posterior additions in some classes (see above §4). The mandibles are jointed and transversely biting. All possess a swinging pair of anterior tentorial apodemes which abduct the mandibles. The Hexapoda also possess a sclerotized head capsule, but differing in detail from that of the Myriapoda, although with a similar segmental preoral composition. The Hexapoda have para-oral unjointed, primitively rolling mandibles, followed by maxillae 1 and 2, the latter fused to form a labium with different structure and muscular connections and support from that of the Symphyla. Two pairs of tentorial apodemes provide the head endoskeleton, but they are rigid, usually fused and never cause mandibular abduction (Chapter 3).

B. *Leg-bases in ancestral Uniramia*

It has been shown above (Chapter 5) how the leg-bases of the Myriapoda have a common type of coxa–body joint and essentially similar leg-rocking mechanism in Chilopoda, Symphyla, and Pauropoda and a built-in leg-rocking mechanism in the Diplopoda associated with the strength required from their legs. The five hexapod classes differ from the Myriapoda and from one another in their leg-base mechanisms and leg-rocking mechanisms used in stepping (Chapter 5). These morphological arrangements are so different and mutually exclusive that one can only suppose that they each arose independently from ancestors with little sclerotized trunks and lobopodial limbs. This implies the existence of multi-legged ancestral Uniramia of at least three kinds,

Broad subdivisions of the Arthropoda

some without head capsules which early organized the simplest of jaws on the second body segment and led to the Onychophora, and two types of multi-legged Uniramia with head capsules of different sorts on which different types of mandibles were differentiated on the fourth head segment. One led to the Myriapoda and the other to the Hexapoda. Since there are mechanical advantages in hexapody (Chapter 5, §3) and since the running on three pairs of legs has been independently evolved in arthropods as remote as some prawns, crabs, and arachnids, it is not an unreasonable suggestion that hexapody probably arose independently five times from multi-legged, soft-bodied ancestors which each elaborated the leg-base mechanism in a different manner, as they independently evolved trunk and limb sclerotization and joints. In other words there never can have been an animal such as one type of ancestral hexapod or insect. These conclusions, based on functional morphology, are summarized in the lower three rows of rectangles in Fig. 6.14.

C. *Series of gaits of the Uniramia*

There is another and surprising feature common to all Uniramia and not shared by other arthropodan groups and it is shown both by the lobopodial limbs of the Onychophora and by the more sclerotized, jointed trunk limbs of Myriapoda and Hexapoda. They all walk or run by a series of gaits in which the phase difference between successive legs and the relative durations of the forward and backward strokes by the legs change harmoniously. A simple example of metachronal rhythm used by a series of onychophoran limbs is shown in Fig. 2.4, the forward and backward strokes being of equal duration approximately. A series of gaits used by the centipede *Scolopendra* are shown with similar conventions in Fig. 6.13, the slower gaits, with longer duration of the backstroke, on the left and the faster gaits with shorter duration of the backstroke (i.e. faster movements) on the right. Such diagrams are called 'gait diagrams' while the dorsal views shown below, called 'gait stills', show the disposition of the limbs at various moments in time during the gaits shown above. Gaits will be considered more fully in the next chapter.

Other segmented animals, such as vertebrates, crustaceans, polychaetes, and arachnids, may always use the same gait (e.g. *Ligia*) or employ little variety in the gaits, or use a set of different types of gait, as in vertebrates. Many depend almost exclusively on different pace durations for changing their speeds of progression. They do not show integrated series of gaits, such as those in Fig. 6.13, which have been found in every uniramian group. So much so that it came as a surprise not to find such a series of gaits in the Arachnida.

D. *Differentiation of habits and uniramian evolution*

The further evolution of the Onychophora and Myriapoda beyond the three lower levels of rectangles in Fig. 6.14 has been intimately bound up with habits. After emergence from

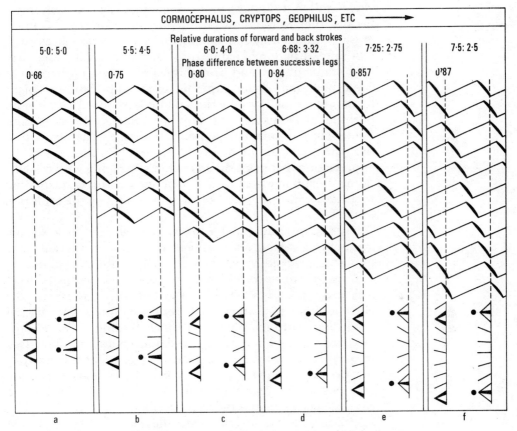

FIG. 6.13. The range of gaits shown by the epimorphic Chilopoda, the slower on the left and the faster on the right. The stepping movements of the series of legs are shown progressing from left to right, the number drawn being sufficient to show a little more than one metachronal wave. The heavy lines denote legs in the propulsive backstroke and the thin lines show the legs in the recovery phase. The lower diagrams show the positions of the legs at the two moments most unlike each other, each being indicated above by the vertical dotted line. The black spots mark legs occupying the common footprints, and their separation represents the length of the stride. (From Manton 1952b.)

the sea, possibly in quiet estuaries leading directly into vegetated damp environments, the little-sclerotized, but soon dry-skinned Uniramia, must have become highly hydrofuge (i.e. unwettable) perhaps as in present-day Onychophora. A first requirement on the land was protection from excessive uptake of fresh water, secondly protection from dessication, and thirdly an ability to find food. That the former is the greater hazard on land is shown by land planarians with wet skins being found, not in damp environments such as woodland, but under stones, etc., in drier surroundings, because they would blow up and burst by osmotic absorption of water which they cannot control in the wetter places.

A differentiation of habits must then have set in. The early inhabitants of the land probably gained shelter wherever it could be found, under large bits of the substratum or plant matter, or by shallow penetration into soft soil with little or no pushing and unaided by any morphological facilitation. Under such circumstances a diversification of habits, within the same environment, could have been initiated. The appreciation that habit diversification, bringing with it facilitating morphology, can occur without any direct adaptation to the environment, is the key to understanding the evolution of the Uniramia.

Early Uniramia capable of finding shelter in a simple manner might have responded in at least four different ways to obstacles standing in their path. The use of exploratory sensory antennae to find narrow crevices or paths of least resistance may have led to the extreme powers of the Onychophora to distort their bodies, without pushing, so penetrating through narrow channels leading to more commodious spaces where larger predators could not follow (see Pl. 4(c)) showing the holes in a card which can be traversed by the *Peripatus* shown). Extreme changes in body shape are only possible to an animal with unstriated muscles and no rigid skeletal elements and the movements must be slow. The Onychophora use an elaborate, flexible connective tissue

FIG. 6.14 (see opposite). A diagrammatic summary of the probable course of evolution of the several groups within the Uniramia as indicated by the evidence of functional morphology and embryology. The inset diagram on the right shows an alternative interpretation of myriapod relationship. It is possible that the Pterygota and Thysanura are both descended from a very early common stage possessing a dorsal coxal articulation with a pleurite. But thereafter the two classes must have evolved independently, with a fixed pleurite in the Pterygota in contrast to the Thysanura with mobile pleurites and distinctive leg structure associated with their peculiar jumping gaits. (From Manton 1973a.)

Broad subdivisions of the Arthropoda

skeleton within the body wall of the trunk and limbs on to which the muscles are inserted. No head capsule is developed, or transversely expanded jaws, or supporting structures other than a longitudinal jaw apodeme (Fig. 3.15(d)).

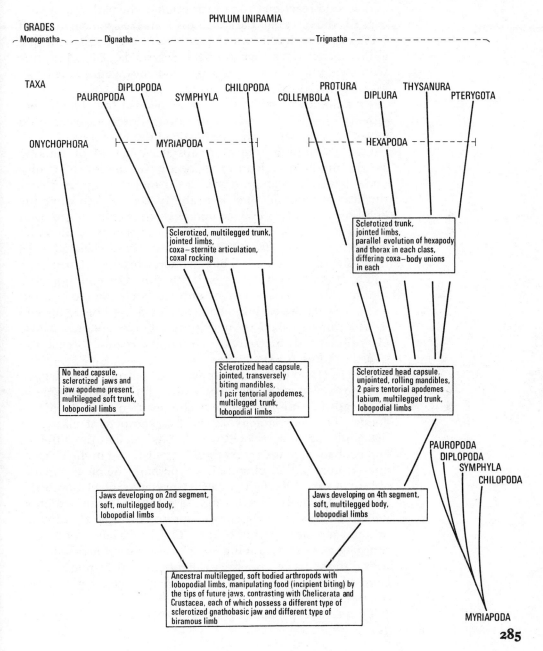

Broad subdivisions of the Arthropoda

The Onychophora, at one time even considered to be intermediate between arthropods and worms, are in fact uniramian in every respect, their morphological peculiarities all being associated with habits.

A second reaction to an obstruction in the path might have evoked the seeking of a route of no resistance, even if tortuous; this could have given rise to the habits, turning ability, gaits, and structure of the Symphyla. A head capsule with swinging anterior tentorial apodemes would now be an asset.

An animal which actively pushed against the obstruction, either with the head end or with antero-dorsal surface, could have led to the structure diagnostic of the Diplopoda.

Lastly, running round an obstacle, instead of walking slowly, might have led to the speedy running of the Chilopoda and their elaborate associated morphology. But centipedes have another and basic ability, that of deforming the body on entering a crevice and then thickening a group of segments, so that a thrust is exerted against the soil, as by earthworms. The morphological features associated with these divergent habits will be considered briefly in Chapters 8 and 9. It is by habit divergencies that the Onychophora and four myriapod classes became differentiated, jointed legs and a sclerotized head capsule of one particular type being formed in the Myriapoda but not in the Onychophora, whose evolutionary advances are in a different direction (see left top quarter of Fig. 6.14).

The hexapod classes presumably arose from multi-legged ancestors, also with a soft body but with a different type of head capsule and postoral head composition from the Myriapoda. The morphology of the five component classes of hexapods is also correlated with habits, such as the different and probably secondary methods of soil-living in the Protura and Diplura. The Collembola are perhaps the most remarkable of hexapod classes in that they preserve and elaborate the use of a haemocoel in forming a jumping escape reaction which has modified the whole anatomy divergently from all other hexapods (Chapter 9). The Thysanura have gone in for jumping gaits and jumping escape reactions of a unique and very different type from the Collembola. The Pterygota have gained speed by the evolution of wings (see right upper quarter of Fig. 6.14).

There were probably many other lines of evolving ter-

Broad subdivisions of the Arthropoda

restrial Uniramia which did not persist to the present day, as suggested in a general way by Sedgwick (1909) in his textbook written for students, as is the present volume. But the terrestrial Uniramia must have had marine ancestors (see Chapter 1) quite different from the Crustacea and Chelicerata and the known early armoured arthropods. The essential clue is provided by the embryological analysis of Anderson (1973) quoted above. Uniramian embryology is utterly remote from that of the Crustacea and Chelicerata (see also Chapter 10, §10). The uniramian theme of development exhibited by the whole phylum can be recognized in the development of certain yolky-egged annelids today. The group Arthropleurida, hitherto known only by the largest terrestrial arthropod, *Arthropleura* (Fig. 1.18) is represented by *Eoarthropleura* from the Lower Devonian (Størmer 1976). These animals could not be close to ancestral uniramians (Manton and Anderson 1977).

6. The haemocoel

The metamerically segmented coelomates destined to give rise to the Uniramia probably already possessed a haemocoel which provided the mechanism for working the lobopodial limbs. The use and enlargement of the haemocoel may even have preceded the formation of good lobopodia. Changes in body shape might have served shallow grubbing into the sea floor and have been an easy way of obtaining protection, as it certainly was for the ancestral Mollusca.

In the Onychophora the haemocoel is essential for the extreme changes in shape of the body and for providing, with the muscles, an isolating mechanism for leg movements (Chapter 4, §5). It has also been shown how shape changes and haemocoelic pressures in Onychophora account for the minute unbranched tracheae, at one time considered to be either a unique or possibly primitive feature, but recently found similarly formed in the centipede *Craterostigmus* which has considerable use for the mechanical attributes of haemocoelic pressures (Chapter 4, §5, Manton 1965, §8(xiii)).

As will be shown later (Chapter 9, §2), the haemocoel is essential for the jumping movements of Collembola and for providing rigidity of body against which muscles can work in burrowing centipedes (Chapter 8, §5B) and in all other arthropods with considerable, but useful, areas of flexible body wall. With the progressive use of the cuticle in providing apodemes, sclerites (see Chapter 3, §2E), and

Broad subdivisions of the Arthropoda

tough flexible portions of the body wall bearing muscle insertions, the mechanical use of the haemocoel has waned, as noted at the end of Chapter 4 (§6E). This use still persists in causing leg extension at leg joints lacking extensor muscles, when little or no heavy work is being done, as on the recovery forward swing of a leg off the ground; and in jumping when legs of a pair work in unison. But always the physiological functions of the haemocoel persist in transporting food round the body, in osmoregulation and excretion. But the glory of the arthropodan haemocoel would appear to belong to the distant past, with *Peripatus* and the burrowing centipedes today serving to stir our imaginations over matters upon which the fossil record will be for ever dumb.

7. Other metamerically segmented invertebrates

The larger groups of extinct and living arthropods have been considered above, but it should be remembered that there are other animals which qualify for inclusion in the old monophylectic Arthropoda, such as the Tardigrada.

A. *The Tardigrada*

It is difficult to assess the status of the Tardigrada because their minute size (Chapter 1, §1H) is associated with very simple organization, as is to be expected in animals so small (Figs. 1.19, 6.15). The lobopodial limbs are moved slowly, much as in Onychophora, the range of swing often being only from a transverse to a posteriorly directed position. The musculature is simple, consisting of widely separated muscle strands (Figs. 1.19, 6.15(c), (d)) as suits a minute animal. The terminal claws on each leg may be only two, as in Onychophora, or there may be more (Fig. 6.15) but they do not form a circlet as claimed for the fossil *Aysheaia*. The terminal mouth region possesses no entognathous jaws, but an elaborate eversible proboscis is present. There seems to be no valid reason for excluding these animals from the Uniramia, because small size could account for all their simplicity of basically uniramian structure (Fig. 6.1). Ramazzotti (1972) considers the Tardigrada to represent a separate phylum but gives no adequate reasons. The resemblances between tardigrades and *Aysheaia* in a terminal suctorial mouth and simple clawed limbs without a 'foot' may be indicative of affinity.

FIG. 6.15 (see opposite). The Tardigrada (see also Fig. 1.19). (a) Ventral view of *Bryodelphax parvulus* to show the general organization. (b) Terminal claws of a leg of *Macrobiotus*. (c, d) Plan of the musculature of *Echiniscus*, (c) in dorsal view, (d) in ventral view. (After Ramazzotti 1972.)

Broad subdivisions of the Arthropoda

B. Opabinia
regalis *Walcott*

It seems fitting to close this chapter by reference to the latest reconstructions of one of the most interesting animals preserved in the Burgess Shale (Fig. 6.16). This animal emphasizes how much we still have to learn about the past history of metamerically segmented invertebrates. *Opabinia* has no antennae, no jointed limbs, and the series of gills situated dorsal to each segmental lateral lobe of the body is not trilobite-like. There are no chaetae or setae. Whittington (1975) considers that *Opabinia* is neither an annelid nor a trilobitomorph arthropod.

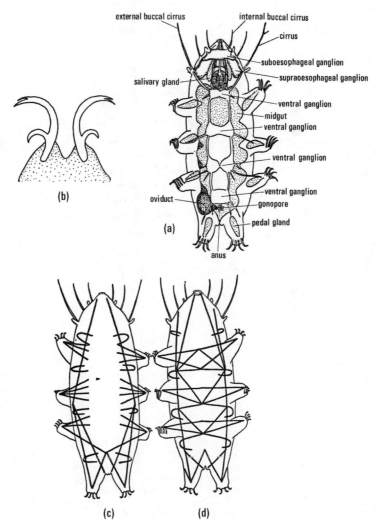

Broad subdivisions of the Arthropoda

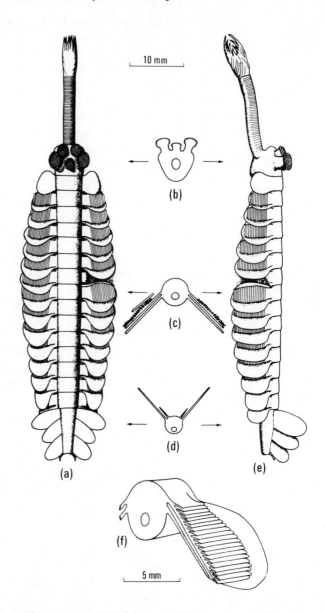

FIG. 6.16. Reconstruction of *Opabinia regalis* Walcott, 1912. (a) Dorsal view, right lateral lobe and gill of segment 7 cut off proximally to show lateral lobe and gill of segment 8. (b), (c), (d) Cross-sections of body at the levels indicated by the arrows, position of alimentary canal indicated. (e) Left lateral view, left lateral lobe and gill of segment 7 cut off proximally to show lateral lobe and gill of segment 8. (f) Oblique view of one segment of the trunk showing left lateral lobe and gill cut through the anterior end; the cross-sectional area is stippled. (After Whittington 1975*b*.)

Broad subdivisions of the Arthropoda

Opabinia was probably benthic in habit; the lateral lobes on each segment may have propelled the animal either along the surface of the substratum or just below it. The upwardly projecting eyes would be suitable to detect changes in light intensity, indicating the close presence of prey or of predators from which escape might be made by sinking into a soft substratum. A long anterior prehensile organ presumably caught soft food and brought it to the mouth, but there is no evidence suggesting that it was a proboscis, as claimed by Sharov (1966), nor any evidence in support of Sharov's claims concerning the ancestral arthropodan status of this legless beast. The lateral processes must have had some rigidity because they are preserved in an identical series. The posterior end of the body shows the lateral processes to be dorso-laterally directed. Possibly the animal ploughed shallowly in the bottom mud, the posterior processes assisting or causing a flow of water over the gills. But however *Opabinia* may have lived, we have clear evidence of a type of animal which, at the present day, we know nothing about. Presumably there were other invertebrates, now extinct, whose morphology is as unique, but they were not suited to meet the pressure of the environment and to survive.

8. The origin of the Arthropoda

It is the policy of this book to stick to the facts, structural, functional, and embryological, and to simple deductions therefrom. The origin of arthropods is not considered in one place, but the relevant evidence, in so far as it goes, is presented throughout the book. It should be mentioned here that the arguments given by Sharov in his *Basic Arthropodan Stock* (1966) are entirely opposed to the facts, except for the chapter on his own excellent work on the fossil Monura. Several errors of fact are present on most pages; his tables are untrue; the corner-stones of his phylogenetic fancies rest (1) on the polychaete *Spinther* being a radially symmetrical derivative of the Coelenterata with lobopodial legs, which it is not (see Chapter 4 and Manton 1967), and thus cannot be ancestral to the Arthropoda, and (2) on *Opabinia*, shown by Whittington to be legless and no arthropod (see above), with a prehensile organ which is not a proboscis. If this masterpiece of misinformation is to be left on library shelves of zoology departments it should be bound up with Anderson's rebuttal of this work (1966).

7 Locomotory mechanisms of the Uniramia

THE probable early differentiation of habits of the Uniramia when they first invaded the land (considered in Chapter 6, §5D) must have brought with it special needs; such as those of speedier running or stronger leg movements serviceable for active burrowing; or changes in body structure facilitating the maintenance of trunk rigidity, or of flexibility; or of distortability; and many other features as needed by the various taxa during their differentiation one from another. The diversification of locomotory mechanisms of evolving uniramian groups went hand in hand with the different advances in structure shown by the various taxa. Each has been dependent upon the other. The locomotory movements are considered in this chapter and the facilitating morphology in Chapters 8 and 9, but only in outline. For further details see *The evolution of arthropodan locomotory mechanisms*, Parts 1–11, Manton 1950–73.

1. General features of locomotory mechanisms

The basic limb movements used in walking were considered in Chapter 2, §2A–C, Figs. 2.2–5. The promotor–remotor swing of the leg usually takes place at the coxa–body junction and is implemented by protractor, retractor, levator, and depressor muscles inserting within the trunk. These extrinsic muscles are assisted by intrinsic levator and depressor muscles situated proximally in the leg and sometimes distally as well; when proximally placed such muscles may function as adductor and abductor muscles, bringing legs of a pair towards or away from one another. Distally along the leg flexor muscles are present at most joints, with functions similar to the depressors. Such flexor muscles may function alone or there may be antagonistic levator muscles also. The way in which the movement of successive leg pairs is co-ordinated in metachronal rhythm was considered in Chapter 2, §2D, Figs. 2.4, 2.5.

Locomotory mechanisms

The locomotory mechanisms of the Uniramia exhibit a series of gaits characteristic of each taxon, as noted in Chapter 6, §5C. Within one series the pattern of the gait, meaning the relative durations of forward and backward strokes of the leg; the phase difference between successive legs; the angle of swing of the leg; and the pace duration change in a harmonious manner to give faster or slower speeds of walking or running in the various animals (e.g. Fig. 6.13).

That such series are remarkable is shown by comparison with other phyla. The Arachnida use only a limited variety of gaits (see Chapter 10) and no long series have been found, as in the Uniramia. Syncarid and peracarid Crustacea use thoracic legs 1 for feeding and legs 2–8 for walking, but their gaits have not been fully investigated. The Isopoda, such as *Ligia*, exhibit the same gait both in water and on land, varying their speeds by changes in the pace duration, as in the land isopods (woodlice). The Decapoda use octopodous and hexapodous walking. The crayfish uses a variety of gaits by four pairs of thoracic legs but with considerable irregularity. Hexapody, as used by the thoracic legs of prawns, brings the same advantage of fanning out the fields of movement as in terrestrial hexapods (Fig. 5.2 right side, and see Ch. 5, §3). Spider crabs attain the same advantage of hexapody in forward walking, while the sideways walking of most crabs entirely separates the fields of movement of the four pairs of walking legs, thus avoiding mechanical interference of one leg by the next.

The mammals also lack long series of integrated locomotory mechanisms. There may be several entirely different types of gait, used alternatively, as in the horse, and great speed is attained by shortening the whole cycle of leg movements (short pace duration). Here the backstroke is of longer duration than the forward stroke during galloping (Gray 1968). By contrast, the usual method of increasing speed in the Uniramia is the employment of fast patterns of gait, in which the backstroke is of relatively shorter duration than the forward stroke and the pace duration is also shortened (e.g. Fig. 6.13). Thus the serial nature of the gaits of the Uniramia is indeed remarkable and is a characteristic feature of this phylum.

Within the Uniramia the locomotory mechanisms can be seen to have evolved along certain definite lines. General features will be considered first, the gaits of the various

2. The use of trunk musculature

The slower methods of walking or crawling seen in arthropods and in polychaetes involve stepping by the limbs or parapodia (see Chapter 4, §4), the trunk remaining straight, with isometric tension in its muscles on the two sides (Pls. 1(c), 5(d)). Increase in speed of crawling in a polychaete is accompanied by horizontal undulations of the trunk (Gray 1939, Clark and Clark 1960, Mettam 1967 and Pl. 1(c)–(e)), which, as they increase in amplitude, lead to swimming. The trunk muscles, contracting alternately on the two sides, increase the efficiency of the parapodia and result in speedier progression. The body is thrown into lateral undulations which progress from behind forwards. The parapodia perform their propulsive backstroke on the outer, exposed, side of each convex curve of the body, the recovery stroke taking place in the concavity. The details differ according to the species. Without parapodia the body undulations would produce backward swimming (Gray 1939). The oligochaetes and leeches also use their body muscles for locomotion, but in different ways. Presumably the use of longitudinal trunk muscles to enhance parapodial effectiveness is a secondary development.

Arthropodan locomotory mechanisms contrast with those of polychaetes in that trunk undulations (Pl. 1(f)), if present, do not contribute directly either to the locomotory movements or to increase in speed of walking or running. The body remains straight, except during changes in direction of progression, and during the casting from side to side of the head end in sensing a path (Fig. 7.1(a), Pls. 4(e), (f), 5(d)–(h)). This trunk rigidity provides a base on which the extrinsic limb muscles can pull, but does not contribute a direct locomotory force.

The appearance of marked body undulations in the Scolopendromorpha (centipedes), during their fastest gaits, and less so in other Chilopoda, is a hindrance to their locomotion, not an asset (Fig. 7.1(b)–(f). Pl. 1(f), and see Chapter 8). Horizontal undulations occur because the two legs of a pair must be used in opposite phase and the positions of the legs is different on each side of the body. These undulations reduce the effective angles of swing of the propulsive legs and therefore reduce the speed potential of the fastest patterns of gait, besides wasting energy in lateral movements of the body during forward progression. It will be shown in Chapter 8, §5C how morphological features in myriapods hinder the appearance of undulations of the body in the horizontal plane. We know nothing of the past history of a lobopodium, which probably arose in the sea in the very remote past, but it is apparent that the Uniramia, and the rest of the Arthropoda, have advanced their locomotory mechanisms by exploiting the rigidity potential of the trunk muscles and the locomotory potential of the intrinsic and extrinsic lobopodial and jointed limb musculature, in contrast to the annelids. Specialized uses of trunk muscles for jumping in air or water exist (Collembola, Thysanura (Machilidae), Copepoda, lobsters, shrimps, etc.), but these are not basic to the evolution either of the Uniramia or of other Arthropoda.

Among the lower vertebrates the axial muscles of the trunk provide the locomotory force for swimming, lateral undulations passing along the body from head to tail. This force is still of locomotory significance in many amphibians and reptiles. But changes in the positions and flexures of the legs of mammals result in the limbs supporting the body from below and limb muscles providing the locomotory thrust, so that no marked lateral undulations of axial structures remain. A further advance, the use of dorso-ventral flexures of the trunk in galloping mammals and in cetaceans swimming, is a new asset. No similar transition in origin of the maximal

FIG. 7.1 (see opposite). The gaits of Chilopoda and the occurrence of undulations of the body in fastest running. The upper diagrams show the range of gaits used by epimorphic Chilopoda; 'gait diagrams' of four successive legs show the relative durations of the forward and backward strokes, the heavy lines indicating the propulsive backstroke, time progressing in the direction of the arrow. Below the gait diagrams are 'gait stills' showing the disposition of the left legs at the moment indicated by the dotted lines. As the relative duration of the backstroke decreases so the distance between successive propulsive legs increases. The lower diagrams (a)–(d) show tracings of photographs of *Scolopendra cingulata* (see another species, Pl. 1(f)) running progressively faster in the gaits shown above. Legs with their tips in contact with the ground and performing the propulsive backstroke are shown in heavy lines, legs off the ground performing the recovery forward swing are shown by thin lines. The points of support of the body against the ground are indicated by black spots. The distances between the back spots show the stride lengths. Legs 1 and 21 are not used in fast running. (e)–(f) Anamorphic Chilopoda with longer legs and therefore using a different type of gait. (e) *Lithobius forficatus* (Pl. 5(g), (e)) running in gait (5·5:4·5) phase difference between successive legs 0·17 approx. (f) *Scutigera coleoptrata* (Pl. 4(j), (k)) using gait pattern (6·35:3·65) phase difference between successive legs 0·135 approx. In (e)–(f) the legs are longer and diverge in the propulsive phase and there is far greater control of undulations of the body than in *Scolopendra* running fast. (From Manton 1965.)

Locomotory mechanisms

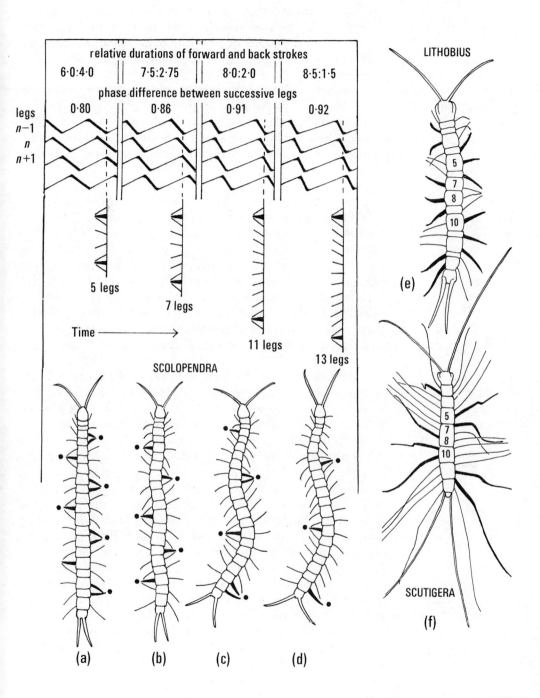

Locomotory mechanisms

locomotory thrust, from axial in worms to appendicular in Uniramia, can be postulated. It seems more probable that two contrasting modes of providing the locomotory thrust were evolved independently in modern Annelida and in Uniramia.

3. Speeds of progression

The speeds of walking or running in arthropods are dependent upon a number of factors which are usually interrelated. Both increases in speed and the use of slow, strong leg movements, are ultimately limited by the physiological capabilities of the animals.

A. *The frequency of stepping*

The pace duration is of importance in controlling speed, quicker stepping leading to faster progression. The proportion of a speed increase due to this factor differs from group to group; some examples are given in Manton (1952a, Table I). Most Uniramia in achieving their maximum speeds increase the angle of swing of the leg and decrease the relative duration of the backstroke (see below) at the same time as reducing the pace duration. But some, with rather inflexible gaits, depend largely on frequency of stepping for changing their speeds as in the millipede *Polydesmus*.

B. *The relative durations of the forward and backward strokes*

These movements of the legs, here called the pattern of the gait, is of considerable importance in changing speeds. The movements of the right legs of an onychophoran, relative to the head, are shown in Fig. 2.4, the thicker lines representing the propulsive backstroke and the thinner lines the forward swing off the ground. Other gait diagrams are simplified here to some extent by the change from forward to backward strokes being shown by sharp angles rather than slight curves in the lines. During most of the paces shown in Fig. 2.4 the forward and backward strokes of the legs are of about equal duration. A decrease in the relative duration of the backstroke gives an increase in the speed at the same or shorter pace durations (see lines representing leg movements in Fig. 7.1(d)). An increase in this relative duration leads to slower progression at the same, or longer, pace durations, and leg movements can also be stronger (see lines representing leg movements in Fig. 7.1(a)). It has been shown in Chapter 5 why strong, slow leg movements and fast, weaker ones, are mutually exclusive and are correlated with different leg structure and uniramian needs.

Locomotory mechanisms

There is no limit to the slowing of locomotion by the employment of progressively slower patterns of gait (Fig. 7.8(b) left side, 7.10(b)). In such gaits all the legs may be simultaneously propulsive over many or all segments. One leg at a time is raised and takes the forward swing (Fig. 7.2(a)). *Macroperipatus* walking very slowly shows a gait with eight propulsive and one recovering leg in some metachronal

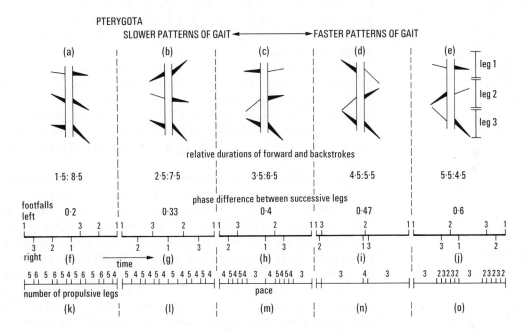

FIG. 7.2. A range of gaits from fast to slow occurring among pterygote insects (Pl. 6) are illustrated by the data assembled under the gait stills (a)–(e). The relative durations of the forward and backward strokes and the phase differences between successive legs for each gait are given below the upper gait stills. These depict the disposition of the legs for the moment in each gait when right leg 3 is at the end of the backstroke. The heavy lines indicate legs in the propulsive phase and the thin lines show legs performing the forward recovery swing off the ground. For convenience the legs are drawn at equal lengths and swinging about the same angle on the body, the range of movement of the tips being shown by the heavy lines on the right of (e). Neither of these features exactly obtains, the legs being of unequal length although taking the same stride (Fig. 5.2). From a fuller series of such diagrams an approximate estimate of a gait can be made from a limited number of still photographs. The actual pace durations in the slower patterns of gait are greater than in the faster. The phase differences between successive legs are those that will give a reasonable but not excessively long or short time interval k, marked on Figs. 7.8(c), 7.9(c), (e). (f)–(o) The heavy horizontal lines indicate one pace length, along which are marked the series of footfalls, from left to right, and the number of propulsive legs at successive moments throughout one pace. (From Manton 1972.)

299

Locomotory mechanisms

waves and the slower gaits used by hexapods have five or six legs in the propulsive phase simultaneously. That slow patterns of gait permit the generation of a stronger leg movement than faster patterns is well shown by the slow gaits of Diplopoda used in forcing their way through soil or decaying leaves (Pl. 7(e)) where 20–22 legs are performing the propulsive backstroke and 8–10 legs are off the ground during their forward swing in each metachronal wave, in contrast to the faster gaits used in a bright light on the surface of the substratum (Pl. 7(d)) where 5–6 legs are propulsive and 6–9 legs in the recovery forward swing, off the ground, in each metachronal wave.

The pattern of a gait, meaning the relative duration of forward and backward strokes, is expressed here by numbers within brackets, the whole cycle of leg movements being taken as 10, see Figs. 7.1–4, 7.8–10. The phase difference between successive legs is given numerically as that proportion of a pace by which leg $n+1$ is in advance of leg n, such as 0·4.

The execution of fast patterns of gait is often dependent upon facilitating morphology (see Chapter 8) which, in the epimorphic Chilopoda, for example, allows fast running with only three legs out of forty in the propulsive phase at any one moment, as in gait (8·5:1·5) (Fig. 7.1(d)). But in animals with few legs, such as hexapods, gait patterns faster than (5·0:5·0) are usually impossible because stability requires a minimum of three legs on the ground at any one moment, arranged round the centre of gravity (Fig. 7.3, where the left leg movements are shown by continuous lines and the right by interrupted lines, the paired legs moving in opposite phase). Some hexapods can manage the unstable moments caused by a backstroke of slightly less than half the pace duration, but the gait pattern, e.g. (5·5:4·5), does not go far in the fast direction (Fig. 7.2(e)). The employment of increasingly faster patterns of gait by the longer legged Myriapoda is limited also by mechanical and physiological features. In *Scutigera*, for example, a faster pattern of gait than the one shown in Fig. 7.1(f) would lead to even more crossing of the legs when swinging forward (thin lines) and an overlap of the equivalent of the propulsive leg 6 by leg 12 right side, leading to mechanical difficulties. Note the well-controlled lateral undulations of the body when running fast (Pl. 4(j)). But there must also be a physiological limit to the speed at which an

Locomotory mechanisms

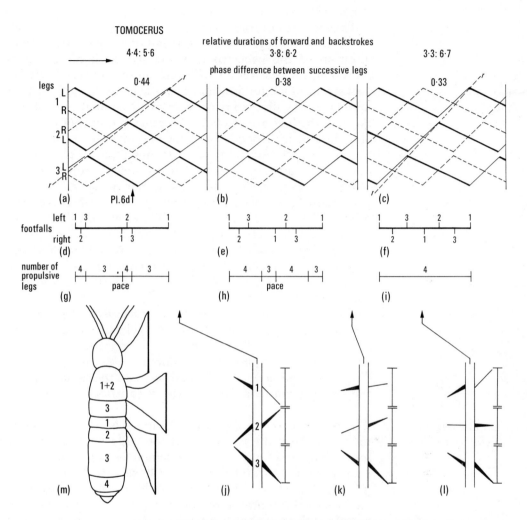

FIG. 7.3. The gaits of Collembola (*Tomocerus*, Pl. 6(a), (d)). Gait diagrams are shown in (a)–(c), the slower on the right; the left leg movements are shown by continuous lines and the right legs by interrupted lines. Gait (a) is seen in Pl. 6(d), the moment being marked by arrow below (a). (d)–(f) Shows the sequence of footfalls of the six legs in the gaits illustrated above. (g)–(i) Records the number of propulsive legs at successive moments during one pace of the gaits shown above. (j)–(l) Gait stills of the disposition of the legs at the moment indicated by the long arrows pointing towards the three gaits shown above. In all stills right leg 3 is at the end of the backstroke, heavy lines indicate a leg in the propulsive phase, and thin lines show the forward swing, the legs are drawn at equal lengths and swinging similarly about their bases, the range being marked alongside on the right. The actual ranges are seen in (m). (m) The range of movement of the right legs of *Tomocerus longicornis* drawn relative to the body, the heavy vertical lines depicting the limb tips during the propulsive backstroke. (From Manton 1972.)

animal can execute its propulsive backstroke.

Thus in most myriapods and hexapods the limit to increase in speed by using fast patterns of gait (decreasing the relative duration of the backstroke) is set by mechanical and by physiological features. A galloping horse contrasts with its backstroke of longer duration than the forward stroke; a limit here to faster galloping may be set by the animal's inability to quicken the forward stroke still further (Gray 1968). In only one class of arthropods have locomotory movements been found in which the ability to perform a rapid forward stroke may constitute an upper limit to speedy running. The Symphyla exploit slow patterns of gait, usually employing gaits with long relative durations of the backstroke (Fig. 7.8(b)) and very rapid stepping. They can run their fastest only in short bursts, suggesting that the legs may be working hard on both forward and backward strokes. Morphological specializations facilitate the rapid forward stroke, and may indicate the difficulties of particular importance in gaining speed (Manton 1966).

The phase difference between successive legs in the Uniramia changes harmoniously with changes in the pattern of gait (see below). Paired legs move in opposite phase except when very slow patterns of gait are the priority, as in diplopods, and here legs of a pair move in the same phase, as they do in Symphyla (Figs. 7.8(b), 7.10(b)).

C. *The angle of swing of the leg*

The promotor–remotor swing of the leg is caused by movements between the leg base and the body in all arthropods other than Arachnida. This movement is controlled by the coxa–body joint in myriapods and hexapods (Figs. 5.3, 5.4, 5.6, 5.7, 5.9–15), but the remarkable legs of the Lepismatidae enables the promotor–remotor swing of the leg to take place mainly at the coxa–trochanter joint (see Chapter 9).

The angle of swing of the leg changes as far as mechanical features permit. An increase in the angle usually accompanies quicker stepping and faster patterns of gait (with short relative durations of the backstroke) and consequently quicker progression.

The effects of changes in the angle of swing of the leg are illustrated in Fig. 7.4. Three patterns of gait are shown, the faster pattern (6:4) in the top three horizontal rows of diagrams, followed by the slower patterns (5:5) and (4:6) in the next two rows. Two phase differences between successive

Locomotory mechanisms

legs, 0·08 on the left and 0·06 on the right, are given for all three patterns of gait. Angles of swing α, β, γ, are indicated in the middle for the several gaits. The propulsive legs are marked by heavy lines, a spot below each marking contact with the ground. The thin lines indicate legs performing the forward swing off the ground and the horizontal lines below them show absence of footfalls. It will be noted that in Fig. 7.4(c), (f), (i) the distance marked x below the 10 legs moving off the ground progressively decreases as the angle of swing increases, so that in (i) the 7 propulsive legs in each metachronal wave together form almost equidistant footfalls along the ground, an advantageous mechanical state giving an even thrust along the body against the ground (Pl. 7(d)–(e). In Figs. 7.4(a), (d), (g) of the same gait pattern but at a slightly larger phase difference between successive legs, the same changes occur, but in (g) there is overcrowding of the forwardly moving legs and the adjacent propulsive legs in

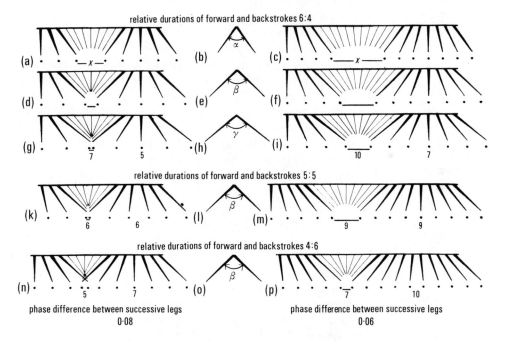

FIG. 7.4. Diagrams showing the effects of alterations in (1) the phase difference between successive legs (number of legs to a metachronal wave), (2) the relative durations of the forward and backward strokes, and (3) the angle of swing of the legs, on the type of gait employed by the Diplopoda. Propulsive legs are shown in black, recovering legs by thin lines. For further description see text. (From Manton 1954.)

303

successive metachronal waves are too close together.

Only the angle of swing β is drawn for the two slower patterns of gait. In pattern (5:5) there are 9 propulsive and 9 recovering legs in each wave and in the (4:6) pattern there are 10 propulsive and 7 recovering legs. Of the four gaits shown, that in (p) is optimal in mechanical advantages because the footfalls (black spots) form an even series all along the ground and there is no undue crowding of the legs off the ground. Gait (k) is undesirable for the same reasons mentioned above for gait (g) and gait (n) is mechanically impossible with crossing propulsive legs. These diagrams illustrate the interdependence of the pattern of the gait, the phase difference between successive legs, and the angle of swing of each leg. Each makes its own contribution, either to increase or decrease in speed of running and to the total force exerted by the legs against the ground at any one moment. Some of these gaits in Fig. 8.4 are used by Diplopoda (see Chapter 8, §3).

Unless the legs are very short (Fig. 5.2, Peripatus), their angle of swing is facilitated by the presence of not too many leg pairs, so that a fanning out of the fields of movement of successive legs becomes possible. There is only a little fanning out of the legs anteriorly and posteriorly in an epimorphic centipede, see Fig. 7.1(a) bottom diagram, but the 14 pairs of progressively longer legs of *Scutigera*, although well fanned out, show considerable overlapping of their fields of movement, as they do in *Ligia* (Fig. 5.2). When the locomotory legs are few, as in scorpions, *Galeodes*, and the earwig *Forficula* (Fig. 5.2) overlapping fields of movement of long legs can be almost or entirely eliminated, making it easy for a variety of leg movements to be executed. Large angles of swing of the leg are usually not found during slow patterns of gait or when the pace durations are long.

D. *Leg length*

The real or effective leg length can influence speed because a longer leg can take a longer stride (Fig. 2.5). The Onychophora elongate their legs when walking fastest by their faster patterns of gait. A jointed leg taking easy strides and moving relative to the body as line B–D in Fig. 4.21 can increase this distance by using a larger angle of swing of the leg, by placing the limb-tip upon the ground a little further forward than the point B and, of necessity, a little nearer to the middle line. Similarly the leg will be raised from the ground behind point D and closer to the middle line. But this

Locomotory mechanisms

will only be possible if the leg jointing permits more acute flexures when half way through the backstroke so that the line A–C becomes shorter. The line B–D is not the stride length, but only that part of one cycle of limb movement during which the limb tip is propulsive and on the ground. During the forward stroke the limb base is carried forwards by other legs in the propulsive phase (see Chapter 2 and Fig. 2.5).

Figures 7.5(a) and (b) show the tracks of the millipede *Callipus*, left, on smoked paper and printed in reverse so that the footprints are black; running more slowly in (a) than in (b). The animal is wearing a boot on one right foot (see Appendix); the black spots alongside show the footfalls of this leg and the distance between them is the stride length. The track (a) made by slow running shows a greater width and shorter strides than track (b) made at speed. The forward curve of the black footprints are made as each tarsus is raised from the ground but carried forward by the moving body. *Callipus* is an unusual type of diplopod which has secondarily become scavenging, carnivorous, and fast-running. Most Diplopoda are not concerned with speedy running but with strong pushing (see Chapter 8, §3A). Figure 7.5(c) shows the track of *Cylindroiulus* wearing a boot on one left leg.

Centipedes are well able to flex their legs strongly when half way through the backstroke and to take long strides which the legs of the fossil myriapod *Arthropleura* (Figs. 1.18, 5.19) certainly could not. Compare the inter-podomere jointing, particularly at the knee, of a centipede and of *Arthropleura* (Figs. 5.3, 5.6, 5.14, with 5.19). In the shorter-legged epimorphic Chilopoda (Fig. 7.1(a)–(d)) the strides are remarkably long for such relatively short legs. The track in Fig. 7.6(a) shows staggered, forwardly directed rows of footprints one stride length apart, made at a slower pattern of gait than in Fig. 7.1(a). At the faster gait pattern (7·25:2·75), Figs. 7.1(b), 7.6(b), all legs of one side use almost the same footprints so forming the common marks one stride length apart. But the propulsive legs are so far separated that the body is not fully supported off the ground and also is thrown into lateral undulations, shown by the median undulating tummy drag along the track. At the even faster gaits, Fig. 7.1(c), (d), stride lengths increase to about 34 mm by a 50-mm animal and the body undulations are greater. Some crowding of the propulsive legs occurs in the concavities of the flank, energy is wasted in lateral movements, stride lengths are

Locomotory mechanisms

shorter than they would be if the undulations could be eliminated, and, although speeds of over 200 mm/s are reached, the unwanted lateral undulations account for a loss of some 8·5 per cent of the potential speed.

The anamorphic Chilopoda, *Lithobius* and *Scutigera* (Pls. 4(j), 5(e), (g)) possess longer legs than those of the Epimorpha (Pls. 1(f), 5(c), (d)) and cannot perform the same type of gait (see below §6C(ii)). A slowish pattern of gait (5·5:4·5) by *Lithobius* (Fig. 7.1(e)) forms a track of spread-out footprints as in Fig. 7.6(c), the black spots marking the stride lengths, but at fast speeds by gait pattern (6·5:3·5) the footprints fall into four groups per stride length as described in the legend for Fig. 7.6(d). The forward scratches by each leg are made as the tarsus was raised from the ground, as in *Callipus* when running fast (Fig. 7.5(b)). Stride lengths of 21 mm and speeds of 280 mm/s have been recorded for this gait by a 25-mm *Lithobius*. The much longer legs of a 22-mm *Scutigera* achieve strides of 33 mm and speeds of over 420 mm/s in gait pattern (6·35:3·65). Each footprint in this track is separate (Fig. 7.5(e)) forming oblique rows, as described in the legend. Leg length in *Scutigera* results in longer strides and much greater speeds than in *Lithobius* and body undulations can be almost entirely eliminated (Fig. 7.1(f), Pl. 4(j) and see Chapter 8, §5C). No other myriapod can run so fast, an

---

FIG. 7.5 (see opposite). Tracks made by myriapods running over smoked paper and printed in reverse, (a), (b), (c), (e), and an onychophoran running over black paper covered by a white powder (d) so that the footprints appear black. (a)–(b) The millipede *Callipus longobardius* (Pl. 4(f)), 42 mm long, wearing a large boot on a right foot. The black spots alongside the mark made by the boot show the stride lengths; (a) running slowly, (b) running fast with longer strides and narrower track. (c) Track of *Cylindroiulus londinensis*, scaled to the same segment volume as for *Callipus*. A more elegant boot is fixed to one left leg; the legs are shorter and the tracks are therefore narrower than in *Callipus*, but the stride length is remarkably long. (d) *Peripatopsis sedgwicki* (another species in Pl. 4(a), (b)), showing the posterior 14 of the 19 pairs of legs used in walking but the few most posterior legs are not in action owing to dislike of the dry powdery substratum; the paired legs are mostly in similar phase and legs of one side use approximately the same footprints in this gait. (e) Track of the centipede *Scutigera coleoptrata* (Pl. 4(j)), 22 mm, recorded as for (a)–(c). Here each leg performing the fast gait (6·35:3·65), stride length 33 mm, speed 420 mm/s, forms an independent footprint, in contrast to (d). The footprints of legs 1 and 14 are marked; legs of a pair move in opposite phase and so the footprints on the two sides are staggered and not level as in (d); the posterior legs are longer and thus oblique rows of footprints are formed. Two successive rows of right footprints are marked by short lines added between the footprints; those of legs 5, 7, and 8 being at times a little out of place, probably these legs were slightly injured. ((a)–(c) From Manton 1958, (d) from Manton 1950, (e) from Manton 1952a.)

Locomotory mechanisms

accomplishment needed for catching spiders and flies for food.

But leg length is of considerable importance to the Uniramia. For a burrowing or soil-living habit, minimal lateral projection of the legs beyond the body is desirable, short legs are needed, and locomotion is slow. But where speedy running is an asset, long legs can provide long strides, but for their use other requirements must be met, such as the control of lateral undulations of the body and avoidance of mechanical interference of one leg by the next. Roughly similar maximum speeds are achieved by *Scolopendra* and

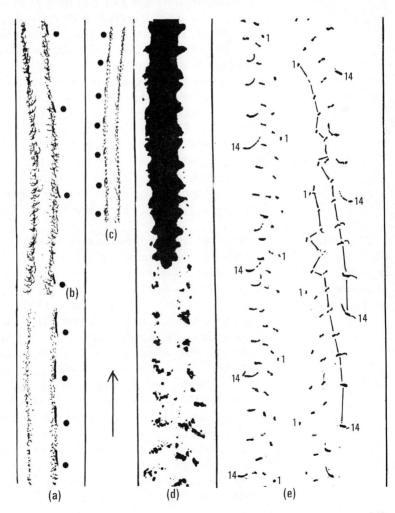

Locomotory mechanisms

other epimorphic centipedes, with short legs, but using very fast patterns of gait with consecutive converging propulsive legs, and by the anamorphic *Lithobius* with longer legs and consequently a different type of gait with diverging successive propulsive legs. These two groups of centipedes, the Epimorpha and the Anamorpha, have evolved their leg lengths and other proficiencies independently as a result of a habit divergence in feeding technique (see §6C(iii) below). Their locomotory mechanisms today represent different methods of achieving much the same fleetness. Only the Scutigeromorpha exceed all the rest in their agility and speed in running; and their long legs have been possible because the animals have also evolved acute eyesight; a superlative antiundulatory mechanism; a special respiratory mechanism; the utmost perfection in mandibular mechanism; but they also need a warm climate (see also Chapter 8, §5C).

The width of a track depends upon leg length, the long strides of centipedes compared with track width contrasts with the much shorter strides relative to track width of the Pterygota (Fig. 7.12(g)). The great differences between the lengths of successive legs in *Campodea* permits overstepping of one leg by the next, giving long strides and fast running and a track as in Fig. 7.12(b) (see Chapter 9, §3). The Hexapoda usually depend upon flight for their maximum speeds but the Myriapoda depend upon their legs.

4. The phase difference between successive and paired legs

These two phase differences have no direct effect upon the speed of progression, but control the direction of travel of metachronal waves of limb movement and mechanical features, such as the disposition of the legs in contact with the ground, in the various gaits. These dispositions avoid mechanical interference of one leg by another during the propulsive backstroke and can produce advantageous arrangements of the points of pushing by limb-tips on the ground. This in turn can give stability to the hexapod and even loading on the legs during the propulsive phase, particularly in myriapods and hexapods; an advantageous condition which sometimes has controlled the exact number of legs used in running (Anamorpha, see below), and the occurrence of footfalls at even intervals of time in the Hexapoda and in other classes. The phase difference changes in an obligate manner with alterations in the patterns of the gaits in the Uniramia (Figs. 7.1, 7.2).

Locomotory mechanisms

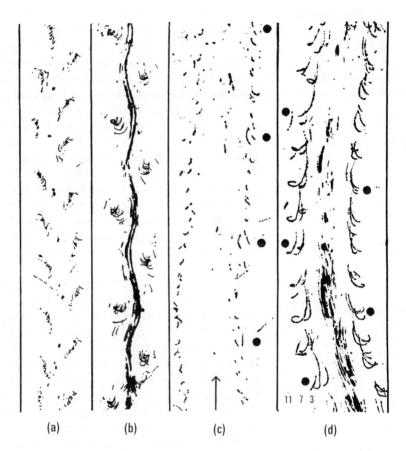

(a)   (b)   (c)   (d)

FIG. 7.6. Tracks of centipedes running over smoked paper and printed in reverse; paired legs in opposite phase. (a)–(b) Tracks of the epimorphic centipede *Cormocephalus pseudopunctatus* (resembles Pl. 1(f)), 50 mm. (a) running slowly by gait (5·5:4·5) and with footprints forming oblique forwardly directed groups, their distance apart representing the stride length; (b) running fast by gait (7·5:2·5) the footprints falling nearly on the same place, as in *Peripatopsis* (Pl. 4(a), (b)) Fig. 7.5(d), but the stride (17 mm) is so long that there is a tummy drag over the ground; the legs being so far apart are unable to hold the body clear of the substratum as in (a). The staggered position of the legs in the propulsive phase (see common footprints) throws the body into imperfectly controlled undulations, more extensive in gait (8·5:1·5) Fig. 7.1(d), where the stride length is 34 mm approx., speed less than 200 mm/s. (c)–(d) Tracks of the anamorphic centipede *Lithobius forficatus* (Pl. 5(g)), 25 mm, recorded as in (a)–(b). The black spots are one stride length apart. (c) Running slowly by gait (5·5:4·5), stride length 14 mm, speed 80 mm/s approx. The footprints form a scattered series but are in fact regular. (d) The fast gait (6·64:3·46), stride length 21 mm, speed 280 mm/s approx. The footprints are now grouped into four rows per stride length, those of legs 3, 7, 11 being marked, bottom left. The next anterior rows are formed by legs 2, 6, 10 and by 1, 5, 9, 13 and by 4, 8, 12, with a repeat of legs 3, 7, 11, level with the next anterior black spot on the left. Legs of a pair are in opposite phase, see staggered black spots. (From Manton 1952a.)

309

Locomotory mechanisms

A. *Metachronal waves and phase differences between successive legs*

Metachronal waves of limb movement pass forwards or backwards over the body as a result of the phase difference between successive legs. In Figs. 7.8–10 lines marked r pass through corresponding points on the gait diagrams of adjacent legs. The line r always passes through the next nearest corresponding point in time, whether it be on the preceding or succeeding leg. All other corresponding points on the same gait diagrams produce lines r which are parallel to those drawn. When leg $n+1$ is less than half a cycle of limb movement in advance of that of leg n, the wave appears to travel forwards over the body, as in Fig. 7.8(a), (b), time progressing from left to right. There is a corresponding wave which travels backwards, but more slowly, and thus does not catch the eye in the living animal. When leg $n+1$ is more than half a cycle of leg movement in advance of leg n the metachronal wave appears to travel backwards, see lines r, Fig. 7.8(c).

Metachronal waves (Figs. 7.8–10) are most clearly apprehended in the many-legged animals and are less easily seen in hexapods, although their presence is just as real (Fig. 7.3(a), (c), lines r). But in the epimorphic Chilopoda (e.g. *Scolopendra*), when all legs of one side occupy the same footprints during the faster gaits, each metachronal wave appears to be stationary relative to the ground and to pass backwards along the body (Figs. 7.8(c), 7.10(d) fast gait, Pl. 1(f)). When the phase difference between successive legs is 0·5 the forward and backward metachronal waves move at equal speeds and neither catches the eye. Sometimes the same animal uses gaits with phase differences both smaller and greater than 0·5, as in Pauropoda, Symphyla, and in *Campodea*. Figures 7.8(b), 7.10(d) show the metachronal waves (lines r) passing forwards in the slow gaits and backwards in the fast gaits.

When the phase difference between successive legs is less than 0·5 of a pace (i.e. leg $n+1$ is less than 0·5 of a pace ahead of leg n), legs or parapodia in the propulsive phase diverge from one another, as seen in the polychaetes and most of the myriapods and hexapods (Figs. 7.8(a), (b) slower gaits, 7.9(b), (c), (d), 7.10(b), (c), Pls. 1(d), (e), 4(j), 5(e)–(g), 6(i), (k), 7(d), (g)). When the phase difference between successive legs is more than 0·5, the propulsive legs converge, as seen in the epimorphic Chilopoda, Fig. 7.8(c), and Pauropoda, medium and fast gaits, Fig. 7.10(d). The contrasting dispositions are shown together in Fig. 7.1(a), (f), the Epimorpha on the left and the Anamorpha on the right and Pl. 1 (cf. (a)–(c) with (f).

Whether legs in the propulsive phase converge or diverge is of mechanical significance. Arthropods avoid crossing propulsive legs. Gait details are always adjusted so that normally no crossing of propulsive legs takes place; if legs are long then crossing occurs on the recovery forward stroke without causing trouble, as in *Scutigera* (Fig. 7.1(f)). The longer legs in Myriapoda, other than the Epimorpha, and in the Hexapoda could not have evolved unless phase differences between successive legs of less than 0·5 were used.

The advantage of using phase differences between successive legs greater than 0·5 in the Epimorpha appears to be associated with advantageous shortness of legs for burrowing (Geophilomorpha) or crevice-living (Scolopendromorpha). Great speeds of running can be obtained from short legs by remarkable shortening of the relative duration of the backstroke, permitting the use of gaits with fast patterns of up to (8·5 : 1·5), which result in only three out of 40 legs being in the propulsive phase at one moment, a remarkable achievement made possible by facilitating morphology (see Ch. 8, §5C). Among the Hexapoda a rare phase difference of just over 0·6 occurs during fast running by the fastest gaits of *Campodea* (Fig. 7.9(e)). The gait (5·5:4·5) is unstable, there being short periods during each cycle of leg movements when only two legs are propulsive and supporting the body (Fig. 7.2(e), (j), (o)). If the phase difference was not increased to more than 0·5 this exceptionally fast pattern of gait for a hexapod would be even more unstable, or impossible to perform, because of a lack of overlap in time of the backstroke of two successive legs (time interval k marked on Fig. 7.9(e)).

The convergences between successive legs in the propulsive phase in, for example, the epimorphic Chilopoda (Fig. 7.1(a)–(d)) shows either two such legs converging, or, where leg n is ending its backstroke, leg $n+1$ is half-way through this backstroke and leg $n+2$ is starting the backstroke, three successive legs converge onto almost the same footprint. The time interval between the footfalls of legs on the ground and the raising of the preceding leg in the same pace is marked k in Figs. 7.8(c), 7.9(c), (e) where the oval marks show successive legs performing the same pace while the rectangles mark successive legs executing successive paces. There is no time interval k complying with the above

Locomotory mechanisms

definition in the gaits of the anamorphic Chilopoda or Diplopoda (Figs. 7.9(a), (b), 7.10(b), (c)) where successive legs simultaneously occupy different footprints (Pls. 4(j), 5(g), 7(e)), or in animals with either a simultaneous change in successive footfalls, as in Collembola (Fig. 7.3(a) shows the gait executed in Pl. 6(d)), or where leg $n+1$ is placed on the ground after leg n is raised in Onychophora and Symphyla (Figs. 7.8(a), (b), 7.9(d), Pl. 4(e)).

The convergence of successive propulsive legs is really determined by the presence of the time interval k. This convergence is seen in *Campodea* at phase differences of both 0·60 and 0·33 (see Manton 1972, Fig. 5(a), (b), (g), (i)) and will occur wherever a moment in time k exists at any phase difference between successive legs, the converging legs occupying the same group of footprints made by the execution of the same pace (one group of footprints see Figs. 7.1, 7.6(b), and Pl. 6(i) right legs 1 and 2). But successive legs occupying different groups of footprints during the propulsive phase diverge, because they are executing successive paces (Pl. 6(i) right legs 2 and 3). Thus convergence and divergence of successive propulsive legs can occur in the same hexapod and in the same gait. The vertical lines u and v in Fig. 7.9(c), (e) pass through a time interval k; the gait stills on these lines show: in u, legs 1 and 2 converging in the propulsive phase, followed by a diverging propulsive leg 3 while in v the converging propulsive legs 1 and 2 are followed by leg 3 moving forwards; moment u is seen almost exactly in Pl. 6(i) right side. The other gait stills in Fig. 7.9(c), (d) show diverging successive right legs; compare the converging propulsive legs in Fig. 7.8(c).

B. *The phase differences between paired legs*

The ease with which various patterns of gait can be performed is dependent upon this phase difference which determines the position of footfalls on the ground. When few legs are in contact with the ground at any one moment, as in fast patterns of gait, the paired legs always move in exactly opposite phase, resulting in support being provided on one side of the body opposite unsupported stretches on the other side (Figs. 7.8(c), 7.9(b), Pls. 1(f), 4(j), 5(d), (g)).

When slow patterns of gait are employed, with the backstroke of longer duration than the forward stroke, many legs are propulsive simultaneously (Fig. 7.10(b)) and the paired legs move in similar phase (Diplopoda and Symphyla),

Locomotory mechanisms

thereby giving an even thrust from the body and no tendency to undulate in the horizontal plane (Pls. 4(e), (f), 5(f), 7(e), (g)). Paired legs move in similar phase in the jumping gaits of *Petrobius* (Pl. 6(f), see below) and in swimming by *Dytiscus* (Hughes 1958).

Hexapods in running and walking use their paired legs in opposite phase (see gait stills in Fig. 7.9(c)–(e) and Pl. 6(d), (e), (g), (h), (i), (k)), although their gait patterns are usually slower than (5·0: 5·0). By this means stability is gained with no necessity for either two, four, or six legs to be in contact with the ground, without the advantageous intermediate numbers of three and five as would occur if the paired legs were moved in similar phase. Hexapodous first instars of Diplopoda similarly use their three pairs of legs in opposite phase as do Pauropoda (Pls. 4(l), 7(h)).

In gaits with equal durations of forward and backstrokes (5·0: 5·0) there is no particular mechanical advantage in either type of phase relationship, which are in fact variable in the 'middle-gear' gait of Onychophora and in the slower gaits of the epimorphic Chilopoda. Note the variable phase differences between legs of a pair in Pl. 5(i), a geophilomorph moving slowly, and in Onychophora (Manton 1950, text-figs. 3–5, Pl. 14, Pl. 15, Figs. 14, 15, Pls. 16, 17).

C. *The effects of the phase differences between successive and paired legs on leg disposition*

These are often quite obvious, as in epimorphic Chilopoda and Diplopoda, but may not be so, as in the Onychophora.

In the centipede *Scolopendra* (Figs. 6.13, 7.1(a)–(d)) the phase difference between successive legs increases along with reduction in the relative duration of the backstroke (right side of the above Figs.). If there was no increase a stable transfer of body weight from one leg to the next behind it could not occur and the fastest patterns of gait would not be practicable (Fig. 7.1(a)–(d)). If the paired legs were not in opposite phase, long stretches of the body would have no support from either side in the fast patterns of gait (right side of same figure), which is also impracticable.

In the Diplopoda the various phase differences between successive legs, which are combined with the different patterns of gait and angles of swing of the legs, control the spacing of the footprints on the ground (Fig. 7.4). The most advantageous combination of these factors for the characteristic slow, strong, diplopodan locomotory movements is as in gait (p), bottom right of the figure, where footfalls, indicated

Locomotory mechanisms

by spots, lie at equal intervals on the ground, there being no gap in the series opposite groups of legs in the forward recovery swing. Thereby a maximum number of legs can be propulsive at once all along the body, with evenly distributed thrusts on the ground. Gait (n) in the above figure is mechanically impossible because propulsive legs cross, gaits (g) and (k) might lead to stumbling unless very exactly performed, while gaits (a), (c), and (f) do not spread the thrust evenly on the ground. The animals' actual choice of gaits are those with even or almost even spread of footfalls (Pl. 7(d), (e)). In fastest diplopodan walking by fastest patterns of gait, when speed rather than strength of pushing is a priority, the even spread of footfalls is sometimes lost. Diplopod paired legs move in similar phase (Pls. 4(f), 5(f)), so ensuring that an even thrust from the two sides is transmitted to the head end or dorsal surface, as suits the several burrowing techniques (Ch. 8, §3A). Lateral undulations of the body, which might be caused by paired legs moving in opposite phase, would be most unsuitable and hinder the generation of a maximum head-on pushing force.

In the Onychophora the three most commonly used gaits are shown in Fig. 7.9(a) with gait stills and side view diagrams of the positions of the legs at one moment during each gait. The mechanical reasons for the onychophoran change in phase difference between successive legs, from a small figure (0·15) in the slower gaits to a larger figure (0·4) in the 'middle-gear' gaits and to a smaller figure again (0·15) in the 'top-gear' gaits, are less obvious than in Diplopoda and Chilopoda above, but they are known (see Manton 1973). Anamorphic Chilopoda also use fairly evenly spaced footfalls, note the distances between the propulsive leg tips (heavy black) in Fig. 7.1(e) particularly right side, Fig. 7.1(f) and Pl. 5(g).

5. The determinants of segment numbers, the order of footfalls, and the loading on the legs

Why do arthropods have different numbers of legs? Often this question can be answered precisely. The total number of trunk limbs in the various taxa is correlated in detail with functional needs. Some of the data concerning Crustacea and Chelicerata have been mentioned in Chapters 3 and 4. The Uniramia will be considered here.

Among the geophilomorph centipedes and the iuliform millipedes (Pls. 5(a), (b), 7(c), (d), (f), (g)) many segments and many limbs are present. In both cases the services rendered by the many legs concern burrowing, but the techniques employed are quite different. A geophilomorph centipede burrows as does an earthworm. Each segment of the body can alter its shape and become long and thin or short and thick. The anterior end of the body elongates and walks into a crevice, in the soil or under a stone. The anterior legs hold on to the substratum while a group of segments behind shorten and thicken, the group exerting pressure on the soil, so widening the crevice. Segments at the front end of the thickening then elongate and walk forwards into the widened crevice while segments from the trunk behind thicken and add themselves one by one to the thickened zone (Pl. 5(c) and see Ch. 8, §5B). The intermittent heaving against the soil may be done by one or more thickened zones and the body must also be capable of traversing gaps in the substratum. A large number of leg-bearing segments, 31–177, is suited to the various detailed habits of the Geophilomorpha and it is probable that the long-bodied British *Haplophilus subterraneus* (Pl. 5(a), (b)) and the Mediterranean *Himantarium gabrielis* can burrow more deeply than the shorter of the British species.

The iuliform millipedes (Diplopoda) also burrow, but by head-on pushing into the substratum, the motive force coming from the limbs. The force from the many limbs is transmitted to the head end where there is a pushing shield of some sort (see Chapter 8, §3A, Pl. 7(c)). The more limbs present the greater will be the thrust exerted at the head end. Up to 66 limb-bearing units lie on the trunk, most of them being diplosegments carrying two pairs of limbs and formed by fusion of two adjacent segments (Pls. 4(h), (i), 7(d), (e)). Diplosegments are deep and short, particularly in the larger species. If the segments were not fused in pairs in this manner the body would have to be much longer if it carried the same number of legs, because intersegmental joints cannot be contrived much closer together than they are between the diplosegments in Fig. 8.3. The habits of the burrowing diplopods do not favour a very long body and the body of moderate length, very strongly constructed, suits the several habits of diplopodan species (see further in Chapter 8).

Fewer trunk segments characterize the scolopendromorph centipedes, some 20 leg pairs being used in running (Pls. 1(f), 5(d)). These animals still hide in crevices and under stones, for which short legs are desirable, but their speciality is fast running, the fastest gait recorded being that shown in Fig.

Locomotory mechanisms

7.1(d). Here only three of the 40 locomotory legs are propulsive at any one moment. If the body was shorter, such a fast pattern of gait could not be used because two legs only supporting the body at any one moment would be unstable. A longer body presumably would make the animal more unwieldy and less capable of rapid turning, etc. It may be suggested that the 20 trunk segments carrying locomotory limbs is about optimal for the needs of the Scolopendromorpha.

The trunks of the anamorphic centipedes, the Lithobiomorpha and Scutigeromörpha, carry only 14 pairs of locomotory limbs. The determinants of this optimal number of segments are less easy to find. It concerns even loading on the legs being a desirable working condition, particularly for fast-moving animals, as are the anamorphic centipedes. Even loading on the legs during running is achieved by a variety of means unless even loading is incompatible with higher priorities. Uneven loading on the legs hardly arises in Diplopoda, where so very many legs are pushing at one moment, but may be an important feature to fast-running anamorphic Chilopoda and to Hexapoda.

Even loading on the legs of *Lithobius* and *Scutigera* occurs when footfalls of the legs are made at equal intervals of time and a constant number of propulsive legs is maintained at all moments. Gait patterns, phase differences between successive legs, and the total number of legs used are concerned in giving the advantage. *Lithobius* (Fig. 7.7(c)) using its fastest gait (6·54: 3·46) at a phase difference of 0·154, and using only 13 of its 14 pairs of locomotory limbs, shows footfalls at equal time intervals (marked along the horizontal line indicating one pace in Fig. 7.7(c)) and always has 9 legs pushing against the ground. If 14 pairs of legs were used in this gait an alternation of 9 and 10 pushing legs would result as shown on the line below. The animals when going all out use only 13 pairs of legs in the above gait and thereby gain even loading on the legs. When running more slowly, gaits as in Fig. 7.7(d), (e) are used. Only gait (e) for *Lithobius* shows even footfall intervals, but by groups of legs. The footfalls in (d) are spread. The number of simultaneously propulsive legs ranges from 11 to 14 in gait (d) and from 10 to 14 in gait (e) when 14 pairs of legs are used. There is no greater uniformity in any feature when 13 pairs of legs are employed and the animals in fact use their 14 pairs in these gaits, gait (d) being

Locomotory mechanisms

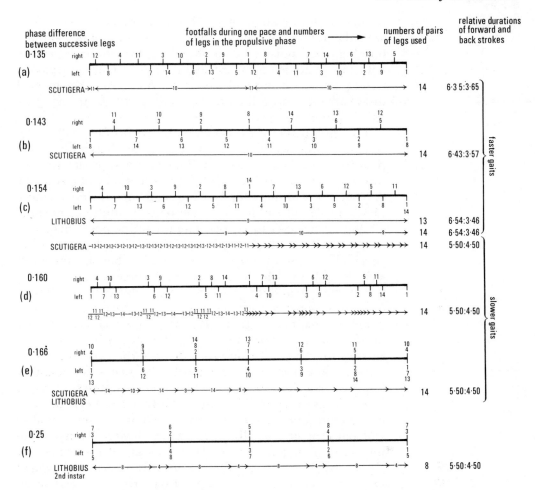

FIG. 7.7. Diagrams showing the order of footfalls and the number of legs in the propulsive phase throughout the duration of one pace (heavy horizontal lines) for a series of gaits, performed by different numbers of paired legs, and arranged in order of increasing phase difference between successive legs, entered on the left. Right and left footfalls are marked along the heavy horizontal lines. The numbers between the arrows gives the numbers of legs which are in the propulsive phase during the time interval between the arrows for *Lithobius* and *Scutigera* as indicated. For further description see text. (From Manton 1952a.)

preferred.

Scutigera (Pl. 4(j), (k)), with longer legs, is incapable of reducing the relative duration of the backstroke as much as can *Lithobius*. The two fastest gaits used are shown in Fig. 7.7(a), (b). Gait (b) gives even footfall intervals and 10

317

propulsive legs at all moments, but by grouped footfalls. The animals prefer (a) where the footfalls are not grouped and the number of propulsive legs is 10, except for two brief periods with 11. There is no advantage in using 13 pairs of legs and the 14 pairs are normally used by *Scutigera*. Mechanical disadvantages to both animals would result from using 15 pairs of running legs.

The early instar of *Lithobius*, with 8 pairs of legs (Pl. 5(h)), can only come near to even loading on their legs with slower patterns of gait than the adult. The closest approach to even loading is shown in Fig. 7.7(f). Thus it is advantageous for *Lithobius* to add more leg-bearing segments during anamorphic growth. The number of segments in adult anamorphic Chilopoda appears to be associated with the performance of mechanically satisfactory gaits which are also physiologically possible. An increase in segment number beyond the adult state would only lead to mechanical inefficiency.

The advantages of possessing only 3 pairs of trunk limbs, or of using only 3 pairs when more are present, have been noted in Chapter 5, §3, and Fig. 5.2. Now we may consider briefly the choice of gaits of the uniramian hexapods, the reasons for the choice, and the effects of the various gaits on stability and leg loading. Usually the needs for stability require three supporting legs at all moments, unless jumping gaits are used, as in Thysanura. Only a few hexapods can manage unstable moments during walking or running and so gait patterns are usually (5·0: 5·0) or slower. Even footfall intervals, giving four propulsive legs at all moments, are shown by Fig. 7.3(c), (f), (i), an advantageous slow gait (3·3: 6·7), phase difference 0·33. Five propulsive legs at all moments are shown by a gait intermediate between those on Fig. 7.2(a) and (b) which also exhibits footfalls at exactly equal time intervals, those in Fig. 7.2(b), (g), are not at exactly equal time intervals. Faster patterns of gait, with obligate changes in phase difference, gives a loss of the even loading on the legs; Fig. 7.2 shows the order of footfalls and numbers of propulsive legs in a series of hexapod gaits. Fast running by the faster patterns of gait, right side of figure, is incompatible with attaining even loading on the legs.

The order of footfalls in Figs. 7.2(b), (g): L_1, R_2, L_3, R_1, L_2, R_3, L_1, changes automatically in Figs. 7.2(a), (f) to: L_1, R_3, R_2, R_1, L_3, L_2, L_1, so that the footfalls of three legs of the same side step down one after another. On the right, Figs.

Locomotory mechanisms

7.2(e), (j), the footfall order is: L1, R3, L2, R1, L3, R2, L1. The various orders of footfalls are the automatic consequence of using the most suitable phase difference between successive legs for each gait. The sequences are interrelated and do not represent isolated adaptations to any special circumstance.

The jumping gaits of the Thysanura will be considered in Chapter 9.

6. The relationships between the gaits used by different Uniramia for walking and running

The gaits of the myriapod and hexapod classes represent specialized developments from gaits such as those of the Onychophora, which in each class have come to serve particular needs. Facilitating morphology has evolved along with the advancement of the gaits (see Chapters 8 and 9). Since the data is considerable, an attempt has been made in Figs. 7.8–10 to indicate the resemblances and differences between the principal gaits found in the Uniramia and their relationships one to another.

A. *The advantages of the retention of a lobopodium; of narrow, pointed limb-tips; and of leg-rocking*

(i) *The lobopodium.* That the Onychophora have not exploited the potentialities of their simple gaits used in their slow modes of walking is due to a habit of overriding importance, the ability to distort the body so greatly that the animals can squeeze through very narrow channels and thereby reach the protection of more commodious spaces out of reach of larger predators (Pl. 4(c), where all holes are passable to the animal). This habit must have been of such survival value that the primitive lobopodium was retained and further developments of walking and running were not required. A lobopodium can be flattened so that it does not project from the body.

(ii) *The narrow, pointed tarsus.* The evolution of a jointed exoskeleton from uniramous lobopodia brings two advantages, hitherto not appreciated. The narrow, pointed limb-tip, whether unguligrade with one or two terminal claws (myriapods and apterygote hexapods), or plantigrade (*Scutigera* and *Pterygota*), enables leg $n+1$ to be put on the ground simultaneously with, or even before, the raising of leg n and often on almost the same footprint (Fig. 7.8(b), (c), 7.9(d), (e)), so transferring the body weight instantly from one leg to the next. This is impossible to the blunt-legged Onychophora (Fig. 7.9(a)). Such improved stepping, made possible by pointed limbs, occurs in all myriapods and hexapods and facilitates their gaits and speedier locomotion (see below) and indeed has made possible the evolution of the hexapod classes.

(iii) *Leg-rocking*. A second advance follows on the possession of jointed limb exoskeleton providing a basal promotor–remotor swing and levator–depressor movement at the next more distal joint (Figs. 2.2, 2.3, 5.3, 5.15). Proximal depressor movements caused by intrinsic and extrinsic muscles can facilitate leg extension in the absence of extensor muscles to several of the distal leg joints, but a rocking movement of the leg assists this movement. The rocking is independent of the promotor–remotor swing and of the levator–depressor movement. The rock is achieved by various ways in different classes (see Chapter 5, §5), but in all it brings the dorsal face of the leg, with its line of dorsal hinge articulations, to a more anterior position during the propulsive backstroke and the reverse on the forward swing.

B. *The gaits of Onychophora*

The Onychophora employ three types of gait, which however merge into one another (Figs. 7.8(a), 7.9(a)). The slow, or 'bottom-gear' gait is used for starting to walk and for gaining speed. The 'middle-gear' gait is used for easy walking once progression has been well established. The 'top-gear' gait gives the best the animals can do towards a fast gait pattern. But this 'top-gear' gait, in the records obtained, is executed at a longer pace duration than the 'middle-gear' gait and the speeds achieved by both gaits were 7 to 9 mm/s in a *Peripatopsis* 66 mm long, with one 10 mm/s record from the 'middle-gear' gait. The 'top-gear' gait appears to be an alternative and perhaps an easier method of walking faster than that of the 'middle-gear' gait. The pace durations in bright light show a very small range in the several gaits, 0·70 to 0·74 s during established walking, and a smaller angle of swing, not a larger one, was recorded in the 'top-gear' gait than in the 'middle-gear' gait. Progression is slow for the size of the animal, a maximum of 9 to 10 mm/s for *Peripatopsis* 66 mm long; compare the 500 mm/s or more found in myriapods and hexapods of shorter body lengths.

C. *The gaits of the Myriapoda and Hexapoda*

In the Myriapoda and Hexapoda any one species uses a series of gaits, but unlike the Onychophora, the range in pace durations is considerable. The faster patterns of gait are associated progressively with larger angles of swing of the leg, shorter pace durations, and greater speeds of running. Thus the 'bottom-gear', 'middle-gear', and 'top-gear' gaits of the Onychophora may remain so designated because they do not correspond exactly with the 'slow', 'medium', and 'fast' gaits of Myriapoda and Hexapoda. The ability to walk about reasonably fast in Myriapoda and Hexapoda is of much more importance for survival than it is to the Onychophora, except for those species living inside decaying logs, etc. Under bright illumination Onychophora show variations in the movements of one or of several legs without causing interference in the execution of the gait pattern by the remaining legs. The Myriapoda and the Hexapoda, using much shorter pace durations and limb-tips which are often very close together, cannot permit such irregularities. The onychophoran facilitating morphology has been described in Chapter 4, §5.

(i) *The gaits of Symphyla and of epimorphic Chilopoda.* Three gaits of the Symphyla and three from the longer series used by the epimorphic Chilopoda are given in Fig. 7.8(b), (c). The relative durations of forward and backstrokes and the phase differences between successive legs are entered above the zigzag lines ('gait diagrams') depicting leg movements progressing from left to right. Below some of these gait diagrams are shown the dispositions of the leg in dorsal view at one moment in time ('gait stills'), heavy lines meaning a propulsive leg with its tip in contact with the ground and thin lines meaning the reverse.

It is seen that the symphylan gaits (e.g. *Scutigerella*, Pl. 4(e)) and those of the Epimorpha (e.g. *Scolopendra*. Pl. 1(f)) form an approximately continuous series in which the relative duration of the backstroke decreases from 7·5 to 1·5 while the phase difference between successive legs changes from 0·25 to 0·92 of a pace. The fastest pattern of symphylan gaits (5·2:4·8) and the slowest of the epimorphan series (5·0:5·0) are closest to the 'middle-gear' gait of the Onychophora, as indicated by the two heavy arrows in Fig. 7.8. The advantage of narrow, pointed limb-tips enables leg $n+1$ of the Symphyla to be put on the ground simultaneously with the uptake of leg n, and not well after this event, as in the blunt-legged Onychophora, and each leg of *Scolopendra* is put on the ground well before the preceding leg is raised. In this way are

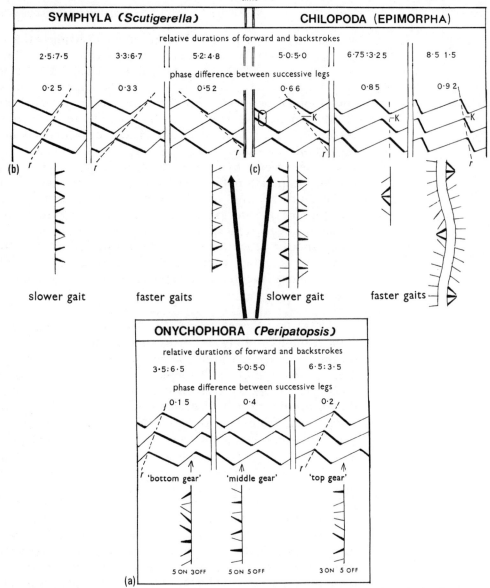

FIG. 7.8 (above), FIG. 7.9 (opposite), FIG. 7.10 (over page). Diagrams depicting the range of gaits found in the various classes of Myriapoda and Hexapoda, shown as gait diagrams to indicate the relationship between these gaits and those of the Onychophora. The latter are repeated on each figure. The relative durations of forward and backstrokes are given above, a whole cycle of movements being taken as ten. The phase difference between successive legs in each gait is given as that proportion of a pace whereby leg $n+1$ is in advance of leg n. As with other diagrams of gaits, heavy lines indicate legs performing the propulsive backstroke and thin lines those in the recovery forward swing. Three successive legs are shown for each animal, forming gait diagrams, time progressing from left to right. Below most of these diagrams are shown gait stills which are diagrammatic dorsal views of either a bilateral or unilateral series of legs performing the given gaits. On some figures the moments in time shown by the gait stills are marked by arrowed or dotted lines, on, or towards the gait diagrams above. In the Diplopoda and

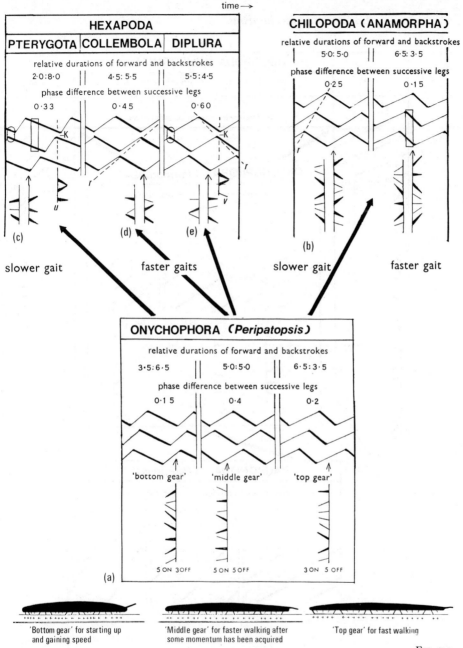

Fig. 7.9

Symphyla, 8(b), 10(b), (c), the paired legs are in the same phase and left legs, if drawn, would be mirror images of the right legs shown on Figs. 7.8(b), 7.10(b), (c). On Fig. 7.9(b), (c), (e), the rectangles mark successive propulsive legs which diverge and are performing different paces, and occupy successive groups of footprints; the somewhat elliptical lines on Fig. 7.9(c), mark successive propulsive legs which converge onto the same group of footprints. For further description see text. (After Manton 1973*b*.)

Locomotory mechanisms

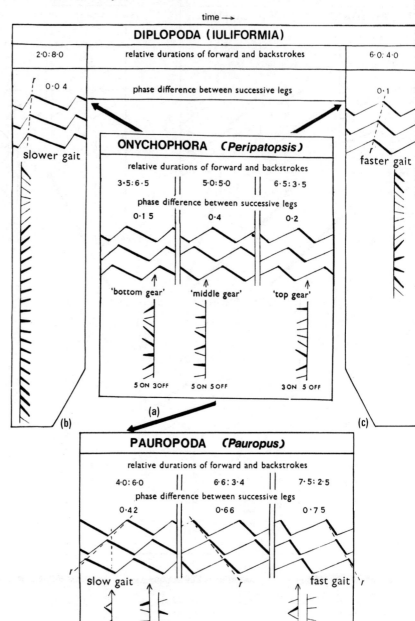

FIG. 7.10. For explanation see pages 322–3.

formed the groups of three legs occupying almost the same footprint (drawn below gaits (6·75 : 3·25) in Fig. 7.8(c)) which are so characteristic of the epimorphic Chilopoda (Pl. 1(f)). In the slow gaits this disposition is almost but not quite achieved and there is slight diversion from it in the fastest gait. Minimal pace durations are of about 0·7 s in the Onychophora, can be reduced to about 0·04 s in the Epimorpha and 0·02 s and less in the Symphyla as the result of striated muscles and the advantage of pointed legs. But there are other morphological features of importance (see Chapter 8).

Symphylan fast running, only possible in sudden darts of short duration, is executed by very rapid stepping and not very fast patterns of gait. This is dictated by the evolution of conspicuous flexibility of trunk (Pl. 4(d)), a higher priority, used in penetrating deeply into soil crevices without pushing. This flexibility is not compatible with fast patterns of gait such as used by the Epimorpha where opposite structural features secure trunk stability (Chapter 8, §4, 5).

(ii) *The gaits of anamorphic Chilopoda*. The range of gaits used by anamorphic Chilopoda (Fig. 7.9(b)) is not as great as in the epimorphic Chilopoda (Figs. 6.13, 7.8(c)). These gaits are executed by longer legs than could be used by the epimorphic Chilopoda. Propulsive leg tips must diverge, unless very small angles of swing are used. The gaits of the Anamorpha resemble the 'top-gear' gait of Onychophora, Fig. 7.9(a), (b) right side, heavy arrow, but are executed with great rapidity.

The gaits of the Epimorpha and Anamorpha represent alternative ways of running fast. The longer legs of the latter require a longer pace duration, 0·07 s in *Lithobius* and 0·08 s in *Scutigera*, in contrast to the 0·04 s in the scolopendromorph *Cryptops*. The maximum recorded speed of *Lithobius forficatus* 25 mm length was 280 mm/s, an achievement of much the same order as the recorded 260 mm/s of *Cryptops anomalans* 38 mm length. *Scutigera coleoptrata* 22 mm length with very long legs is the fleetest of all myriapods, achieving a recorded 420 mm/s, but can certainly run faster.

(iii) *Divergence in feeding habits of epimorphic and anamorphic Chilopoda, as determinants of the choice of gaits*. A divergence in feeding habits, gaits, and associated structure must have constituted a parting in the directions of early chilopodan evolution. The short-legged Epimorpha can penetrate deeply

Locomotory mechanisms

into crevices and the remarkable flattening of their head and feeding organs (Pl. 5(g) shows slots passable to the animal) enables them to feed within as well as outside crevices. They can deal with prey which is protected by heavy cuticle. Small, soft animals can be swallowed whole, but the efficiency of the feeding mechanism centres round the ability to seize, paralyse, and tear lesions into the bodies of armoured arthropods by means of paired large and strong poison claws (Chapter 3, §5B(iii), Fig. 3.22). The soft internal parts of the prey are scooped out by softish mouth-parts aided by external digestion; the movements of mandibles, maxillae 1, and maxillae 2 are associated with this mode of feeding. Among Epimorpha, all Geophilomorpha and many Scolopendromorpha are blind; in the others there are only four pairs of simple ocelli which do not appear to give acute vision; these animals do not apprehend their prey at a distance. All are adept at dealing with prey in confined spaces.

The Anamorpha, by contrast, pounce upon spiders and flies with alacrity. Their vision is usually acute and their catching movements swift. These longer-legged Chilopoda can start up in their fastest patterns of gait, but the Epimorpha start running in their slower gaits and gradually increase speed by transition to faster gaits. This transition does not take long, but briskness off the mark in the Anamorpha must be of extreme importance in catching flies. The form of the more lightly built anamorphan poison claws (Figs. 3.22, 8.17) provides swifter weaker movements than those of the Epimorpha. Prey once immobilized by the Anamorpha is cut up by a much more powerful mandible, pieces being swallowed whole; compare the small weaker mandibles of the Epimorpha with those of *Lithobius* and *Scutigera* (see Chapters 3 and 8, Figs. 3.22, 8.17). The management of so huge a mandible as that of *Scutigera*, housed within the head capsule, is a peak of evolutionary achievement. Small prey can, of course, be swallowed whole by any chilopod.

The hunting and feeding methods of the anamorphic Chilopoda must have led to increasing leg length, suited to produce very swift movements, and thus imposing the use of the anamorphic type of gaits. The need for speed obtained by longer legs has meant an increasing need for perfection in anti-undulation mechanism and of other morphological advances (see Chapter 8, §5C). The gaits of the Anamorpha

Locomotory mechanisms

cannot be comprehended without reference to facilitating morphology and the habits other than running, with which they have evolved.

(iv) *The gaits of Pauropoda.* The slowest gaits used by the short-legged Pauropoda (Fig. 7.10(d) left side) are seen in the early instars with fewer legs than the adult. But slow and fast gaits can be employed by the adult. Leg $n+1$ is put on the ground simultaneously with the raising of leg n; otherwise the gait is similar to the 'middle-gear' gait of the Onychophora, but the pace duration is much shorter, 0·2 to 0·25 s. As the number of legs increases to the adult 9 pairs, so the ability to execute fast patterns of gait ensues. The phase difference between successive legs increases as the relative duration of the backstroke decreases, but not so steeply as in the Epimorpha, and thus the three successive legs converging on to one footprint, so characteristic of the Epimorpha, see Fig. 7.8(c) gait (6·75 : 3·25), are never present, leg n and leg $n+2$ being off the ground when leg $n+1$ is halfway through the backstroke. However, the nature of the gaits of the Pauropoda bears a general resemblance to those of the Scolopendromorpha. The adult is very unstable when using its fastest gait, with few legs in contact with the ground, and is readily blown over by air currents. But the normal environment is draught-free within decaying logs, etc. Undulations of the body are completely controlled (see Chapter 8, §6).

(v) *The gaits of Diplopoda.* The closely set, not very short, legs of the Diplopoda necessitate divergence of successive propulsive legs, as in anamorphic Chilopoda, the positions resulting from phase differences between successive legs of less than 0·5. Some features concerning diplopod stepping have been mentioned in §3C above. Faster and slower diplopod gaits are shown in Fig. 7.10(b), (c). They bear obvious resemblance to the 'bottom-gear' and 'top-gear' gaits of the Onychophora, Fig. 7.10(a). The great speciality of the Diplopoda is the execution of extremely slow patterns of gait (Fig. 7.10(b)) in which very many legs are simultaneously propulsive in long metachronal waves (Pl. 7(e)), paired legs being in similar phase. By this means a maximum propulsive thrust is exerted by the legs for use in burrowing, either by head-on shoving or by pushing with the dorsal surface (for diplopodan burrowing techniques see Chapter 8, §3A). Such gaits could have been readily derived from the 'bottom-gear'

Locomotory mechanisms

gaits of the Onychophora, where the legs are used usually in similar phase.

The prerequisites for slow, strong gaits of diplopods are very many leg pairs and intersegmental joints which do not telescope. The leg force must be transmitted as fully as possible to the pushing surfaces at the head end or at the dorsal or antero-dorsal parts of the tergites. Neither of these features is an easy morphogenetic acquisition (see Chapter 8). The faster gaits, Fig. 7.10(c), with shorter metachronal waves and groups of fewer legs simultaneously in the propulsive phase, are seen as the animals walk over a surface and are not serviceable for burrowing (Pl. 7(d)). Diplopods without very large numbers of legs (Pl. 6(n)) cannot exert so great a head-on pushing force against the substratum as those with very many legs, but they have other accomplishments, such as the ability of the Oniscomorpha to roll into a sphere (Pl. 3(e)), and the strong pushing with the dorsal surface by Polydesmoidea and Nematophora (Pls. 4(h), (i), 5(f)).

(vi) *The effect of body size on locomotory forces.* The utilization of strong, slow gaits for burrowing by the Iuliformia is only effective over a certain size range. The force put out by the legs is proportional to the transverse sectional area of the muscles concerned, but the weight of the animal, as it becomes larger, increases in terms of cubes. Thus with increase in size, more and more of the leg force is needed to carry the body weight leaving less available for pushing. The largest known Iuliformia are near the theoretical limit in size for effective burrowing and many of these species in fact burrow little except into soft material.

It is quite easy to measure the maximum force which can be exerted by a millipede. Most species will not push with the head end to order, but when harnessed by an adhesive strap from the dorsal surface (Pl. 4(h)) attached by a lead to a suitable sized pan, they make the most willing 'cart-horses'. The pulling force exerted by the legs cannot be very different from the pushing force normally used in burrowing and the maximum pulling force can be determined by the addition of weight into the pan. Rough damp paper forms a suitable substratum. The actual force exerted by the animal can be ascertained by attaching the same maximum loaded pan (on the same damp paper) by a fine thread over a pulley to another pan hanging freely in which gram weights can be added

Locomotory mechanisms

sufficient just to shift the 'cart'. The method is serviceable for ascertaining differences between species.

Comparisons between different sized millipedes with roughly comparable leg lengths are shown in Fig. 7.11 where the force exerted by each species per weight of the animal is plotted against the cube-root of the volume of a diplosegment, thus making allowance for Iuliformia with different numbers of legs. It is seen that the larger species, to the right, exert the smaller pushing force. The vertical arrow marks the size of the largest known species of iuliform millipede and any larger size would clearly reduce the burrowing ability to almost nil. *Blaniulus* on the left is a very small millipede with consequently a large force/weight, but the area of application of this force to soil particles is small and this little beast must be near the lower limit in size for effective burrowing, unless soil particles are small. The most effective burrowers today have sizes between those of *Blaniulus* and *Ophistreptus*, but the latter is a poor burrower.

Some comparisons may be made with the pulling ability of small vertebrates. If a mouse or a lizard of equal weight to one

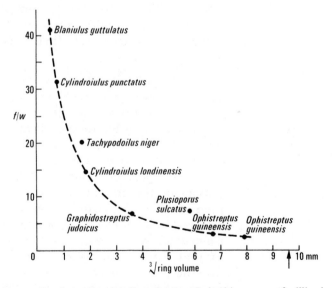

FIG. 7.11. Graph showing the effect of size on the pushing power of millipedes. The pushing force exerted (f)/body weight (w) is plotted against the $\sqrt[3]{}$ *ring volume*, and points for iuliform species with approximately comparable leg length are joined by the dotted line. The greater force exerted by *Tachypodoiulus* and *Plusioporus* is correlated with their longer legs and their points lie outside the curved dotted line. For further description see text. (After Manton 1954.)

of the larger iuliform millipedes be harnessed to a 'cart' it is found to be able to put out but a quarter of the pulling force of the millipede. But millipedes can only be efficient over a certain size range and this range is quite different from that of vertebrates. Vertebrates can become much larger and exert considerable forces from their limbs, but only using two pairs, not the many small limbs of a millipede. Functional matters set a limit to the smallness in size of a vertebrate but this size greatly exceeds the lower limit in size of a iuliform millipede.

(vii) *Conclusions concerning gaits and facilitating morphology of the myriapod classes.* The co-ordination of the data show- the gaits of every myriapod class could have been derived from one or other of the simple gaits seen in the Onychophora. But each myriapod class has advanced its selected type of gait far beyond an onychophoran-like level and each one along different lines. It is suggested that ancestral millipedes at a soft-bodied stage used gaits much as in Onychophora today. The Onychophora cannot have been ancestral to any myriapod, because the head morphology is so different. The trunk skeleto-musculature of each myriapod class is so profoundly bound up with the gaits employed (Chapter 8) as to indicate that in each class the one evolved with the other.

(viii) *The gaits of Hexapoda.* Some hexapod running and walking gaits are shown in Fig. 7.9, one of the slowest gaits of a pterygote in (c), a faster gait in a collembolan in (d), and the fastest gait recorded for a dipluran, *Campodea*, in (e). All are limited to medium and slower patterns because of the need for stability by a six-legged animal. Three legs simultaneously upon the ground and disposed about the centre of gravity of the animal are essential to give stability at all moments. All leg pairs move in exactly opposite phase. Rarely, short unstable moments can be managed, as in *Campodea*, when momentarily two legs only are on the ground. In the gaits shown in Fig. 7.9(a), 'bottom-gear', 'middle-gear', and 'top-gear' gaits respectively show five, four, and three of the six legs in the propulsive phase at the moments drawn below. The most important difference between these hexapod gaits and those of the Onychophora lies in the improved stepping in which leg $n+1$ is put on the ground simultaneously with the raising of leg n (d), or as is

Locomotory mechanisms

more usual, well before this event, (c) and (e), and the more rapid stepping. Hexapodous running would not be possible with the onychophoran type of stepping in which leg $n+1$ is put on the ground well after leg n is raised (Fig. 7.9(a)) because of the wide blunt leg tips.

Successive propulsive legs of hexapods diverge, except when momentarily changing footholds (Pl. 6(i), right side, leg 2 has been put down and leg 1 is about to be raised), a necessity for legs which are long and well fanned out, so spreading their fields of movement. The range in gaits used by hexapods extends from the slowest possible pattern to those shown in Figs. 7.2, 7.3, 7.9(c)–(e). The gaits could have been derived from the 'bottom-gear' and 'middle-gear' types of gait seen in Onychophora (heavy arrows in Fig. 7.9). The advantage of the hexapodous condition (Chapter 5, §3) provides great freedom of movement by the three pairs of long legs which do not interfere mechanically with one another because they are well fanned out. Gait changes over a limited field are easy. The three leg pairs must be situated close together, thereby minimizing undesirable body undulations caused by paired legs moving in opposite phase (Pl. 6(d), (e), (g), (h), (i)).

There are many specialized methods of using hexapod legs, the quadrupedal walking of *Protura* (Pl. 6(k)) and many butterflies being examples. All the various ways in which hexapod legs are used, Figs. 7.2, 7.3, 7.9, occur within the limitations of normal hexapod gaits used in walking or running. But in each of the five hexapod classes there are constant ways in which the legs are used and the differences between the classes reveal themselves clearly in the tracks (Figs. 7.12, 7.13(h), (g) and legends and Chapter 5, §3). The general similarities between the gaits of the five hexapod classes are due to the limitations set by using only three pairs of legs.

7. Jumping in Myriapoda and Hexapoda

Jumping may be carried out by legs, by the body, and by special jumping organs. Regular series of jumps or hops result in a method of steady progression in the Thysanura and in a few millipedes. Isolated large jumps, executed at irregular intervals and in various directions, upwards, forwards, backwards, or sideways, form escape reactions making the animal difficult to follow by a predator. Only a few of the many examples of jumping will be mentioned here.

Locomotory mechanisms

A. *Jumping gaits of Thysanura*

Some very unusual hexapod running is shown by the jumping gaits of the Machilidae in which each leg pair jumps in similar phase. As in hopping and jumping animals of other phyla, the momentum of the jump carries the animal through unstable moments. In the Machilidae only one pair of legs is in contact with the ground at a time during the fastest jumping gait, shown by the black rectangles in Fig. 7.13(a). The backstroke starts a little before the leg strikes the ground and the recovery swing lies between successive rectangles. Leg 1 (black rectangle) leaves the ground long before legs 2 jump and similarly for legs 3 performing the same pace (marked on the left), so that the footprints are staggered in a forward direction, as in the track shown in Fig. 7.13(h). The order of jumps is always the same, legs 1 jump, then legs 3 and then legs 2 (marked by circles on the same figure) and the phase differences between successive legs are always 0.33.

The photo of *Petrobius* in Pl. 6(f) shows a slightly slower jumping gait, about (5.0:5.0) in pattern (see rectangles marked by an interrupted line in Fig. 7.13(a)), giving less forwardly staggered groups of footprints and the number of propulsive legs in contact with the ground changing from 2 to 4 at successive moments throughout each pace (Fig. 7.13(g), (d), cf. (h) and (c)) with 2 propulsive leg pairs at all moments. The blurred image of the animal in Pl. 6(f) is intentional to show a differentiation of fast-moving and slow or stationary parts at an exposure time of 1/60 s. The body is moving

---

FIG. 7.12 (see opposite). Diagram showing footprint positions, from tracks made by hexapods running over smoked paper and performing their fastest gait. The tracks are scaled to a roughly common thoracic segment volume in order to show comparisons between species. Actual tracks are seen on Pl. 6(d)–(g). The median line between the right and left footprints measures two stride lengths and the numbers 1–3 indicate the positions of the three paired thoracic footprints. (a) Collembola, *Tomocerus longicornis* (Pl. 6(a),(d)). The change in footfalls is almost simultaneous as shown in Fig. 7.3 and therefore the footprints lie at the same transverse level, see also Pl. 6(d), where left leg 3 is on the ground in between the footprints of legs 1 and 2. (b)–(c) Diplura (b) campodeid (Pl. 6(e)), in which the differential length of the thoracic legs is so great that leg 2 is put on the ground in front of leg 1 before the latter is raised and leg 3 is put on the ground in front of leg 2 before that leg is raised; (c) japygid, in which the three legs are so short that the three footprints are backwardly and not forwardly staggered. (d)–(e) Thysanura, (d) *Thermobia* sp. (Pl. 6(g)), paired legs in opposite phase and forwardly staggered footprints alternate in position, (e) *Petrobius brevistylis* (Pl. 6(f), (m)), in which the paired legs move in similar phase and the forwardly staggered series of footprints, as in *Thermobia*, are formed by jumping gaits. (f)–(g) Pterygota, (f) the beetle *Cylindronotus laevioctostriatus* (Pl. 6(i)) and (g) *Dolycoris baccarum*, a hemipteran, both showing the typical disposition of pterygote footprints contrasting with all others. (After Manton 1972.)

Locomotory mechanisms

forward fast. The tarsi of legs 1 are stationary, on the ground, forming a clear double claw-mark on the right, plainly seen also in the track showing three previous footfalls of legs 1. Legs 2 are moving forward so rapidly that only a blurred field

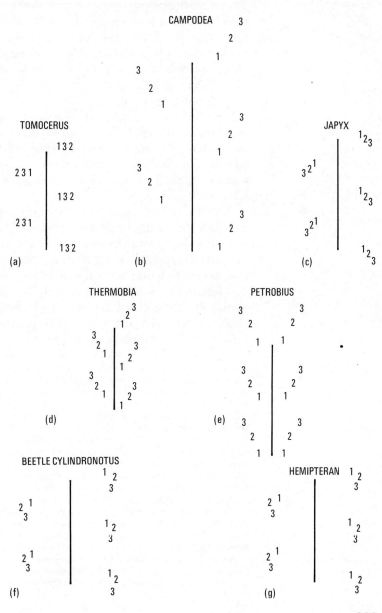

333

Locomotory mechanisms

of movement is seen; they have just vacated the spot-like footprint to the side and a little in front of the double claw-marks of legs 1. Legs 3 are also off the ground, having vacated the most lateral footprints level with the posterior abdomen and will be put down lateral to the two footprints of legs 1 and 2 level with the posterior part of the thorax; but these legs 3 are almost sharp, showing that they were moving forward less fast than the body and preparing for their next jumping backstroke. These remarkable jumping gaits are only practicable on upper surfaces. On the underside of stones normal slow hexapod gaits with legs in opposite phase are used, the tarsi serving for clinging.

The Lepismatidae use their legs in opposite phase, but the movement of each leg, employed singly, is a jumping movement, forming forwardly staggered footprints as in the Machilidae, but alternating on the two sides, see tracks d and e in Fig. 7.12. The leg action and morphology are unique (see Chapter 9).

The dipluran *Campodea* is the only other hexapod known to produce forwardly staggered groups of footprints (Fig. 7.12), but here the great difference in length of successive leg pairs permits the footfall of leg 2 to take place well in front of leg 1, and in slower gaits, but not in the faster ones, just before leg 1 has left the ground, and similarly for legs 3 and 2, Pl. 6(e) shows the fast gait pattern (5·5 : 4·5), phase difference 0·6 with slight irregularity at the moment of exposure to very bright lights; and see Manton 1972, Fig. 5).

FIG. 7.13 (see opposite). *Petrobius brevistylis,* jumping gaits (Pl. 6(f), (m)). Both legs of a pair are in the same phase, cf. all other hexapods. (a) Three alternative gaits are shown on a composite gait diagram. Horizontal rectangles indicate the period of time during which a leg pair is in contact with the ground. The backstroke starts a little before the leg strikes the ground, and the recovery forward stroke, unmarked, lies between successive rectangles. Time progresses from left to right as shown by the horizontal arrows. The relative durations of the periods of time during which each leg is off and on the ground in each gait are given above, the black rectangle being the fastest gait (6·37:3·3), the rectangle marked by interrupted lines being the intermediate gait (5·0:5·0), and that marked by dotted lines being the slowest gait (3·3:6·7). All have a phase difference of 0·33 between successive legs, so that footfalls, marked by large rings, occur at equal intervals of time, see (b). (c)–(e) Show the number of propulsive legs at successive moments throughout a pace in the three gaits. (g)–(h) The plan (g) of four successive groups of footprints from the track shown in Pl. 6(f), length of the animal 10 mm, stride of jump for each leg 6·4 mm, pattern of gait (5·0:5·0) approx. (h) Plan of four successive groups of footprints from a track made at greater speed, scale to correspond with body length 10 mm, stride or jump length 8·3 mm, pattern of the gait between (6·0:4·0) and (6·7:3·3). The four paces shown on the tracks correspond with the four paces marked by oblique lines on (a). (After Manton 1972.)

Locomotory mechanisms

B. *Jumping millipedes*

Millipedes of the genus *Diopsiulus*, found in Sierra Leone, all show a characteristic jumping behaviour. The animals jump in the direction in which they run. A millipede jumps 2–3 cm, lands, slides, runs forward, jumps again, lands, slides, and so on. Evans and Blower (1973) have analysed this behaviour by high-speed cinematography. Essentially the jump is an escape reaction. A sudden and rapid humping of the anterior part of the body behind the head raises the centre of gravity (black spots in Fig. 7.14). The humps become a loop which is then thrown upwards and forwards, dragging the posterior half of the animal after it. The head and anterior segments remain stationary upon the ground until they are overtaken

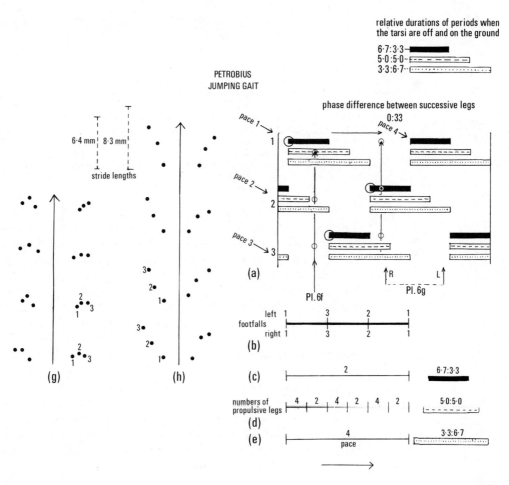

Locomotory mechanisms

by the rapidly moving body loop, when they are also pulled forwards and upwards. The millipede lands in a U-shaped position and normally falls on its side. It swings the head forwards after coming to rest and then runs forward before the next jump. This remarkable behaviour is probably a modification of the normal defensive rapid spiralling of the body (Pl. 4(o)) shown by most millipedes (see Chapter 8).

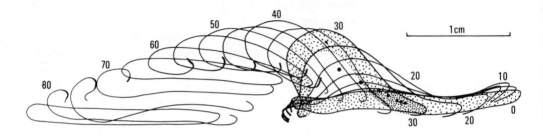

FIG. 7.14. Side view of a jump, from right to left, of the millipede *Diopsiulus regressus*, based upon successive frames of a ciné film taken at 2000 frames s^{-1}. Every tenth frame is shown, and the figures are in ms after the start of the jump. For most frames only the dorsal edge of the profile and a head and tail are shown (the head is indicated by a thickened line). The initial position (at 0 ms) and the position at take-off (30 ms) are shaded. The successive positions of the centres of gravity of each profile shown between the 0 and 30 ms positions are indicated by large black dots. They cover the acceleration distance. In this particular jump the millipede rotated 180° about its longitudinal axis between take-off (at 30 ms) and landing (at 75 ms). (After Evans and Blower 1973.)

C. High jumping and escape reactions of hexapods

The high jumping of locusts and grasshoppers is mediated by one pair of very large legs, the jumps being isolated events. Comparable but very high jumping by fleas using legs 3 is associated with much complex morphology (Rothschild, Parker, and Sternberg, 1972). High jumping escape reactions are used by the Machilidae, the motive force coming, not from the legs, but from the abdomen (Chapter 9). The click beetle ousts the flea in its jumping power, projecting itself into the air to a height of 30 cm from a position lying on its back, the mechanism resembling that of a mousetrap (Evans 1972). In Collembola the high jumping escape reactions are implemented by the terminal springing organ, probably derived from a pair of legs. The whole anatomy of the jumping Collembola is profoundly modified in association with the jumping mechanism, the motive force for which comes from the trunk musculature working on the hydrostatic skeleton (Chapter 9). Many adult and larval hexapods use catapult principles in projecting themselves upwards, such as click beetles which start by lying on their backs (Evans 1972). Various dipteran larvae such as cheese skippers (Sepsidae) and some fruit-fly larvae, e.g. *Dacus* (Oldroyd 1964) arch the body until the mandibles grasp a papilla at the caudal end and when muscle tension has built up, the jaws let go and the body straightens and jumps, like a spring unbending.

A little detail of only two contrasting examples of jumping are given below and their morphological basis will be found in Chapter 9.

(i) *High jumping by Machilidae.* When the seashore bristletail *Petrobius* is disturbed it jumps high off the ground landing some 3–10 cm away. The jump may be repeated several times, but the animal soon tires. This predator-baffling reaction lands the animal with the head pointing in any direction. *Petrobius* may turn and face its original direction of movement, or it may run off by the normal jumping gait in any direction. The jumping action consists of a strong tail beat which causes an abdominal take-off by depression of the abdominal coxal plates against the ground. This movement swings up the anterior part of the body, including the centre of gravity. The details of both forward and backward jumps have been analysed by high-speed cinematography (Evans 1975). The pre-jump position with the tail cocked is shown in Pl. 6(m), with the abdomen strongly concave dorsally and a slight convex curvature of the thorax. This is the energy-storing position. A sequence of profiles of a small forward jump are shown in Fig. 7.15(a), the tail beat in the earlier frames being very marked.

It is suggested that the jumping results from two movements spreading from either end of the animal: 1) from the posterior end there is a relaxation of dorsal longitudinal muscles allowing a rapid straightening of the abdomen and a tail beat lasting 2–8 ms; 2) from the anterior end, there is an increasing ventral curvature of the thorax which often leads to complete rolling up. Movement (1) swings up the anterior end of the body, thus raising the centre of gravity, but the final direction of the jump, forwards, backwards or up, depends on the rapidity of movement (2), as shown in Fig. 7.15(b), (c).

Locomotory mechanisms

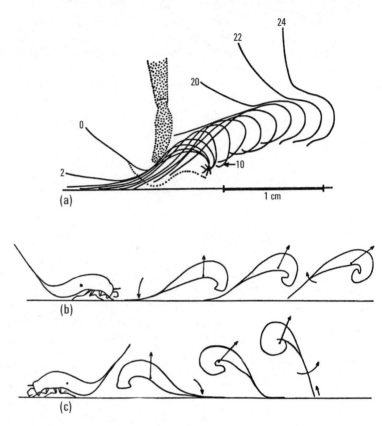

FIG. 7.15. High jumping in *Petrobius* (Pl. 6(m)). (a) The start of a small forward jump in side view (left to right). Dorsal profiles of even-numbered frames up to 24 are shown. The ventral surface of the *Petrobius* in frame o (stationary) is dotted, as is the brush which provided the stimulation for the jump. Frames are 1·05 ms apart. (b)–(c) A model of the way in which forward (b) and backward (c) jumps may take place in *Petrobius*. The bristletail is seen in side view. In both forward and backward jumps a rapid tail beat with abdominal depression serves to raise the centre of gravity, but the ventral curvature of the thorax takes place at different speeds. In (b) it occurs relatively slowly, thus allowing an anterior displacement of the centre of gravity leading to a forward jump. In (c) it proceeds rapidly so that the centre of gravity is finally displaced posteriorly, thus leading to a backward jump. The centre of gravity is in the 2nd abdominal segment. (After Evans 1975.)

(ii) *High jumping by Collembola.* When woodland litter is disturbed Collembola such as *Tomocerus* (Pl. 6(a), (d)) jump high into the air, repeatedly and in any direction, a very effective escape reaction. Again the mechanism of the jump has been unknown until recently. The existence of a jumping organ at the end of the abdomen, formed perhaps by the

Locomotory mechanisms

partial fusion of a pair of legs into a median manubrium and distal paired furca, is well known but the hydrostatic mechanism of movement of the organ is unique and comes as a surprise (Manton 1972).

The collembolan jumping organ is elaborately hinged to the 5th abdominal segment along the heavy line in Fig. 7.16(c) and seen end on as a spot in (a) and (b). Elaborate flexible and stiff parts of the cuticle enable the organ to swing

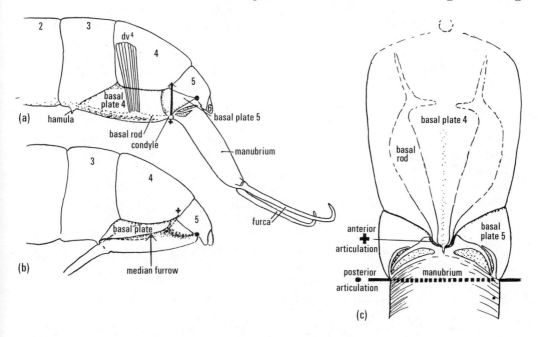

FIG. 7.16. The jumping organ of the collembolan *Orchesella villosa*. (a) Lateral view of the posterior end of the body with the springing organ extended, as at the end of a jump, drawn diagrammatically showing the limits of the sclerites; the midventral abdominal furrow is not shown. The posterior hinge between manubrium and body is marked by a black spot. The mobile anterior articulation between the arm of the basal rod on the body and the manubrium is marked by a cross. The arrow shows the direction of movement of this anterior articulation when the manubrium swings forwards to its resting position. (b) Lateral view of the posterior end of the body with the springing organ flexed forwards, almost to its resting position when the hamula impinges into the base of the furca (dentes). The point marked by a cross in (a) and the ends of the basal rods of the basal plate 4, are infolded and raised to the point marked by the cross in (b) corresponding with the arrowhead in (a). (c) Ventral view of the abdomen with the manubrium directed posteriorly as in (a), and on a larger scale. The axis of the posterior articulation between the manubrium and the body is shown by the heavy dotted line. The anterior articulation, marked by a cross as in (a) and (b), is shown in detail. The stippled cuticle indicates the slots of lesser sclerotization in the proximal part of the manubrium. (After Manton 1972.)

away from the body striking the ground with the furca. Almost all the force used to swing the springing organ against the ground is hydrostatic in origin, generated by slight but strong contractions of the longitudinal and deep oblique muscles of the trunk combined with devices which prevent expansion of the cuticle elsewhere. The return of the organ to its resting position is done by muscles, the point marked by a cross being pulled upwards as shown in Fig. 7.16(a), (b). A slot at the base of the furca or dentes fits over a ventral abdominal projection, the hamula, thus enabling the strong muscles used to put the springing organ away in its ventral groove, to relax (Manton 1972, §9A). The basic anatomy of Collembola must have evolved along with the hydrostatic needs for the springing organ dating from the earliest differentiation of the group (see Chapter 9, §2).

8. Neural co-ordination of arthropod movement
(by Professor G. A. Horridge, F.R.S.)

For some years it has been apparent that the co-ordination of movement in the Annelida and Arthropoda is primarily based on fixed patterns of interaction in the central nervous system. At first, however, progress was slow and until lines of thought escaped Sherrington's influence, which stressed reflex interactions, and a fresh start was effectively made by von Holst, before electrophysiological techniques provided the mass of recent confirmatory information.

Retzius, Zawarzin, and others trained in the last century, had shown that in arthropods the sensory cell bodies are peripheral, the motor neurons are relatively few per muscle, and both sensory and motor axons run all the way from the central nervous system to periphery. They showed that the central ganglia are composed of (a) motoneuron dendrites, (b) sensory axon arborizations which usually do not spread beyond the ganglion, and (c) side branches of interneurons. They showed that most interneurons spread over many segments with arborizations in each ganglion on the way, and that relatively few interneurons are restricted to one ganglion. It follows from this, but has been rarely said, that segmentation in the nervous system is largely an effect which is imposed on motor and sensory neurons by the peripheral muscles and sensory structures. The main integration of the animal is done by long interneurons which are not related to segmental boundaries. Anatomical details will be found in Bullock and Horridge (1963).

In annelids and arthropods the basic locomotory wave is

controlled by a central mechanism in which excitation travels forwards along the nerve cord and generates a sequence of movements that are caused by patterns of impulses in groups of motoneurons in successive ganglia. As shown by numerous experiments, including the famous one in which an earthworm is severed and then sewn together with a thread, waves affecting all segments can be initiated by any segment. The more advanced the arthropod the more likely is the basic central pattern to be modifiable by peripheral sense organs, but it seems at present that all specialized movements of all arthropods are basically centrally determined patterns that are derived from fixed connections between interneurons, and from interneurons to motoneurons. This applies alike to flight, song generation, jumping, swimming, eye-stalk movements in decapod Crustacea, respiratory rhythms, and especially to locomotion. For details, see papers by Ikeda and Wiersma (1964), Mendelsohn (1971), and Pearson, Fourtner and Wong (1973). Even so, feedback from peripheral sense organs, or lack of it where limbs are amputated, can so modify the phase, direction, or strength of individual movements that they appear to be reflexly controlled. Really, they are reflexly modified on a background central pattern, and so-called reflexes and postures in these animals are fixed patterns that are centrally controlled and triggered by appropriate peripheral stimuli, although endlessly elaborated in detail by environmental loads.

The velocity of the locomotory wave along the body is much slower than that of the slowest impulses along long interneurons. Therefore not these long neurons, but others that are restricted to single segments, must link and originate the sequences in each segment. The long interneurons activate and inhibit and may keep other ganglia informed of activity originating in each ganglion. Present evidence, derived from tedious recording from identified motoneurons and interneurons in insects and crustaceans, is that local feedback circuits and single neurons act as oscillators within each ganglion. Many fixed circuits are present in each ganglion to deal with groups of joints and different behaviour patterns. In the locomotory wave along the body activity in each ganglion sets off the appropriate oscillators in the next anterior ganglion with a latency that is relatively long and modifiable by sensory feedback, so that phase relations are controlled in relation to loading.

Electrophysiological analysis of the locust shows that the following details are highly significant (Burrows and Horridge 1974):

(1) Position of joints affects little the sequence of the motoneuron impulses in the same or other appendages.

(2) Motion of joints, via proprioceptors, has a small effect on subsequent movements of the same or other joints, except through forces acting via the ground.

(3) Reflex effects to the contralateral side and to other segments are very weak, except via forces acting through the ground.

(4) Loading of joints and resistance to spontaneous movement has a very strong effect on subsequent efforts made by the same or other joints.

From this it follows that recording of joint positions or movements will never give significant information for the analysis of locomotory mechanisms in terms of activity in central neurons, because forces and loads, not postures or excursions, are the factors accepted by the central mechanism in so far as it is influenced at all by reflex control.

Patterns of movement are generated by the patterns of connection between identifiable interneurons which drive motoneurons. The interaction which produces the pattern is essentially between interneurons.

Posture, in all its forms, is primarily based on a central 'set' within the nervous system and hardly at all determined by reflexes. Posture is determined by background levels of impulse rates in slow motoneurons, and these in turn are controlled by central neurons that are little influenced by joint movements or positions.

Rhythmic movements, especially those of locomotion, are 'switched' on or off by the activity of 'command' neurons which run for considerable distances along the cords and have branches in the appropriate ganglia. They drive stereotyped behaviour but the pattern of movement has its origin in the local ganglion: the command fibre activates but does not produce the pattern.

The way in which the oscillators generate pattern in the individual ganglia is not understood but there are strong indications from recordings of interneurons that the control of the pattern and rhythm lies with a few local interneurons which oscillate relatively slowly without the participation of nerve impulses. The data come from insect respiration, insect flight neurons, cicada song, crustacean mouth-part movements and cockroach walking, and so may represent a fundamental mechanism.

The conclusion from these recent findings on the nervous control mechanisms is that the interesting interactions are those that occur between interneurons and determine the exact form of pattern generated. Research is now at the stage of mapping these. It is already apparent that the evolution of progressively more complex movement patterns is the evolution of more individually unique and complex interneuron connections. A deeper level of understanding, as to how one pattern rather than another is dominant in appropriate circumstances, is not yet approached experimentally.

8 Habits and the evolution of the Myriapoda together with the basic skeleto-musculature of the Uniramia

1. Introduction

IN this chapter an attempt will be made to account, shortly, in terms of function, for the variety in overall body shapes characteristic of the many myriapodan taxa, besides an indication of the significance of some of their particular structural details. Such an assay on a functional basis has never been possible before. It is of importance for a better understanding of myriapodan evolution and also gives an indication that there is much significance in the overall shapes in other, less well understood, taxa.

The four myriapod classes, the Diplopoda, Chilopoda, Symphyla, and Pauropoda have in common: jointed, sclerotized body with paired limbs on most trunk segments; a common type of leg-rocking mechanism (Chapter 5); a particular type of head capsule; transversely biting, jointed mandibles which are abducted by swinging anterior tentorial apodemes (Chapter 3); a basically dignathan head with or without posterior additions (Chapter 6). The five hexapod classes differ from the Myriapoda in all these features, and in being hexapodous and not multi-legged.

The trunk segmentation of arthropods is established early in embryonic development, and it is the series of mesodermal somites which evokes the more superficial segmental structures, the limbs, the ganglia, and visible ectodermal features. We have found sound and reliable information concerning the laying down of the series of trunk segments in all arthropodan classes (Anderson 1973), but we must look to other disciplines for the meaning of the many externally visible divergencies which exist in the serial repetition of certain, mainly dorsal, features which appear in later developmental stages in myriapods in particular (§2B below).

2. Basic form of the skeletomusculature

First it is useful to review the component parts in which exoskeleton and endoskeleton is laid down and the basic form and functions of the major constituents of the musculature.

A. *The skeleton*

The exoskeleton consists of the surface cuticle, areas of which when little sclerotized form flexible arthrodial membrane at the intersegments and flexible flanks of the body (white on Fig. 8.1(a)). There is always some thin surface sclerotization which bends readily but does not stretch. Localized sclerites are formed where the cuticle is sclerotized or tanned to increasing extent, enhancing both the rigidity and the elastic properties, so that a bent sclerite will return to its original shape when deforming forces are removed, as in Chilopoda. Calcification may be the main stiffening agent producing a rigid non-deformable cuticle, as in Diplopoda.

Typically there may be a segmental series of dorsal tergites, ventral sternites, and one or more pairs of lateral pleurites (Figs. 8.1(c), 8.10, 5.12). The posterior edge of one tergite usually overlaps the anterior edge of the succeeding tergite. The sclerites together may envelop the whole body, as in Geophilomorpha and iuliform Diplopoda (Figs. 8.1(c), 8.3, Pls. 5(a)–(c), 7(c)–(e)). The margins of each sclerite may be permanent or the shape may change (Fig. 8.2), and the edge may not always be near the points marked q, v, f, and g (see below). Pleurites and other sclerites similarly change in shape so that the overall shape of a segment can alter considerably. The sclerites of a segment may fuse together forming a complete or almost complete ring, as in Diplopoda (Fig. 8.3, Pls. 4(i), 7(c)–(e)), or there may be considerable areas of flexible pleuron, as in Symphyla, Chilopoda, and Pauropoda (Figs. 8.1(a), 8.7, 5.10, 5.11). Pleurites in the form of paired procoxa and metacoxa often support the leg base when legs project laterally (Fig. 8.1(a), (c), 8.10). Many additional pleurites may be present (Fig. 8.1(c), 5.12).

The endoskeleton consists partly of apodemes projecting as folds of cuticle into the anterior of the body and used for muscle insertions (see prophragma of diplopods and spine apodeme of *Japyx*, Figs. 8.3, 9.6).

There is also a connective tissue endoskeleton, as in Onychophora, below the ectoderm. It is thin in most areas, but can be thickened into solid tendon which projects far into the body and, like the apodemes, carries much musculature (Fig. 8.2, tendons in black).

Habits and the evolution of the Myriapoda

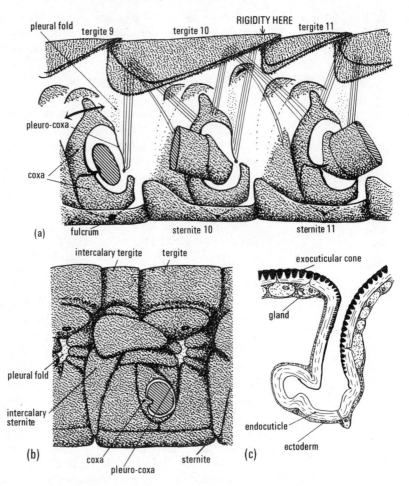

Fig. 8.1. Lateral views of some trunk segments, head end to the left, to show different degrees of heavy sclerotization. (a) Lateral view of three segments of *Lithobius forficatus* where the tergites and sternites are separated by flexible pleuron (white), the sternites are equal in length but the tergites are alternate in length; the coxa is expanded up the side of the body with some of the muscles shown which cause the rocking movement. The trochanter of leg 9 is cut at its base to show the strong anterior hinge between coxa and trochanter. The posterior part of the coxal ring is incomplete and ends in a small knob. (b) Lateral view of a geophilomorph centipede to show how the surface armour is complete. Yet the entire shape of a segment can change because the edges of the sclerites are mobile as shown in (c). Here exocuticular cones of heavy sclerotization set in the softer underlying endocuticle allow plenty of flexures in any direction. The basic plan of the sclerites in (b) is the same as in (a) except that intercalary tergites and sternites are well formed, the former being capable of sliding right over the tergite behind and there are more pleurites in (b). Note the complete investment of the coxa by surrounding sclerites. (From Manton 1965.)

In the Myriapoda and Hexapoda a metameric series of essentially transverse and intersegmental connective tissue endoskeletal units is formed along the trunk, as in trilobites and crustaceans (Chapter 3, §2E(ii)). Each may become very complex, especially in the apterygote hexapod thorax; greatest elaboration occurs where each unit becomes divided into lateral components, and these may sink forwards into the segment in front to different degrees, forming the series of segmental tendons within the lateral and sternal longitudinal muscles (Figs. 8.2, 8.13, 9.3, 9.4, 9.6, 9.10(a), 9.14, simplified

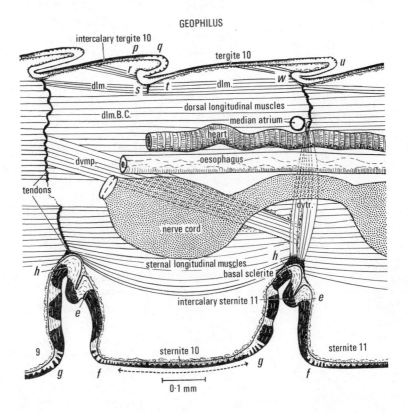

FIG. 8.2. Longitudinal half of *Geophilus carpophagus* viewed from the middle line with fat body and some muscles removed. The mobile edges of the intercalary and principal tergites and of the sternites are marked (see text). The segmental tendons attached to the anterior part of the intercalary tergite above and intercalary sternite below are shown in black and they carry the dorsal longitudinal and sternal longitudinal muscles on either face; they also carry the antagonistic deep oblique muscle *dvmp.* and deep dorso-ventral muscle *dvtr.* as shown. The lateral longitudinal muscles are obscured by the sternal longitudinal muscles (see Fig. 8.9). (From Manton 1965.)

tendons being shown in black). The main tendons of the dorsal longitudinal muscles probably belong to the same tendon series (Figs. 8.2, 9.14, etc.), but have become displaced posteriorly.

Muscles are inserted on to sclerites via tonofibrils, as in Crustacea (Fig. 3.11(a)) but not enough bulk of muscle could be inserted in this manner and derivatives of intersegmental tendons may sink into the body and even become detached from the surface connective tissue layer; see the ventro-lateral segmental tendons of chilopods (Fig. 8.13, tendons in black). Tendons in the Uniramia anchor muscles marginally or on to the faces of sclerites, or even on to flexible cuticular membrane near to sclerites, as is mechanically suitable. The details of all these features are correlated with habits.

B. The principal myriapodan divergencies in exoskeleton

These are: 1) the extra tergites on at least segments 4, 6, and 8 in the Symphyla; 2) there are well-formed intercalary tergites (Symphyla) and intercalary sternites as well in the geophilomorph centipedes (Figs. 8.2, 8.7, 8.8); 3) there are poorly-defined intercalary tergites in the scolopendromorph centipedes; 4) there is heteronomy in tergite lengths in the Scolopendromorpha, Lithobiomorpha, and Scutigeromorpha; the degree of heteronomy increasing in the scutigeromorph direction with long tergites in all these groups over legs 7 and 8 (Figs. 7.1, 5.7, 5.9, Pls. 1(f), 4(j), 5(e), (g)); 5) the Pauropoda lack tergites on every other segment (Fig. 5.11); and 6) in the Diplopoda most of the apparent trunk segments are diplosegments bearing two pairs of legs, two pairs of tracheal systems, two pairs of ostia to the heart and initially two pairs of mesodermal somites (Figs. 8.3, 8.5, 8.6, Pls. 5(f), 7(e)). There are other divergencies existing in the most anterior and most posterior segments of the body of many Myriapoda but these will not be considered here.

It has been suggested in Chapter 6, §5D that a differentiation of habits in early terrestrial Uniramia has led to the evolution of the structural features distinguishing the four classes of myriapods, and these include the segmental divergencies listed above as well as features of the head, which are also associated with habits, and considered below.

C. Muscular systems and their functions

(i) *Superficial muscles.* These muscles, e.g. Fig. 8.9 (*s.ob.a., s.ob.b., pms., plcx., tmi., ptm., pam.*), Fig. 8.3(a) (muscles on segments marked 3 and 4), control the alignment between

Habits and the evolution of the Myriapoda

successive segments and between well-separated tergites and sternites, and, when well developed, are capable of holding intersegmental flexures against strong outside forces. Some of the superficial pleural muscles form part of the leg-rocking mechanism in Chilopoda, Symphyla, and Pauropoda (see Figs. 5.7–11, *tcx., pct.l.,* etc.).

This category is found wherever there is marked trunk flexibility. These muscles are well represented in epimorphic Chilopoda, extremely well developed in the Diplopoda, present in Symphyla, fairly well formed in the thorax of the Diplura, and absent from the rigid bodies of Pauropoda except for one leg-rocking muscle, Fig. 5.11(b) *pct. tep.*, and in normal form in jumping Collembola.

(ii) *Longitudinal muscles.* In their simplest form they extend from each anterior marginal apodeme (Fig. 8.3(a), segments marked 1 and 2, muscles *ret.dor., ret.par., fl.inf.long., mus.lac.v.*) or from tendons of the major segmental sclerites to the next tendon (Fig. 8.2, tendons shown in black between muscles *dlm.B, C* and sternal longitudinal muscles). Long sectors of these muscles crossing more than one intersegment serve special purposes in certain groups (Fig. 8.13, *dlm.A, dlm.B,* Fig. 8.14, *dpl.* segment 8 to 10).

Dorsal longitudinal muscles. These muscles, with the laterals (when present) and sternal longitudinal muscles, give stability and cohesion to the whole body. In Geophilomorpha (Fig. 8.9) and to a lesser extent in Scolopendromorpha, shortening of the muscles produces thickening of the body, so that a burrowing thrust can be exerted upon the surrounding medium. In the jumping Collembola body shortening is of importance for the jumping mechanism as it is dorsally in the abdomen of the Thysanura Machilidae (see Chapter 9). Other specialized functions concern anti-undulation mechanisms.

Dorsal longitudinals are always present and usually well developed. The details differ considerably from group to group and they are always associated with function.

Lateral longitudinal muscles. These muscles, with the dorsal and sternal longitudinals, complete a cylinder of longitudinal muscles round the body which is particularly useful in causing body thickening and shortening used in burrowing by epimorphic Chilopoda (Fig. 8.9) and in hydrostatic jumping by Collembola (see Chapter 9).

Lateral longitudinals, as separate entities, are present only in the Onychophora, Diplopoda (*ret.par.*), epimorphic Chilopoda, and Collembola. They are clearly fused with the sternal longitudinals in anamorphic Chilopoda and are not distinguishable from them in other Myriapoda and Hexapoda.

Sternal longitudinal muscles. They (Figs. 8.9, 8.13), with the dorsal longitudinals, and the lateral longitudinals when present, maintain stability of the whole body and particularly control the sternites which carry the legs in Myriapoda, Diplura, and Protura.

These muscles are present in almost all groups, but are small or absent from some Diplopoda, e.g. Figs. 8.3, 8.4, *mus.lac.v.* (or retractor ventralis). They are usually termed ventral longitudinal muscles when lateral longitudinals are not recognizable.

(iii) *Deep oblique muscles*. This category (Figs. 8.9, 8.13, *dvmp., dvma.,* 9.3, 9.6, 9.7) has two main functions. The tension strongly shortens and stiffens the body and provides part of an effective anti-undulation mechanism. Wherever these muscles are present, so are their antagonists, the deep dorso-ventrals. The deep oblique and deep dorso-ventral muscles are not required in Pterygota where thoracic segments are well integrated by the exoskeleton and endoskeleton. Deep oblique muscles are absent in Symphyla (Fig. 5.10(a)) and weakly developed in Diplura (Figs. 9.6, 9.7), groups whose habits require great flexibility of trunk and thorax respectively. They are well developed and roughly similar in form in Pauropoda, Chilopoda, and Thysanura (Figs. 5.11(b), 8.9, left side 8.13). They occur in Diplopoda (Figs. 8.3, 8.4 *lev.ap.post.*) in modified form, sometimes an involvens inferus muscle being present additional to levator apophysis posticae, particularly where the ventral skeleton is free and needs stabilizing (Manton 1961 (text-fig. 35), 1973 (Appendix)). The deep obliques are enormously elaborate in the machilid abdomen where they are used in jumping (Figs. 9.13, 9.14).

(iv) *Deep dorso-ventral muscles*. These are the antagonists of the deep oblique muscles, and since they have only to extend relaxed muscles and do no outside work, they are less bulky than the deep obliques (see muscle *dvtr.* in Figs. 5.11(b), 8.9, 8.13). Additional functions in the Chilopoda and Diplura

Habits and the evolution of the Myriapoda

concern stabilizing the coxal base by the *dvc.* complex (Manton 1965).

Deep dorso-ventrals are present wherever there are deep obliques. They are absent in Symphyla and simple in *Campodea*, but in several sectors in the japygids. They are present where there is much trunk flexibility as in the dipluran thorax (Figs. 9.6, 9.7, *dv., dv.1., dv.2., dv.3*).

(v) *Conclusions concerning trunk musculature.* All the bewildering complexities of the trunk musculature of the Uniramia fall into line on the supposition that the musculature of the Onychophora, Myriapoda, and Hexapoda has been derived by easy morphogenetic changes in association with functional needs from ancestral musculature much as in the Onychophora today.

Onychophoran-like superficial circular and oblique musculature could have separated into the component sectors of the superficial oblique system, very different in detail in the various classes. Longitudinal muscles, already in dorsal, lateral, and sternal series, could easily have become divided into segmental units as the trunk exoskeleton evolved, giving anterior segmental apodemes or tendons of insertion for these muscles. An onychophoran-like muscle sheet 3 could have divided into the deep dorso-ventral and deep oblique muscles (see Chapter 4 and Figs. 4.25, 4.26).

Full details concerning the functional components of the trunk musculature of the Uniramia will be found in Manton 1973 (Appendix).

3. The Diplopoda

A marked ability to push, using the motive force of the legs, is as diagnostic of the Diplopoda as is the possession of diplosegments. This, and the ability to curl the body into a tight or a loose spiral, are the two habits which must have been of major importance in the differentiation of the group. An ability to burrow by pushing into the substratum, using a motive force derived mainly or entirely from the legs, is a habit with which the morphology of this class has evolved. Usually diplopods are slow moving. They are found in damp, but not sodden, environments; in small spaces under stones, in woodland leaf litter and in leaf mould, in decaying logs, etc. They feed on large quantities of decaying wood, decaying leaves, and rarely on living plants. Each species has its own particular microhabitat and food preferences.

A. *Diplopodan burrowing techniques*

Three types of burrowing are practised by different diplopodan orders. Each mode of burrowing is associated with its particular type of morphology but a simple form of head-on pushing into the substratum appears to be the habit of basic significance for the class and accounts for the evolution of diplosegments. This habit has been perfected in modern Iuliformia, §(i) below, and has given place to the two other derived methods of burrowing, (ii) 'flat-back' pushing, and (iii) wedge burrowing which will now be considered.

(i) *The iuliform type of head-on pushing and the evolution of diplosegments*. The burrowing shown by typical Iuliformia probably represents the present-day perfection of a type of burrowing which has led to the evolution of the many diagnostic diplopodan features. Head-on pushing by the cap-like collum (a dorsal sclerite) covering the intucked head, enables the animals to penetrate soil and decaying wood (Pl. 7(c)). The cuticle is calcified, stiff, and inflexible. The incompressibility of the joints between the diplosegmental skeletal rings enables the propulsive force exerted by the many legs to be transmitted forwards with the least loss of energy. The form of the skeleton is shown in Fig. 8.3. The overlap at a joint is exactly circular, enabling a little rotation on the long axis of the body, essential when an animal is walking itself out of the spiral protective position (Pl. 4(o)) and on other occasions when the substratum is not flat. The tergo-pleural arch covers most of the body and in extant Iuliformia is fused with the sternite (Fig. 8.3). The anterior margin of the tergo-pleural arch, or of the complete ring, is turned inwards at right angles to the body surfaces to form a strong apodeme, the prophragma, on both sides of which are inserted the dorsal and lateral longitudinal muscles (Fig. 8.3(a), see segments marked 1 and 2).

On dorso-ventral flexure the anterior margins of the sternites all along the body are pulled upwards into the preceding segments, as occurs during enrolment, or less great dorso-ventral bending at other times, so that the ventral surface is shortened into a tight zigzag formed by tilted sternites alternating with intersegmental arthrodial membrane as shown in Fig. 8.3(a), cf. 3(b), the dorsal overlap of the tergites being reduced.

Intersegmental joints with the flexibility shown by Fig. 8.3(a) could not be constructed closer together. Two single segments, with one pair of legs to each, occupying the same length as one diplosegment shown on Pl. 7(c)–(e) would be mechanically impossible. Herein lies the functional explanation of diplosegment evolution. Fusion together of most of the trunk segments in pairs to form diplosegments enables each to be deeper and shorter than single segments could be and a long unwieldy body composed of single segments would thus be avoided when many legs are a priority. It has already been shown how the strongest gaits are only possible to a diplopod where many legs are present (Chapter 7, §6C(v) and Pl. 7(c)–(e)) and therefore the possession of many diplosegments is suitable for a bulldozer type of burrowing. The general shape of the iuliform diplopods, that is the ratio of length to width of the body, is not exactly the same in all species, but there is a general similarity. Smaller genera roughly resemble the larger ones such as *Ophistreptus* (Pl. 7(c), (d)) in body proportions, although the smaller genera usually possess fewer, longer diplosegments than the larger, for the reasons considered in Chapter 7, §6C(vi). The force put out by the limb-musculature/body-weight becomes smaller as size increases and so it is understandable that species of larger size increase their number of diplosegments as far as is mechanically practicable, so increasing the number of propulsive legs; decrease in diplosegment length maintains the same type of overall shape. Doubtless the optimal iuliform length of body possessed by the various species is not uniform but varies with the exact habits and environmental niches occupied.

During head-on pushing the ventral surface of a iuliform diplopod is held as close to the substratum as the S-shaped bends (Fig. 5.3(a)) of the leg will permit. The range of swing during the propulsive backstroke is small and thus a maximal force can be exerted by the limb-tip against the ground. Pl. 7(c) shows the millipede *Ophistreptus* walking on a flat surface by a slow-pattern gait, with many legs in the propulsive phase (marked by black spots) and but few swinging forwards off the ground (marked by a horizontal line) a gait used during bulldozer-like pushing: the head tucked under and the collum tergite forming the anterior pushing surface. A stronger pushing gait is shown in Pl. 7(e) with many more simultaneously propulsive legs. When walking fast, iuliform millipedes use a fast-pattern gait (Pl. 7(d)) with shorter metachronal waves, few legs in the propulsive phase and more in the recovery forward swing than in Pl. 7(c). The body is also held further from the ground, permitting longer strides, larger angles of swing, and a narrower track as shown by the differences between the tracks and stride lengths of another type of diplopod shown in Fig. 7.5(a) and (b), a fleet carnivore, *Callipus*. Diplopod gaits are considered in Chapter 7, §6C(v).

The narrow head of the coxa of iuliform diplopods, sunk in a tight socket in the sternite near the middle line (Figs. 5.3(a), 8.3(a), (c)), is a provision permitting maximum leg length, fairly wide angles of swing (Fig. 7.4), and great strength due to the absence of a large expanse of arthrodial membrane at the joint (cf. the enormous expanse of coxa–body arthrodial membrane in centipedes, Figs. 5.7, 5.8, 5.9, 8.1(a)). The ventral position of leg origin, in contrast to all other myriapod classes, results in a minimal or no leg projection beyond the flanks, as is suitable during bulldozer-like burrowing. There

---

FIG. 8.3 (see opposite). Diagrams of the skeleto-musculature of a iuliform millipede *Poratophilus punctatus* to show the principal features associated with head-on pushing used in burrowing and acute dorso-ventral bending used in assuming the protective spiral position. Muscles have been removed in whole or in part of certain diplosegments and the inner air face of the exoskeleton is black. Muscles are lined, the inner edge of each prophragma is white, and the bladders of the stink glands are also white. The paired ventral tracheal apodemes seen in Figs. (a)–(c) are white. (a) Median view of a longitudinal half of some segments of the body in acute dorso-ventral flexure. The sternites are pulled upwards anteriorly into the preceding segment so that the ventral surface forms a close zigzag of sternites and arthrodial membrane, cf. (b). The tracheal pouch apodemes appear in lateral view to the left of each sternite. The pair of legs borne by each sternite are cut short. For further description see text. (b) Lateral view of the ventral parts of two successive diplosegments drawn from the body when it is longitudinally straight. Note the position of the sternites and of the legs. The arthrodial membrane between successive sternites which appertains to the intersegment is covered by the striated posterior part of each tergite. (c) Horizontal ventral half of three diplosegments viewed from above to show the ventral structure. The body shows a small horizontal flexure, and different numbers of muscles have been cut back progressively from the upper part of the figure to the lower. In particular note the white tracheal pouches linking the sternal longitudinal muscle (*mus.lac.v.*). The levator apophysis posticae (*lev.ap.post.*) is much foreshortened, cf. (a). The extrinsic promotor and remotor coxal muscles are shown on the lower segment attached to and covering the tracheal pouch apodemes. (d) Horizontal dorsal half of the body (not flexed), with musculature progressively removed towards the third diplosegment shown. (e) Oblique lateral view of the skeleton of one diplosegment; the internal cuticular face is black and the prophragma lying at right angles from the anterior margin of the tergo-pleural arch is shown. The small ventral sternite, fused to the tergo-pleural arch carries two pairs of legs. (f) Anterior view of the same showing in particular the prophragma, sternite and anterior tracheal pouch. (g) Ventral view of the skeleton of one diplosegment on which the left legs have been removed showing the small coxal sockets in the sternite. Note that the sternite is a little forwardly expanded which is a help in assuming the dorso-ventrally flexed position in (a). (From Manton 1954.)

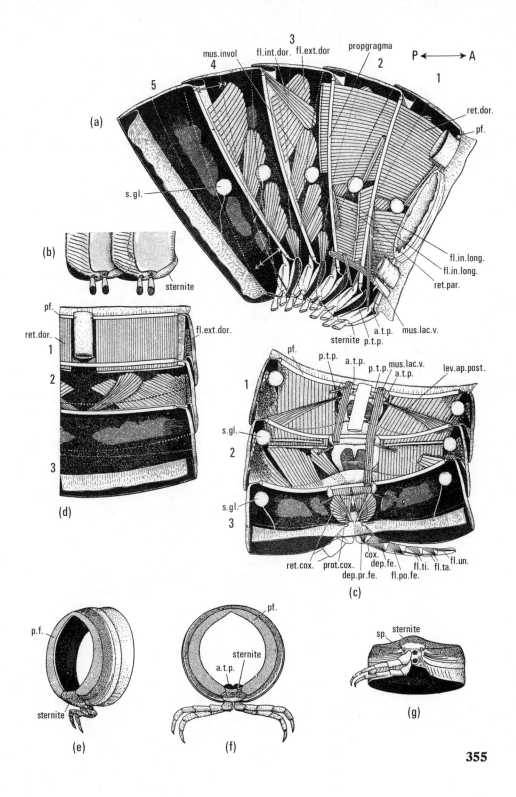

is no lateral leg projection in the millipedes walking in Pls. 4(g), 7(f), the legs being invisible from above, but millipedes which do not burrow well by the above method, or which burrow differently, have longer, laterally projecting legs, as in Pls. 5(f), 7(g).

The superficial oblique musculature between diplosegments is well developed (Fig. 8.3(a), segments marked 3 and 4, and see Fig. 8.5(b)) and maintains alignment between one diplosegment and the next in whatever position may be momentarily suitable during burrowing or walking, including rotation of one diplosegment upon the next. The dorsal and lateral longitudinal muscles, retractor dorsalis and retractor paratergalis, are very stout and hold the body firmly during transmission of the force from the legs to the head end (Fig. 8.3(a), segments marked 1 and 2). The sternal longitudinal muscles are quite abnormal because extreme shortening could not be effected if they extended between successive tilted sternites, since striated muscles cannot shorten to the degree required. Instead these muscles are small or absent and extend between the tracheal pouch apodemes as shown in Fig. 8.3(a), (c), *mus.lac.v.*, or retractor ventralis.

Of importance in bulldozer-like burrowing is the even width of the whole body, so that penetration of the soil or of decaying wood by the head end prepares a path wide enough for the whole trunk to follow. Typically there is no anterior tapering of the body, as shown by diplopods practising the next two types of burrowing, (ii) and (iii) below. The anterior segments of all diplopods are less deep ventrally than those behind, leaving space for the head end in the enrolled position. Of importance too is the nature of the diplopod cuticle which is hard, stiffened largely by calcification and is unyielding. The chilopod cuticle contrasts in being stiffened largely by sclerotization and the sclerites, although firm, possess considerable flexibility and elasticity.

(ii) *The polydesmoidean type of dorsal-surface or 'flat-back' pushing.* The 'flat-backed' millipedes, Pl. 5(f) and Fig. 8.4, are relatively shorter than iuliform millipedes; they possess fewer, longer diplosegments, each forming a united skeletal ring which is produced laterally into an expansion, the keel, projecting from the posterior part of the diplosegment. The legs are longer than in the Iuliformia and very strongly constructed (Figs. 5.16–18).

The 'flat-backed' millipedes are adepts at widening crevices which tend to split open along one plane, such as under bark or in layered decaying leaves. Below the keels the stout legs can find almost complete cover (Figs. 5.3(b), 5.16). The tapering head end is stuck into a crevice; the legs, flexed under the keels, straighten a little and together deliver a thrust from the wide dorsal surface, which widens the split, in decaying leaves for example, and enables the animal to penetrate deeper. The splitting of the immediate environment in one plane leaves lateral space for the longish legs, which in walking project far (Pl. 5(f)), and can be housed in a widened split. No such leg projection is practicable to a iuliform millipede, unless its habits are modified towards better running ability and lesser burrowing ability, as in the British *Tachypodoiulus niger*, and large species such as *Plusioporus sulcatus*, where the legs project markedly beyond the flanks.

Polydesmoidean legs arise ventro-laterally and not midventrally from the sternite as they do in most millipedes (cf. Fig. 5.3(b) with 5.3(a)). The initial difficulties attendant upon the iuliform S-shape of the leg are thereby avoided, since the keels from the dorsal surface permit the legs to project beyond the flanks proper and still retain protection during burrowing. The ventral surface can be held close to the ground, as seen in the side-view of *Polydesmus* walking in Pl. 4(i). Thus the effect of leg straightening during pushing is to raise the dorsal surface much more easily than could be done by legs inserted on the body as in the Iuliformia.

The polydesmoidean legs are stouter and stronger, relative to diplosegment size, than are those of the Iuliformia and can put out greater forces. These forces can readily be measured in a comparative way by harnessing millipedes of various kinds to sledges carrying weights and ascertaining the maximum load which each animal can shift. The force in grams required to shift the load must then be measured (see Appendix) and the weight of each animal also recorded. Millipedes make amiable 'cart-horses' because pulling is not very different from their normal pushing. It is not possible to induce a maximum pushing effort from a series of live diplopods. The pushing (pulling) powers of Polydesmoidea are two to five times as great as those of Iuliformia possessing the same diplosegment volume but different numbers of legs. That these forces are large for this type of animal is shown by

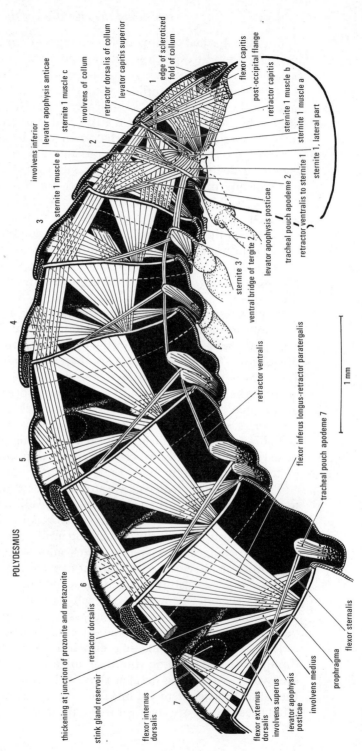

FIG. 8.4. Internal view of a longitudinal half of the anterior segments of an immature male *Polydesmus angustus*; the large numbers show the tergites of the trunk segments and parts are shown of legs 1–3 from trunk segments 2–4. The well-formed prophragma carrying the dorsal longitudinal muscles and levator apophysis posticae is shown; the dorsal musculature is weaker than in Fig. 8.3; and the weak and small ventral longitudinal muscle (retractor ventralis) is present only in the anterior segments. Segments 2–4 carry only one pair of legs and one pair of tracheal pouches. The flexor sternalis muscles are well developed and used in spiralling. The keels from the dorsal sides of the tergite are seen in Pls. 4(i), 5(f). (From Manton 1961.)

comparing the pulling forces exerted by lizards, mice, and hamsters of the same body-weight as large Iuliformia; the vertebrates put out only 0·2 to 0·3 of the forces exerted by the millipedes (but see Chapter 7, §6C(vi)).

Polydesmoidean extrinsic leg muscles are long and strong and arise both from the coxa and from the prefemur. The extreme modifications of the joints at the proximal and distal ends of the trochanter have been described in Chapter 5, §6C, Fig. 5.18, together with the unique existence of synovial cavities in the arthrodial membrane at the leg joints of the Polydesmoidea (Chapter 5, §6D, Fig. 5.17). All these features contribute to leg strength and ease of flexure and extension during 'flat-back' pushing. The depressor femoris and depressor preformis muscles straighten the flexed leg and so raise the body. They are housed in dilatations of the leg podomeres (Figs. 5.16, 5.18, *dep.p.fe* and *dep.fe.*). In addition the large species, such as *Platyrachus* spp., possess the rare additional extensor muscles from the distal hinge joints, *lev.po.fe.* and *lev.fe.*, working at poor mechanical advantage from apodemes arising at the hinges themselves (Fig. 5.16). With such leg morphology and size it is not surprising to find that Polydesmoidea can run some three times as fast as Iuliformia of comparable diplosegmental volume, the only satisfactory way of comparing the one with the other. However it is burrowing ability and not speedy running that is of greatest moment to these animals. If speed was a priority faster gait patterns would be expected (for gaits see Chapter 7, §6C(v)).

The exact shape of polydesmoidean diplosegments is related to dorsal surface pushing. The anterior part of each entire diplosegment is cylindrical and rotates freely in its circular articulation with the preceding diplosegment. The absence here of keels enables flexure of one diplosegment on the next to take place in various directions. The dorsal cuticle is rendered particularly rigid in an engineering sense by the posterior part being domed and this region and the dorsal faces of the keels are not smooth but lumpy with small curvatures, all of which promote mechanical rigidity suitable for delivering the strong dorsal thrust without buckling.

Thus the speciality of the Polydesmoidea is no longer head-on bulldozing, as in their probable ancestors and in present-day Iuliformia, but the development of 'flat-back' pushing as a particular exploitation of the wedge principle in burrowing, but entirely different from that described below (iii) for Nematophora and Colobognatha. The combination of a strong upward thrust by the dorsal surface and a forward thrust from the wedge-shaped anterior end of the body contrasts with most present-day Iuliformia. The tapered anterior end of the Polydesmoidea, the smaller head capsule, and small collum sclerite are clearly more serviceable for transmission of an oblique and upward thrust than the direct forward one of the Iuliformia.

(iii) *The nematophoran and colobognathan type of wedge-pushing*. These two orders, and the Colobognatha in particular, have elaborated yet another burrowing technique. Both orders possess a tapered anterior end to the body (Figs. 8.5, 8.6(f), Pl. 4(g), (h) and many, but not all, possess keels projecting over the legs (Pls. 4(h), 7(g)). The sternites alone, or both sternites and pleurites, are united only by arthrodial membrane to the tergites or tergo-pleural arches (Fig. 5.5). The intersegmental joints are not incompressible antero-posteriorly as they are in the Iuliformia and Polydesmoidea. The prophragma from the tergites is well developed (e.g. *Dolistenus*, Fig. 8.5(a)). Each segment can be pulled forwards into the segment in front to some extent. When the legs of a segment are firmly holding on to the substratum and the legs of the following segment are off the ground, strong and elaborate trunk musculature telescopes the more posterior segment into the more anterior one and pulls the posterior sternites, with their appended legs, forwards to an even greater extent, because the sternites are capable of sliding back and forth within the tergo-pleural arch (Chapter 5, §4D). The anterior end of the body is strongly tapered (Pls.

FIG. 8.5 (see opposite). The colobognathan millipede *Dolistenus savii* to show the muscular complexity associated with the colobognathan type of wedge-burrowing (see Chapter 8, §3A(iii), cf. Figs. 8.3, 8.4). The large numbers denote trunk segments and the smaller numbers 1 and 2 denote the anterior and posterior coxae of a diplosegment respectively. (a) Internal view of a longitudinal part of head and anterior trunk segments. The ample folded ventral arthrodial membrane between the sternites is white bearing black spots. The internal margin of the prophragma apodeme is white, as in previous figures. Note the very large size of the tracheal pouch apodemes (white) carrying the sternal longitudinal muscles which are so large anteriorly where the telescoping of the interior segments is of such importance in burrowing (cf. Figs. 8.3(a), (c), 8.4). Segments 2–4 carry one pair of legs only. (b) Lateral view drawn as a transparent object to show the enormous size of the superficial muscles contributing to the telescoping of one segment into the next aiding the burrowing technique, those in the anterior segments spreading both forward and upwards as shown. (From Manton 1961.)

Habits and the evolution of the Myriapoda

4(g), (h), 7(g)) and thus tergo-pleural arches of increasing size are dragged forward, so widening a crevice. Most of this wedge-burrowing is done by the tapered anterior end of the body; the anterior segments of *Dolistenus* can become caked with mud when the animals force their way into shelter under a slab of rock.

The bodies of most Nematophora are relatively longer than those of Polydesmoidea and more like the Iuliformia. Some Colobognatha are short bodied, as is the British *Polyzonium*

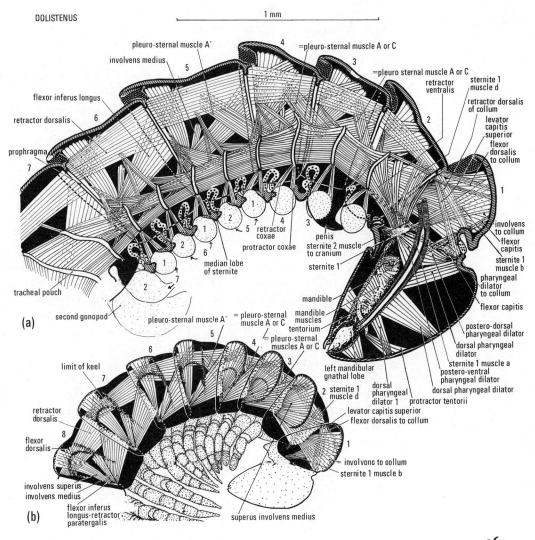

(Pl. 4(g)) and North American *Brachycybe* but others are long bodied, as is the Mediterranean *Dolistenus* (Pl. 7(g)) and the South American *Siphonophora* (Pl. 7(f)). The Colobognatha and Nematophora have other proficiencies besides wedge-burrowing with which their several body shapes are associated. However, the wedge-burrowing of all these millipedes is associated with free ventral sternites and often free pleurites. The strength of the legs is not outstanding in contrast to the Iuliformia and Polydesmoidea, but the strength of the telescopic segmental movement and sliding sternites provided by trunk musculature is great and provides the motive force for the wedge-burrowing, in contrast to the types of burrowing (i) and (ii) described above.

When the forward movement of the ventral parts of the tergo-pleural arch is momentarily greater than that of the dorsal tergite (Fig. 8.5), humping results in *Dolistenus*. A wave of humping travels forwards over the anterior segments of Nematophora and of Colobognatha as they push, the degree of humping increasing as the animals push more strongly. Such a hump is seen in *Polymicrodon* harnessed to a weighted sledge in Pl. 4(h), where a strong, slow gait pattern is also in use, 20 consecutive legs forming the propulsive group in one metachronal wave instead of 5 used in fast walking; 8 or 9 consecutive legs are off the ground, performing the forward swing, in both gaits. In *Polyzonium* (Pl. 4(g)) there is no humping during free walking but harnessed to a weighted sledge marked humping occurs as in *Polymicrodon*. In *Siphonophora* no humping occurs, but the anterior part of each diplosegment is of markedly smaller diameter than the posterior part so that the pulling of one segment into the next can effect a local shortening of up to 20 per cent in body length.

The action of sliding sternites used in wedge-burrowing by Nematophora and *Siphonophora* is well seen in a ventral view through glass of the nematophoran *Callipus* (Pl. 4(f)), actually running fast, but the movements are the same during burrowing. *Callipus* is a secondary fast runner (see below, §3C(i)). Propulsive limb-tips are marked by white spots on the left of the photograph and the greater distances between the bases of the propulsive compared with the forwardly moving legs off the ground show the sternite movements. They slide forwards, packing the legs close together, during the forward stroke while the sternites slide backwards during

Habits and the evolution of the Myriapoda

the propulsive backstroke, so, with the effect of metachronal rhythm, enhancing the range of the propulsive leg movement, while the sternal musculature provides additional strength to the leg movements. In the siphonophoran slowly walking, shown in ventral view in Pl. 7(f), there is far less sternite sliding, although it can be detected, and there is greater difference in the phase relationship of the paired legs than in Pl. 4(f), because of the extensive lateral turning movements during walking.

The sliding sternites, so important in wedge-burrowing as well as in walking, are moved by many muscles; see the enormous sternal longitudinal muscles (retractor ventralis) of *Dolistenus* in Fig. 8.5(a), unlabelled, extending between the large tracheal pouches (labelled on left) which arise from each sternite in contrast to the very small corresponding muscles in the Iuliformia and Polydesmoidea, Figs. 8.3, 8.4. This muscle in the Polydesmoidea is present as a minute strand in anterior segments only. Other muscles in *Dolistenus* (Fig. 8.5(a), (b)) are modified for wedge-burrowing, particularly in the anterior segments. The lateral longitudinal muscles (flexor inferus longus united with retractor paratergalis, labelled on left of Fig. 8.5(b)) are substantial and obliquely placed. Involvens superus and involvens medius become very large and fused in the anterior segments, compare the left-hand and right-hand ends of the external view in Fig. 8.5(b). These are the muscles most concerned in wedge-burrowing, pulling one segment forward into the preceding segment whose legs firmly hold the substratum, so increasing the diameter of the body and widening a crevice. Note too the thickness of the legs in Fig. 8.5(b), which could not be stouter if an ability for tight enrolment was to be maintained. Other muscles concerned in the movement of sternites are shown in Fig. 5.5. All the muscular features associated with wedge-burrowing are modifications of the basic diplopodan musculature present in all diplopodan groups. The spectacular correlations with function do not represent the evolution of unique muscles, but only modifications of basic ones.

Thus the colobognathan and nematophoran types of wedge-burrowing, used also for pushing under stones, is entirely different from that of the polydesmoidean antero-posterior incompressibility of diplosegments. The colobognathan habit speciality appears to have accompanied the structural evolution of the group, but it is probable that this type of wedge-burrowing is a secondary evolution from ancestors with basic diplopodan pushing abilities.

The three main types of burrowing found in diplopods are not entirely distinct. There are a few Iuliformia with tapered anterior ends and some Polydesmoidea without tapering. There are endless variations in detailed habits and in associated structure, but the habits considered above appear to have been of major importance in the direction of the structural evolution of the diplopods.

B. *Diplopodan spiralling techniques*

A second habit of evolutionary importance to Diplopoda is the ability to enrol in a tight spiral, Pl. 4(o). The ability to flex the body dorso-ventrally, a nodding movement, results in the tight spiral or enrolled position, assumed when resting in the soil and as a protective reaction at any time, shielding antennae and legs. The anterior single segments, legless or with one pair of legs, are dorso-ventrally less deep than the rest, so making space for the intucked head and a tight centre on dorso-ventral flexure for the spiral position of the body. The Polydesmoidea with their long diplosegments cannot achieve as tight a spiral as can the Iuliformia and Nematophora, but all diplopods use this protective reaction to the best of their several abilities. The Polydesmoidea curl round their nests of eggs, protecting them until they hatch, and for this their loose spiralling ability is sufficient.

Enrolment in the Iuliformia is achieved by a pulling together of the ventro-lateral parts of successive diplosegments by the retractor paratergalis muscles; and by a raising of the anterior ends of the tilted sternites by the flexor inferus longus and levator apophysis posticae muscles, all shown on Figs. 8.3(a), (c). The slender sternal longitudinal muscle *mus.lac.v.* must be of some assistance in approximating successive tracheal pouch apodemes. The contrasting much more limited lateral flexure of the body is shown in Fig. 8.3(c).

As noted above in §3A(i) and shown in Fig. 8.3(a), there is mechanical difficulty in achieving sufficient ventral compression with a body cylindrical in shape and diplosegments short and deep. A simple evasion of the difficulty is the elimination of the ventral part of the cylinder by reducing the body to a half-cylindrical shape and keeping the axis of the nodding movement between one diplosegment and the next at the equatorial level. Enrolment of such a body needs little or no ventral compression and is mechanically easy. The half-cylindrical shape has been independently evolved many times and may be a permanent or a facultative feature. It has also been evolved in a parallel manner among the Crustacea by some Isopoda (Pl. 3(d), (e)). Fig. 8.6(d), (e) shows a partly enrolled *Glomeridesmus mexicanus* (Limacomorpha) and the skeleton of one diplosegment which is almost half-cylindrical in shape. The same is seen (h) *Oniscodesmus fuhrmanni* (Polydesmoidea). But in the 'pill-millipedes' (Oniscomorpha), Pl. 3(e), Fig. 8.6(g), the raising of the free sternites and pleurites to give the half-cylindrical shape is done on enrolment by muscles; some are indicated by rods in Fig. 8.6(c) for *Sphaerotherium*, the tightly enrolled position being shown in (a). Where pleurites and sternites are free there are more deep oblique muscles than in the Iuliformia and deep dorso-ventrals are also present (Manton 1954, 1961).

There are mechanical devices which keep the axis of the nodding movement in the desirable ventral position, not least of these being the overlap of the pleurites from before backwards while that of the sternites is in the opposite direction (Fig. 8.6(c)) and the staggered position of the two legs, two sternites, and one pleurite, all paired, of each diplosegment, as shown. The Isopoda which can enrol, such as *Armadillidium* (Pl. 3(d)), also use imbricating tergites overlapping from before backwards and sternites overlapping from behind forwards which similarly fixes the axis of the nodding movement low down on the half-cylindrical body. Strong and complicated musculature in *Glomeris* both raises the ventral sclerites horizontally to the equatorial level and pulls them together longitudinally. A large paired apodeme projects low down from the prophragma of each tergal arch which carries upward and downwardly directed muscles. A large tendon, arising from the tergo-pleural junction, carries muscles to the diplosegment in front (Fig. 8.6(b)). A further device giving perfectly fitting enrolment with no undue dorsal expansion, when the ventral parts of the segments are pulled tightly together, is provided by a ridge on the underside of the tergal overlap which snaps into a corresponding groove on the anterior part of the following tergite. The differences between the tergal faces of a walking and enrolled *Glomeris* are seen in Fig. 8.6(a) and Pl. 6(n). This tight enrolment is achieved at a price; lateral bending is limited and the power of rotation of one ring upon another has been almost abandoned and burrowing must be done with few legs. The remarkable convergence between 'pill-millipedes. and crustacean isopods capable of similar enrolment, but by somewhat different means, is shown in Pl. 3(d), (d1), (e), (e1) (see legend).

The 'pill-millipedes', as well as showing such perfectly fitting enrolment, are reasonable burrowers by the bulldozer technique in spite of the small number of trunk segments. The body does not taper anteriorly, the collum segment is small and the pushing shield is formed by the second tergite (Pl. 6(n)). The maximum possible use is made of the legs, of which there are only 17 pairs in *Glomeris*. They do not project beyond the flanks and do not insert, as is usual in diplopods, in strong sternal sockets. The mid-ventral body surface in *Glomeris* is formed by a flexible pedigerous lamina surrounding the leg bases, the sternites being more lateral in position (Fig. 8.6(b)). But from the antero-lateral face of each transversely flattened coxa extends a stout strut articulating with the corresponding very large tracheal pouch apodeme. Normal promotor–remotor coxal movements take place

---

FIG. 8.6 (see opposite). Diagrams of the skeleton of millipedes capable of assuming a spherical enrolment. (a)–(c) *Sphaerotherium dorsale* (Oniscomorpha), (a) rolled up in side view. The heavy dotted line indicates the position of the union between tergites and pleurites and the lightly dotted line shows the margin of the pushing shield (tergite 2) over which the lateral tergite wings of succeeding segments fit; (b) oblique anterior view of the cuticle of 2 diplosegments showing: two tergal arches, each with prophragma and a pair of apodemes; two movable pleurites; visible parts of 4 sternites, 4 pairs of legs; 4 pairs of tracheal pouches, the spiracle being lateral to each leg base. Muscles are shown diagrammatically by rods (for muscle labelling see Manton (1954) Fig. 4); a plan of the parts is shown in (c). The marked tendon arising from the tergite at its junction with the pleurite carries muscles which pass anteriorly to the tergite in front, inserting on the tergal face and prophragmal apodeme; these muscles pull the ventral parts of the skeleton together and keep the axis of the nodding movement at an equatorial level. (c) Ventral view of the skeleton shown in (b). The single pleurite and two sternites appertaining to the lower diplosegment are stippled and staggered in position, a device which helps to keep the axis of the nodding movement at an equatorial level. The legs are cut off and are set in flexible sternal membrane but articulated directly with the tracheal pouches. The three most median black spots mark the median and outer limits of the tracheal pouches while the right hand black spot marks the flexible junction between pleurite and tergite above which extends the very large apodeme from the prophragma. (d), (e) *Glomeridesmus mexicanus* (Limacomorpha). (d) Posterior view of the skeleton of a diplosegment showing protection afforded to the legs, the prophragma of large size, and the approach towards a half-cylindrical shape. (e) Lateral view of a dead specimen showing the manner of rolling up; the posterior parts can cover the head end as far as the arrow. (f) *Polyzonium germanicum* (Colobognatha). Anterior end of the body in lateral view; note the tapered shape in contrast to the Iuliformia in Pl. 7(c), (d). (g) Anterior view of a diplosegment of *Sphaerotherium giganteum* for comparison with (b). (h) *Oniscodesmus fuhrmanni* (Polydesmoidea). Anterior view of a diplosegment. (From Manton 1954.)

Habits and the evolution of the Myriapoda

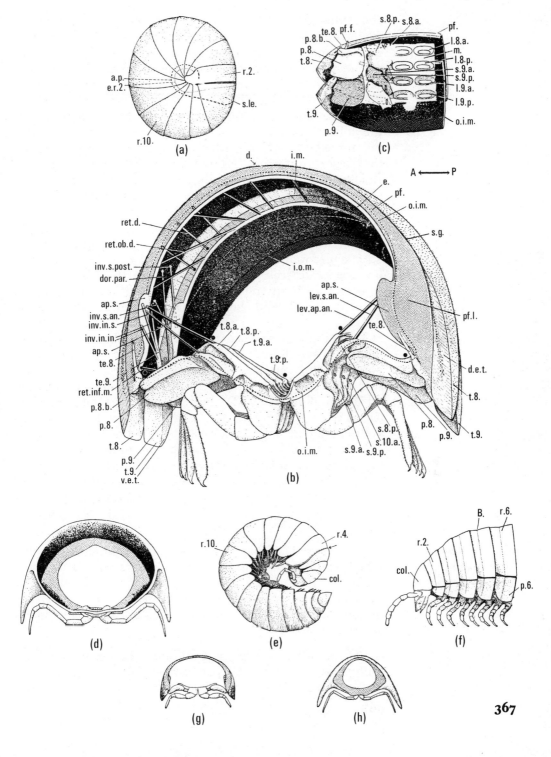

about this hinge. Stout, wide, sternal longitudinal muscles extend horizontally between one tracheal pouch apodeme and the next and cause antero-posterior sliding of these apodemes and of their appended legs during stepping, exactly as is caused by sliding sternites of Nematophora and Colobognatha (Pl. 4(f)). These movements greatly enhance both the stride length and the force exerted by the legs against the substratum. A harnessed *Glomeris* can put out a pulling force/body-weight, which is greater than exerted by the larger Iuliformia but less great than put out by the smaller Iuliformia. This ability enables *Glomeris* to execute effective shallow burrowing into its particular type of woodland soil.

The achievements of the 'pill-millipedes' in respect of tight enrolment, burrowing, and the facilitating morphology are spectacular indeed. The Oniscomorpha probably have been long separate from other diplopods in their evolutionary history. The position of their gonopods differs from those of other diplopods and the pushing head shield is differently placed.

The Siphonophoridae (Colobognatha) show yet another method of spiralling. They are half-cylindrical in shape, good dorso-ventral muscles pulling up the free pleurites and sternites (Fig. 5.5). The sternites slide antero-posteriorly during walking (Pl. 7(f)) but, unlike the Oniscomorpha, lateral trunk flexibility is great and the body is long. The legs are short, being entirely covered by the tergal arch. *Siphonophora hartii* and *S. portoricensis* rest in a tight or an irregular horizontal coil upon a ceiling, a very unusual feat. Dorso-ventral spiralling is also practised on the ground. No other diplopods have been found with such marked lateral flexibility.

C. *Secondary diplopodan achievements*

The habits and facilitating morphology just described represent basic features possessed by the class and further advances therefrom. Two examples will now be considered showing more marked secondary deviations from basic diplopodan habits which have resulted in contrasting habits and morphology.

(i) *The Lysiopetaloidea (e.g. Callipus)*. These nematophoran millipedes show a partial abandonment of typical bulldozer-like pushing, slow movements, and herbivorous feeding.

They have secondarily gone in for faster running, scavenging-carnivorous feeding and have good rock-climbing ability. *Callipus longobardius* is found on rock surfaces inclined at any angle in its north Italian habitat. *Callipus* retains the narrow coxa–body articulation near the middle line, but with the usual vertical axis of swing characteristic of the Nematophora, but not of the Iuliformia (Figs. 5.3(a), (b), 5.4, and further figures in Manton 1958*b*). This feature permits the ventral surface of the body to be held close to the substratum at all times (cf. Iuliformia, Chapter 5, §4F, Fig. 2.3, and Manton 1958*b*) and facilitates the maximum angle of swing of the leg and longest stride lengths, see the contrasting tracks in Fig. 7.5(b), cf. 7.5(a)). Leg joints and musculature of the typical iuliform diplopod are suited to produce strong, slow movements (Fig. 7.5(c), cf. 7.5(b) and the faster walking in the Iuliformia needs the body to be held well up from the substratum as shown in Pl. 7(d). Lysiopetaloid modifications promoting faster movements (Chapter 5, §4B, F) and the convergent resemblances to chilopods are the long intrinsic and extrinsic leg muscles, which cross more than one intersegment, and the long tarsus which is also divided. The Lysiopetaloidea use their sliding sternites to the full in gaining long strides (see the more conspicuous movements shown by Pl. 4(f) than Pl. 7(f), the latter being a slow mover). These morphological features result in fleetness far exceeding that recorded for any other millipede; *Callipus* can run twice as fast as *Polymicrodon* and the latter is fleeter than a iuliform millipede of corresponding size.

Such running achievements suit the pursuit of prey and may be the travelling of considerable distances to find it. As might be expected, these secondary achievements have resulted in a loss of the original basic diplopodan ability to burrow by head-on pushing. There are no keels, and the long legs project too far from the flanks to permit anything but the feeblest burrowing in loose soil. Harnessed to a sledge, *Callipus* will not pull readily, an indication that no longer is either head-on pushing or wedge-burrowing one of its accomplishments. *Callipus* is as near to a fleet carnivore as can be secondarily arrived at by a diplopod, but even so, its achievements are inferior to those of the fast running Chilopoda, although the rock-climbing ability, not conspicuous in Chilopoda, may be a substantial asset in the environments inhabited by *Callipus*.

(ii) The Pselaphognatha (*e.g. Polyxenus*). These millipedes are even further removed from basic diplopodan type than are the Lysiopetaloidea. They are fleet for their size, but their outstanding achievement of survival value is the ability to live upside down on glass-smooth ceilings of rocky places, such as flint surfaces in rock walls. They leave their hides to run out for their algal food and return again to their ceilings. Closely placed cast cuticles still adhering to the ceilings indicate the proficiency of *Polyxenus* for passing much time in such places.

Polyxenus is an example of an animal with many accomplishments. It lives also among lichens, on walls, and under bark and has been found in places as diverse as the surface soil of woodlands, among the roots of coastal plants, and on the leaves of willows. But it is probable that the ability to live in colonies on protected ceilings in confined spaces, too smooth for most predators to follow, is the habit of greatest survival value under adverse conditions. This is the habit which is associated with the extraordinary morphology and locomotory movements of *Polyxenus*.

Polyxenus lagurus is about 3 mm in length (Pl. 6(j)); it is rightly classed as a diplopod, although this has been questioned in the past when the proficiencies of the animal were not understood. At times it has been regarded as an ancestral type of diplopod; or possibly intermediate between Chilopoda and Diplopoda, because the legs in propulsive phase converge as in epimorphic centipedes (see side view of *Polyxenus* in Pl. 6(l) and note the convergence of propulsive legs in Scolopendra Pl. 1(f) and Fig. 7.1(a)–(d) opposite the black spots). *Polyxenus* can neither push nor roll up, but the basic diplopodan legacy is shown in the limbless collum segment, the next three segments being single, with one pair of limbs each, followed by diplosegments, and the position of the gonopores. The body is short, with only 13 pairs of legs, and the internal anatomy is decidedly diplopodan.

The shortness of the body of *Polyxenus* is probably associated with its flattened form and need for manoeuvreability in confined spaces. Protection against flooding of crevices as well as against predators is essential. The body is very lightly constructed, without calcification in the cuticle. The short legs are completely protected by overhanging rosettes of large spines and similar spines are situated in dorsal rows (Pl. 6(j), (l)). These spines are large, extremely light, and hollow (Pl. 6(b) shows transverse sections of some spines). The spines and the whole cuticle are extremely hydrofuge so that the animal is unwettable. The spines are protective and the lateral lobes of the body bearing the spine rosettes and the terminal bundles are all mobile. This spine armature replaces the normal diplopod protection by a rigid cuticle and enrolment. The diplosegments are not rigid; they can telescope into one another and enhance spine mobility. The internal anatomy and musculature is basically diplopodan although modified in connection with spine mobility and with flexibility of the ventral pedigerous lamina bearing sternites and legs (see Manton 1956).

The short limbs end in complex adhesive arrangements at the tip of the tarsus. A soft adhesive lobe is pressed on to a ceiling by the two terminal claws forming a fork around it. The leg is short and the tarsus is divided, as in fast runners (*Callipus* above and fast-running centipedes, Figs. 5.3(c), 5.6(c)) and, as in these myriapods, there are long extrinsic and intrinsic leg muscles indicative of fast movement, and the observable large angle of swing. The coxae are immobile and flat, the promotor–remotor muscles operating from the prefemur and femur. Very close adherence to the substratum is thereby gained, making possible the spine protection.

Polyxenus has achieved some measure of fleetness from its short legs by using a large angle of swing, a slightly quicker pace duration than in most diplopods, and a slow type of gait (4·0:6·0), phase difference between successive legs 0·8, or slower, in which many legs (from 7 to 13 at one moment) adhere to a ceiling. A fast pattern of gait with few legs adhering to a ceiling at one moment would be impracticable. The only practicable manner of using these features is the employment of a large phase difference between successive legs, greater than 0·5, as in Scolopendromorpha, the limb-tips converging during the propulsive phase as shown in Pl. 6(l). The use of a normal diplopod phase difference between successive legs of less than 0·5 (Figs. 7.8(c), 7.10(b), (c)) is impracticable to *Polyxenus* because it would lead to crossing of propulsive limb-tips and other mechanical disadvantages (see Manton 1958, text-fig. 5). Thus the gaits show no true resemblances to those of Chilopoda.

The morphological and locomotory features of *Polyxenus* are fully explicable on the habit which appears to be of greatest survival value. The Pselaphognatha cannot be regarded as anything other than divergent diplopods in which the original accomplishments and structure have been considerably modified towards new habits and proficiencies. The Pselaphognatha are not primitive diplopods nor are they intermediate between diplopods and chilopods.

D. *Conclusions concerning the Diplopoda*

The essentials concerning the evolution of the Diplopoda and of its component taxa, presented above only in outline, can now be comprehended for the first time. The presence of diplosegments, unique among extant arthropods, and of very many other structural features (some 32 are listed in Manton 1954, pp. 352–3 and more are considered in Manton 1958*b*, 1961) are explicable in terms of the exploitation of burrowing habits, enrolment, and a vegetarian diet with some deviations therefrom. The diplopodan taxa considered above have been selected to illustrate how their very diverse general shapes have become meaningful for the first time. It follows that the general shapes of other non-diplopodan taxa should be equally meaningful when we have the right functional information to apply to them. This concept is of particular importance concerning taxa whose relationships are obscure, such as the Pycnogonida, considered in Chapters 6 and 10, and the centipede *Craterostigmus*, see below §5D.

4. The Symphyla

The symphylan habits of survival value concern their ability to seek a passage through the soil, etc., by flexing and turning (Pls. 4(d), (e)) into narrow, tortuous existing channels without pushing and without actual body deformation (Manton 1966) such as takes place in Onychophora (Pl. 4(a)–(c) and Manton 1958*c*). A second habit, probably of survival value against certain predators, is not spiralling as in Diplopoda, but the ability to change direction of running very suddenly and often, to turn in a hairpin bend or run off in a different direction. Such tactics hinder pursuit by certain predators which cannot turn with ease.

Symphyla are found in woods, sometimes on the soil surface but able to penetrate deeply into humus formed by decaying wood. They also occur in fields, some characteristically deep in the soil. They feed on decaying vegetable matter but of softer composition than eaten by many diplopods.

The excellent developmental studies of *Hanseniella* by Tiegs (1940, 1945) leaves no doubt about the formation of the extra tergite on segments 4, 6, and 8 and on further segments

Habits and the evolution of the Myriapoda

in some other genera; these sclerites are a duplication of the normal single dorsal sclerites on just these segments. The function of these extra tergites is now known. The sclerites of Symphyla are lightly sclerotized and thin, the sternites are also thin and separated by intercalary folds of soft cuticle, and the soft flanks of the body (white on Fig. 8.7) are extensive. Symphyla are delicate animals depending upon flexibility (Pl. 4(d)), not on pushing power.

A. *Extra tergites of the Symphyla*

The position of the extra tergites on segments 4, 6, and 8 of *Scutigerella* are shown in lateral view in Fig. 8.7, the body slightly flexed dorso-ventrally in (a) and markedly so in (b); a tracing of a photograph of *Symphylella* crawling round the

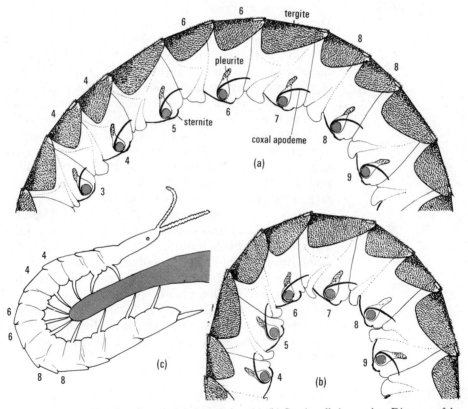

FIG. 8.7. Symphyla in lateral view. (a), (b) *Scutigerella immaculata*. Diagrams of the left side of the body to show the tergites (mottled), pleuron (white), coxal apodemes (black), and sternites (black). In (a) the body is slightly and in (b) markedly dorso-ventrally flexed. (c) *Symphylella* sp. larval stage crawling round the edge of a leaf, from a photograph. (From Manton 1966.)

edge of a thin leaf (c) shows how the extra dorsal hinge-lines, formed by the extra tergites, are used. They provide a contribution of 75° to the total 180° flexure of the trunk. Remarkable intersegmental flexibility also exists in the horizontal plane, Fig. 8.8(b), (c). The postero-lateral tergal lobes slide over one another in the concavity of the bend and on the convexity they give protection to the dorso-lateral surfaces which would otherwise be exposed by such great flexures. Extra tergites are also present in the chilopod *Craterostigmus*, where they contribute to the extreme secondary flexibility of the trunk (see §5D below).

B. *Intercalary tergites of Symphyla*

In addition to the extra tergites, all tergites of Symphyla show transverse intercalary tergites differentiated by transverse hinge-lines anteriorly from the main tergite (Fig. 8.8(a)). Intercalary tergites are very mobile and can even fold back over the tergite behind, as shown by intercalary tergite 3 and anterior intercalary tergite 4 on Fig. 8.8(a), posterior intercalary tergite 4 being in the normal position. The presence of the intercalary tergites aids the general flexibility of the body. They occur wherever marked trunk flexibility is required, e.g. in some Chilopoda and in the japygid and proturan thorax (Fig. 5.12). In the chilopod *Craterostigmus*, using marked antero-posterior movements of the head for its peculiar feeding mechanism, the intercalary tergite on the poison claw segment can similarly fold back over the main tergite behind (Manton 1965, Fig. 83(a)).

There is reason to suppose that neither the Diplura nor the Protura are ancestral to any other type of hexapod (Manton 1972, and Chapters 6 and 9). They share with the Symphyla and Chilopoda the separation of the tergites and sternites by a wide pleuron which bears separate sclerites in various numbers and arrangements (Chapter 5, Fig. 5.12). The differently contrived dorsal and ventral intercalary features are associated with the need for flexibility in all these classes. They can be only partial retentions of primitive features on which are superimposed structures used for advanced and particular purposes.

C. *Musculature and other features of Symphyla*

In association with the marked trunk flexibility of Symphyla the massive superficial pleural muscles serve this need (marked on Fig. 5.10(a), *pct*.). Equally striking is the entire absence of the rigidity-promoting deep dorso-ventral and deep oblique muscles, and the long sectors of the dorsal

Habits and the evolution of the Myriapoda

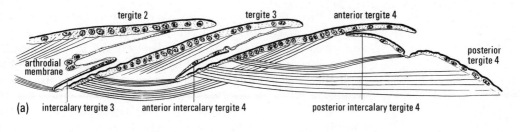

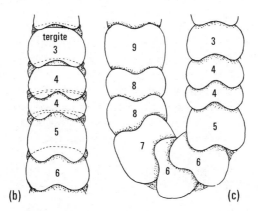

Fig. 8.8. *Scutigerella immaculata*. (a) Longitudinal section to show the intercalary and principal tergites, marked sclerotization being indicated by heavy black and the more flexible cuticle and arthrodial membranes are bounded by thin lines. The lateral and the submedian series of dorsal longitudinal muscles both comprise two sectors. There are no long dorsal muscles crossing more than one trunk intersegment as there are in many Chilopoda (Figs. 5.10, 8.13). (b) Tergites of segments 3 to 6, drawn from a photograph of a living animal with the body held straight. (c) Tergites of segments 3 to 9, drawn from a photograph of a living animal with the body strongly flexed in the horizontal plane. The tergites are drawn as if their margins were constant; it is possible that slight rolling at the margins takes place as in many Chilopoda (Figs. 5.10, 8.13). (From Manton 1966.)

longitudinal system (see below).

These Symphyla are faced with having to use a locomotory mechanism within the limitations imposed by the trunk flexibility, a higher priority than normal fast running by the employment of fast-patterned gaits. This they do in a remarkable manner. They are fast running on the soil surface (Pl. 4(e)) ceaselessly sensing the ground with very rapid antennal movements and gaining safety from pursuing arachnids of small size, which cannot turn with ease. Symphylan trunk flexibility makes it impossible for them to use gaits with fast patterns in surface running, for which trunk rigidity is required, because few legs would be in contact with the ground at one moment in such gaits (Fig. 7.8(c)). Instead Symphyla use slow or moderate patterns of gait in which stability is gained by the many legs on the ground simultaneously and speed is gained by rapid stepping (Fig. 7.8(b)) with short pace durations. Alone among fleet Uniramia, fast running may be limited by an inability to execute the forward leg stroke with greater speed. Leg morphology and leg muscles show many peculiarities which assist a rapid forward swing (Manton 1966).

D. *Conclusions concerning Symphyla*

All the functional and morphological evidence concerning the symphylan trunk shows peculiarities of structure which harmoniously fit together with the way the whole works and with the needs of the animals. Symphylan evolution has been bound up with the habits named above, and along with habit perfection have evolved the facilitating skeleto-muscular system. All are based upon the simple metameric segmental series marked by the trunk limbs and revealed by developmental studies. There are no diplosegments as there are in Diplopoda.

5. The Chilopoda

There are four classes of centipedes: the Geophilomorpha and Scolopendromorpha (Epimorpha), Pls. 1(f), 5(a)–(c), which hatch with approximately the adult number of segments; and the Lithobiomorpha and Scutigeromorpha (Anamorpha) Pls. 4(j), (k), 5(g), which hatch with few segments (Pl. 5(h)) and add leg-bearing segments at subsequents moults; several sub-adult instars follow before the adult number of segments is acquired.

That the Chilopoda are fast-running carnivorous Myriapoda is well known, but now the morphology of the head, trunk, and limbs can be shown for the first time to be correlated with these proficiencies. The habits which appear to have directed the evolution of centipedes are not many. They have led to: (i) legs capable of a speedy backstroke, which is as diagnostic of Chilopoda as are the differently constructed legs of the Diplopoda providing their strong, slow burrowing movements; (ii) a body capable of dorso-ventral flattening and of some shortening and extension, in contrast to typical incompressible Diplopoda. *Lithobius* on Pl. 5(g) can easily pass through both slots in the white card, as *Haplophilus* can traverse the slot seen in Pl. 5(a).

Capabilities (i) and (ii) above are primarily used in

obtaining food and shelter by traversing both open spaces and those which are wide enough to take the leg-span, but dorso-ventrally confined, as between rocks, between stones and dry soil, under bark, etc.

Superimposed upon these two basic capabilities shown by all centipedes, are two outstanding proficiencies, the fast running exercised by the Scutigeromorpha and by the less advanced Scolopendromorpha and Lithobiomorpha, and the earthworm-like burrowing of the Geophilomorpha, in which a thrust against the soil is delivered from the cuticular surface of a group of segments which become short and thick (Pl. 5(c)). The movements were briefly described in Chapter 7, §5.

A. *Structural features correlated with basic habits common to all Chilopoda*

A long stride and a rapid backstroke by the legs is facilitated by the lateral leg insertion on the body; the ventral surface being held near the substratum (not practicable for most diplopods); the axis of swing on the body being roughly vertical (Fig. 5.3(c)); and a well-developed leg-rocking mechanism being situated between coxa and body (Figs. 5.8, 5.9(a) and Chapter 5, §§4B, F, 5A, C). The coxal base is very wide, the dorsal part being expanded up the flanks giving good leverage for the rocking muscles. There is an abundance of extrinsic leg muscles (Fig. 4.29, (cf. Diplopoda Fig. 5.3(a)) and long as well as short intrinsic muscles (Fig. 5.6). A series of dorsal tergites, ventral sternites, and lateral pleurites, arranged differently in the various classes, provides both stability and flexibility of the body (Fig. 8.1(a), (c)).

Carnivorous feeding is dependent upon the strong poison claws (Figs. 3.22(a), 8.17(b), (d)), formed by the first trunk limbs, and the entognathous mandibular mechanism (Chapter 3, §5B(iii), same Figs.). A dichotomy in ancestral hunting methods has presumably led to the differentiation of epimorphic and anamorphic centipedes (Chapter 7, §6C(iii)). Remarkable head flattening involving all feeding limbs, leaves space for manipulation of food in shallow crevices, particularly in the Epimorpha and Lithobiomorpha.

B. *The structural features of the Geophilomorpha associated with burrowing*

After the head end of a geophilomorph centipede has walked into a crevice in soil or under stones, a group of segments becomes short and thick, exerting a heave against the soil. Such a thickening is shown in Pl. 5(c) near the anterior end of a glass bridge o–o against which the animal is pushing.

Segments at the anterior end of the thickening become slender and walk forwards, normally into the crevice widened by the heave, while segments behind the thickening become thick and add themselves to it. The body-surface armour, however, remains remarkably intact throughout these changes.

Each intercalary tergite slides over the tergite behind (Figs. 8.1(c), 8.2); buckling of both can be effected by the dorsal musculature (Fig. 8.9); and sternites and pleurites change their shape enormously by marginal sinking into the body, or the reverse, Fig. 8.1(b). These furling and unfurling movements are made possible by cones of greater sclerotization being set in flexible endocuticle in the marginal zones (Blower 1951) so providing great strength and flexibility. The apparent tergite and sternite margins in Fig. 8.2 may be near p,

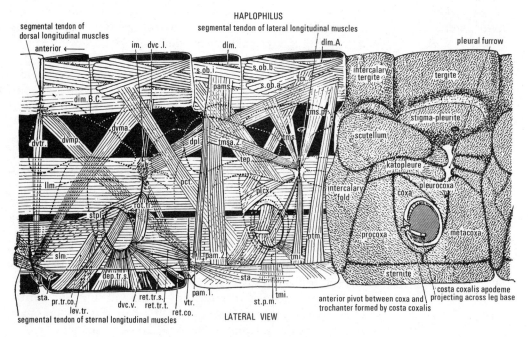

Fig. 8.9. Lateral view of three segments of the common British *Haplophilus subterraneus* (Geophilomorpha) with the legs cut off at the trochanter, to show, right: the superficial sclerites (mottled), whose shape can change momentarily by furling and infurling of the edges (see Fig. 8.1(c)), middle: the superficial musculature, and left: the deeper musculature and extrinsic leg muscles viewed from the body surface (N.B. most of the other figures of trunk musculature are viewed from the sagittal plane). The principal muscles only are shown, there are many smaller pleural muscles. (From Manton 1965).

Habits and the evolution of the Myriapoda

q, f, g, or at considerable distances on either side of these levels.

Shape-changes of the body are also mediated by different movements on the two sides when the body flexes horizontally. Intercalary sternites open out or fold away as shown in Fig. 8.10. Most of the burrowing is done by the anterior third of the body where the legs, used in pulling the body forwards and holding it during a burrowing heave, are much stouter than they are posteriorly. Anteriorly also are the carpo-

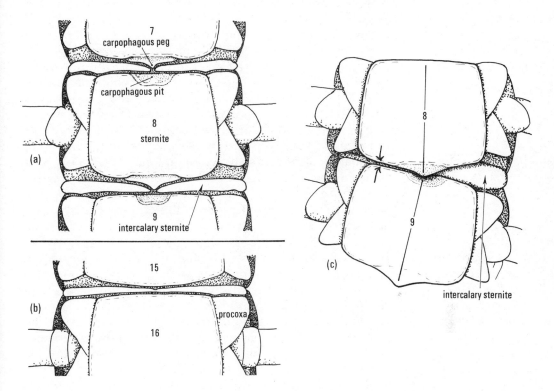

FIG. 8.10. Ventral views of the exoskeleton of *Geophilus subterraneus* (Geophilomorpha). The legs are cut off close to the body. (a) Sternites of leg-bearing segments 7, 8, and 9; the intercalary sternites are separate in the mid-ventral line and the carpophagous peg from the posterior margin of sternites 7 and 8 fits into the carpophagous pit on the anterior median part of the following sternite when the body flexes dorso-ventrally. (b) Sternites 15 and 16 where there are no carpophagous structures and the intercalary sternites are united mid-ventrally. (c) Sternites 8 and 9 moderately flexed in the horizontal plane so that the carpophagous structures interlock; the intercalary sternite on the left side of the diagram has entirely folded inwards as shown by the two arrows while the corresponding intercalary sternite on the other side of the body is expanded but still meets the adjacent sternites so maintaining the integrity of an entire surface armour. (From Manton 1965.)

phagous pits and pegs of certain genera. These structures are at maximal development about leg-segments 7–8 in *Geophilus carpophagus*. When burrowing commences at an angle to the general surface of the substratum, the anterior segments must flex dorso-ventrally as well as execute burrowing heaves of thickening. The carpophagous structures interlock as the intersegmental folds of cuticle deepen and the pegs and pits prevent undue infolding and damage to internal organs. Geophilomorpha with no carpophagous structures may be expected to burrow at right angles to a surface less easily.

Geophilomorph musculature shows correlations with function. The dorsal, lateral, and sternal longitudinals are very stout and supply the main force giving the burrowing heave against the soil, aided by the deep oblique muscles *dvmp*. The deep dorso-ventral muscles *dvtr*. act as antagonists to both in restoring the elongated shape. All the components of the dorsal longitudinals are short, stout, muscles crossing only one intersegmental joint. There is an abundance of superficial pleural muscles (Fig. 8.9). Such strong antagonistic muscles can only operate in the presence of the incompressible haemocoelic fluid, since chilopodan cuticle lacks the overall rigidity that is provided by diplopodan cuticle.

C. *Facilitation of fast running in Chilopoda by cuticular and muscular features*

Faster running is practised by the Scolopendromorpha and Anamorpha using fast patterns of gait with very short pace durations. Such a rapidity of stepping is not seen in the Geophilomorpha. The massive longitudinal muscles of Geophilomorpha maintain trunk rigidity when using fast patterns of gait with few legs in contact with the ground and slow stepping, but other measures are needed by the fast runners.

The Scolopendromorpha possess considerable flexibility of trunk, used in fitting themselves under cover, but they need facultative rigidity during fast running, as do the Anamorpha. The Cryptopidae (Pl. 5(d), gait (6·0:4·0)) with flexible, narrow trunks, cannot employ the fastest patterns of gait used by the Scolopendridae (Fig. 7.1(d)). The gait patterns of the former are moderately fast, up to (7·5:2·5) (cf. Fig. 7.1(b), Pl. 1(f) with (d)), and speed is gained by remarkably rapid stepping.

The simplest use of small intercalary tergites and intersternite infolding during fast running is shown by the narrow-bodied Cryptopidae (Manton 1965, Fig. 16). Good dorso-ventral bending could not take place without a hinge between intercalary and principal tergites. The Scolopendridae use faster gait patterns with as few as three propulsive legs at one moment and for this greater trunk rigidity is required. Well hinged intercalary tergites would be unsuitable and sufficient mobility is provided by the incomplete hinge lines situated anteriorly on the tergites (Manton 1965, Figs. 35–7). The Anamorpha with longer legs have an even greater need to control unwanted flexibility and intercalary tergites are here absent and there are no sclerotized intercalary sternites (Fig. 8.1(a)).

(i) *Tergite heteronomy in Scolopendromorpha and Anamorpha*. The control or limitation of unwanted yawing during fast running is essential to all fleet chilopods. Widely separated single legs, or groups of propulsive legs, always moving in opposite phase, tend to throw the body into horizontal undulations (Fig. 7.1) and these are controlled in large measure by the different degrees of tergite heteronomy and the associated changes in the segmental musculature. The movement of the head and antennae in sensing a path wide enough for the legs to follow also tends to start trunk undulations (Fig. 7.1, Pl. 1(f)).

Cinematography shows that the most stable region of the body lies at legs 7–8 in both Scolopendromorpha and Anamorpha: see line 7 on Fig. 8.11, showing the position of the mid-posterior margin of this segment as the animal ran from below upwards in (a)–(e). Segments posterior to this level show increasing undulations.

The recording of the flexibility which actually exists along the body can be made from good quality photographs of animals flexed in the horizontal plane, resting or otherwise. In Fig. 8.12 the middle of such bends on the trunk are marked by an arrow. In the geophilomorph centipede *Orya* and in the diplopod *Ophistreptus* the line joining the points form an even curve. In *Lithobius* the small dots show the degrees of flexure at the anterior end of the short tergites and the large dots show the much smaller flexures at the anterior end of the long tergites, X on Pl. 5(e) marks the visible rigidity at the anterior ends of the long tergites in a running *Lithobius*. There is greater tergite heteronomy in *Lithobius* than in the Scolopen-

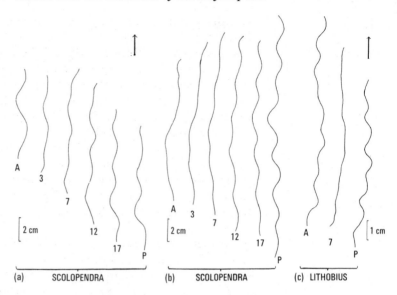

FIG. 8.11. Each line, plotted from cinematograph films, shows successive positions relative to the ground, of the head end, A, of the mid-posterior margin of tergites 3, 7, 12, and 17, and of the posterior end of the body, P, of *Scolopendra morsitans* and of *Lithobius forficatus*. In (a) the head executes marked sensory casting aside of the antennae, speed 180 mm/s. In (b) the head was maintained in the direction of motion, speed 200 mm/s. In (c) the head executed conspicuous casting aside of the antennae, speed 90 mm/s. The animals were running in the direction of the arrows from the bottom of the page towards the top. (From Manton 1965.)

dromorpha and plottings for the latter show less rigidity at the anterior end of the long tergite (see arrow marking rigidity in Fig. 5.9(a)), and it is least in *Cryptops* where heteronomy is very slight. In *Scutigera*, Fig. 8.12, the flexibility at the anterior end of the short tergites has been eliminated by their small size and position under the edge of the long tergites. The short tergites are invisible on Fig. 5.9(b) and Pl. 4(j), (k). The general rigidity has been increased by fusion of the two large central tergites which now cover legs 6–9 (Fig. 5.9(b), 7.1(e), (f)). The effect of increased tergite heteronomy is shown in Pl. 4(j), where there is little undulation on the body during fast running (cf. Pl. 1(f)), an absolute necessity for the long-legged, swiftly moving *Scutigera*. This acquisition of facultative rigidity is an evolutionary triumph and contrasts with the flexibility of the trunk displayed during cleaning.

(ii) *Musculature and heteronomy*. The basis of the control of yawing associated with tergite heteronomy is not far to seek.

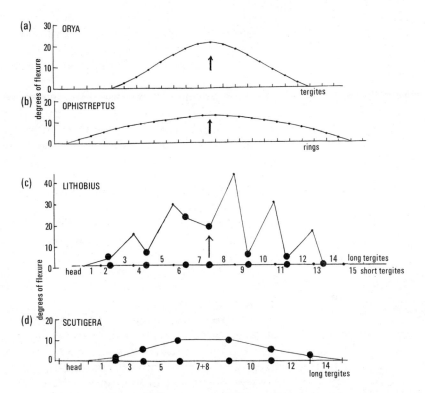

FIG. 8.12. Graphical representations of the flexures between the successive tergites from photographs of animals showing acute bends of the body in the horizontal plane when moving freely or resting. Degrees of lateral flexure are shown by the ordinates and the length of successive tergites are marked on the abscissae. The heavy dots on (c) and (d) show the points of flexure at the anterior ends of the long tergites, and the vertical arrows mark the middle of each bend of the body. (a) The geophilomorph centipede *Orya barbarica*, middle part of the body, in a hairpin bend, 16 segments effecting a total flexure of about 180°. (b) The iuliform millipede *Ophistreptus guineensis*, middle part of the body, in a hairpin bend, 23 diplosegments effecting a total flexure of about 180°. (c) *Lithobius forficatus* in a hairpin bend of about 190° by segments 1–13. (d) *Scutigera coleoptrata* in as acute a bend as is possible to the animal, about 35° by segments 1–14. The small dots marking intersegments with greatest flexibility in *Lithobius* are in effect totally eliminated by *Scutigera* owing to the overlap of the long tergites by the short tergites so that only the points of maximally controlled flexure, the heavy dots, remain. For further description see text. (From Manton 1965.)

Every segment possesses a similar complement of muscles all along the body. The segments at the head end and at the posterior extremity of the body are not being considered here because their peculiarities do not illuminate the composition of the major part of the trunk. The serial segmental complements of muscles are uniform in number of main muscles, but sometimes one muscle may possess more than one sector (e.g. *dpl.* and *dpll.*, *dvmp.a.* and *dvmp.b.* (Manton 1965, Figs. 51 and 56)). The latter is a common phenomenon associated with function. A deep oblique muscle, with much force to generate, but no complexity of movement to create, may be unusually large, as in Collembola (Chapter 9, Fig. 9.3(a), muscle *ob.*, abdominal segment 3) and its functions in the thysanuran abdomen dictate subdivision into a number of sectors with varied dorsal insertions in association with more complex movements (Chapter 9, Figs. 9.13, 9.14, muscles 24, 25A, 25B, 26, which correspond with *ob.* of Collembola and *dvma.* of Chilopoda). The presence or absence of more than one sector of a muscle is correlated with functional needs and provides no evidence of partial elimination of segments along the trunk.

Three features of the musculature of fast-running Chilopoda contribute towards this accomplishment by providing facultative rigidity, controlling the tendency to yaw, as well as rigidity preventing sagging of the body on to the substratum between widely spaced propulsive legs. (i) Well-developed deep oblique muscles *dvmp.*, *dvma.*, and their antagonistic deep dorso-ventral muscles *dvtr.*, (Fig. 8.13); these muscles are entirely absent in Symphyla where flexibility of trunk is the priority. (ii) The single short sectors of the dorsal longitudinal muscles crossing one joint in Geophilomorpha (Figs. 8.2, 8.9) are represented by a complex system composed partly of short muscles, *dlm.*, *dom.*, and more bulky long sectors, muscles *dlm.A.*, *dlm.B.*, *dlm.C.*, crossing more than one intersegment, Fig. 8.13. (iii) A shifting of the dorsal insertions of many trunk and extrinsic leg muscles off the short tergites and on to the long tergites, in progressive measure as heteronomy in tergites length increases. The shifting of the dorsal insertions of chilopod muscles is shown diagrammatically in Fig. 8.14.

The extra rigidity at legs 7–8 shown by Figs. 8.11, 8.12, and Pl. 5(e) is neatly contrived (Fig. 8.14 for *Lithobius*). These two long tergites receive muscles respectively from the short tergites in front and behind, so that between them they carry more muscles than do any two normal successive segments. This zone of greatest rigidity lies at a workable distance behind the head with its antennal movements, giving a maximal damping-out of undulations. The process is

Habits and the evolution of the Myriapoda

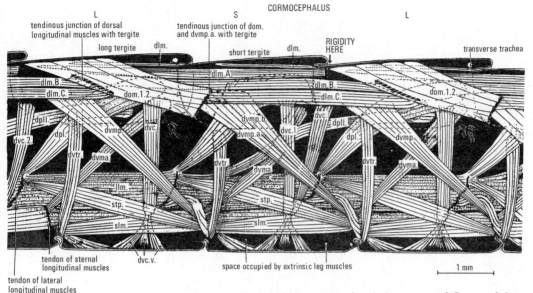

FIG. 8.13. Showing the principal muscles of typical segments of *Cormocephalus calcaratus* (Scolopendromorpha) with long (L) and short (S) tergites. Internal median view of the right half of the body after removal of the viscera, fat body, and nerve cord; head end to the left. Dotted lines show the lateral extent of the tergites. The dorsal musculature is foreshortened. The superficial (short) dorsal muscles *dlm.* are not shown here being obscured by the deep longitudinal (long) dorsal muscles *dlm.A*, *dlm.B*, and *dlm.C*; the deep oblique muscles *dom.1.2*, corresponding with *dom.1* and *dom.2* of *Cryptops*, are shown. The segmental tendinous junctions, anchored to the tergites, are indicated diagrammatically in black, those between the *dlm.B* and *dlm.C* muscles being covered by *dom.1.2* in this aspect. Heteronomy is shown by the short and long segments in the dorsal muscles *dlm.A*, *dlm.B*, *dlm.C*, *dom.1.2*, and by the deep dorsal-ventral and deep oblique muscles *dpl.1*, *dvc.(dvc.1, dvc.2)*, *dvmp.*, *dvtr*. Note muscle *dvc.2* and *dpll.* giving lateral stability to the anterior ends of the (L) tergites. Muscle *vtr.* arises widely on the tendon of sternal longitudinal muscles and inserts on the tendon of the lateral longitudinal muscle. Muscle *dvmp.* in the (L) segments and sector *dvmp.b* in the (S) segments insert on the tergites, but sector *dvmp.a* in the (S) segments inserts on the tendon of the *dom.–dlm.* muscles. (From Manton 1965.)

completed by the fusion of these tergites together in the Scutigeromorpha (Fig. 5.9(b)) and by the absence of a long posterior part of the body in the Anamorpha (Fig. 7.1(e), (f)), so that the long tergites over legs 7 and 8, already differentiated in the Scolopendromorpha (Fig. 7.1(a)–(d)), become central on the body in the Anamorpha. It has already been shown that segment numbers are controlled by other features, such as even loading on the legs during fast gaits (Chapter 7, §5).

A summary is given in Fig. 8.15 of the general manner in

385

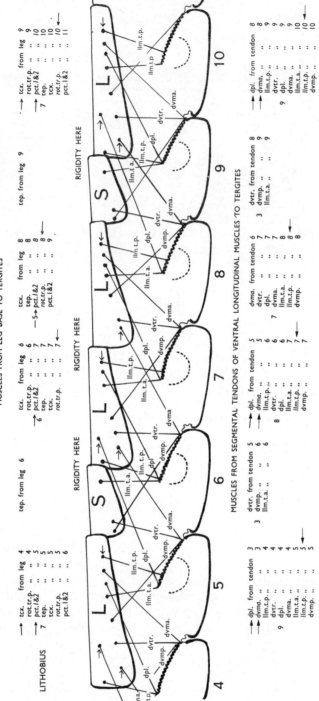

FIG. 8.14. Diagram showing the disposition of the principal deep dorso-ventral and deep oblique muscles on tergites 5 to 10 of *Lithobius forficatus* in order to depict the comparison between the muscle insertions of typical consecutive long and short tergites, such as those on segments 5 and 6, and the insertions present on the consecutive long tergites on segments 7 and 8. The muscles arise from segmental tendons (black zigzag lines) of the ventral longitudinal musculature (fused tendons of the sternal and lateral longitudinal muscles). Below each sternite are listed the deep dorso-ventral and deep oblique muscles inserting on the corresponding tergite and above are listed the extrinsic coxal muscles and those from the katopleure and procoxa which insert upon each tergite. The arrows and italics indicate muscles whose dorsal insertions have migrated, and the arrows indicate the direction from which they have come. A simplification in the diagram shows muscles *pct.1* and *pct.2* as one muscle; and muscle *dpl.* in the (S) segments, which divides into three sectors, only one being shown here. The dotted area shows the positions of the proximal rim of the trochanter in each segment. (For further description see Manton (1965) Fig. 34.)

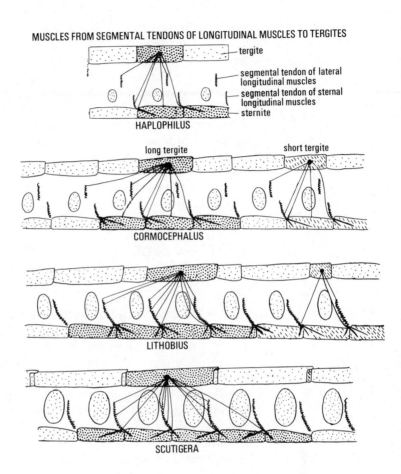

FIG. 8.15. Diagrams showing the linkage of the tergite to the sternite by muscles via the segmental tendons of the longitudinal muscles (black zigzag lines) and the coxal apodemes (not shown) in examples from the four chilopodan orders. One tergite of the geophilomorph *Haplophilus subterraneus* and one long tergite of each of the other animals, together with their associated sternites, are marked by heavy stippling. Note that one tergite of *Haplophilus* is indirectly linked to two sternites, in *Cormocephalus* to three sternites, in *Lithobius* to four sternites, and in *Scutigera* to five sternites, thus promoting trunk stability because each long tergite is similarly connected. One short tergite of *Cormocephalus calcaratus*, *Lithobius forficatus*, and *Scutigera coleoptrata* is marked by interrupted hatching and its corresponding muscular connections shown. Each line represents one muscle. From left to right the muscles attached to one stippled tergite are: *Haplophilus* (six muscles), *dpl.*, *dvma.*, *dvtr.*, *dvc.*, *llm.t.*, *dvmp.*; *Cormocephalus* (seven muscles), *dpll.*, *dpl.*, *dvc.2*, *dvma.*, *dvtr.*, *dvc.*, *dvmp.*; *Lithobius* (eight muscles), *dpll.* (two sectors) from preceding segment, *dpl.* from own segment, *llm.t.p.*, *dvtr.*, *dvma.*, *dvmp.*, *llm.t.a.*, *llm.t.p.*; and *Scutigera* (twelve muscles), *dpl.*, *dvma.*, *dvtr.*, *llm.t.a.*, *llm.t.p.*, *dpl.*, *dvma.*, *dvtr.*, *dvmp.*, *llm.t.a.*, *llm.t.p.*, *dvmp.* (From Manton 1965.)

which trunk rigidity is promoted in increasing measure as faster gaits and longer legs are employed by the four main orders of centipedes. In the Geophilomorpha there is no heteronomy in tergite lengths and every tergite is linked with the segmental tendons and sternites as shown in Figs. 8.9, 8.15, while at the bottom of the latter figure each long tergite is linked with five successive sternites instead of with two. The extrinsic muscles from the legs also migrate progressively on to the long tergites (Manton 1965, Fig. 33). Overall rigidity is thus well provided for in the longest legged, fastest runners, the Scutigeromorpha.

An attempt has been made in Fig. 8.16 to show how some of the long and short sectors of dorsal longitudinal muscles cooperate with the deep oblique muscles $dvmp$. Tension on a muscle is shown in black and by stippling. In Fig. 8.16(a) $dlm.A.$ and $dlm.B.$, extending between two long tergites, and $dvmp.5$ cannot do other than flex the body dorso-ventrally, aided by tension on the lateral and sternal longitudinals. In Fig. 8.16(b) the tension on the black muscles cannot do other than straighten the body, here drawn in the flexed position. Corresponding diagrams for *Scolopendra viridicornis* possessing much greater heteronomy in tergite length are given in Manton 1965, Fig. 39, where it is shown how the greater heteronomy and the further differentiation in the dorsal musculature provide even greater efficiency for both the movements shown. This is not a complete analysis of the co-ordination between muscle systems, but just an example of the inevitable effect of the muscle changes associated with heteronomy.

D. *A secondary chilopodan achievement:* Craterostigmus

Two examples were given above (§3C) of secondary diplopodan achievements by taxa with markedly divergent habits and morphology from what must be regarded as basic diplopodan structure. Even great doubt has reigned as to the systematic position of *Polyxenus* until the functional morphology of this animal was worked out. A similar situation existed concerning the centipede *Craterostigmus* from Tasmania and New Zealand. This animal was regarded by systematists, working on the external form of dead specimens, as an aberrant lithobiomorph, largely because it possesses 15 pairs of legs as in *Lithobius* (Pl. 5(g), 7(i)). But note the scolopendromorph gait of *Craterostigmus*, with groups of three legs converging towards the same footprint, marked by black spots, as on Pl. 1(f) and Fig. 7.1(a)–(d), entirely unlike the gaits of *Lithobius* (Pl. 5(g) and Fig. 7.1(e)) where the propulsive legs diverge (marked by black spots on Pl. 5(g)).

The gait pattern of *Craterostigmus* (5·5 : 4·5) is slow for a scolopendromorph; only one leg moves forwards between successive groups of propulsive legs on each side of the body (cf. Fig. 6.13). Moreover, under extreme stimulation of bright light for cinematography the animal showed itself to be

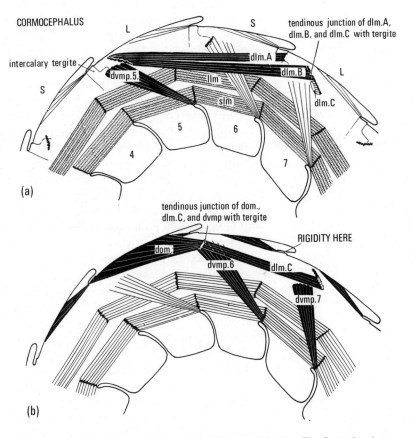

Fig. 8.16. Diagram of part of the trunk of *Cormocephalus* (see Fig. 8.13) showing dorsal muscles *dlm.*, *dlm.A*, *dlm.B*, *dlm.C*, deep dorso-ventral muscles *dvmp.*, and lateral and sternal longitudinal muscles *llm.* and *slm.*, viewed from the sagittal plane, the long and short tergites being marked L and S. Strong tension on a muscle is indicated in black or by stippling. Muscle tensions in (a) tend to flex the body and in (b) tend to straighten the body. For further description see text. (From Manton 1965.)

incapable of faster patterns; an attempt at faster running led to uncontrollable lateral undulations of the body and crossing of propulsive legs and was abandoned in a fraction of a second. Slow running and great flexibility are the trunk peculiarities of *Craterostigmus* and are associated with many structural modifications (described in Manton 1965), but exactly how this animal uses its remarkable flexibility in its natural way of life we do not fully understand because so little is known of its habits.

In Tasmania *Craterostigmus* females have recently been found in the same localities during certain months of the year only, under bark guarding their egg masses until they hatch, as does a normal scolopendromorph. No males were present with the females. Where they all go in other months we do not know because they have eluded diligent searching. Males have been obtained in Tasmania many years ago and males and females of two species are known from New Zealand.

The extreme trunk mobility of *Craterostigmus* is based upon normal scolopendromorph morphology with a degree of tergite heteronomy characteristic of this group. But every long tergite (on segments 3, 5, 7, 8, 10, and 12) has become secondarily divided into two at the levels marked by the arrows on the left of Pl. 7(i). Other flexibility-promoting skeletal features are: sternites which can fold; strongly developed intercalary tergites; a collapsible upper pleuron; and particular features concerning the procoxa and legs. The animal has extreme powers of body flattening and penetration of shallow crevices without pushing. The musculature is profoundly modified, but on a scolopendromorph basis. The dorsal muscles serve local telescoping and shuffling together of tergites; the stabilizing long sectors of dorsal musculature *dlm.A.*, *dlm.B.*, and *dlm.C.* (Fig. 8.13) are absent; the many muscular features normally inhibiting lateral undulations during running are absent (for details see Manton 1965).

Craterostigmus in captivity in England refused all live and dead food which has been readily accepted by European, Asian, and African centipedes and only accepted live Tasmanian termites. The feeding mechanism is peculiar. The poison claws are very large and conspicuous in dorsal view of the head, as they are in *Lithobius* and less so in *Scolopendra* (Pls. 1(f), 5(g), 7(i)). This is due to size, the structure of the poison claws resembling that of the Scolopendromorpha. The mandible of *Craterostigmus* is very small and entirely unlike a lithobiomorph or a scutigeromorph mandible (Figs. 3.22, 8.17(a), (b)) and more closely resembles the shape and size of a mandible of a geophilomorph centipede (Fig. 8.17(c), (d)). The poison claws can be protracted in front of the head to a grater extent than in other centipedes and are used expertly to pull termites out of cracks. Feeding is accompanied by external digestion and extensive licking by the hypopharynx, the head capsule sliding forwards and backwards as recorded for no other centipede. The latter movement is caused by muscles, but the protraction can only be done by hydrostatic pressure of haemocoelic fluid forced forward by trunk musculature.

Craterostigmus has been observed by Professor V. Hickman to make tunnels through compacted decaying wood and this is done by means of the very mobile and protrusible poison claws, movements presumably also used in extracting termites from confined spaces, but no pushing by the body surface is done as in the Geophilomorpha. This spade-and-shovel method of burrowing is yet another myriapodan technique not found in diplopods or in other chilopods, although utilized in hexapods by mole crickets and others.

The feeding movements and the extreme mobility of the trunk (for details see Manton 1965) are probably associated with the almost onychophoran-like tracheal system. The large crater-like spiracles situated on the segments bearing the original long tergites, which gives the animal its name, supply not a few large tracheal trunks which branch internally, as in other centipedes, but a very large number of minute tracheae which do not branch and which pass as bulky sheaves of minute tubes to their various destinations. Air-filled tubes, 2–6 μm in diameter, can resist deformation by changing haemocoelic pressures in *Craterostigmus* and in Onychophora. Tracheal systems of other centipedes also are modified in detail to suit their requirements. The scolopendromorph *Plutonium*, living deep in rock crevices, has an opposite type of tracheal modification; large sinuses are present which probably act as reservoirs removing CO_2 from the tissues in the limited and unchanging air conditions in deep, confined rock spaces. The minute tracheae of *Craterostigmus* do not represent a primitive state or show evidence of onychophoran affinity but merely an adaptation to the use of hydrostatic forces in promoting shape changes. *Craterostigmus* incidentally brings the tracheal system of the

Habits and the evolution of the Myriapoda

Onychophora into a new light.

Reference has been made above to the serviceability of certain overall body shapes which characterize diplopodan taxa such as the Iuliformia. The same ratio of body length to width is variously maintained. The long segments of *Craterostigmus* facilitating mobility of trunk, and their smaller number than in other Scolopendromorpha, results in about the same overall proportions (cf. Pl. 1(f) with 7(i)). It is probable that the shape suits the partially understood habits of *Craterostigmus*.

There are plenty of little understood problems among myriapods. We do not understand the full significance of hatching with the adult number of segments in epimorphic centipedes (Geophilomorpha and Scolopendromorpha) and hatching with a small number which is subsequently increased in the anamorphic centipedes (Lithobiomorpha and Scutigeromorpha). There are three pairs of legs in the newly hatched millipede and in *Pauropus* (Pls. 4(l), 7(h)); eight leg pairs in Symphyla, the 8th being rudimentary; and seven in *Lithobius* (Pl. 5(h)). *Craterostigmus* confounds us by brooding its egg mass, as in Scolopendromorpha, but the young hatch with 12 pairs of legs (Pl. 4(n)) not the full adult 15 pairs or the smaller number possessed by *Lithobius* hatching from singly abandoned eggs, rendered inconspicuous by sticky eggshells being rolled in soil.

We can but conclude that this interesting centipede *Craterostigmus* must once have enjoyed all the anti-undulation features associated with fast running as in a typical scolopendromorph. But now it has reversed its principal needs which are flexibility of trunk and not rigidity. This aberrant scolopendromorph has probably been long separated from the Scolopendridae and is quite unrelated to the Lithobiomorpha. The dichotomy in feeding habits which was probably associated with the differentiation of the epimorphic and anamorphic centipedes (Chapter 7, §6C(iii)), places *Craterostigmus* squarely with the Epimorpha but in a separate group the Craterostigmidae parallel with the Scolopendridae and Cryptopidae.

E. Conclusions concerning Chilopoda

It has been shown in outline above and in Chapter 7, but in considerable detail in the original accounts (Manton 1952*b*, 1965) how the form of the body; the number of segments; the details of structure of the cuticle; the type and disposition of

Habits and the evolution of the Myriapoda

the trunk sclerites; and various arrangements of sectors of short and of long musculature are bound up with habits and with functional needs. All must have evolved together and could have resulted from early habit divergencies outlined in Chapter 6, §5D.

The progressively stabilizing effect of tergite heteronomy; of successive large tergites over legs 7 and 8; and of the associated muscle changes is demonstrated beyond question and further evidence is provided by the remarkable centipede *Craterostigmus*.

The various groups of centipedes do not form a series showing increasing perfection of various habits. Each taxon has its different detailed proficiencies. The Geophilomorpha are artists in the earthworm-like burrowing technique; among the Scolopendromorpha the Cryptopidae achieve speed by little tergite heteronomy but very fast stepping; while the Scolopendridae, with stouter, wider bodies and greater heteronomy in tergites and muscles, achieve speed by these features which permit the use of very fast gait patterns, impossible to the Cryptopidae. The Lithobiomorpha show an alternative way of achieving speed of running and are proficient at penetrating crevices, but without pushing.

The Scutigeromorpha are outstanding in their many perfections and are in no way simple primitive centipedes as their dead bodies have suggested to the earlier systematists. Acute vision from closely grouped optic units permits the location of spiders and flies; and rapidity off the mark, speed of running, and quick movements of lightly constructed poison claws ensure the capture of such fast-moving prey. Not least of the scutigeromorph perfections is the enormous size and strength of the entognathous mandibles which enable the cutting up and swallowing of pieces of hard food. The mandible is shown in Fig. 8.17(a) on the right where it moves within the head capsule, contrast with the moderate sized mandible of a scolopendromorph shown in Fig. 3.22 and the relatively minute geophilomorph mandible in Fig. 8.17(d). The latter puts the whole head inside the prey and scoops up the soft tissue. The light build of the scutigeromorph poison claw (left side of Fig. 8.17(a)) and its basal articulation are not primitive and contrast with the more slowly moving, but massively strong, poison claws shown by the Scolopendromorpha, Fig. 3.22.

The Scutigeromorpha are indeed highly advanced and not

primitive centipedes. Their long legs with elaborate jointing and 34 extrinsic muscles per leg (cf. the 19 muscles in a scolopendromorph, 13 in a geophilomorph, and 2 or 4 in a diplopod) testify to the same conclusion, as does their

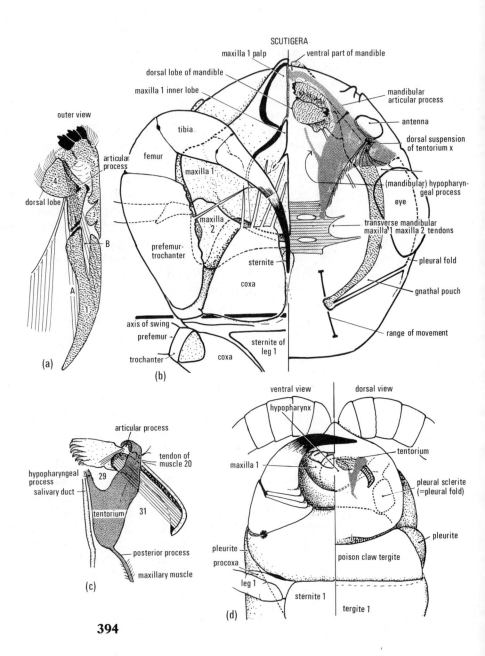

extreme tergite heteronomy and muscle connections perfecting the anti-undulation mechanism without which such long legs and speedy running would not be possible.

Consideration has been given in Chapter 7, §6C(iii) to the importance of a probable habit dichotomy between the ancestors of epimorphic and anamorphic centipedes. The ancestors of the latter, gaining greater leg length and exploiting quickness off the mark in starting to run, by their characteristic type of gait, has led to catching swiftly moving prey. Eyesight became acute, poison claws not too heavy, and mandibles very large culminating in the cutting up and swallowing of pieces of large, hard prey. The epimorphic centipedes, on the contrary, went in for longer series of gaits by shorter legs; speed, although great, is attained gradually through a series of gaits. Food is frequently stumbled upon by accident and reliance is placed on strong cutting and tearing by the poison claws, the soft parts only of large prey being eaten. Such a dichotomy in habit could have led to the several structural and habit proficiencies of the main chilopodan taxa. There have also been reversals of habit and of structure, an extreme example being *Craterostigmus*; some Lithobiomorpha are blind and therefore cannot apprehend their prey from a distance and pounce. The conclusions concerning centipede evolution outlined here could not have

FIG. 8.17 (see opposite). Diagrams to illustrate the superlative advances of the mandible of *Scutigera* (a), (b), compared with other centipedes. The mandible of the scolopendromorph *Cormocephalus* is shown in Fig. 3.22 and that of the geophilomorph centipede *Orya* here (c), (d). (a) Head of *Scutigera coleoptrata* drawn to the comparable scale used for the figures of *Cormocephalus* and of *Orya*. Ventral view of the head of *Scutigera* on the left; and dorsal view on the right; drawn as a transparent object to show the mandible (mottled) and swinging anterior tentorial apodeme (black, red in Chapter 3). The musculature is omitted, the transverse segmental tendons appertaining to the mouth parts are shown by horizontal ruling (pink in chapter 3), and the curved heavy line shows the range of movement of the posterior end of the mandible. Note the great size of the mandible and the anterior tentorial apodeme which causes abduction compared with *Cormocephalus*. (b) Isolated right mandible to show the sclerites and intrinsic muscles in lateral, inner view on the same scale as in (a). Intrinsic mandibular muscles A and B and extrinsic mandibular adductor 20 (unlabelled) are shown. (c) Similar diagram of the head of geophilomorph centipede *Orya barbarica*, the ventral aspect on the left and the dorsal on the right, drawn as a transparent object. Note the minute size of the mandible and of the anterior tentorial apodeme compared with *Scutigera*, but note also that both structures are of the same general type as in *Cormocephalus*, another epimorphic centipede. (d) Isolated mandible and tentorium of *Orya barbarica* on a larger scale than (c) in order to show the details. The anterior tentorial apodeme, marked tentorium, is stippled and not black. Intrinsic mandibular muscles 29, 31 are marked, and extrinsic mandibular adductor 20 arises from the marked tendon. (From Manton 1965.)

been obtained from a study of dead animals. The functional approach has been essential.

6. The Pauropoda

These minute animals (Pl. 4(m) and the first instar in Pl. 4(i)) form a fitting conclusion to the analysis presented above. As in Symphyla, there is little structural variation within the class. Again Tiegs (1947) has demonstrated by his fine embryological studies how the trunk of *Pauropus sylvaticus* is established by a series of mesodermal somites, ganglia, etc., corresponding to the adult series of legs. Tergites develop on every other segment and spread anteriorly and posteriorly, but do not quite touch one another (Fig. 5.11(a)). These animals normally live in draught-free environments inside decaying logs, etc. They move about rapidly for their size. They use fast patterned gaits with few legs in contact with the ground at any one moment (Fig. 7.10(d)). They show great rigidity of body and no flexibility. It comes as no surprise to find that the flexibility promoting superficial oblique muscles are entirely absent (cf. Symphyla) and that the rigidity promoting dorso-ventral and deep oblique muscles are well developed (Fig. 5.11(b)) as in Chilopoda, Collembola, Thysanura, etc. (Figs. 8.13, 9.3, 9.6, 9.7, 9.13). The absence of alternate tergites represents a more extreme form of heteronomy than is reached by the Chilopoda and only the long, stability-promoting, sectors of the dorsal longitudinal muscles are present, extending between the tergites on alternate segments. There is no special rigidity provided by long central tergites on successive segments, there being no casting sideways of the head in running and the legs and body being short. The whole organization of sclerites and muscles forms an end term to the known modifications suited to the use of fast gait patterns (Manton 1966).

7. General conclusions on myriapod evolution

The trunk morphology of the myriapod classes is based upon a common plan, but the class differences are so great as to suggest parallel evolution of these classes from soft-bodied ancestral Uniramia with a common head type. These conclusions could not have been reached without the completion of an extensive functional and comparative study of the principal structural features of both heads and trunks of Onychophora and Myriapoda.

The outline suggested for myriapod evolution in Chapter 6, Fig. 6.14 is amply supported by the detail considered in this chapter. Embryology and accurate morphology recently described for the first time, combined with a study of the habits and accomplishments of the various groups of myriapods, show that the trunk morphology has a common basis. Serial segmentation of the body has been modified in various ways which facilitate the special accomplishments of the animals concerned, the series of legs indicating the metameric segmentation of the major part of the trunk, all segmental divergencies in sclerites being secondary. A basic type of head structure is similarly modified in association with habits. Not one of the groups considered in this chapter could possibly have been ancestral to any other. All represent independent divergencies from ancestors with many legs, flexible bodies, and a particular type of head capsule and head limbs.

9 Habits and the evolution of the hexapod classes

1. Introduction

AN abundance of fresh and revealing information on the hexapods is now available from the field of functional morphology, but this information could not have been derived from the hexapods alone. The interpretation of their structure rests in part on modern studies on the hexapods themselves, but the work accomplished on the rest of the Uniramia, and on other arthropods, also has been essential. A comparative viewpoint of the data is usually needed before sound conclusions can be reached.

The fundamental differences between the structure and modes of action of the coxa–body junctions in the five hexapod classes have been described in Chapter 5. These differences are so deep-seated as to suggest independent evolution from multi-legged ancestors, having in common a head type quite different from that of myriapods (Chapter 6, §5, Fig. 6.14). The abundance of different theories of insect origin now can be set aside because they all entail the assumption of functionally impossible evolutionary stages, supposedly leading from one type of ancestral hexapod. But no such ancestor could ever have existed (Chapter 7, §6C(viii)).

With increasing competition to survive, perhaps since the Carboniferous when conditions may have been easier, or the fauna less dense than it is today, speedier locomotion or other methods of achieving protection may have become useful. The hexapod classes appear to have met this challenge in the first place by the parallel evolution of hexapody, bringing with it the advantages considered in Chapter 5, §3, Fig. 5.2, Pl. 6. Thereby they avoided all the different myriapodan solutions to competitive life on land, based on habits and the associated structural advances (Chapter 6, §5D, Chapter 8). Later it is probable that competition and selection pressure made simple hexapody an insufficient remedy for immediate problems of survival, just as the early simple habit differentiations of the Onychophora and four myriapod classes also in themselves became insufficient. Perfection of habits and the correlated structure of Onychophora and Myriapoda led to the present-day proficiencies of these classes, advances in the sense considered in Chapter 1, which fit them to live better in the same or in a variety of habitats. In the end the hexapods too needed more advanced accomplishments for their survival. Speed became desirable, unless small size and a secondary reduction in leg length enabled refuge to be taken in the soil, inside decaying logs, and such places. Greater speed of running or perfection of escape reactions, were obtained by the apterygote Collembola, the dipluran Campodeidae, and by the Thysanura, all unguligrade, but with distinctive locomotory mechanisms. Many Collembola secondarily lost their jumping escape mechanisms and became small in size and shorter in the leg as they gave up surface living and sub-surface hiding and descended more thoroughly into the substratum (Pl. 6(c)). The dipluran Japygidae and the Protura took to the same sort of habits, but with marked structural correlations, see below §§3 and 4. The Pterygota developed flight in the adult state as their solution to the speed problem; they are usually plantigrade and they have not employed the myriapodan types of running, different pterygote species being fast or slow moving. The ability to evolve the power of flight was dependent upon the previous establishment of a certain type of coxa–body junction, differing from all other hexapods and myriapods in possessing a fixed, immobile pleurite, to which the dorsal proximal rim of the coxa was articulated. Without this flight could not have evolved.

It has been suggested in Chapter 5, §3 and Chapter 7, §5C(viii) that the superficial similarities between the gaits of all uniramian hexapods are due to the limited number of ways in which six legs can be used; in other words these similarities are convergent. What is also new is the appreciation that each class of hexapods has elaborated its own particular way of using six legs, all five classes supposedly having started from multi-legged, soft-bodied ancestors in which different types of coxa–body junctions were being evolved. These joints served somewhat the same purposes but are mutually exclusive and could only have evolved from a soft-bodied lobopodial leg.

Habits and the evolution of the hexapod classes

The head structure of the hexapod classes, all with uniramous mandibles biting with the tips, suggests the existence of a common type of head capsule at a time when the trunk was soft-bodied and multi-legged (Chapter 6, Fig. 6.14). However, this type of head contrasts with all myriapod heads in possessing two pairs of fixed tentorial apodemes, not an anterior pair alone as in myriapods; hexapod anterior tentorial apodemes never swing within the head capsule, nor are they involved in causing mandibular abduction as in myriapods (Chapter 3, §5). The basic mandibular movements in hexapods is a rolling one directly derivable from the promotor–remotor swing of a walking leg, while the basic myriapod mandibular movement is transverse biting by jointed mandibles. These differences are decisive. Secondary transverse biting used for hard-food feeding has been evolved in the Pterygota and some Thysanura, and secondary, protrusible entognathous mandibles have been independently evolved several times in the apterygote classes. The details concerning entognathy differ in the several classes. The means whereby a protrusible mandible, and sometimes maxillae 1 also, are obtained and used are not the same in the Collembola, Diplura, and Protura (Chapter 3), and contrast in detail with the entognathous Chilopoda, siphonophorid Diplopoda, ectoparasitic Crustacea, and ectoparasitic Arachnida where stylets are present within a forwardly projecting part of the body (Chapters 5, 6, §3, Figs. 6.5–9, and Manton 1964).

The one-time notion that insects as they are today have descended from one ancestral insectan group is now untenable; and since the term 'insect' becomes ambiguous, because it may or may not imply the pterygotes alone, it is more satisfactory to use the term hexapod with no phylogenetic implications. Hexapody also occurs in Crustacea and Arachnida. The five class names for the uniramian hexapods are precise, four of them can be described as apterygote (wingless) for convenience but the 'Apterygota', as a taxon, has no more meaning than the 'Entognatha', both terms used abundantly even in the most recent literature, but having no more than descriptive significance. We may now go on to consider some of the advances and special achievements of the several hexapod classes with which their morphology is associated. A longer and fuller account is given in Manton (1972).

2. The Collembola

The Collembola are the oldest of known hexapods. Fossil remains of *Rhyniella praecursor* Hirst and Maulik occur in the Devonian, but it was already an advanced collembolan with fused prothoracic and mesothoracic tergites, of service for the jumping escape reaction.

The habits and habitats of modern Collembola have been summarized by Kühnelt (1961). Their most important locations are as follows. (1) The soil surface and porous superficial layer, in litter and moss, under logs, etc. Here the species are large, 3–4 mm; ocelli and good springing organs are present and the legs are long, e.g. *Tomocerus* (Pl. 6(a), (d)) and *Orchesella*. (2) The upper soil layers where the species are smaller, 2 mm; heavier in build; with a shortening of the springing organ and the legs; e.g. *Folsomia*. (3) The deeper soil layers, where legs are short; springing organs absent, ocelli absent, and body worm-like and white. Small size enables easy passage through minute cracks and spaces, but no active burrowing is possible, e.g. *Tullbergia, Friesea*. Greater resistance to desiccation is shown by the species living at upper levels, but all require humidity, and even *Tomocerus* can go down to 15 cm in the soil during dry conditions. The food is very various in these three habitats, ranging from decayed plant material, fungal mycelia and spores, animal food such as fly pupae, fly larvae, dead flies, other Collembola, etc.; and each species has its own particular preferences which determine to some extent the microhabitat and locomotory activities. (4) Above the soil live the Symphypleona, with globular bodies; small thorax; long legs; and long antennae. They are found on damp wood; *Alacma fusca* climbs trees at night in Britain and feeds on epiphytes, and *Sminthurus viridis* can be a pest of crops in Australia.

Both running and jumping are well displayed by the larger and longer-legged Collembola living on the soil surface and in the upper porous layers. In the deeper niches both accomplishments wane, locomotory movements become slower and jumping absent.

Precision stepping takes place in Collembola in spite of a hydrostatic mechanism used in jumping. *Orchesella villosa* Geof. and *Tomocerus longicornis* Müller under bright illumination almost always use their same fast gait (4·4: 5·6) at a phase difference between successive legs of about 0·44 (Fig. 7.3(a)). But much slower patterns of gait are employed when the animals are pottering about in dark and dimly-lit conditions. Two of these gait patterns are shown in Fig. 7.3(b), (c), bringing with them the usual advantages of more even loading on the legs, Fig. 7.3(f), (h), (i). The change in support by successive legs on the same side of the body is almost simultaneous and therefore the track shows groups of footprints of legs 1–3 in transverse rows, Fig. 7.1˚2(a), Pl. 6(d). The stable moments marked k in Figs. 7.8–10 are absent, but since the moments of support at the change over of successive legs are staggered between the two sides of the body, particularly in Fig. 7.3(c), stability is maintained. The speeds achieved are not great, some 25 mm/s by *Tomocerus longicornis*, but the speciality of greatest evolutionary significance for these Collembola is a superb jumping escape reaction.

A. *The collembolan jump*

The collembolan jump was shortly described in Chapter 7, §7, Fig. 7.16. When Collembola, equipped with well-formed jumping organs, are disturbed they jump readily and repeatedly, but tire easily, as do jumping Thysanura. The successive jumps take the animals in any direction, 10–20 cm upwards and landing a little distance away. The usual disruptive or cryptic colouration makes the jumping movements difficult to follow by the eye; the exercise provides protection against predatory birds, etc., hunting for food in woodland litter and such places.

The jumping mechanism of the Collembola is found to be basic to the evolutionary differentiation of the whole group; the lack of the full jumping mechanism in Collembola inhabiting environments (2) and (3) above being secondary. The jumping mechanism involves modifications of the entire morphology, features which must date from the time of the original differentiation of this class from ancestral soft-bodied Uniramia.

Peculiarities of the cuticle, endoskeleton, tendons, and musculature maintain an almost constant body volume by not permitting a billowing out of flexible or of infolded cuticle under momentary increases in internal hydrostatic pressure which cause the propulsive jumping thrust against the ground. This thrust is not provided directly by musculature, as has hitherto been supposed.

Tergo-pleural cuticular arches envelope most of the thorax, leaving very little pleuron exposed; Fig. 9.1(a) shows the flexible pleuron, stippled, between head and leg 1, also the arthrodial membrane, stippled, at the coxa–trochanter joints,

Habits and the evolution of the hexapod classes

the body being in a natural position; Fig. 9.1(b) shows a potash preparation of cuticle stained with chlorazol black, the thorax is distended and torn open showing the ventral as well as the lateral aspect, stippling here marking the stained, sclerotized areas; compare with the large expanse of pleuron in the Diplura and Protura, Fig. 5.12. There is only one pair of pleurites associated with the metathoracic legs (Fig. 9.1(a), *pl*, (b), *pleurite*). There are no sub-coxal segments, or remains thereof, either in Collembola or in any other hexapod or myriapod, contrary to the statements in the literature (see Manton 1972). The coxae of the mesothorax and metathorax are completely divided into two segments in the former and partially divided in the latter, giving functional advantages (see below). The rest of the pleural cuticle is considerably sclerotized, as shown by suitably stained sections, and unstretchable.

Alone among hexapods and myriapods, the Collembola lack a coxa–body articulation. In other arthropods the hinge and pivot joints, or their derivatives, situated between the limb podomeres and between coxa and body, possess imbricating sclerotizations separated by short links of arthrodial membrane between the actual articulations, the arthrodial membrane progressively increasing in expanse away from each articulation (Chapter 4, §6, Figs. 4.27, 4.28, and Chapter 5, §§4, 5, 6, Figs. 5.6–9). These arrangements confer both strength and localization to the movements at the joints, the amplitude depending upon podomere diameters at the joint and on the degree of the expanse of arthrodial membrane (see above references). These cuticular articulations, other than perhaps that of the very elaborate chilopodan anterior coxa–trochanter pivot (Figs. 4.29(a), 5.8), cannot resist a force which tends to slide the imbrications apart in the plane of the general surface of the cuticle; such articulations could not withstand a sudden increase in internal hydrostatic pressure such as accompanies the collembolan jump. Special arrangements exist at all these joints in Collembola which prevent such dislocation and thereby prevent any momentary increase in internal volume at the joints which would hinder the momentary transmission of maximal hydrostatic force to the springing organ.

(i) *Collembolan musculature associated with jumping*. The entire musculature of the body is concerned with jumping.

Habits and the evolution of the hexapod classes

The longitudinal musculature of the Onychophora in dorsal, lateral, and sternal sectors, Fig. 4.25, forms a complete cylinder round the trunk which can control the shape of the body, i.e. whether it is short and thick or long and thin (Pl. 4(a), (b), the same animal), the weaker antagonists being the superficial dorso-ventral muscles. Appropriate tension on all this musculature could also shorten the body and increase the internal hydrostatic pressure but the latter is not obligatory, as has been stated by ill-informed textbooks. Among the Myriapoda only the Geophilomorpha possess such a cylinder of dorsal, lateral, and sternal longitudinal muscles, here of service in producing very strong longitudinal body shortening (Pl. 5(c)), braced by the deep oblique muscles, which deliver the burrowing heave against the soil (Chapter 8, §5B, Fig. 8.9). The completeness of the muscular cylinder and of the surface cuticular armour, which changes shape according to momentary needs (Fig. 8.1(b), (c)), and the whole component musculature of trunk segments not executing the burrowing heave, together enable a maximal local thrust to be exerted against the soil with no waste of pressure due to outbillowing of body surfaces elsewhere.

The Collembola, alone among hexapods, possess such a cylinder of dorsal, lateral, and sternal longitudinal muscles. Lateral longitudinal muscles are absent from other hexapods. This cylinder of muscles (Figs. 9.1(c), 9.2, 9.3(b), 9.4), reinforced by massive deep oblique muscles (Fig. 9.3(a), *ob*), particularly in abdominal segment 3, cause exceptional body rigidity with no flexures in any direction. These muscles cause strong, slight, and sudden trunk shortening which

---

FIG. 9.1 (see opposite). Collembola. (a) Diagram of the thorax and leg bases of *Tomocerus longicornis* Müller (Pl. 6(a), (d)). The surface scales of the body have been removed in order to show the segmentation: **ab** marks the coxa of leg 1; **a, b,** show the two coxal segments of leg 2 and similar lettering marks the partially divided coxa of leg 3; **pl** marks the pleurite of 3; stippling marks the flexible arthrodial membrane at the coxa–trochanter joint and between head and thorax. (b) *Orchesella villosa* Geof. Cuticle of the anterior part of the body after treatment with potash and staining with chlorazol black. The more sclerotized parts of the cuticle appear grey (stippled) in contrast to the less sclerotized parts which remain clear. The cuticle is torn on the animal's right-hand side so that it is opened out ventrally to show the middle line. Labelling of the coxa is the same as in (a). (c) *Tomocerus longicornis* Müller (Pl. 6(a), (d)). Diagrammatic reconstruction of the dorsal longitudinal musculature of the thorax and abdomen. The middle line is dotted, short muscles, crossing one intersegment marked (S), are shown on the left, the outer (O) and the inner, more median (I) series being numbered according to the segments. Long muscles crossing two intersegments, or clear derivatives therefrom, are shown on the right. (From Manton 1972.)

Habits and the evolution of the hexapod classes

drives the haemolymph into the cavity of the springing organ causing the jump, other muscles being concerned with the release of the organ from its resting position. Muscles are concerned with the return of the springing organ to the ventral abdominal groove, a major one being an enlarged sector of the lateral longitudinal musculature (see Manton 1972, Fig. 17). Some intrinsic muscles within the median

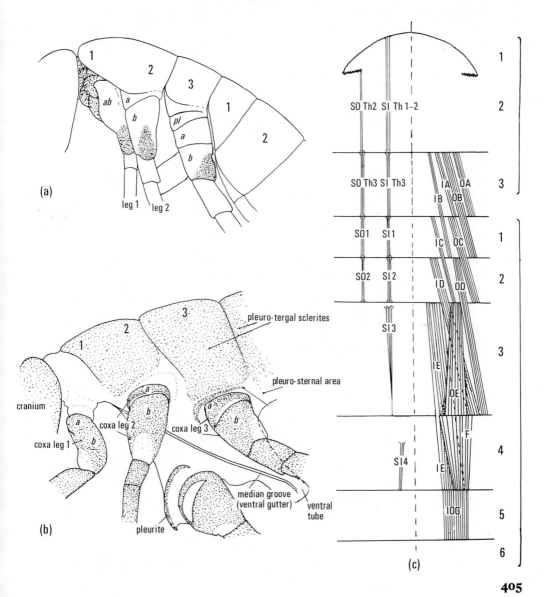

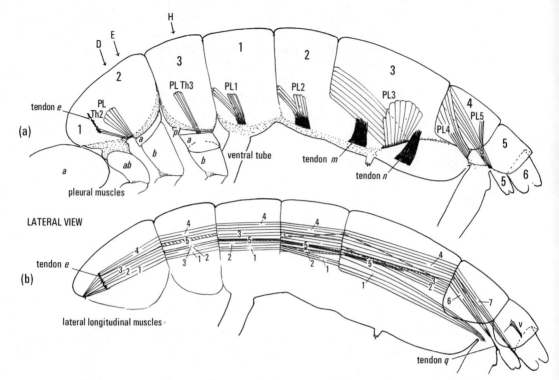

FIG. 9.2. *Tomocerus longicornis* Müller. Diagrammatic reconstructions of the trunk musculature. The body curvature is a post-mortem effect, the trunk being straight in life. Tendons are shown in black. The ventral surface of the body is drawn as if the abdomen was circular in transverse section, the ventral groove and laterally projecting folds being opened out unnaturally, as is impossible to the animal, in order to show the tendons. (a) Superficial pleural muscles. Tendons m and n bear no muscles, the lower ends of these tendons are sub-median, on the lateral wall of the ventral furrow, and they pass almost horizontally outwards to the lateral body wall. (b) The lateral longitudinal muscles. Tendon e marks the fused junction of the protergite and mesotergite. Sectors 1–5 are recognizable on many segments. Sector 1 of abdominal segment 3 forms the main flexor muscle of the springing organ. (After Manton 1972.)

manubrium of the springing organ reinforce the movement of the furca against the ground, while a tendon without muscles traverses the whole length of the manubrium, so preventing over-inflation of the furca, Fig. 7.16 (for details see Manton 1972).

differ from those of the Geophilomorpha in possessing many long sectors (Figs. 9.1(c), 9.3(a), (b)) crossing more than one intersegment. We have already seen the long muscles in the legs of fast runners (Chapter 5, §4B, Figs. 5.4, 5.6) where speed of action is required. The Geophilomorpha can take

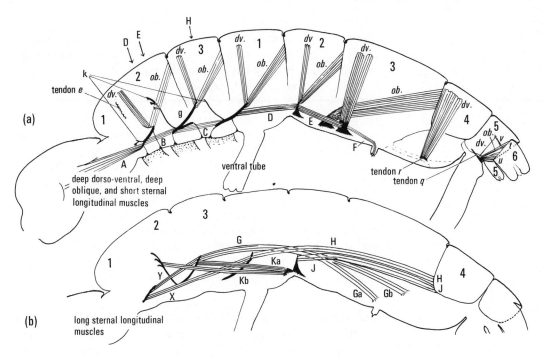

FIG. 9.3. *Tomocerus longicornis* Müller. Diagrammatic reconstructions of the musculature with the segmental thoracic endoskeletal tendon systems much simplified and shown in black as are the ribbon tendons in the abdomen. (a) The deep dorso-ventral muscles *dv.* and deep oblique muscles *ob.* are shown with the short sternal longitudinal muscles B–F, each traversing one segment only. The thoracic endoskeletal plates **g** are anchored directly by processes **k** to the tergite of the meso- and metathorax and indirectly by process *k* and a muscle on the prothorax. The large ribbon-like tendinous insertions of the muscles *dv.* and *ob.* in the abdomen, shown in black, insert ventrally on either side of the ventral groove. (b) The long sectors of the sternal longitudinal musculature, traversing more than one segment. Only those parts of the tendinous endoskeleton bearing insertions of these muscles are shown. Muscles Ga and Gb insert posteriorly in a shallow furrow between tergite 3 and basal plate 4. For further description see Manton (1972).

their time about generating a maximal thrust against the soil from short muscles with large transverse sectional areas, as can the Diplopoda, but the Collembola need to work at speed in order to obtain the best jump. Note the complex arrangement of the long sternal longitudinal muscles, extending between the segmental thoracic endoskeletal complexes of connective tissue and the unique ribbon tendons of the abdominal deep oblique and deep dorso-ventral muscles (Figs. 9.3(c), 9.5, 9.10(a), 9.11(a), 9.12, and below). Note also the posterior termination of the longitudinal muscles, particularly the sternal, which end at the proximal margin of 'ventral plate 4' shown in Figs. 7.16(a), 9.3(b), so leaving the posterior end of the abdomen to receive the blast of haemolymph which lets fly the springing organ.

It has been shown in Chapter 8 (§2C(iii), (iv)) how the deep obliques and their antagonists the deep dorso-ventral muscles promote body rigidity and this they do in a superlative manner in the Collembola. But with the large size of these muscles in the abdomen have been developed enormous ribbon-like tendons, black in Fig. 9.3, from the subectodermal connective tissue layer. Very large ectodermal cells packed with tonofibrils unite these tendons with the cuticle. Many small cones, or rope-like tendons, anchor each elaborate thoracic endoskeletal unit with the cuticle, e.g. Fig. 9.4(a), (b), *e*. Such ribbon tendons have not been found elsewhere and are associated with the jumping mechanism. The superficial oblique muscles are considered below, §(ii).

(ii) *Exoskeletal and endoskeletal features associated with the collembolan jumping mechanism.* However good the muscular cylinder and deep categories of muscles may be, it is essential that all billowing-out of cuticle should be avoided for maximal jumping efficiency. Trunk rigidity is promoted by the very short intersegmental arthrodial membranes which permit little movement; by the presence of dorso-lateral sclerites forming tergo-pleural arches (Figs. 9.1(a), 9.4) and not simple flattish dorsal plates as in Chilopoda, Symphyla, Pauropoda, Diplura, and Protura (Figs. 5.3(c), 5.9, 5.10(a), 5.11(a), 5.12); by the fusion of these units on the prothorax

FIG. 9.4 (see opposite). *Tomocerus longicornis* Müller. Thick transverse sections at the levels D, E, H shown on Figs. 9.2 and 9.3. Parts of the segmental endoskeletal tendon systems are shown in black and marked **g**, the tonofibrilar ties of this skeleton to the cuticle are marked **e, f, q**, and other ties are unlabelled. The dorsal longitudinal muscles are marked by letters, the lateral longitudinal muscles by numbers, and the sternal longitudinal muscles by letters. Only some of the limb extrinsic muscles are shown in each section. The endoskeletal coxal plate within the coxa is cut transversely to the plate in each section. (a) Section passing through the middle of coxa 2 showing the separate coxal segments **a** and **b** and the tie **r** between the mesothoracic tendinous endoskeleton and coxal segment **a**. (b) Section passing through the middle of the third coxa showing united coxal cuticle **ab** on the median side and divided coxal cuticle into **a** and **b** on the outer side. Connective-tissue tie **r** corresponds with that in coxa 2 and an additional tie **p** links the pleurite laterally with the metathoracic endoskeleton. (c) Section through the thorax 1–2 intersegment between the legs showing the anterior face of the coxa of leg 2 with separate coxal segments **a** and **b** and the extrinsic coxal promotor muscles. Muscle abbreviation are: *abd.* abductor; *add.* adductor; *co.* from coxa; *dep.* depressor; *lev.* levator; *pr.* protractor; *ret.* retractor; *sus.* suspensory muscle. For further description see Manton (1972).

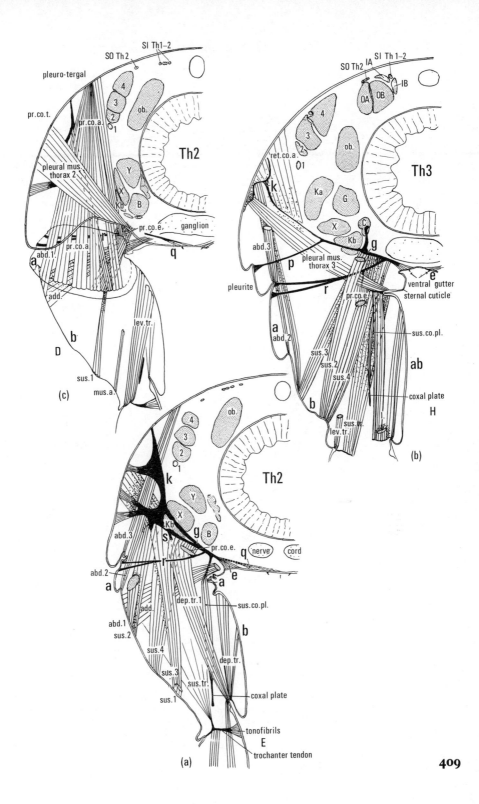

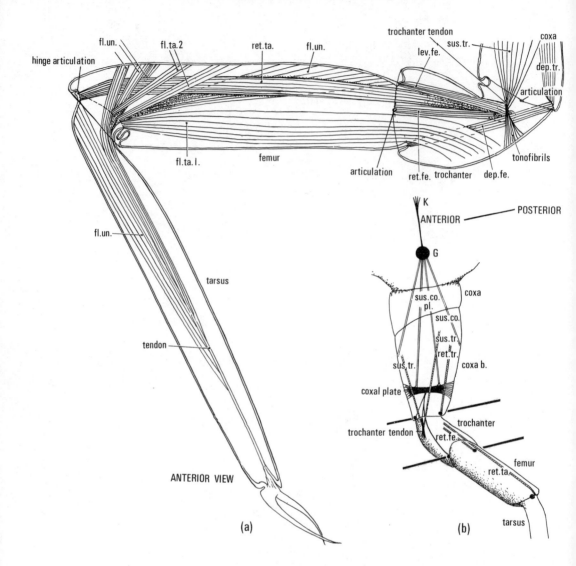

FIG. 9.5. *Tomocerus longicornis* Müller. (a) Diagram to show the suspensory system of left leg 2; G represents the mesothoracic tendinous endoskeleton linked to the tergite by the tie K and bearing the coxa and coxal plate by suspensory muscles. There is no coxa–body articulation. The positions of interpodomere articulations are shown by black spots, with a horizontal axis of articulation marked by heavy lines at the coxa–trochanter and trochanter–femur joints. The femur–tarsus hinge joint possesses a dorsal articulation as shown. The muscles *ret.fe.* and *ret.ta.* arise too close to an articulation to cause any flexure, they stabilize the joint against dislocation caused by momentary increase in hydrostatic pressure. (b) Anterior view of leg 2 with the coxa–trochanter joint strongly flexed. (From Manton 1972.)

Habits and the evolution of the hexapod classes

and mesothorax; and by the small expanses of flexible, but well sclerotized cuticle elsewhere which, however, are elaborately controlled. The flexible membrane above leg 1 and between leg 1 and the head (stippled in Fig. 9.1(a)) is required for the great mobility of leg 1. Outbillowing under pressure is prevented by three pairs of rope-like ties of connective tissue anchored to this cuticle, one above another, and inserted on to the elaborate prothoracic connective tissue endoskeleton. The two bulges left between these three ties on either side are the so called sub–coxal segments of the literature! (For details see Manton 1972.)

The next most important provision against unwanted outbillowing of the body wall is a complete diversion of function of the superficial oblique muscular system. These muscles promote horizontal trunk flexibility in epimorphic Chilopoda, in Symphyla, and in Hexapoda specializing on mobile thoracic segments, e.g. Japygidae. The collembolan superficial obliques, one or two pairs per segment, are shown in Fig. 9.2(a). They are enormous in the abdomen, arising ventro-laterally from their own ribbon tendons. Nearby, in abdominal segment 3 and from ventral plate 4, huge ribbon tendons, without muscles, maintain the integrity of the ventral abdominal groove which houses the resting springing organ. The pleural muscle of the metathoracic segment is shown in Fig. 9.4(b) marked 'pleural muscle thorax 3'.

The ribbon tendons of Collembola (Figs. 9.2(a), 9.3(a)) represent a remarkable component of their skeletal system (for details see Manton 1972). The thoracic segmental endoskeletal units, marked in black g in Figs. 9.3, 9.4, are much more complex than drawn and are also formed by connective tissue; note the connective tissue rope-like struts marked p and r uniting each endoskeletal unit with the pleurite over leg 3 and with the proximal coxal segment a by a double insertion on both mesothorax and metathorax. The process k to the tergo-pleural arch is very large in the mesothorax and anchored there by an abundance of tendinous fibrils.

(iii) *Features of collembolan legs associated with running and abdominal jumping.* Of equal importance in control of outbillowing are the special and unique features of the legs. The coxae are long and ventrally directed, the diagram in Fig. 9.5(b) summarizing the principal innovations. The endoskeleton of one thoracic segment, represented by *G*, is linked to the tergo-pleural arch by the process *k*, as shown in Fig. 9.4. The whole coxa is suspended from *G* by a pair of anterior and posterior suspensory muscles, the posterior being marked *sus. co.* Across the distal part of the coxa, spanning the lumen, lies a flat coxal plate formed by compact connective tissue and anchored to the anterior and posterior coxal cuticle by fans of fibrils. This coxal plate is cut transversely in Fig. 9.4(a), (b), as indicated. From this plate pass a pair of suspensory muscles *sus. co. pl.* inserting on the thoracic endoskeleton. There are coxal promotor, remotor, adductor, and abductor muscles as well, but the coxal suspensory system just outlined holds the coxa to the body and allows freedom of movement caused by coxal extrinsic muscles, in a manner impossible to any normal coxa–body articulation if it were under internal pressure. There are other extrinsic coxal muscles; those marked *sus. 1–sus. 4* in Fig. 9.4(a), (b), (c) assist in maintaining strong coxal suspension from the endoskeleton and some of these sectors arise from the coxa–trochanter arthrodial membrane (stippled in Fig. 9.1(a)) and so prevent outbillowing of this large expanse of flexible membrane, which is needed by the wide movements between coxa and trochanter.

The two pivot articulations at the coxa–trochanter and trochanter–femur joints and the hinge joint between femur and tarsus are also stabilized and prevented from dislocation under internal pressure. Tendons extending from the coxal plate hold the trochanter and prefemur at their anterior pivot articulations, black lines in Fig. 9.5(b), while muscles *ret. tr., ret. fe.,* and *ret. ta.* cannot cause leg flexures from points of origin so close to the articulation but can prevent longitudinal dislocation of each joint. Further, there is an elaborate dorso-ventral tendon situated proximally within the trochanter and strongly anchored ventrally by a cone of tonofibrils and more simply linked with the dorsal proximal margin of the trochanter (Figs. 9.4(a), 9.5(a), (b)). From this tendon arise enormous suspensory muscles *sus. tr.* inserting on the coxa (Figs. 9.4(a), (b), 9.5(a), (b)) and the tendon on its distal face carries the *lev. fe.* muscles (Fig. 9.5(a)) from the femur. A somewhat similar trochanter tendon is present in the jumping Machilidae where the coxae are also pendant from the body and great differences in flexure are employed between coxa and trochanter.

Finally the subdivision of the mesothoracic coxa into two segments and the partial subdivision of the metathoracic coxa can now be shown to be correlated with functional needs and the many previous interpretations of these features can be set aside. The undivided coxa of leg 1, marked *ab* in Figs. 9.1, 9.2(a), the two segments *a* and *b* on the mesothoracic coxa, and the union between these two segments on the median face of the metathoracic coxa are clearly shown on these figures and in figure 9.4. The cuticles of *ab* and of *a* and *b* are identical in structure and colour and quite distinct from that of the small amount of exposed pleuron above. The musculature of legs 2 and 3 is essentially similar and there is no reason to suppose that coxal regions *a* and *b* are not homologous one with another on legs 2 and 3. In both the proximal coxal margin is supported by the endoskeletal strut *r*. The much larger leg 3 (Pl. 6(a)), which has to support the body before and after the abdominal jump, and which carries out the whole of its propulsive thrust posterior to the leg base (Fig. 7.3(m)), is also supported by the partially encircling pleurite lying just in front of the leg base. This pleurite is stabilized by strut *p* from the endoskeleton (Fig. 9.4(b)).

All these features appear to have a two-fold explanation. The prothoracic coxa is undivided since there is ample pleural flexible membrane lying above it permitting the wide movements executed by this shortest of the legs, the pleural membrane itself being stabilized by the three struts of connective tissue mentioned above. Complete or partial subdivision of the long pendant coxae on legs 2 and 3 permit easier adductor–abductor movements in that the displacements can be shared between the coxa–body and intracoxal joints. This movement is needed most by legs 2 where the intracoxal joint is complete. Legs 3 need greater strength, provided by the solid median coxal face and the supporting pleurite.

Secondly, leg length and disposition of extrinsic leg muscles is greatly affected by the form of the coxae. Most of the promotor and remotor muscles arise from the proximal margin on leg 1 and from the proximal margin of the distal segments *b* on legs 2 and 3 (Fig. 9.4(c), *pr. co. a.*, *pr. co. p.*, and part of the retractor muscles *ret. co. a.* as shown, see also Manton 1972, Figs. 18, 22, and 25). If these muscles arose from the proximal margin of segment *a* on legs 2 and 3, the muscles would be much shorter, so that the same degree of muscular contraction would cause a smaller displacement of the coxae. But on leg 1 length of these muscles is obtained by a fanning out to the head and second tergo-pleural arch; the muscles do not converge, as over legs 2 and 3, so no subdivision of coxae 1 is needed. There is clearly a thoracic space problem in housing the many thoracic muscles (see Manton 1972, Fig. 18).

B. *Conclusions concerning the Collembola*

An understanding of how the collembolan body works substantiates a simple interpretation of their coxae and of the absence of free pleurites other than one supporting the third legs, and refutes the mass of speculations concerning transformed hypothetical pleurites and sub-coxal segments. The detailed correlations of body structure with the jumping mechanisms (for full details see Manton 1972) indicates an exploitation of the hydrostatic forces which must have been in normal use by lobopodial soft-bodied ancestors. There is no indication that any other class of hexapods ever passed through a jumping ancestral stage with the facilitating morphology outlined above. In other words, the idea that Collembola represent simple primitive insects is entirely outdated. The small number of abdominal segments, six, suits the jumping mechanism; Collembola could not use a springing organ effectively if it was situated at the end of a long abdomen. Evolutionary shortening of the abdomen probably occurred as the jumping organ became effective. There is no reason to suppose that the ancestors of other hexapods shortened their bodies and then elongated them again to about 14 segments or so. The presence of more leg podomeres in a collembolan would mean more joints to be stabilized. There is no sound reason to suppose that Collembola ever possessed more leg podomeres than they do at present, although sometimes there is complete or partial fusion of the trochanter and femur, providing extra stability. Other hexapods possess one more podomere. The whole trunk and leg morphology of Collembola could have been derived directly from lobopodial soft-bodied ancestors in which hexapody was developed anteriorly while the posterior end of the body was pushed along by its sub-terminal limbs, moving in unison, which later became the springing organ and hamula.

The secondary loss of jumping in habitats (2) and (3) above is accomplished by progressive regression of the springing

Habits and the evolution of the hexapod classes

organ, diversion of its extrinsic muscles to other purposes, and enlargement of trunk intersegmental arthrodial membranes, so providing a flexible body capable of wriggling through small spaces (Pl. 6(c), cf. (a)). The median ventral tube present on the first abdominal segment, which can be inflated by hydrostatic pressure, is provided with retractor muscles only and is another unique collembolan organ used for absorbing water. Moisture is taken up by extrusible coxal organs in many myriapods and hexapods, but the collembolan ventral tube is unlike all of these.

3. The Diplura

This class of wingless hexapods (Pl. 6(e), (h)) is another group which has been regarded as 'primitive insects', but there are other interpretations when structure and function are considered. The entognathy of dipluran heads is quite unlike those of Collembola and Protura and has probably been independently acquired (Chapters 3 and 6). The dipluran leg-rocking mechanism differs from those of all other hexapods (Chapter 5, §5B(i)).

The Campodeidae are widely distributed but are most abundant deep in highly porous soil in which they negotiate crevices by small size and flexibility of body, but not by pushing. The Japygidae live in deep humus, in ground litter, and under stones. Some japygid species attain very much larger size than do the campodeids. The japygids do not press against the soil by the body surfaces, but burrow by using existing crevices, great flexibility of trunk and very strong leg movements shifting soil particles. The Campodeidae are fairly fast runners on the soil surface, a 4·5-mm *Campodea* achieving up to 54 mm/s, owing to their long legs and long strides, facilitated by the great differences in leg length between the three pairs (Pl. 6(e)). Thus, overstepping of one propulsive leg by the next produces a track of forwardly directed footprints, as already noted (Chapter 7, §6C(viii), Fig. 7.12(b)), not found in other hexapods. The short legs of the japygids, relative to segment length, are suited to their method of burrowing and cause backwardly staggered groups of footprints, Fig. 7.12(c), as in no hexapods other than *Thysanura*. The japygids apparently dislike surface locomotion and alone among hexapods investigated (Manton 1972) would not maintain steady surface locomotion for speeds to be recorded with certainty; their gaits are much the same as in *Campodea* (Fig. 7.9(e)) although strides and speeds are very much smaller. Japygids excel at moving in tortuous spaces by virtue of thoracic flexibility and by 'scrabbling' movements of their strong, short legs. The wide distribution of the Campodeidae is probably due to their considerable speed for their size and consequent ability to find suitable moisture conditions with rapidity. The Japygidae are more limited in their distribution in spite of remarkable structural facilitations towards crevice-living. Both groups are carnivorous.

A. Structural facilitations to crevice-living in Japygidae

The outstanding proficiencies of the Japygidae are determined by flexibility of body, by their unique telescopic antennae, and by their strength of leg movement.

The thoracic segments are narrower than those of the abdomen, permitting the short legs to be held close into the body without much lateral projection except when walking. There is great flexibility in the thorax (Pl. 6(h)) permitting flexures which may amount to the head being able to turn postero-laterally, then over the dorsal surface to reach the other side, without movement of leg 3 standing on the ground. The intersegmental regions are narrow and flexible but there is intrasegmental thoracic flexibility as well, which takes place between intercalary and principal tergites and between the lateral sclerite expansions marked intercalary sternite, presternite, and sternite in Fig. 5.12(a). Some of these sclerites can bend over one another giving somewhat the same effect as in the Geophilomorpha, but by different means (see Manton 1972). The coxal promotor–remotor axis of swing is marked by the heavy line and the position of the pleurite, and lateral expansions of the sternite, stabilize the leg base, in spite of so much intrasegmental flexibility. The sclerites of *Campodea* are much the same, but less extensive in area. The intrinsic dipluran leg-rocking mechanism (Chapter 5, §5B(i)) may be correlated with this intrathoracic flexibility of body leading to required coxal stability. A simple labelling of pleurites, sternites, and intercalary sclerites is given here because the complexities of much entomological labelling appear to be needless; there are no sound reasons for considering pleurites to be homologous from one class to another.

The antennae in Pl. 6(h) are of different lengths, an appearance not due to foreshortening. Each antennal segment can telescope into the next, the shortening so produced being

Habits and the evolution of the hexapod classes

either local or all along the antenna. The antennae are highly sensory and used to feel soil spaces before a route is selected. There is ample intersegmental musculature within the antenna capable of causing shortening. Imms (1939) first saw the central blood vessel within the antenna which dilates into a large sinus within each antennal segment, but he did not understand the significance of these sinuses or know about antennal shortening in japygids. It is probable that each sinus is capable of dilatation or contraction by the tension of musculature within the walls of the vessel, acting against blood pressure, and that dilatation can cause antennal extension, either local or entire. Jointed musculated antennae are present in myriapods and in Collembola (Pls. 4, 5, 6(a)–(e), 7(d), (f), (g)), the movements of the former being less spectacular than those of Collembola (Pl. 6(a)). The antennae of butterflies and the beetle in Pl. 6(i) are unjointed, except at the base, from which they are moved, but no other example of telescopic antennae has been found in any other hexapod or myriapod class. The telescopic antennae of japygids must be regarded as a very special advance correlated with their burrowing habits.

The strength of the leg movements and the flexibility of the thorax is mediated by the muscles and endoskeleton shown in Fig. 9.6. As might be expected, the dorsal longitudinal system is composed of short muscles crossing one intersegmental or other joint. No long rigidity promoting muscles, such as *dlm. A, dlm. B,* and *dlm. C* of Scolopendromorpha (Fig. 8.13) exist. Superficial pleural muscles, the flexibility promoting category, are represented by *superficial dorso-ventral muscles a* and *b* and others in Fig. 9.6.

It is the ventral endoskeleton and musculature which is the most remarkable. From the posterior median margin of each thoracic sternite arises a large backwardly projecting 'spine apodeme' or 'spina'. A median cuticular endoskeletal plate in *Campodea* (Fig. 9.7(b)) corresponds with the spine apodeme of the japygids. The spine apodeme is supported by two infoldings across the sternite forming, with the spine apodeme, a Y-shaped structure; the anterior ends of the arms of the Y support the coxa–sternite articulation, a very strong provision, supporting the very strong leg movements, in a unique manner, cf. centipede costa coxalis Figs. 5.8, 5.9(a).

The ventral longitudinal musculature is complex, forming many sectors, *vlm. 1–vlm. 6,* extending either between successive sternites or between successive paired lateral endoskeletal tendons marked on Fig. 9.6. There are no long sectors, such as shown by the rigidity promoting sternal longitudinals of Collembola (Fig. 9.3(b)). The bulkiest of the japygid sectors is *vlm. 2* where length and muscle thickness is provided by insertion on the spine apodeme, see mesothorax in Fig. 9.6. This arrangement provides very strong flexure at the ventral thoracic intersegmental joints. The spine apodeme also carries the insertion of the enormous coxal retractor muscle *ret. co. 6,* responsible for the strong retractor leg movements. The stabilizing, rigidity-promoting, deep oblique muscles *dvm. d.* are small, compared with those of Collembola where rigidity is of paramount importance (Fig. 9.3(a)), but the antagonistic deep dorso-ventral muscles *dv., dv. 1–dv. 3* are elaborate, see metathorax in Fig. 9.6, where their contraction must cause elongation of relaxed *dlm. 2* and *ret. co. 6* muscles as well as body flattening and lateral flexure, unusual additional features for this category of muscles.

The functions of the spine apodeme have not previously been ascertained and much ingenuity is found in the entomological literature purporting to prove that this structure is a primitive insectan feature. But functional studies of all hexapod groups fail to reveal the existence of a comparable apodeme in any of them. Even in the undoubtedly related Campodeidae, a quite different type of endoskeleton is present near the site corresponding with the origin of the spine apodeme, supporting both limb extrinsic and ventral longitudinal muscles. It may be presumed that the elaborate campodeid endoskeleton and appended musculature suits their rapid leg movements. The sternite is very narrow; cf. Fig. 5.13(a) with 5.12(a) where the wide sternite is seen in lateral view, though in ventral view it appears much as in a centipede in Fig. 4.29. There is no more reason to regard the spine apodeme of the Japygidae as a primitive feature than the corresponding endoskeletal plates and median connections of the sternite in Campodeidae. Both suit different needs. The spine apodeme and its muscles are superlative features associated with japygid achievements and have nothing to do with a primitive status of any kind. The so called spine apodeme in certain hexapod groups is no more than the site of insertion of muscles on the posterior margin of a sternite or on the flexible membrane associated with it.

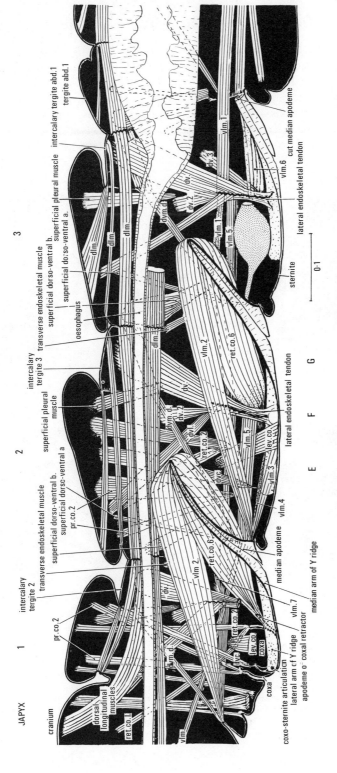

FIG. 9.6. Japygid sp. View from the sagittal plane of the principal muscles of the three thoracic segments numbered above. Fat body and vascular system are omitted and endoskeletal tendons are shown in black. The ventral Y-shaped apodemal ridge consists of a median infolding of cuticle from which arises the median spine-like apodeme, the paired arms appear in internal view, stippled white, and anteriorly support the coxa–body articulation. On the metathorax the spine apodeme is cut away together with other muscles in order to show the ventral longitudinal muscular system more fully. (For further description see Manton 1972.)

419

Habits and the evolution of the hexapod classes

B. Structural facilitations to crevice-living in Campodeidae

The Campodeidae are more lightly built than the Japygidae, their musculature is simpler in the trunk; compare Fig. 9.6 with 9.7(b). Their rapid running is served by the many slender and long leg extrinsic muscles, Figs. 9.7(a), 5.13(a), such as seen also in centipedes. Figure 4.29 shows the many leg extrinsic muscles inserting on the sternite; many others insert elsewhere. The Campodeidae do not possess huge wide

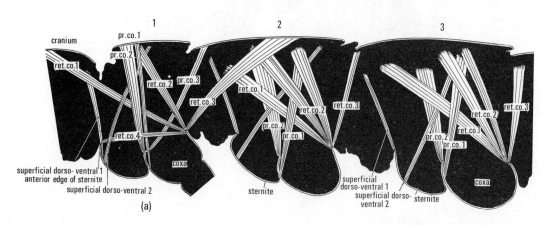

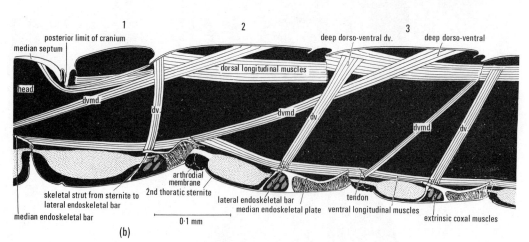

FIG. 9.7. *Campodea* sp. (Pl. 6(e)). Longitudinal views of the thoracic musculature. (a) Internal view of the right half of the thorax, the muscles and body wall shown in (b) being cut away to expose the extrinsic limb muscles and the superficial dorso-ventral muscles. The origins of the coxal protractor muscles *pr.co.* and retractor muscles *ret.co.* are shown, but are not exactly the same on successive legs. (b) Internal aspect of the right half of the animal showing the dorsal and ventral longitudinal muscles, the deep dorso-ventral *dv.* and deep oblique *dvmd.* muscles and the endoskeleton. (From Manton 1972.)

Habits and the evolution of the hexapod classes

muscles such as the japygid *ret. co.* 6, and the very wide campodeid coxae are compatible with a very narrow sternite (Fig. 5.13(a)), a weaker coxa–body insertion than present in japygids, but suited to fast, weaker leg movements. The combination of living predominantly deep in porous soils, at any rate during the day-time, and the ability to run fast by relatively long legs, is perhaps unexpected, but these features are incompatible with large size, since leg projection would be too great for soil living. The basic structure of Campodeidae resembles that of the Japygidae, although the median endoskeleton in the posterior part of the thorax is quite different from the spine apodeme complex of the Japygidae (Figs. 5.13(a), 9.6, 9.7(b)); for further details see Manton 1972.

C. Conclusions concerning Diplura

A claim to a primitive insectan status for the Diplura fails as it did for Collembola, but for different reasons. The entognathy and mandibular mechanism of the head differs from other entognathous hexapods and is far removed from what must be regarded as the basic hexapod type of mandible and its mechanism, but could have been derived therefrom by functional intermediate stages. The leg mechanism, including leg-rocking, differs from other hexapods and does not lead towards the leg base or leg-rocking of other groups. The trunk sclerites on the thorax are concerned with flexibility of body, as they are in the Protura, but the details differ. Neither the form of the pleuron, nor the presence of intercalary sclerites, nor the presence of a spine apodeme in the Japygidae alone can be upheld as primitive dipluran features. They are all associated with present day functional needs and advancements.

The skeleto-musculature of the Campodeidae and Japygidae are correlated in detail with their crevice-living habits and associated accomplishments. It is probable that trunk flexibility was an early feature associated with crevice-living and that the intrinsic leg-rocking mechanism evolved in dependence upon it, leaving the leg-base as stable as possible for strong or fast promotor–remotor movements, in spite of intrasegmental thoracic flexibility. Probably the japygid legs became a little shorter than in the dipluran ancestor as their burrowing abilities increased, because such short legs would hardly account for the differentiation of hexapody and the thorax. The legs of the Campodeidae certainly became longer, their speciality centring about lightness of build and the ability to move with rapidity to suitable micro-habitats providing the right moisture and prey organisms. The japygids abandoned running proficiencies in favour of very strong leg movements used in burrowing, along with telescopic antennae, a unique feature among hexapods and myriapods. But these proficiencies, without the ability or inclination to move well over the surface of the ground, have inevitably restricted the geographical distribution of the japygids which so contrasts with the ubiquitous presence of campodeids in temperate climates.

Thus within the Diplura we do not find primitive hexapods, but we now can start to appreciate the significance of their internal and external body form, correlated as is usual with habits. It has been shown in Chapters 3 and 8 why the Symphyla can have had no connection with uniramian hexapods and cannot be ancestral insects. The Diplura cannot be related to Symphyla and are not advanced Symphyla. Both taxa have in common much trunk flexibility (Fig. 8.7, Pl. 4(d), Pl. 6(h), showing conspicuous intrathoracic flexure, cf. rigid thorax of a beetle, Pl. 6(i)) associated with crevice-living and deep soil penetration in the two classes.

4. The Protura

Little new work has been done on the functional morphology of this class of minute litter and soil living hexapods (Pl. 6(k)). They have no antennae. The first pair of legs is forwardly directed at the sides of the head and may be mainly sensory, or of use in pulling the animal forward by remotor flexures in the vertical plane. Legs 2 and 3 are locomotory, by a slow quadrupedal gait, pattern (2·0:8·0) phase difference 0·25. The order of footfalls is R.2, L.3, L.2, R.3, R.2, resulting in three propulsive legs at most moments but four such legs during four brief periods during each pace. There are other examples of quadrupedal hexapods, such as seen in some butterflies walking about on flower heads, but quadrupedal walking is clearly secondary in hexapods. The entognathy of Protura differs in detail from those of other entognathous hexapods.

Protura possess intersegmental and intrasegmental flexibility of the thorax, marked by arrows in Fig. 5.12(b), (c) where lines of lesser sclerotization traverse both tergites and sternites. The coxa–sternite articulation, where the sternite is narrowest (Fig. 5.12(b)) brings the coxa well into the flanks of the body, reducing leg projection, as suits living in confined spaces. A median longitudinal apodeme strengthens the divided sternal plate and the coxal union with the body is reinforced by a pleural articulation at the dorsal coxal margin (Fig. 5.12(c), black spots). Pleurite 1 can slide against a facet on pleurite 2, so providing another unique leg-rocking mechanism.

The Protura afford no evidence of a primitive hexapodan state, but instead, of yet another line of hexapods with their own particular solutions to the problems of life common to some other classes as well. The presence of dorsally divided tergites, ventrally divided sternites, ventral coxa–sternite articulations, lateral leg projection, and flexible pleuron supported by a variety of pleurites are all features associated with trunk flexibility and occur in scolopendromorph and lithobiomorph centipedes in particular as well as in the hexapodous Diplura.

5. The Thysanura

In both the Machilidae and Lepismatidae (Pl. 6(f), (g), (m)) the leg movements are basically jumping ones, in contrast to the walking or running leg movements of other hexapods and myriapods. Particular thysanuran structural features are associated with these jumping gaits. In addition, the Machilidae employ high jumping escape reactions implemented by the abdomen, as in no other hexapods, while the extreme speed and turning ability which can be achieved for short periods by the Lepismatidae serves as their escape mechanism.

In the jumping gaits of *Petrobius*, serviceable on upper surfaces, the legs of each pair jump in unison and successively. In *Lepisma* legs of a pair jump in opposite phase and make possible their fast progression on walls as well as on more or less horizontal surfaces (see Chapter 7, §7A, Figs. 7.12(d), (e), 7.13). On the underside of rocks *Petrobius* uses normal hexapod gaits, with paired legs in opposite phase, and slow patterns giving many adhering legs at one moment, the jumping gaits with legs of a pair in unison here being unserviceable.

A. *Structural features associated with jumping gaits of Thysanura*

The coxae in the Machilidae are pendant from the body and in the Lepismatidae they are ventro-laterally inserted (Figs. 9.9–11). The paratergal expansion from the lateral margins of the tergites enclose each thoracic segment as far or beyond the leg base, leaving little exposed pleuron (Figs. 9.8, 9.9).

Habits and the evolution of the hexapod classes

A coxa–body articulation lies antero-dorsally on the coxal margin which articulates with a mobile pleurite; postero-ventrally the coxal margin is surrounded by flexible pleuron and sternal body wall as in no other hexapod order so far described.

When structure and function are considered together, the leg base, pleurites, and leg structure of neither the Machilidae nor the Lepismatidae measure up towards a primitive condition of anything. They lead only towards thysanuran perfections of habits, which differ from those of other classes of hexapods.

(i) *Coxal movements and associated structure in Thysanura.* The thorax of *Petrobius* is shown diagrammatically in Fig. 9.8(b). Under the tergal lobes lie two closely placed pleurites over leg 1 and a single horizontal pleurite over legs 2 and 3. The position of the coxa–pleurite articulation is indicated by black spots, a section through the mesothoracic coxal articulation being shown in Fig. 9.9(b). Except at this point the pleurite and coxa are not closely opposed, there being ample arthrodial membrane or flexible pleuron between them (Fig. 9.9(d)). The mesothoracic and metathoracic pleurites approach the coxae more closely posteriorly, but there is no articulation here. Small adductor–abductor movements take place between the coxa and pleurite. Anteriorly the pleurites are strongly slung from the tergal cuticle by a large bunch of tonofibrils passing above the opening to the paratergal lobe (Fig. 9.9(a)). Posterior to the coxa–pleurite articulation a large pleural apodeme passes into the coxa (Fig. 9.9(c), (d)) penetrating into the body itself and carrying massive long levator and depressor muscles from the trochanter (*dep. tr., dep. tr. l.* are labelled). Shorter muscles from the trochanter insert on the proximal faces of the coxa but these muscles are still long owing to the length of the coxa (Fig. 9.8(a), *lev. tr., dep. tr.*). The base of leg 1 is a little different (see Manton 1972).

FIG. 9.8 (see opposite). *Petrobius brevistylis* Carpenter (Pl. 6(m)), leg structure. (a) Posterior face of left leg to show the muscles and joints. The pivot joints (dicondylic, one pair of articulations on either side) between coxa and trochanter and between femur and tibia are marked by white and by black spots respectively; neither joint is equatorial, they are situated towards the ventral and the dorsal side of the leg respectively giving great range of movement on the opposite side. (b) Lateral view of the thorax to show the position of the pleurites, the large overhang by the paratergal lobes, and the points of closest union of coxa and pleurite (marked by black spots). Two pleurites x and y lie above leg 1. (From Manton 1972.)

Habits and the evolution of the hexapod classes

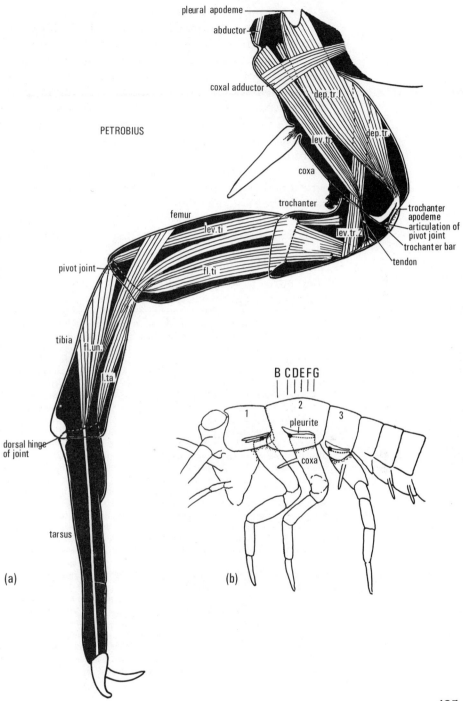

No promotor–remotor movement takes place at the coxa–pleurite junction of legs 2 and 3. This movement occurs between pleurite and body, each pleurite swinging out about its anterior inextensible tie with the tergite (Fig. 9.9(a)). The free posterior end deforms the flexible pleuron. No such promotor–remotor movement has been found in any other group. Coxal and pleural protractor muscles of *Petrobius* are shown in Fig. 9.9(b), (c), the coxal and pleural retractor muscles lying posterior to the plane of the sections (see Manton 1972).

In *Lepisma* three pleurites lie in front of the leg base, positioned as shown in Fig. 9.10. They are the trochantin, katopleure, and anopleure of the entomologists but these names indicate no homology with chilopodan pleurites or with those of other classes. The three do not form a pavement of sclerites, but the more anterior overlap the next behind. The katopleure ends ventrally as shown in Fig. 9.10, but the anopleure curves posteriorly below the trochantin and overlaps it as seen in the section passing through the anterior part

FIG. 9.9 (see opposite). Thick transverse sections through the leg bases of Thysanura. (a)–(d) *Petrobius brevistylis* Carpenter. (e)–(g) *Lepisma saccharina* L. (Pl. 6(g), (m)), the capital letters indicate the levels of the sections, marked on Fig. 9.8(b). (a) Section passing immediately anterior to the sclerotized edge of the single mesothoracic pleurite, a thick cone of tonofibrils pulls on the pleurite and terminates above on the basement membrane below the tergal ectoderm. The paratergal lobe here arises as a double-layered outfolding of the body without an internal cavity. (b) Section showing the anterior coxal articulation with the pleurite; the stout pleural protractor muscles 91, 92 passing postero-ventrally, their coxal origins being seen in levels (c), (d). The coxal protractor, 101, arises from the anterior external margin of the coxa posterior to this level. (c) Section showing the origin of the pleural apodeme from the pleurite. The strengthened lateral coxal margin (not a trochantin) is now separated from the pleurite by an expanse of flexible pleural cuticle. On the pleural apodeme insert the long depressor muscles from the trochanter (*dep.tr.l.*). (d) Section passing through the middle of the coxa, the pleural apodeme is invested by the *dep.tr.l.* muscles and dorso-laterally the small muscle 116 is probably a stabilizer. (e) Section below passing through the coxal articulation with the pleurite (trochantin) and above lying a little more anterior. The oblique dorso-ventral line marks the axis of swing of the coxa on the body but there is no ventral articulation at the proximal coxal rim with any pleurite, the trochantin and anopleure are cut ventral to the coxa and a thin plate in the coxal wall lies below the thick rim carrying the dorsal articulation. Above, the section passes through the katopleural apodeme formed from the floor of an intucking below the paratergal lobe, and the apodeme carries many muscles. (f) Section a little more posterior than (e) passing through the coxa where the lateral apodeme tucks in and carries adductor muscles 106. Dorsal to the coxal rim lies flexible pleuron and no articulation of the coxa with any sclerite. (g) Transverse section through the anterior part of the coxa of *Lepisma* showing the pleurites above and below the anterior edge: katopleure, trochantin, and anopleure; some muscles are shown and the great width of the sternite below the coxa boxing it in. The femur of leg 1 is flexed dorsal to the coxa and lies below sternite 2. (For further description see Manton 1972.)

Habits and the evolution of the hexapod classes

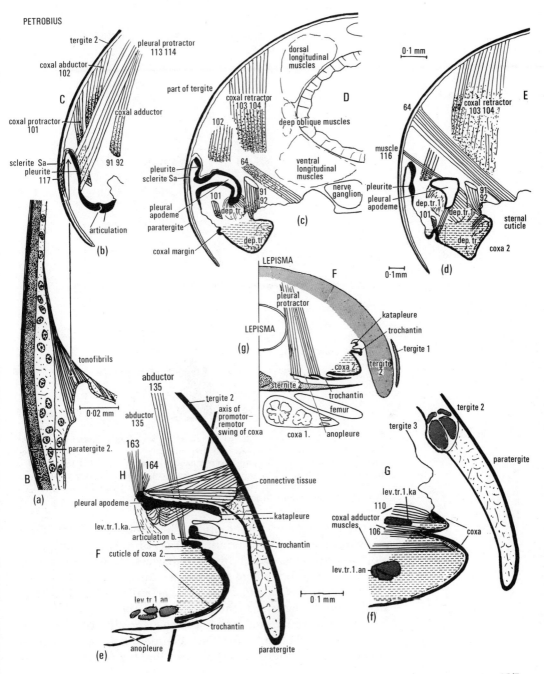

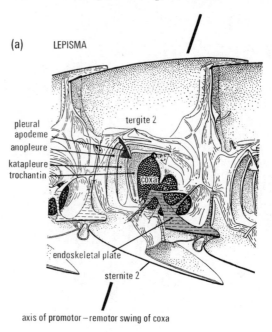

(a) LEPISMA

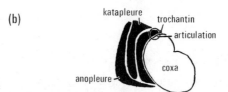

(b)

FIG. 9.10. *Lepisma saccharina* L. (a) The inner surface of the mesothoracic cuticle. The tendinous endoskeletal plate and its connectives are marked by interrupted lines. The close union forming an articulation between coxa and trochantin is marked by a circle. (b) The limits of the three pleurites are indicated by the black zones but in reality they overlap and do not form a pavement, see Fig. 9.9(g). (After Barlet 1951, from Manton 1972.)

of the coxal face, Fig. 9.9(g). The mobility of the three pleurites producing different degrees of overlap along their dorso-ventral edges and different degrees of movement away from the main body is of importance for the leg mechanism. As in *Petrobius*, a coxa–pleurite articulation lies antero-dorsally with pleurite 1, the trochantin (Fig. 9.9(e)). There is no other articulation, none antero-ventrally between the trochantin and coxa, as has been claimed, and none postero-dorsally where there is nothing but flexible pleuron close to

Habits and the evolution of the hexapod classes

the coxa (Fig. 9.9(g), (e)). A groove across the lateral face of the coxa carries an intucking and tendon bearing adductor muscle 106. Less thick, but sclerotized cuticle unites this intucking with the anterior part of the coxal rim, as shown in Fig. 9.9(e), (f). The proximal rim of the coxa is the upper black cuticular piece in both sections. The thin part of the coxa ventral to its rim here is probably concerned with the vibrations caused by the very rapid leg action.

The pleural membrane of *Lepisma* forms flexible bulges round the upper margins of the pleurites; see the two direction lines marking the trochantin in Fig. 9.9(e), the lower only indicating the actual sclerite. From the katopleural flexible bulge of pleuron a stout apodeme passes into the body, as shown in Fig. 9.9(e), the upper part of this thick section being more anterior in position than the ventral part. The actual sclerotization of the apodeme represents only the floor of the intucking. This apodeme carries massive levator muscles from the trochanter, as in *Petrobius*, but also muscles from the anopleure. The internal end of the apodeme is much more strongly tied by muscles to the tergite than it is in *Petrobius*; cf. Fig. 9.9(e), muscles 163, 164 and those passing laterally to the base of the paratergal lobe, with Fig. 9.9(d) muscle 116. These muscles in *Lepisma* are correlated with the very rapid and wide telopod movements.

The movements of the coxa of *Lepisma* are unique, as is the disposition of the leg on the body. Each coxa, arising from the interrupted line marked on the prothoracic coxa in. Fig. 9.11(a), (b), is inclined posteriorly close to the thorax. The sternites are very large and project posteriorly as free median pointed plates overlapping the anterior end of each following sternite, as shown by the median view of a lateral half of the cuticle of the body in Fig. 9.10(a) and by Fig. 9.11. Each paratergal lobe envelops the coxa which thus comes to lie in an oblique antero-posterior groove situated between the paratergal lobe and the sternite (Figs. 9.9(g), 9.11). The basic promotor–remotor coxal movement thus comes to be a slight depression of the coxa away from the body and the raising of it towards the body. The adductor–abductor movement of the coxa is shown by the double arrowed line on the animal's left mesothoracic coxa in Fig. 9.11(b). In practice these two movements, entirely different in other arthropods, are here merged into a coxal promotor–adductor and an antagonistic remotor–abductor movement, both small

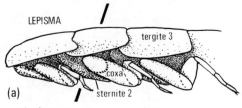

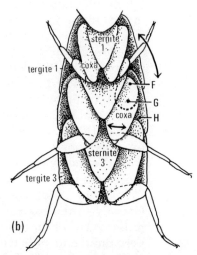

Fig. 9.11. *Lepisma saccharina* L. (a) Lateral and (b) ventral views to show the positions of the legs and their movements. The three sternites slope postero-ventrally and carry postero-median and antero-lateral flanges, as shown. The lateral flanges lie ventral to and against the coxae. The coxal origin from the body is roughly indicated by the heavy dotted line at the anterior end of left coxa 2 in both figures. The coxae arise ventro-laterally and lie in grooves between the paratergal lobes and lateral flanges of the sternites. The coxae are flattened and slope postero-ventrally. Owing to this position the small promotor–remotor movements of the coxa about its axis marked in (a) combines with the adductor–abductor movement of the trochanter on the coxa to give the stepping movement, a condition found in no other hexapod. The morphology of the coxa–trochanter joint is such that the trochanter and femur swing forward dorsal to the coxa and most extensively on leg 1; see also Fig. 9.9(g). (From Manton 1972.)

but strong and rapid and facilitated and supported by the vertical array of strongly musculated imbricating pleurites (Fig. 9.9(g)). The very large and strong coxae house the large, fast-moving muscles that cause the main stepping movement which takes place here from the coxa–trochanter pivot joint and not mainly at the leg base as in other Uniramia; no such remarkable leg action has been found anywhere among arthropodan groups.

Habits and the evolution of the hexapod classes

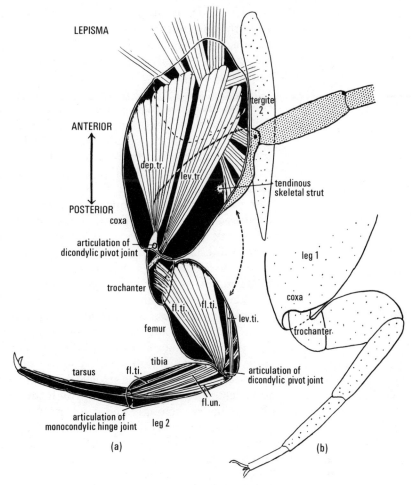

FIG. 9.12. *Lepisma saccharina* L. Leg morphology and movement. (a) Left leg 2 in ventral view, the overlapping sternal flange has been removed but the paratergal lobe is in place. The leg lies in its natural position, the coxa sloping posteriorly and slightly downwards from the body (see Fig. 9.11). The morphologically anterior face of the coxa becomes the apparent ventral face. The union of coxa and body is shown by the dotted line. The intrinsic leg muscles are drawn but only an indication is given of the extrinsic muscles. The trochanter articulates into a postero-dorsal concavity on the coxa, the parts being so shaped that the femur when elevated lies against the dorsal posterior face of the coxa, between the coxa and body. The heavy dotted outline and the stippled face of the telopod shows the elevated position. The levator–depressor movement of the trochanter on the coxa swings through a very large angle, over 90°, as shown by the contrasting positions on (a) and the arrowed, interrupted line. This movement of the trochanter becomes the largest component of the promotor–remotor stepping of the limb-tip. (b) Outline of part of the left prothoracic leg in the same view as (a). The coxa is similarly backwardly directed but the head of the trochanter is even more curved than on leg 2, allowing the more forward stepping movement of leg 1. (From Manton 1972.)

(ii) *Telopod movements and associated structure in Thysanura.* In both *Petrobius* and *Lepisma* the coxae are remarkably long; their cavities house bulky muscles from the trochanter (Figs. 9.8, 9.12). Certainly in *Lepisma* and probably in *Petrobius*, the jumping leg movement is mainly due to telopod action, whereby the leg strikes the ground with maximum velocity at the beginning of each jump. The coxa–trochanter joint is different from all others so far investigated. It permits movement through a wide angle; see Fig. 9.12 where the interrupted arrowed line between the extreme leg positions marks the movement. The coxa–trochanter joint is very narrow in *Lepisma* and fairly so in *Petrobius*, in a way resembling the very strong coxa–sternite joint of a iuliform diplopod where the arthrodial membrane is also minimal. There is no wide expanse of intersegmental arthrodial membrane in *Lepisma*, such as is present in chilopod legs simply for speed (Figs. 5.6–9). In both *Petrobius* and *Lepisma* (Figs. 9.8(a), 9.12) there is a very solid apodeme from the trochanter, passing into the coxal cavity and bearing massive depressor muscles. But the axis of movement through the coxa–trochanter anterior and posterior pivot articulations is far from the equator of the joint and near to the ventral face. This position promotes easy wide levator movements (the recovery non-propulsive stroke), while the massive depressor muscles pull on the apodeme which forms a lever swinging the trochanter strongly about its articulation.

In *Petrobius* the rod-like trochanter tendon anchored to the ventral face of this podomere has already been mentioned. It bears a large levator muscle and swings about its tonofibrillar base so that the levator muscles always pull at the same angle on the tonofibrils of origin; the tendon would swing, and never be at an angle to the *lev. tr.* muscles as is drawn in Fig. 9.8(a); the drawing serves to emphasize the point. The arcuate sclerite at the femur–patella joint of the spider leg serves a similar need but by entirely different means (see Chapter 10, §5D(ii), Fig. 10.5). A less elaborate but stout tendon is present in the trochanter of *Lepisma*, a smaller animal than *Petrobius*.

The proximal end of the trochanter in *Lepisma* is strongly curved (Fig. 9.12(a), (b)) and to different degrees in the three pairs of legs.

The result of these features in *Lepisma* is the most extraordinary leg movement here recorded. The stepping–

jumping movement takes place obliquely nearly in the horizontal plane (Figs. 9.11, 9.12(a), (b)). There is no conspicuous upstanding knee, although foreshortening in Fig. 9.11(a) exaggerates the knee. Also remarkable at the coxa–trochanter joint are the cuticular details which allow the femur to flex not only up to but dorsal to the coxa, as shown by dotted outlines at the overlaps of these podomeres in Fig. 9.11(a), (b); and see section in Fig. 9.9(g), coxa and femur cut below sternite 2. This astonishing provision contributes to the very large angle of swing of the telopod. The rapidity of the series of jumping actions by each leg results in the great speed achieved by *Lepisma*.

B. *The mechanism of the high escape jumps of* Petrobius

These escape jumps by *Petrobius* were described in Chapter 7, §7C(i), Fig. 7.15. They are isolated events and implemented by the abdomen. The tergo-pleural arches envelop the body and overlap the sides of the ventrally situated coxal plates which bear the reduced telopods (Fig. 9.13(b), (c), Pl. 6(m)). The tergo-pleural arches have very strong anterior margins, Fig. 9.14, and easily telescope over one another to give a dorsal concavity of the abdomen in the cocked position before a jump (Fig. 7.15). The musculature is complex. The dorsal longitudinals are twisted round one another like ropes, Fig. 9.14 muscles 1–5 and 8–11, as they are in jumping Malacostraca, Fig. 4.20. Segmental tendons lie between the segmental sectors of the ventral longitudinal muscles 15 and are anchored to the coxal plates by muscles 19. These tendons are linked to the intersegmental tendon system which is shown diagrammatically as black zigzag lines on the figures, but are much more complex and all linked with the tendons of the dorsal longitudinal muscles and so to the tergites (Figs. 9.13(d), 9.14). The ventro-lateral parts of these tendons carry a normal deep oblique muscle 22 and the extraordinarily elaborate additional sectors, muscles 23, 24, 25A, 25B, 26, which cross several intersegments and loop round one another in a manner reminiscent of the jumping Malacostraca (Fig. 4.20). The normal simple antagonistic deep dorso-ventral muscle is duplicated as muscles 20 and 21. The abdomen cannot flex horizontally and superficial pleural muscles, the horizontal flexure-promoting category, so well developed in Diplura, Symphyla, and many Chilopoda, are absent.

The mode of action of this complex abdominal musculature appears to be: first the creation of the dorsal trunk concavity by contraction of the rope-like dorsal muscles during which muscle tension builds up, followed by contraction of the elaborate and bulky deep oblique system which strongly and suddenly straightens the abdomen, beating the coxal plates against the substratum so causing the jump.

C. Conclusions concerning the Thysanura

The coxa–body junctions of *Petrobius* and of *Lepisma* and the structure and mode of action of their legs contrast with all other hexapod classes, although there are considerable differences as well as similarities between *Petrobius* and *Lepisma*. The basic promotor–remotor swing of the legs, normally so uniform, is entirely differently contrived in the Thysanura from that of other classes. The mobile pleurites form integral parts of the leg mechanism in contrast to the fixed pleurite of the Pterygota (Fig. 5.14(a) and see §5 below). The pleurites and legs of *Lepisma* form together as unique a leg mechanism as exists among myriapods and hexapods.

The crescentic pleurites of *Lepisma* cannot be the remains of complete rings round the leg base. The absence of pleurites behind the coxa is essential, here as on legs 3 of Collembola, which are also backwardly inclined, their single pleurite forming an antero-lateral crescent (Figs. 9.1(a), (b), 9.4(b)). The leg base can thus deform the pleuron on the remotor movement, so giving a maximum angle of swing. Even the coxa is attenuated posteriorly in the fast running Scolopendromorpha and Lithobiomorpha (Figs. 5.7, 5.8, 5.9), permitting the same effect. The metacoxal pleurite, when present at all, behind the leg base in these centipedes, leaves plenty of room for pleural deformation on the backstroke. Only in the Geophilomorpha is a close investment of

---

FIG. 9.13 (see opposite). *Petrobius brevistylis* Carpenter (Pl. 6(m)). Skeletomusculature of the abdomen. (a), (b) Diagrammatic transverse sections through abdominal segment III, at the levels indicated by the arrows on (c). In (a) abdominal tergite III and coxo-sternal plate 3 complete the circumference; the dorsal anterior margin of tergite 4 is also cut where it carries muscle 20, see (d) above. (b) Transverse section at a more posterior level, see (c), (d) where the whole of the anterior end of the exoskeleton about abdominal segment IV is also cut, coxo-sternal plate 3 now lying free from the body and covering the anterior end of coxo-sternal plate 4. (c) Ventral view showing the nature of the overlap of the exoskeletal structures. Line Z marks a supposed junction between coxa and sternite; it is a cuticular fold shown in section (e). (d) Longitudinal median view through a partially flexed abdomen to show the nature of the overlaps of the parts on dorsal shortening and the layout of the trunk musculature. The two components of the endoskeletal system, the coxal tendon and the lateral endoskeletal tendon complex, are marked on the right, but the latter is much more complex than shown. (From Manton 1972.)

Habits and the evolution of the hexapod classes

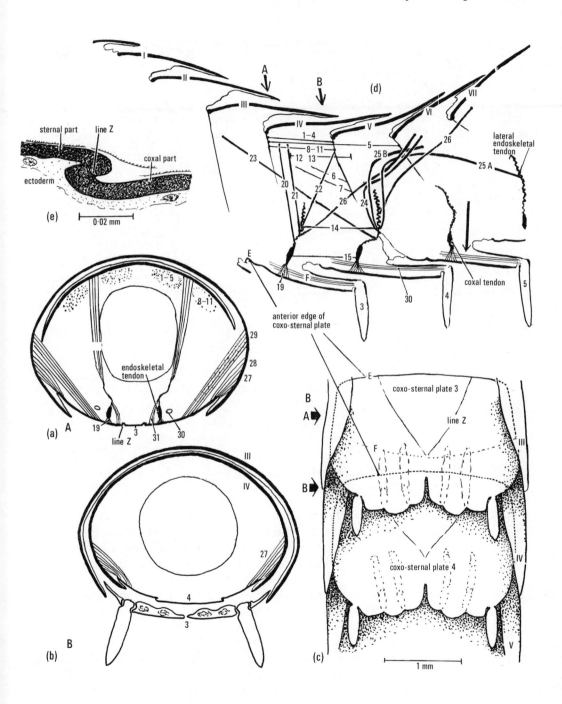

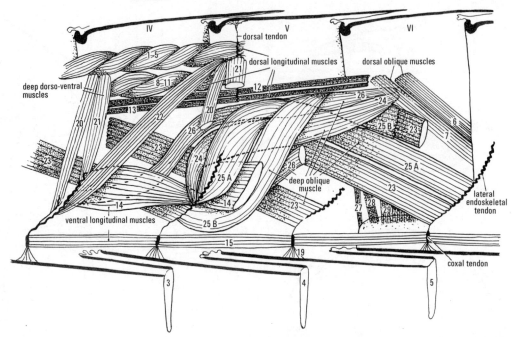

FIG. 9.14. *Petrobius brevistylis* Carpenter. The musculature of the abdomen viewed from the middle line in segments IV–VI redrawn from data kindly provided by Dr. G. E. E. Scudder. Tendinous endoskeleton is simplified and shown diagrammatically in black as are the more sclerotized parts of the cuticle. The dorsal longitudinal muscles comprise the twisted muscles 1–5 and 8–11 and untwisted muscles 12 and 13. There are two ventral longitudinal muscles, 14 and 15. The normal dorso-ventral muscle 20 is reinforced by another muscle 21. The normal deep oblique muscle 22 is reinforced in enormous complexity by muscles 23, 24, 25A, 25B, and 26. Muscles 27–9 pass from the anterior part of the coxal plate to the tergite above. (From Manton 1972.)

the leg base by pleurites round the posterior as well as the anterior coxal margin needed for supporting the strong burrowing movements of the body and legs (Fig. 8.1(c), cf. 8.1(a); see Tiegs 1955, 'subcoxa-pleurite' muscles of *Ctenolepisma* and their antagonists). The musculated, imbricating pleurites of *Lepisma* (Fig. 9.9(g)), moving rapidly in support of the rapid, small, but strong coxal movements, reflect the need for anterior support by the whole leg of most myriapods and apterygote hexapods with digitigrade tarsi. The fusion of the procoxal pleurite with the coxa in the centipede *Craterostigmus* provides coxal support in another manner for this exceptionally flexible animal (Manton 1965).

Habits and the evolution of the hexapod classes

The anterior components of pivot joints are usually larger and stronger than the posterior ones (Fig. 4.27(d), (e), showing the large anterior pivot articulation; the posterior ones are small). Sometimes the posterior components of pivot joints are absent, as at the coxa–trochanter joint of scolopendromorph and lithobiomorph centipedes (Figs. 5.7, 5.8, 5.9(a)), thus allowing the same pleural deformations by the leg base on the remotor swing, as in *Lepisma*. All such useful deformations in fast runners would be prevented by encircling pleurites.

It is the complete lack of understanding of how leg mechanisms work and the lack of appreciation that a mechanism in *Lepisma* exists at all which has led to the entomological theories concerning homologous pleurites from group to group and a claimed primitive status for pleurites and leg base in *Lepisma*. The dorso-lateral pleurites of Symphyla and anamorphic Chilopoda are largely concerned with assisting or creating the leg-rocking mechanism, as they do in the Protura, by different means (Figs. 5.7–9, 5.12(a), (b)). The form of all pleurites in the Diplura and Geophilomorpha promotes trunk flexibility (Figs. 8.1(b), (c), 8.9). In all these arthropods the form and number of pleurites is only comprehensible in terms of function since all serve particular needs.

The head capsule of Thysanura, with typical hexapod fixed anterior and posterior tentorial apodemes, but with no entognathy, contrasts with other apterygote classes. There is a tendency for these apodemes to fuse together in the Lepismatidae, as secondary transverse biting of harder food than eaten by *Petrobius*, is developed. This is a partial parallel to the more efficient changes towards hard-food feeding accomplished by the Pterygota. But the primitive rolling action of the jaws of *Petrobius* is combined with a very neat hydraulic method of collecting the scraped-up algal cells (see Chapter 3, §5C(ii)), which is not a primitive feature.

Among modern Thysanura we have nothing but hexapody and a basic type of head capsule in common between them and the other apterygote classes and also the Pterygota. The leg mechanisms alone in all these classes are all so different that any one does not appear capable to have arisen from any other. The fossil record is of little help in further assessment of the status of the Thysanura. There are the beautifully preserved Monura from the Lower Permian, but they are close to modern Thysanura, although possessing more obvious external segmentation in the head cuticle and other slightly more primitive features in the abdomen; compare *Dasyleptus* in Fig. 1.7(b) with *Petrobius* in Pl. 6(m). The Monura do not span the gap between Thysanura and other hexapod classes and are undoubtedly not far removed from thysanuran ancestors.

6. The Pterygota The adaptive radiation of winged insects, or Pterygota, on land equals that of the Crustacea in the sea and fresh water. The Pterygota show adaptations to every conceivable terrestrial habitat and food, and their habits are of infinite variety. These adaptations are too many to be considered here: one can recall the innumerable modifications of mouth parts to suit particular feeding habits; the modifications of limbs, serving for instance prehension in the mantis, burrowing in mole crickets, jumping in grasshoppers, swimming in water beetles and water boatmen, etc.; and the several ways in which one or two pairs of wings are used. Yet where the origin of the Pterygota lies we do not know. Because the Thysanura lack entognathy they have, with no adequate reasons, been regarded as at least pointers to the origin of the Pterygota. Recent textbooks even figure the Monura and Thysanura as antecedents of the Pterygota forming one evolutionary line. But let us look at the evidence.

A. *The origin and evolution of the Pterygota*

The fossil record discloses Devonian Collembola, already advanced in the jumping habit and associated morphology (§2 above) which could have had nothing to do with the origin of the Pterygota or of any other class of hexapod. The fossils from the Carboniferous indicate an abundant and probably rapidly evolving pterygote fauna, with species often of large size such as the Palaeodictyoptera, dragonfly-like, with a wing span of up to a metre. There are abundant remains of wings of many kinds and sizes, but there is no conclusive evidence as to where and how wings came into existence. It can be assumed that paired thoracic expansions first formed gliding organs. Some Palaeodictyoptera, such as *Stenodictyon*, possessed small prothoracic dorso-lateral expansions as well as large mesothoracic and metathoracic wings, with wide bases, probably used for gliding. A narrowing of the wing base and added complexity in structure could lead to the ability to fold the wings when not gliding, an obvious asset possessed by most winged insects today. Simple flapping flight presumably came after the ability to fold the wings. Some Pterygota are poor fliers with slow or weak wing movements, while highly advanced flight mechanisms use wing beats of frequencies of up to 1000/s in the Hymenoptera and Diptera, other orders showing flight proficiencies of intermediate kinds. The mechanisms are not the same in each order. The most advanced of such flight mechanisms involve rapidity of muscle contractions far exceeding the frequency of nerve impulses arriving at the muscles. Automatic muscle actions at high frequency have been evolved, along with special flight-muscle histology; with 'click' mechanisms of importance in dipteran flight; with elastic, energy-storing units; and with specializations far removed from normal ambulatory skeleto-muscular systems.

It is not intended to provide an outline of the abundant information which we now have concerning the wonderful advances shown by flying insects. There are many accounts of insect flight mechanisms in recent textbooks. We are concerned here with the evolutionary aspects of insect flight mechanisms.

(i) *The origin of flight mechanisms.* Wings develop ontogenetically as double layered outgrowths of the body wall on the mesothorax and metathorax. In the 'exopterygote' or 'lower' orders of pterygotes, where a series of nymphal stages precedes the adult sexually mature animal, the wings increase in size at successive moults but become functional only after the last nymphal moult. In the 'endopterygote' orders a series of larval stages without wings is succeeded by a non-feeding pupal stage during which extensive reorganization of the body takes place, even to almost entire replacement of the tissues. Wings and flying muscles develop during the pupal stage and the latter are thus formed directly and serve the needs of flight without precursor muscles in the larval stages. Tiegs (1955) pointed out that in acquiring the ability to fly the Pterygota must have used existing thoracic musculature of non-flying ancestors. Thus it became of importance to discover just how flight muscles develop throughout the nymphal stages of Orthoptera and Homoptera and to compare the functional morphology of the musculature of species with non-flying females and flying males in the adult stage. But first we must note some basic structural features of the Pterygota.

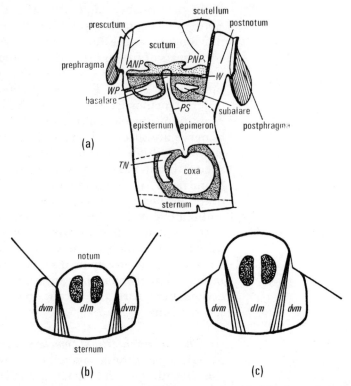

FIG. 9.15. Pterygota (Pl. 6(i)). (a) Diagrammatic lateral view of a typical wing-bearing segment, *ANP.*, anterior notal process; *PNP.*, posterior notal process; *PS.*, pleural suture; *TN.*, trochantin; *W.*, wing (cut through); *WP.*, pleural wing process. (Redrawn after Snodgrass 1935, from Pringle 1957). (b), (c) The classical diagram showing, in transverse section of the thorax, the mechanism of the indirect flight muscles according to the scheme of Chabrier (1822). (b) Top of upstroke. (c) Bottom of downstroke. *dlm.*, dorsal longitudinal muscle; *dvm.*, dorso-ventral muscles. (Redrawn after Magnan 1934, from Pringle 1957).

(ii) *Thoracic structure.* Unlike the hexapods so far considered, the thoracic segments bearing the wings possess a well sclerotized thoracic wall which is continuous round the body (Fig. 9.15(a)). There is no flexible sternum as in Collembola and Machilidae, and no flexible pleuron as in Diplura, Protura, both Machilidae and Lepismatidae among the Thysanura, and in most Myriapoda (Figs. 5.3, 5.7–9, 8.1, 8.7, 9.1, 9.4, 9.9). Flying pterygotes possess a deep, sclerotized pleuron surmounted by a few small epipleural sclerites at the wing base (Fig. 9.15(a)). Each mesothoracic and metathoracic segment is enclosed roughly by a sclerotized cuticular box with top and sides deformable under muscle pressure. Apodemes or phragmas, corresponding with the large prophragma of diplopods (Figs. 8.3–6), are formed to different extents in the various species at the anterior and posterior ends of a segmental thoracic box, the dorsal longitudinal and other muscles, inserting on either face, as is usual. The wings arise dorso-laterally at the junction of tergite and pleuron.

(iii) *Wing movements and musculature.* In many pterygotes the up and down stroke of the wings can be represented (Fig. 9.15(b), (c)) as the effects of the antagonistic muscles shown. The dorso-ventral muscles *dvm.* deform the tergum, pulling it downwards and so elevating the wings, and the longitudinal muscles *dlm.* from the anterior and posterior margins of the thoracic segments compress the box in the opposite direction, so causing elongation of the relaxed dorso-ventrals and depression of the wings. Various adjustor muscles at the wing base control detailed movement, dealing with tendencies to roll, pitch, or yaw. In some of the most highly perfected flying mechanisms this basic deformation of the thoracic walls has been almost dispensed with; the movements become very small and flight-muscle contractions almost isometric and therefore functioning with maximal physiological efficiency.

We have to turn to functional morphology of living animals and to their ontogenetic development for information as to how flight could have originated. A study of simpler, less efficient flying mechanisms than those of wasps and flies is needed, as well as the functional morphology of non-flying pterygotes. The growth, musculature, and functions of the musculature throughout the nymphal stages to the flying adult show that most of the flight muscles are directly derived from ambulatory limb muscles, without cessation of function (Tiegs 1955). With the sudden ability to fly after the last nymphal moult some muscles become more specifically concerned with flight. Much information is yielded from a comparison, not only of the musculature of nymphal stages, but of the ensuing adults when the female is flightless and the male flying, as in certain grasshoppers. There is no doubt that the large muscles in the flying male, often possessing flight-muscle histology, are indeed used for flight. In some cases direct electrical stimulation of the pleural muscles with supposed both ambulatory and flight functions, has produced wing depression. In general, the flight muscles undergo considerable enlargement by longitudinal cleavage of their fibres at the last nymphal ecdysis. Muscles of similar homology are not always modified to serve the same functions in the flying mechanism, a feature of muscles occurring elsewhere in the Arthropoda. Changes in insertions of homologous muscles are listed for Diplopoda in Manton 1961 (pp. 457–8) and shown by Fig. 8.14 here for *Lithobius*; but functions can change along with shifts of insertion or of origin.

We may now consider just those muscles which are concerned with flight in some Orthoptera, other muscles being omitted. There are four categories of such nymphal muscles: (1) tergal muscles; (2) dorso-ventral muscles; (3) pleural muscles; and (4) wing-adjustor muscles.

Tergal muscles. Muscle *mdl.* in Fig. 9.16(a), (c) corresponds with part of the dorsal longitudinal muscles *dlm.* of all myriapods and hexapods, as in Figs. 5.10, 5.11, 8.2, 8.3, 8.5, 8.9, 8.13, 9.1(c), 9.6, 9.7(b). Muscle *mdl.* is small in *Blattella*, a feeble flier, but with the formation of phragmas (apodemes) at the anterior and posterior margins of the segment, the muscle becomes much larger as in the grasshopper *Acridopeza*, a strong flier, forming a strong indirect *wing depressor* (Fig. 9.16(c)). A small oblique tergal muscle *o.t.* in *Blattella* forms the main *wing depressor* and in the male grasshopper *Acridopeza* it could be a weak indirect *wing depressor*.

In this category of muscles we see the transformation of the *dlm.* muscles from the main rigidity or stability promoting muscles of other hexapods, but regarded by Tiegs as flexor promoting in the non-flying nymphs, to undoubted *wing depressors* in the male grasshopper *Acridopeza*. In *Blattella* the skeletal changes occurring at the last nymphal ecdysis, when slight phragmas are first formed, impart a new action to

---

FIG. 9.16 (see opposite). Orthoptera (Pterygota). To show the four categories of mesothoracic and metathoracic trunk muscles concerned with flight. (a), (b) The cockroach *Blattella germanica* L. (a) View from the median plane of flight muscles of mesothorax and metathorax and leg bases. (b) Mesothorax with the more median muscles removed. (c), (d) The male grasshopper *Acridopeza reticulata* Guérin. (c) View from the median plane of mesothorax and metathorax and leg bases. (d) Mesothorax with more median muscles removed. (e) Mesothorax of a young nymph drawn to the same scale. The numbers preceding the abbreviations mean that the muscles arise in the second or third segment of the leg; the muscle abbreviations mean: $bas._{1-3}$ with basalar insertions; $cx._{1-2}$ with coxal insertions; *m.d.l.* median sector of dorsal longitudinal muscle; *o.t.* oblique tergal muscle; *pl.* with pleural insertions; *s.* with sternal insertions; *sub.* with subalar insertions; *t.* with tergal insertions; *tr.* with origin on trochanter. (From Tiegs 1955.)

Habits and the evolution of the hexapod classes

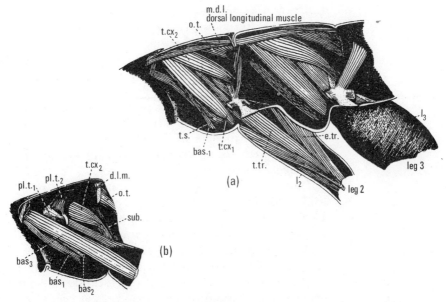
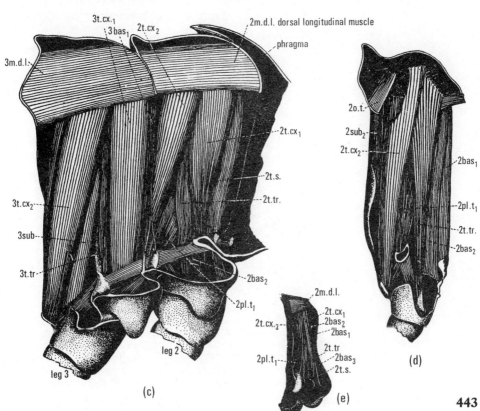

Habits and the evolution of the hexapod classes

functional nymphal muscles converting the tergal muscles exclusively into indirect *wing depressors*. With the dorso-ventral spread of the phragmas in other species, the insertion of the *mdl.* muscles spreads along them, as shown diagrammatically in Fig. 9.17(c), (d), the muscles here being marked *dlm*. The oblique tergal muscle *o.t.* similarly increases in size and its insertion spreads along the phragma as shown in Fig. 9.17(c), (d) marked *adm*. The position of this phragma leaves much of this muscle in the mesothorax in a dorso-ventral direction, so forming the principal wing levator while the *mdl.* muscle forms the chief *wing depressor*.

Dorso-ventral muscle. These muscles are present in the middle of a segment as the *dvc.* complex in Chilopoda and Diplura (Figs. 8.13, 9.6). This muscle or tergo-sternal *t.s.* (Fig. 9.16(a), (c)) is small in *Blattella* but larger in *Acridopeza* and very large in the aphid in Fig. 9.17(a) forming an *indirect wing levator*. Extrinsic leg muscles inserting on the tergite and contracting together form indirect levators to the wings; they are the coxal promotor and remotor, *cx. 1* and *cx. 2*, and the

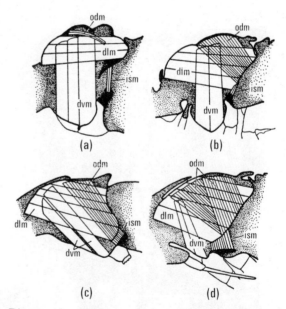

FIG. 9.17. Diagrams showing the arrangement of the mesothoracic indirect flight muscles in Homoptera. (a) *Aphis fabae*; (b) *Prylla mali* (Redrawn from Weber 1930). (c) *Cyclochila australasiae* (Cicadidae); (d) *Erythroneura ix* (Jassidae). *dlm.* dorsal longitudinal muscle; *dvm.* dorso-ventral muscle; *ism.* oblique intersegmental muscle; *odm.* oblique dorsal muscle. (Not to scale; redrawn after Tiegs 1955, from Pringle 1957).

Habits and the evolution of the hexapod classes

depressor from the trochanter, *t. tr.* (Fig. 9.16). The coxal remotor muscle in *Acridopeza* is one of the strongest wing levators.

Pleural muscles. The lower families of Orthoptera show that the wing depression was initially done by pleural muscles attached to the epipleural sclerites just below the wing base (Fig. 9.15(a)) and only in grasshoppers and others do the median sectors of the dorsal longitudinal muscles become of importance as the main wing depressors. The pleural muscles inserting on the epipleural sclerites and contracting together give strong wing depression. Those inserting on the basalar sclerite are (Fig. 9.16): *bas. 1*, a coxal promotor; *bas. 2*, a coxal abductor; and *bas. 3*, a depressor of the trochanter. A pleural muscle inserting on the subalar sclerite also gives wing depression: *sub.* a coxal remotor and strongest direct flight muscle.

Wing adjustor muscles. There are only two in *Blattella* (Fig. 9.16). Tergo-pleural muscle *pl. t. 1* arises from the pleural apodeme and inserts on the scutum above the wing base, it tilts the wing up. Tergo-pleural muscle *pl. t. 2* of similar origin but inserting on thin cuticle above and around the allula and presumably flexes the wing. Other more complex adjustors are present in different species.

(iv) *Coxa–body articulation and wing evolution.* Most of the above muscles could not work unless the coxa and pleuron were firm. Rapid contraction of extrinsic leg muscles inserting on to the tergite and pleuron could not be used effectively for flight unless the coxa–body articulation be firmly situated in incompressible trunk sclerotizations. The arrangement of hard parts in the pleuron itself, shown diagrammatically in Fig. 9.15(a), is essential.

The coxal articulation of the Pterygota is quite unlike that of other hexapod classes. In the Diplura and Protura the principal coxal articulation lies on the ventral proximal rim of the coxal margin, as in most Myriapoda, and above the coxa the pleuron is flexible and compressible (Figs. 5.9, 5.12, 5.13, 8.1(a), 8.7). In the Collembola there is no coxa–body articulation, the coxa being slung by internal ties in a complex manner (Fig. 9.5). The proturan coxa is articulated with a pleurite as well as a sternite, as is that of the Symphyla, but the pleuron is flexible and compressible in both. In the Thysanura alone is the dorsal proximal rim of the coxa

Habits and the evolution of the hexapod classes

articulated with a pleurite, leaving the ventral coxal margin free in flexible membrane, but the details are very different from those of the Pterygota. The whole thysanuran leg-base mechanism is dependent upon mobility of pleurites, not upon their rigidity, as in the Pterygota. The promotor swing takes place between pleurite and body in Machilidae and much of it likewise in the Lepismatidae, a unique position. Contrast the coxal articulation of the prothoracic leg of a locust shown in Fig. 5.14(a). The pleurite bearing the coxal articulation is firmly fixed to the tergite above and to an apodeme from the sternite below; and the sternite is rigid, not flexible as in Collembola and Thysanura. Much of this pleurite comes to form an internal ridge situated dorso-ventrally on the pleuron of the mesothorax and metathorax.

If functional continuity has been maintained throughout ancestral stages leading up to non-flying orthopteran ancestors, as indeed it must, then for duplication of muscle functions, the same muscle serving the purposes of ambulation and of flight, a thysanuran-like coxa–body junction would seem to be entirely untenable. The later, independent growth of some flying muscles at the end of nymphal life, or the progressive use of phragmas for flight muscle insertions, does not alter the situation. The thysanuran coxa–body junction of neither the Machilidae nor the Lepismatidae could provide appropriate morphology for nymphal stages in which existing nymphal muscles, without cessation of function, become flight muscles. The thysanuran coxa–body junction is associated in detail with jumping thysanuran gaits, such as are performed by no pterygotes.

The leg-base mechanism of the Pterygota is unlike those of all other hexapods. The dorsal coxa–pleurite articulation (Figs. 5.14(a), 9.15(a)) is extremely strong while the ventral arthrodial membrane round the coxa is ample and permits a variety of movements. There is the basic promotor–remotor swing about an oblique dorso-ventral axis passing through the coxa–pleurite articulation. Sometimes there is an extra coxal articulation with the trochantin, a pleurite situated in the arthrodial membrane just anterior to the coxa (Fig. 5.14(a)). The axis of the promotor–remotor swing can be fixed in position by this second articulation. Sometimes, as in a stag beetle, the leg smoothly swings in an apparent slot in the even, hard contours of the body, fitting as tightly and closely as a revolving door in a wall. Hairs along the pleural

Habits and the evolution of the hexapod classes

margin of the slot keep out unwanted particles. The coxa can often perform small abductor and adductor movements as shown by the heavy arrows below the leg base in Fig. 5.14(a), and also some twisting about the long axis; note the difference in the position of the coxal lobe marked x on the pair of coxae in this figure. There is no rocking mechanism in the pterygote leg, but instead there are paired antagonistic muscles to the pivot joints (black spots surrounded by circles in Fig. 5.14(a)). This feature, and the plantigrade tarsus with simple hinge joints dorsally between the tarsal segments, are probably responsible for the relatively shorter strides taken by pterygotes in contrast to the apterygote classes (Fig. 7.12, cf. (f), (g) with (a)–(e)).

It would seem therefore that the apterygote classes with unguligrade tarsi have achieved speedy running, in those species which practice this, by means of long strides and all their other facilitations. The Pterygota, on the other hand, have avoided all the apterygote types of advancement and at first must have relied on plantigrade stability and a very stable coxa–body junction, giving the ability to walk more easily on inclined surfaces and perform stronger leg movements than contemporary apterygotes. This may have compensated for any immediate urge for speed. But when this need was caused by environmental pressure, the solution lay in modification of existing structures for flight.

7. Conclusions concerning the hexapod classes

Many conclusions have already been mentioned in the Introduction to this chapter and in §§2B, 3C, and 5C.

We are led to the postulate that there never can have been any one type of ancestral hexapod or insect capable of giving rise to the five hexapod classes.

The functional morphology of the leg base alone is so different and mutually exclusive in the five classes that one can conceive of their evolution only by independent descent from multi-legged ancestors acquiring sclerotization, jointing, and increase in leg length in parallel. But since the hexapods all have fundamental similarities with one another in head structure in contrast to the head capsule, mandibles, and tentorial apodemes of the Myriapoda, it is probable that the multi-legged ancestors of the hexapods already had a common type of head before they developed hexapody, longer legs, and class distinctions in leg-base mechanism. Flight itself could only have arisen from non-flying ancestors of the Pterygota and not from any of the other hexapods which possess a quite different and unsuitable type of leg base and pleuron.

The Thysanura in particular have been favoured as possible pterygote ancestors because they are the only apterygote class lacking entognathy. But their pleural and leg-base structure is bound up with the unique jumping gaits used by this class, which are not found at all in the Pterygota or in any other hexapod class. Thysanuran pleural structure does not lead towards that of a non-flying pterygote nymph. Tiegs (1955) noted that thysanuran musculature differs considerably from that of the Orthoptera. The probable evolutionary history of the hexapod classes was summarized in Chapter 6, §5, and see Fig. 6.14. It is unlikely that ancestral pterygotes wrapped the thorax in protective paratergal lobes serving an essentially flexible pleuron needed for thysanuran unique locomotory mechanisms and more likely that the pleuron became sclerotized directly in the pterygote ancestors. Whether paratergal lobes can ever be equated with the more dorsally situated wing projections is most uncertain. If this similarity is regarded as common ground between the ancestors of Thysanura and Pterygota it is an extremely tenuous one which might belong to the distant past when none of the characteristic and contrasting features of the pleuron, locomotory mechanism, leg base and musculature in the two classes had been initiated.

Functional morphology and ontogenetic studies tells us much more about the past history of hexapods that can be obtained at present by any other methods. This sound evidence enables a drastic sifting out of fiction from facts in existing entomological theories. It is possible that the giant Carboniferous winged hexapods bore no more relation to the ancestors of the modern Pterygota than the dinosaurs did to the mammals. But however that may be, we have no sure indication as to the origin of non-flying pterygotes, although we know much more about them than we did.

10 Locomotory habits and the evolution of the Chelicerata and Pycnogonida

1. Introduction

As already shown, modern work in several fields is being of great help in piecing together the past history and evolution of the various groups of Arthropoda, but the available new information is not evenly spread. We have much more detailed knowledge of the correlations between structure and function in the Uniramia and Crustacea than in the Chelicerata.

2. The Merostomata

For the extinct Merostomata we can only guess at the correlations between habits and structure in the light of modern forms. An early divergence between predominantly bottom-living shallow burrowing in the Xiphosura and bottom-living swimming habits in the Eurypterida has been mentioned in Chapter 1. The similarity between the fossilized tracks of the eurypterids *Mixopterus* (Fig. 1.12) and a modern scorpion is striking (Pl. 7(a), (b)), indicating similar movements, but the former took much shorter strides and probably had a less perfected leg-rocking mechanism than modern scorpions (see below §5D(i)).

Limulus and related genera of the Xiphosura can be studied alive; we know their leg movements and methods of feeding (Chapter 3, §3A) and of burrowing (Figs. 1.9, 2.6, 3.12, 3.13). The terminal spine levers the cephalic shield (prosoma) down into the mud or sand while the sixth pair of prosomal limbs shovel out the substratum in a posterior direction by strong extensor–remotor movements, the wide terminal setae on these legs fanning out against the substratum (Fig. 3.12(c)). The fields of movements of these legs is separated from those of legs 2–5, so permitting flexibility in performance (Fig. 2.6(c)). The more anterior legs grasp at food organisms, living or dead, in the substratum. The form of the prosoma suits the burrowing habit used in search of prey such as lammelibranchs; the curvatures and the lateral expansion of prosoma and mesosoma, with the longitudinal grooves providing mechanical strength against deformation of the exoskeleton, all contribute to burrowing efficiency. The biramous mesosomal limbs, used for swimming and respiration, are well protected and boxed in. An understanding of the functional significance of the shape of *Limulus* and of the mode of action of its morphology is suggestive as to the modes of life of some of the early merostome-like fossil animals.

The leg-joints of *Limulus* with their several movements are shown in Fig. 10.4(f), (h) where black spots indicate the positions of the articulations and the horizontal or vertical heavy lines show the axis of movement at each joint. The patella is fused with the tibia, strengthening legs used for digging, and the knee-joint between femur and patella–tibia rises steeply under protection of the overhanging carapace. Leg movements of *Limulus* differ from those of all other extant chelicerates (i.e. the Arachnida) and resemble those of Crustacea and the Uniramia. The promotor–remotor swing of the coxa on the body is basic to the stepping movement of the whole leg, as in Crustacea, Myriapoda, and Hexapoda, where the principal levator–depressor movement occurs at the coxa–trochanter pivot joint, that is distal to the basal promotor–remotor swing, with further levator–depressor movements at the trochanter–femur joint. Hinge joints lie more distally along the leg. This order in the positions of the main types of joints and of the movements they cause resembles those of other arthropodan groups (Figs. 4.5, 5.14, 5.15). The transverse adductor–abductor movement of the coxa and its gnathobase, used in feeding, is a movement at right angles to the walking promotor–remotor swing and differs from the mode of action of the gnathobase of Crustacea (Chapter 3, §3A, Chapter 6, §2, Figs. 2.6, 10.4(g)). For Pycnogonida see §9 below.

3. The invasion of the land

Although Uniramia today are basically terrestrial, the Crustacea have contributed to the land fauna. Isopoda (woodlice), a few Amphipoda, and even some Copepoda live in damp places out of water; there are land decapods in very limited variety and there are plenty of freshwater Crustacea: the Branchiopoda, Copepoda, Branchiura, Ostracoda, and a few from other crustacean groups. The primary habitat of

Locomotory habits

Crustacea is the sea. The great chelicerate invasion of the land, leading to modern arachnids, was another event of the greatest importance (see Chapter 1, §1C(iii)). It must have taken place entirely independently from the uniramian invasion of the land.

The origin or origins of the Arachnida, on land, is not shown in any detail by the fossil record. Silurian scorpions, aquatic and terrestrial, have been mentioned in Chapter 1; see drawings of *Waeringoscorpio* and modern scorpions in Figs. 1.13, 1.14(d). Whether the arachnids are descended from scorpion-like eurypterids lacking the conspicuous specializations shown by many of these animals (Fig. 1.11) we do not know. Spiders and Pedipalpi occur in the Carboniferous and all the principal orders of modern Arachnida are present as fossils in the Tertiary rocks.

4. The Arachnida

That the Arachnida and Uniramia lie in separate phyla is certain from their general anatomy, their embryology (Anderson 1973), and from the arachnid use of gnathobasic chewing. Yet they can only take in fluid food obtained by chewing and external digestion. The hexapods have a great variety of food and feeding techniques, but the taking in of fluid food after gnathobasic chewing is not one of their accomplishments.

The Uniramia possess a head bearing sensory antennae anteriorly. The Arachnida possess a prosoma (Figs. 1.9–13), as in the aquatic Merostomata, bearing preoral chelicerae. Arachnid postoral pedipalps with chewing gnathobases (Fig. 3.14) and often distal chelate grasping organs are used as shown for the scorpion in Fig. 1.13. Four pairs of walking legs follow. The prosoma is covered dorsally by a tergal shield or carapace which usually is entire, extending over all leg-bearing segments (Pls. 7(a), (b), 8(a), Fig. 10.1(a)) but sometimes there are separate tergites on the more posterior leg-bearing segments. The united prosomal sternite is usually small (Fig. 10.1(b)) or absent and the leg bases large.

The more posterior part of the body is variously constructed. An opisthosoma is always present, separated from the prosoma by a constriction in spiders but not demarcated by a waist from the prosoma in scorpions, mites, and harvesters (Opiliones). The adult segmentation of the opisthosoma, as present in scorpions, is absent in other arachnids (spiders, ticks, mites, harvesters, etc. Figs. 1.13, 10.1, 10.7,

451

Locomotory habits

10.8, Pl. 8(a)–(c), (e)–(j)), several posterior somites being embryonic. A narrow, segmented, terminal flagellum is present in Uropygi and some Pedipalpi (whip scorpions), the anus being at the base of the flagellum, as it is in *Limulus* in respect of the terminal spine. The body form of the various orders of Arachnida shows the common theme evident throughout the Chelicerata but possesses greater range in detail than seen in the marine chelicerates.

5. Arachnid locomotory mechanisms associated with terrestrial life

It is beyond the scope of this chapter to present more than an outline of some of the correlations between arachnid structure and function which have most recently been appreciated. The locomotory mechanisms suiting arachnid land life contrast with those of the Uniramia more than might be expected. Although most arachnids possess four pairs of legs posterior to the pedipalps, some species have become hexapodous in habit and this they do in a manner different from the uniramian hexapods.

Arachnid and xiphosuran prosomal walking legs probably represent the endopod of originally biramous limbs seen today in the branchial limbs of *Limulus*. Many malacostracan Crustacea have formed a walking endopod from a biramous limb as also have the trilobites.

Ancestral Arachnida probably left the sea already possessing good sclerotization and longish legs, in contrast to the Uniramia which probably emerged on land with soft bodies and short legs (Chapter 6). Unlike the multi-legged Uniramia, the Arachnida have probably never possessed more than the five postoral locomotory limbs seen in *Limulus*. The differentiation of a pedipalp reduced the ambulatory leg pairs to I–IV, as usually numbered. The legs of modern arachnids, relative to body size, are on the whole longer than those of myriapods and hexapods. The short legs of small Acari and of the deep-living Collembola are probably secondarily so in both groups.

The Arachnida and the Pycnogonida differ from all other arthropodan groups in showing no promotor–remotor swing of the coxa on the body. The arachnid leg-joints between the podomeres provide a different order of the various movements used in stepping from those found in other arthropodan groups (Figs. 10.2(b)–(g), cf. (a) and Fig. 5.15); leg-rocking mechanisms, analogous to those of myriapods and hexapods, exist but they are intrinsic in arachnid legs. Only

Locomotory habits

the Diplura outside the Arachnida have an intrinsic leg-rocking mechanism, but of a different kind (see Chapter 5, §5B(i), Fig. 5.13). The legs of Pycnogonida are considered in §9 below.

A. *Arachnid coxae*

The evolution of long legs brings with it the problem of obtaining sufficient coxal stability. This is gained in Arachnida in a manner contrasting with those of the Uniramia and Crustacea. Except to some extent in spiders, no proximal arthrodial membrane allows mobility of the coxae; the cuticle at the coxal margins is minimal in extent and well sclerotized, the coxa thus being firmly fixed on the prosoma. It is advantageous for the trochanter to project laterally and not ventrally, so that the body can be held close to the ground. Coxae inserted ventrally on the prosoma are often boat-shaped, the prow being a ring of firm cuticle which carries the trochanter just clear of the body laterally, the rest of the coxa being open dorsally to the haemocoel. The upper margins of the boat-shaped coxae II–IV of an amblypygid are united on either side by scanty but strong cuticle (Fig. 10.1) and coxae III and IV of scorpions are fused together on either side. Reduction in size of the sternite accompanies its disuse for insertion of extrinsic coxal muscles. In the Solifugae there is no sternite between coxae III and IV, which are fused in pairs mid-ventrally (Fig. 10.3), some intersegmental flexibility of the prosoma being retained. Only in spiders does the coxa possess some mobility on a well-formed sternite; the coxae are cylindrical in shape but the ventral face is considerably longer than the dorsal face.

A horizontal fanning-out of the coxae on either side is usually evident, so spreading the fields of movement of successive legs and reducing the possibility of mechanical interference of one with another (see scorpion and spider in Fig. 5.2 and of Amblypygi in Fig. 10.1(a)). In mites the coxae form almost flat plates upon the body in some species but with a projecting ring of cuticle carrying the trochanter. The legs here are well separated by the anterior position of coxae I and II and the posterior position of coxae III and IV.

B. *Arachnid leg positions*

Arachnids hang down from their legs, as do most arthropods, unless the legs are very short. When legs are long the knee must project in some direction when the legs are not outstretched. Normally the knee rises upwards beyond the

Locomotory habits

dorsal level of the body, as in the centipede *Scutigera* (Fig. 5.14(b), Pls. 4(k), 8(h)) or the metathoracic legs of a grasshopper. In Opiliones and spiders, as in *Limulus*, the knee is directed upwards (Fig. 3.12, Pl. 8(h)); there is no particular knee in Acari but the dorsal hinge articulation between the distal podomeres is directed upwards.

The legs of *Limulus*, arachnids, myriapods, and hexapods tend to be flattened a little on their presumed anterior and posterior faces, leaving a dorsal surface which carries the series of hinge articulations. But in some Arachnida, notably scorpions and Pedipalpi, the dorsal margin of the leg is permanently displaced so that the anterior face is always visible from above. This condition is probably not due to any actual twist of the leg but to changes in its ontogenetic organization about the long axis between certain proximal and distal limits, marked by black spots on the amblypygid leg shown in Fig. 10.1(a) (see further below). In the scorpions the leg organization about the long axis is such that there appears to be a permanent slight twist of the leg at the coxa–trochanter joint so that the flattened anterior face is visible dorsally, the normal narrow dorsal edge being tilted backwards (see the leg beyond the trochanter in Fig. 10.2(d), Pl. 7(a), (b)).

This change in the position of the dorsal face of the leg,

FIG. 10.1 (see opposite). Pedipalpi (amblypygid) sp. (Pl. 8(a), (b)) moulted cuticle, arthrodial membranes white. (a) Dorsal view showing the coxal cavities open to the haemocoel and the sternite, the carapace having separated from the rest of the cuticle at the upper level of the pleuron. Right leg III only is drawn in entirety, but in two pieces, which join as shown by the arrowed line. Between the large black spots the leg organization is such that the face normally considered to be dorsal is backwardly directed so that the knee lies in this position at the joint between femur and patella. The hinge articulations at the femur–patella and tibia–tarsus 1 joints are seen in morphological lateral view, while the axis of the pivot articulation between tarsus 1 and tarsus 2 lies at right angles, in the plane of the paper. (b) Ventral view of the same cuticle. The position of the costa coxalis (an apodeme) is shown only on leg III, but is similarly present on coxae II–IV. The coxa of leg I is invisible ventrally. The coxae of legs II–IV are well fanned out and immovable. The coxa–trochanter joint carries anteriorly the strong articulation between the trochanter and distal end of the costa coxalis. Posteriorly this joint is provided with a Y-shaped sclerite f spanning the arthrodial membrane. Little if any rocking occurs at this joint, which provides the main levator–depressor movements of the leg. The trochanter–femur joint forms a vertical pivot providing the promotor–remotor swing of the leg. The dorsal articulation in (a) is small with large femoral emarginations on either side and the ventral articulation is stout, strong, a little postero-ventral in position and supported by a short costa fading in the trochanter. Anteriorly the joint carries crescentic sclerites as shown. (c) The tarsal joints of leg III in lateral view, setae omitted. The promotor–remotor swing here takes place at the trochanter–femur joints, cf. Figs. 10.3, 10.4. (From Manton 1973*b*.)

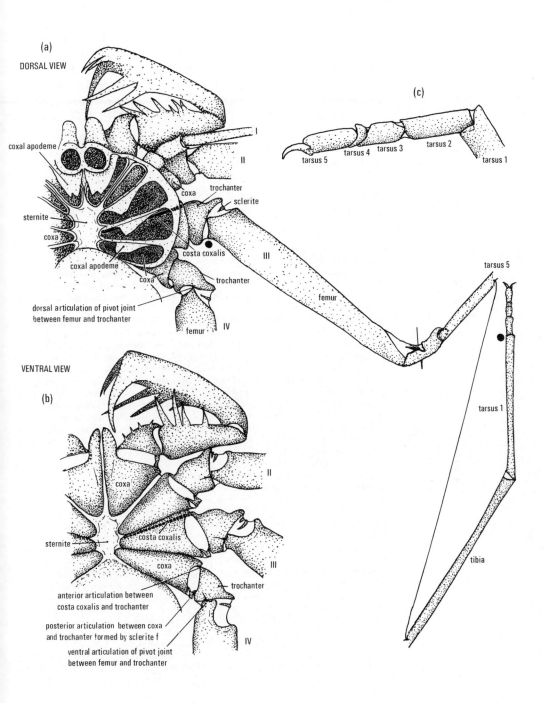

Locomotory habits

often bearing the dorsal hinge articulations, in scorpions and amblypygids is reflected in the position of the knee. In scorpions, the knee when flexed does not project upwards but folds againt and over the body (Fig. 10.4(j)). Thus there is no hindering leg projection during burrowing. In the Amblypygi the legs are organized as if twisted by almost 90° at the trochanter–femur joint (Fig. 10.1), and the knee projects backwards, rising little above the body (Pl. 8(a), (b)). Thus the flattened amblypygid trunk and the long legs, with movements keeping them close to the substratum, are all suited to crevice penetration. In the Solifugae the knee is mainly upwardly directed (Fig. 10.3) but the curvature of the proximal podomeres and the set of the prefemur–postfemur

---

FIG. 10.2 (see opposite). Joints on walking legs III of *Limulus,* of arachnids, and of a pycnogonid (Figs. 1.10, 1.16(a), Pl. 8) are shown, comparatively and diagrammatically. Each joint is viewed proximally and end-on with the anterior face of the leg on the left and the posterior face on the right, as if the leg was extended from the body in the transverse plane. The concentric lines represent the proximal and distal overlapping margins of successive podomeres at each joint. When the coxa is immovably fixed to the body it is shown as a stippled area, the outline indicating the ventral coxal face. In the spider the coxa is less firmly fixed on the body than in other arachnids and can rock and undergo a promotor–remotor swing to some extent. The positions of articulations are marked by large black spots. At pivot joints possessing imbricating articulations, (i) on anterior and posterior faces or (ii) on the dorsal and ventral faces of the leg, the axis of movement at the joint is marked by a heavy line passing through these articulations and the resultant movements, promotor–remotor or levator–depressor, are marked *pr., re., lev.,* and *dep.* respectively. A hinge joint is shown in heavy black, usually forming an extended transverse and dorsal articulation, with or without conspicuous paired dorso-lateral articular components. The axis of movement at a hinge joint is shown by the heavy line passing through the hinge articulation; here muscles can flex the leg ventrally but cannot extend the leg as at a pivot joint. Hatching across parts of the margins of the podomeres indicates strong emarginations in the proximal–distal directions which cannot be shown in the plane of these diagrams. Some joints permit a rocking movement of one podomere on the next, as indicated by the curved, arrowed interrupted lines. In *Galeodes* at the trochanter 2–prefemur joint the emarginations are so extreme as to leave the axis of movement between the two articulations in a longitudinal position on the leg, the black spot here indicating an end-on view of the axis through the most proximal articulation (see Fig. 10.3). The flexible sclerite between coxa and trochanter f in scorpion, spider, and Amblypygi, spanning the wide arthrodial membrane between the two podomeres, permits the rocking movement, arrowed interrupted lines, on the coxa–trochanter articulation in scorpion and spider. (a) *Limulus*, see also Fig. 10.4(f), (h), where the whole leg is shown as well as similar joint diagrams. (b) Opiliones, (Pl. 8(g)) the more distal tarsal hinge joints are omitted. (c) Acari, (Pl. 8(e), (f)) the more distal podomere joints not included because they are inconstant in number and no differentiated knee is present. (d) Scorpion, (Pl. 7(a), (b)) *re.* signifies retractor movement at tibia–tarsus 1 joint; the tarsus 1–tarsus 2 joint is better seen in Manton (1958*a*, Fig. 4(c)–(e)). (e) Spider (Pl. 8(d)). (f) Pedipalpi, a species of amblypygid (Pl. 8(a), (b)). (g) Solifugae, *Galeodes* sp. (Pl. 8(e)). (h) The pycnogonid *Decalopoda antarcticum*. (After Manton 1973*b*.)

Locomotory habits

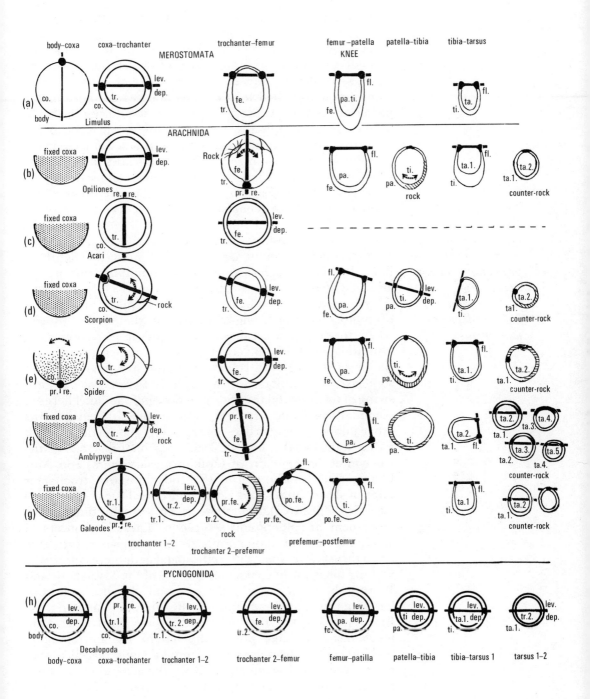

Locomotory habits

joint on the more posterior legs tend to bring the knee over the body.

As shown in Chapter 5, stepping movements of well fanned-out limbs of Uniramia and Arachnida result in progressive flexure during the backstroke of anterior limbs from an outstretched forward position, while those of the posterior limbs show progressive extension during the backstroke from an initially flexed position. A leg moving forwards and backwards equally about its base, situated in the middle of a fanned-out series of limbs, shows a backstroke consisting first of progressive flexure and followed by progressive extension (Fig. 5.2). These features have their effects upon arachnid leg mechanisms.

C. *Arachnid leg movements: the promotor–remotor swing; leg-rocking; and levator–depressor movements*

The usual immobility of arachnid coxae is presumably secondary since mobility is present in *Limulus* and slightly so in spiders (Fig. 10.2(a), (e)). Coxal immobility means that arachnid stepping must be differently contrived from that of all arthropods with mobile coxae. That stepping involves the promotor–remotor swing of the leg on the body is usually taken for granted because it is found in vertebrates, in Uniramia, and in polychaetes using their parapodia for walking (Pl. 1(b)–(e)). In arachnids the stepping, except to some extent in spiders, does not involve a promotor–remotor swing of the coxae on the body as it does in *Limulus*. There are secondary promotor–remotor movements situated at some other joints in some orders, as shown by the vertical axes of swing indicated by heavy black lines on Fig. 10.2 at the trochanter–femur joint of Opiliones (harvesters) and Amblypygi (see also Fig. 10.1(a), (b)) and at the coxa–trochanter joint of Acari (mites and ticks) and of Solifugae (*Galeodes*) (see also Fig. 10.3), but there is no joint producing this movement in spiders and scorpions. These two orders of arachnids depend upon simple flexor–extensor movements and a thrust developed within the distal podomeres forming a rocking mechanism analogous to those of the Myriapoda and Protura between body and coxa and the intrinsic rock between trochanter and femur in the Diplura (Chapter 5, §5, Figs. 5.8, 5.13). Other arachnid orders employ rocking mechanisms situated proximally on the leg and counter-rocking contrivances are always more distally placed so that the clawed tarsi remain squarely on the ground throughout the backstroke (see below). As in the Uniramia, the details of

Locomotory habits

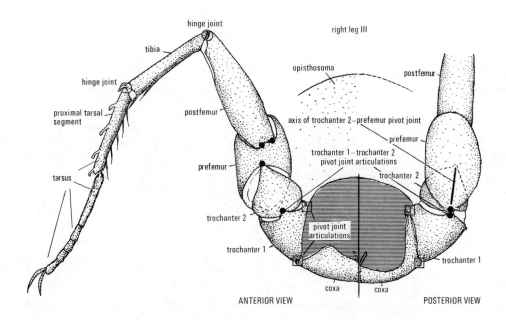

FIG. 10.3. Solifugid sp. (Pl. 8(c)), drawing made from a potash preparation, arthrodial membrane white. Transverse piece of the body bearing walking legs III to show the cuticle and joints. Anterior view on the left with the dorsal level of the more posterior opisthosoma indicated and posterior view on the right. On the left the resting cuticle brings the basal leg podomeres to a vertical position. The paired coxae are fused mid-ventrally. Trochanter 1 articulates with the coxa by strong midventral sclerotizations and mid-dorsally there is a wider but less sclerotized close union between these podomeres; both junctions are marked by a surrounding square. This coxa–trochanter pivot joint, with a vertical axis of swing, provides the promotor–remotor swing of the leg, usually residing at the coxa–body articulation in Myriapoda and Hexapoda. Trochanter 1–trochanter 2 joint possesses strong articulations at similar levels anteriorly and posteriorly marked by black spots on either side of the diagram. The axis of swing at this joint is horizontal and longitudinal, so providing levator–depressor movements. Such a joint is normally present at the coxa–trochanter joint in Myriapoda and Hexapoda. Trochanter 2 is long on its anterior face but the posterior face is so much shortened that the posterior trochanter 2–prefemur articulation is confluent with that of trochanter 1–2. The prefemur is strongly bulged posteriorly, so that the axis of swing at the pivot joint between trochanter 2 and prefemur becomes longitudinal to the leg, the movement about this axis giving leg-rocking. A prefemur–postfemur hinge joint, shown on the left, lies obliquely on the anterior face of the leg fitting into a deeply emarginated prefemur. The postfemur–tibia 'knee' joint possesses a very strong and wide dorsal hinge (ringed) and is much emarginated ventrally. The tibia–tarsus dorsal hinge articulation (ringed) is normal. The several tarsal segments permit compensatory rocking so that the claws remain on the ground squarely during the backstroke. The promotor–remotor swing takes place at the coxa–trochanter joint, the rock at the trochanter 2–prefemur joint, and the compensatory counter-rock at the tarsal joint, cf. Figs. 10.1 and 10.4. (From Manton 1973*b*.)

459

Locomotory habits

the rocking mechanisms differ from group to group; the positions are shown by the double arrowed interrupted lines on Fig. 10.2 for some arachnid orders.

To understand the way in which the various arachnid rocking mechanisms work it is useful to look again at simple hinge and pivot joints as seen in *Limulus*, see above (Fig. 10.4(f), (h)). These are formed in the same way in all arthropodan groups owing to their being the simplest mechanical solutions to joint formation. More advanced and particular joints are peculiar to each of the several arthropodan groups. A pivot joint, with close articulations of one kind or another situated on opposite faces of the leg, with ample arthrodial membrane elsewhere on the joint, results in movement of the distal podomere about an axis passing through the articulations; see the coxa–trochanter joint of an amblypygid in Fig. 10.1(b) permitting free levator–depressor movement while the trochanter–femur pivots permit free promotor–remotor movement.

Rocking movement in arachnids is usually developed at pivot joints where one of the two pivot articulations is suppressed, as at the coxa–trochanter joint of chilopods (Figs. 5.8, 5.9), but unlike this particular joint which permits levator–depressor movements only, the coxa–trochanter joint of the spider and scorpion has strong but freely moving anterior pivot articulations which allow rocking of the posterior margin of the trochanter, surrounded by ample arthrodial membrane, back and forth as indicated by the double arrowed interrupted lines on Fig. 10.2(d), (e). In both spider and scorpion a flexible sclerite *f* traverses and supports the expanse of arthrodial membrane situated at the posterior side of the joint. In the scorpion a principal and slightly oblique levator–depressor movement also takes place about the axis shown at the coxa–trochanter joint on Fig. 10.2(d). In the Opiliones the vertical axis of the promotor–remotor swing at the trochanter–femur joint is not fixed in one place dorsally; the femur can rock a little about its ventral pivot articulation, thus rocking the more distal part of the leg (Fig. 10.2(b)). In the spider there are three rocking joints situated at the coxa–body, coxa–trochanter, and patella–tibia positions (Fig. 10.2(e)). Counter-rocking takes place distally at the tarsus 1–tarsus 2 joint and is differently contrived in other orders.

Levator–depressor movements take place at pivot joints

Locomotory habits

with horizontal axes of swing; see horizontal heavy lines indicating these axes on Fig. 10.2. Active depressor or flexor movements occur at hinge joints as shown. Leg straightening (extension) is indirectly accomplished at hinge joints by proximal depressor muscles, rarely by hydrostatic pressure.

D. *Arachnid leg mechanisms*

The variety in positions of joints producing the promotor–remotor swing (vertical axes); the levator–depressor movements (horizontal axes); and the rocking movement in arachnids contrast with the greater uniformity in distribution of leg-joints giving these movements shown by myriapods, hexapods, and *Limulus* (the latter uses no rocking). Four examples of different arachnid leg mechanisms are given below, while some others are summarized in Fig. 10.2. Whether each has arisen from a scorpion-like mechanism is uncertain, although that is possible; we do not know either the origin or interrelationship of the several arachnid orders which are each considerably different from one another (see below §10).

(i) *Stepping movements and leg mechanisms of scorpions.* As noted above, scorpions use flexor–extensor movements of the leg joints and rocking movements, but no promotor–remotor swing of the coxa–body or at any other joint. There is very little movement of coxae I and II and coxae III and IV are fixed on the body. A rock takes place, as shown above, at the coxa–trochanter joint and the counter-rock at the tarsus 1–tarsus 2 joint. The stepping of the left leg IV is shown in lateral view in Fig. 10.4(i) and in dorsal view in (j). Outlines are given of the leg at the beginning and at the end of the backstroke. Typical pivot joints (Fig. 10.4(b), (d)) with antagonistic muscles lie at the trochanter–femur, femur–patella, and patella–tibia junctions, marked by black spots. Muscles at these joints cause the leg to flex and extend and could cause some stepping but with the knee rising unsuitably high during flexure.

A rocking movement of the coxa–trochanter joint (see arrows) swings the dorsal face of the leg in a posterior direction during the forward stroke, thus turning the tarsus of leg IV further forward than would be possible to a simply flexed leg and at the same time keeps the knee well down over the dorsal face of the body, giving little projection, and therefore the least hindrance during burrowing. Leg I is most

Locomotory habits

fully flexed at the end, not at the beginning, of the backstroke (see Pl. 7(a), (b)) and the stride is similarly increased by leg-rocking. During the backstroke of left leg IV (Fig. 10.4(i), (j)) the rock aligns the femur–patella wide dorsal hinge to take the thrust following from leg extension caused by depressor muscles; and the counter-rock helps to tuck the knees well in

FIG. 10.4 (see opposite). Diagrams illustrating the two basic modes of stepping, (i) by a promotor–remotor swing of the leg on the body, as in *Limulus* (f), (h) and Uniramia Diplopoda (a), (b), (c), Crustacea are also similar; and (ii) by leg-rocking on a fixed coxa in scorpion (i), (j). Black represents arthrodial membrane, hatching indicates a cut body surface, trunk sclerites of scorpion are mottled, and podomeres are stippled. Axes of swing of the joints are shown by heavy lines passing through the points of articulation or close union, which are usually marked by black spots as in Fig. 10.2. (a) The promotor–remotor swing of the coxa on the body in ventral view, exemplified by a polydesmoidean diplopod (Pl. 5(f)); the close articulation of coxa and body is seen at the proximal left-hand edge of the coxa. (b) Shows the coxal in the sternite, the coxa being cut short, aspect similar to that in (a). (c) Vertical longitudinal section through three successive legs, set in a diagrammatic sternite, as in (a), the coxae being cut short and their axis of swing on the body being marked by black spots which represent the end-on view of the heavy lines drawn in (a), (b). The promotor–remotor swing brings the dorsal face of the coxa to a more posterior position and the ventral face to a more anterior position while the remotor swing does the reverse, so that the rock and the promotor–remotor swing are combined. (d) Lateral view of a typical dicondylic pivot joint between podomeres, the axis of swing passing horizontally across the leg through the black spot. Antagonistic levator and depressor muscles (arrowed lines) cause levator and depressor movements of the distal podomere as shown. (e) Lateral view of a typical hinge joint between podomeres. The close union may be in the form of a mid-dorsal articulation or a transversely elongated articulation situated across the leg with or without particular dorso-lateral condyles. A single flexor muscle (arrowed lines) causes flexure at the joint; extension towards the straight position is achieved by remote muscles or by hydrostatic pressure, see text. (f) Anterior view of left leg III of *Limulus* and cut prosoma (Figs. 1.9, 3.12). The promotor–remotor swing of the coxa takes place about the axis shown, the only close articulation being situated between the coxa and the pleurite, marked by a black spot. Two succeeding pivot joints and two hinge joints lie along the leg as shown by the black spots. (g) The coxa, drawn as in (f); the dotted outline shows the abductor swing about the coxa–pleurite articulation, this movement being used in feeding but not for locomotion. (h) Diagrammatic transverse views of the joints across the leg of *Limulus* at the positions indicated by the arrowed lines. The method of presentation of these diagrams is the same as described for Fig. 10.2. (i), (j) Shows the stepping movement of left leg IV of the scorpion (Pl. 7(a), (b)) (i) in lateral view and (j) in dorsal view. There is no coxal promotor–remotor swing as in (a–c). The positions of the leg at the beginning and end of the backstroke are shown. The coxa projects postero-laterally and is fixed to the body. From the outstretched position with the tarsus on the ground in a posterior position, three movements bring the tarsus forward: flexure at the distal leg joints, levation at the trochanter–femur joint, and clockwise rocking at the coxa–trochanter joint. The propulsive backstroke utilizes the reverse movements: counter-clockwise rocking at the coxa–trochanter joint, depression at the coxa–femur joint, and extension at the more distal joints. Counter-rocking at the tarsus 1–tarsus 2 joint enables the tarsal claws to remain squarely on the ground. The manner in which the anterior face of the leg is exposed during the forward swing is more clearly seen in dorsal view (j) and arrows across the leg positions indicate the direction of rocking. (From Manton 1973*b*.)

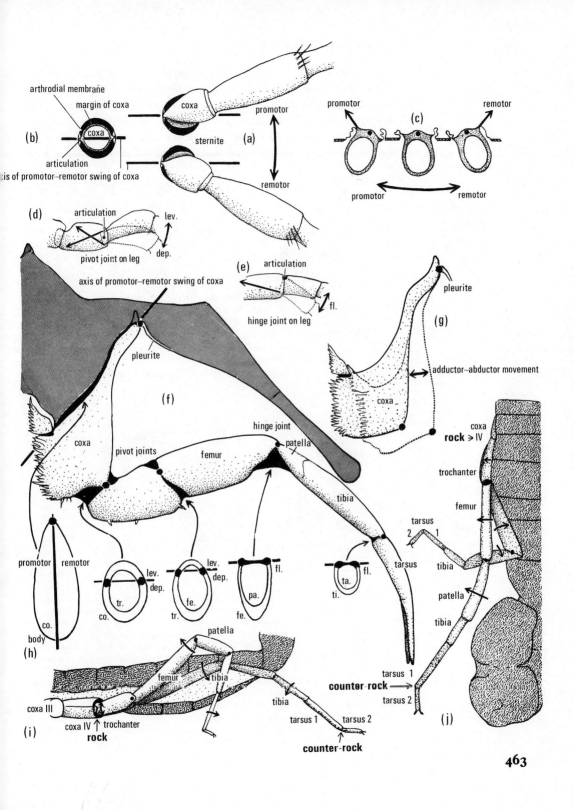

Locomotory habits

over the dorsal surface of the animal. (For muscles see Manton 1958a, Fig. 2). The tarsus 1–tarsus 2 joint is complex and permits counter-rocking in the opposite direction to that taking place at the coxa–trochanter joint, thereby enabling the tarsus 2 claws to remain squarely on the ground without participating in any rock of the telepod (Manton 1958a, Fig. 4).

The coxa–trochanter joint of the scorpion provides

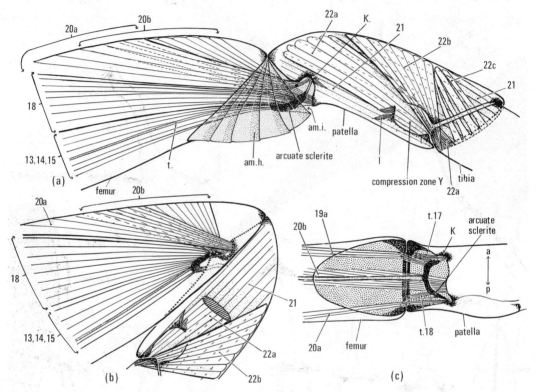

FIG. 10.5. Skeleto-musculature at the femur–patella and patella–tibia joint of leg IV in the spider *Ciniflo ferox* (Pl. 8(d) shows another species). The almost black zones denote heavy, amber-coloured sclerotization. Extreme extension in (a), flexure in (b), posterior views, and (c) ventral view. (a) The arthrodial membrane between femur and patella is extensive. Within it are several hoops of light sclerotization, *am.h.*, and a heavily sclerotized arcuate sclerite (marked) which swings in a patella notch K on either side. A blind infolding *am.i.* of the membrane lies distal to the arcuate sclerite. Normal flexor muscles 20b pull on the ventral lip of the patella. Additional flexor muscles 17, 18 pull from the lateral parts of the arcuate sclerite (tendons of origin of muscles 17, 18 shown in (c) and muscle 18 shown in (a), (b), and muscles 13–15 from a tendon arising on the arcuate sclerite, the tendon is shown in (b)). The postero-ventral compression zone on the patella, *Y*, is marked and the lyriform sensilla *l* in the cuticle. (From Manton 1973b.)

levator–depressor movements to the leg about the axis shown in Fig. 10.2(d) as well as rocking. This axis is oblique, so turning the whole distal part of the limb a little; see the postero-dorsally directed dorsal surface and the similar positions of the axes of the more distal pivot and hinge articulations in Fig. 10.2(d) and Pl. 7(a), (b) where the anterior faces of the legs are fully visible except when approaching the outstretched position (Pl. 7(a), right leg IV, Pl. 7(b), right leg III, left leg IV). The anterior pivot articulation in the scorpion is very strong and is supported by a costa coxalis as in Fig. 10.1(b) for Amblypygi. Condyles from the trochanter on either side of the main articulation carry strong muscles. The posterior component of the pivot articulation is reduced to a mobile sclerite f. Thus stepping takes place without any direct promotor–remotor swing of the leg.

The contrasting stepping of a diplopod using an oblique axis in the transverse plane of the body giving promotor–remotor swing at the coxa–body joint is shown in Fig. 10.4(a)–(c), cf. 10.2(a)–(c). The coxa of *Limulus* swings similarly (Figs. 10.2(a), 10.4(f), (h)). The leg of *Limulus* beyond the coxa possesses two pivot joints followed by two hinge joints, marked by black spots (Fig. 10.4(f), (h)), the diagrams below showing the joint construction and movements. Such a leg of simple type could have been the forerunner of a scorpion leg, except for the fusion of patella and tibia, which in *Limulus* strengthens the leg beyond the knee, as suits a burrowing habit. The adductor–abductor coxal movement used in chewing is shown in Fig. 10.4(g) and takes place at right angles to the walking movement (see Chapter 3, §3A).

(ii) *Leg mechanism of spiders*. In spiders the coxa retains some of the primitive promotor–remotor swing on the body about a close mid-ventral union between the coxal rim and the sternite. There is no dorsal articulation, as seen in *Limulus*, and the coxa can also rock slightly on its ventral articulation. The levator–depressor movement of the leg takes place at the horizontal trochanter–femur dicondylic pivot articulation, Fig. 10.2(e), not at the coxa trochanter joint as in scorpions. The coxa–trochanter joint is constructed essentially as in scorpions with a strong anterior pivot articulation; the posterior side is free, the articulation being represented by sclerite *f*. The joint permits a little rocking but no conspicuous levator–depressor movements. The trochanter of spiders is very short and its muscles and distal joint are considerably modified in providing a wide range of strong levator–depressor movements; mid-ventrally from the trochanter a sclerotized lip from the ventral distal margin projects into the femur, instead of the normal podomere overlap in the reverse order (Fig. 10.2(e), and Manton 1958*a*, p. 171).

In the spider a further contribution to leg rocking is provided at the patella–tibia joint (Fig. 10.5(a)). Here there is little difference in podomere diameters but considerable elaboration in shape at the margins. A dorsal hinge firmly unites these podomeres. The proximal postero-ventral edge of the tibia can be pulled up into a 'compression zone Y' of flexible cuticle on the posterior face of the patella. This small movement rocks the ventral face of the leg a little posteriorly during the backstroke and can assist the extension of the tibia–tarsus joint lacking extensor muscles. The compression zone Y is larger on leg II than on leg IV and probably the rocking here is greater. The patella–tibia joint is very strong and the lyriform sensilla in the cuticle are doubtless of importance in maintenance of stability by their response to tensions building up in the cuticle so near to the knee where leg flexures can be maximal (Fig. 10.5(a)).

The counter-rocking at the tarsus 1–tarsus 2 joint of the spider is achieved by different morphology from that of the scorpion and cannot be shown adequately on Fig. 10.2(e).

The Arachnida excel in cuticular specializations within their arthrodial membranes in a manner not found among the Myriapoda and Hexapoda. The mobile, bendable sclerite *f* of the coxa–trochanter joint occurs somewhat similarly in scorpions and spiders and facilitates the rocking movement of the coxa–trochanter joint.

Other sclerites lie at joints producing wide angles of movement; some simple ones are seen at the trochanter–femur joint of Amblypygi (Fig. 10.1(b)) situated on the anterior side of the joint. A more elaborate arcuate sclerite occurs at the knee of spiders where an extreme range of movement takes place at the adjacent podomeres which are very much emarginated ventrally, so exposing a large area of arthrodial membrane when the leg is outstretched (Fig. 10.5). The range of flexure at the femur–patella joint is shown in

Description of plates

The black or white spots alongside leg tips on some of the photographs indicate limb tips in contact with the ground. Short black or white lines alongside some photographs indicate legs off the ground and performing the forward swing. The relative durations of forward and backward strokes of the limbs are indicated respectively by the numbers within brackets whose sum is 10. The phase difference between successive legs is given as that proportion of a pace by which leg $n+1$ is in advance of leg n. All the photographs depict live animals, usually performing activities of evolutionary significance to the taxa in which they belong, except for the X-ray of one fossil animal.

PLATE 1. (a) *Cheloniellon calmani* Broili, Devonian. X-ray of the anterior end in ventral view; antennae 1 are marked A on either side: antennae 2 lie posterior to antennae 1, the base only being within the field on the left; the labrum lies between. The cusped gnathobases of the anterior trunk limbs lie on either side of the middle line. X-ray by Stürmer, personal loan. For reconstruction of entire animal see Fig. 1.8. (b–e) Polychaeta walking in sea water; (b, c) in dorsal view; (d, e) in ventral view through a glass-bottomed tank. (b) *Hermonia hystrix* (Savigny) 43 mm, walking fast on smoked paper held down by weights showing the track left by the parapodial setae which hold the animal clear of the substratum. Each transverse row of marks is made by every fourth parapodium, i.e. one group consists of the 'footprints' of parapodia 1, 5, 9, 13, 17, etc., and is followed by the next posterior group consisting of the footprints of parapodia 2, 6, 10, 14, 18, etc. The distance between successive transverse rows is the distance between the footfalls of two successive legs, as seen in a ventral view. Speed 31·6 mm/s, pace duration 0·9 sec, and stride length 20 mm. The resemblance of the track to that of *Lothobius*, using a quite different gait, Fig. 7.6(d), has no phylogenetic significance. (c) *Neanthes fucata* (Savigny). Part of the body of the worm crawling on the substratum. There are only three segments to each metachronal wave. (d) *Alentia gelatinosa* (Sars) 55 mm. Ventral view walking slowly on glass. (e) The same, walking faster. In both (d) and (e) there are 13–14 segments to a wave of metachronal movements, two or three successive parapodia being in contact with the ground, marked by dots, gait (8·3:1·7) approx., phase difference 0·08. Body undulations caused by trunk musculature are greater in (e) than in (d) giving larger effective angular movements of the parapodia, greater distances between successive footfalls, longer strides, and greater speeds, but the gait characteristics are little changed. (f) *Scolopendra morsitans*, 88 mm, running fast by gait (7·8:2·2), phase difference between successive legs 0·09. There are 9 legs approx. per metachronal wave, the propulsive legs being in the concavities of the body, converging on to almost the same footprints (see Fig. 7.6(b) for a track made by almost the same gait), the forwardly moving legs off the ground being on the convexities of the body, cf. the opposite in (d) and (e).

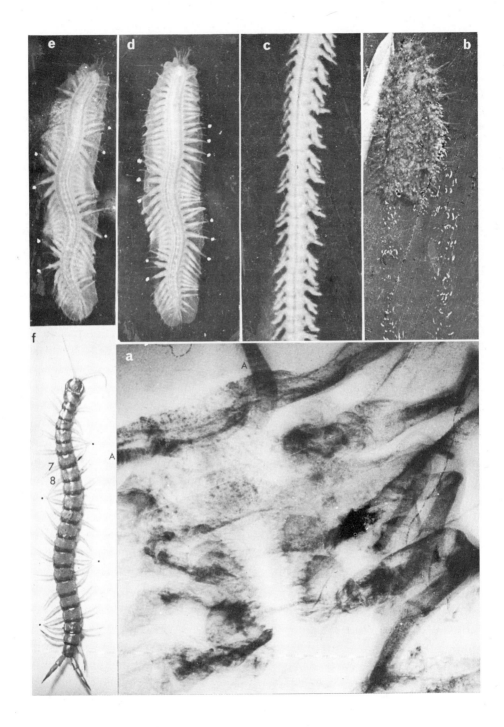

PLATE 2. (a) The branchiopod crustacean *Triops cancriformis* swimming near the bottom and displaying the ventral surface. The antenna 1 and long biramous antenna 2 with the median labrum are plainly seen. The mandible and maxillae are too close to be distinguished, but the trunk limbs with their large endite lobes are clear (see Fig. 4.16). The carapace, clearly seen on the left, lies over the anterior and middle parts of the body. The caudal furca from the telson extends beyond the field. (b) The branchiopod crustacean *Sida crystallina* (Cladorcera) in side view. The long biramous swimming antennae 2 project far out of the field. The short antennae 1 are seen below the black compound eye, across the tip of the rostrum. Black eggs lie in the ovary and the empty 'brood pouch' between dorsal carapace and body wall is clear. The ventro-lateral margins of the carapace below the limbs and extending to the limbless abdomen and caudal fork are transparent but just visible. Thoracic legs 3–6 show enlarging interlimb spaces, closed distally by the backwardly curved limb tips (the suction phase of filtering). Legs 1 and 2 have obliterated their interlimb space which is open distally. Some stalked epizoic protozoa are attached to the head dorsally. The abdomen can be flexed forwards within the carapace. (c) The barnacle *Balanus* sp. Crustacea Cirrepedia. Two specimens in posterior view are 'fishing' with outstretched casting net formed from the elongated thoracic limbs (see Fig. 4.13). The opercular valves are open to allow the spread of the 'net'. A massive base of the barnacle is formed by a pallisade of fixed calcareous plates cemented to a rock. In the foreground are two smaller barnacles with the opercular valves almost closed, the fishing limbs being withdrawn. The fine setules on the cirri forming the casting net are scarcely visible.

PLATE 3. (a) An ostracod crustacean, *Cypria ophthalmica* 0·5 mm, living, in side view, anterior end to the right. Antenna 1 and antenna 2 project forwards from between the two valves of the shell (carapace) and other limbs project ventrally. The pale central mark is the insertion on the shell valve of the transverse adductor muscle; the dark mass is the viscera and the black spot above and to the right is the median eye. (b) A copepod crustacean *Cyclops vicinus* 2 mm in dorsal view, living female, showing long antenna 1, median eye, cephalothorax followed by free segments of thorax, and narrow abdomen ending in a telson bearing a caudal fork; a pair of egg masses are attached to the first abdominal segment. (c) A copepod crustacean *Cyclops viridis* 3 mm in lateral view. Antenna 1 is upwardly directed and foreshortened; antenna 2 projects downwards, the mouth parts, maxilliped (thorax 1 limb), and first swimming limbs project antero-ventrally; the remaining three pairs of swimming limbs project ventrally and the minute 6th thoracic limbs are clearly visible. The cephalothorax incorporates two thoracic segments; the limbless abdomen ends in a telson and caudal fork, the branches appearing superimposed. (d, d1) The isopod crustacean *Armadillidium* sp. 15 mm almost tightly enrolled and walking respectively. Note that the head forms part of the smooth sphere and that thoracic tergite 1 and 2 are fused and bear pointed postero-lateral terminations which complete the sphere when the animal is enrolled. (e, e1) The millipede (Uniramia, Diplopoda) *Glomeris marginata* 14 mm, tightly enrolled and walking respectively, showing remarkable convergent similarity with *Armadillidium*, but note that the head and collum tergite are entirely covered by the large telson on enrolment, the two following trunk tergites being fused, so forming a pushing shield during burrowing, and completing the enrolled sphere with evenly curved margins, as shown on the upper and middle part of (e) cf. (d). (f) The syncaridan crustacean *Anaspides tasmaniae* 38 mm, progressing slowly to the right over a piece of white rock in a pool on the summit of Mount Wellington, Tasmania. Antennae 1 and antennae 2 project antero-laterally, the flat short exopod of antenna 2, used in swimming, being visible. The following thoracic legs 1–6 are not clearly distinguishable, but some gill lamellae (excites) are visible on the animal's right side (see transverse section in Fig. 4.5(a)). Thoracic legs 7 and 8 are backwardly flexed, ventral to the pleopods. All 6 abdominal limbs (pleopods) are visible; pleopods 1–3 are performing the backstroke and pleopods 4–5 are moving forwards, the series in metachromal rhythm, contributing to swimming (see Ch. 2, Figs. 2.4, 2.8, 2.10); the 6th pair of abdominal limbs are outspread beside the telson forming the tail fan which, with the anterior exopods of antennae 2, control direction of swimming. (g) *Phthisica marina* 10 mm a peracaridan amphipod, Crustacea Caprellidae, holding on to a hydroid stem with the 7th pair of thoracic legs. Note the large tarsal claws on the free 8th pair of thoracic legs and their resemblance to the tarsal claws of the pycnogonid *Nymphon* and of *Palaeoisopus* in Figs. 1.16, 6.6. All limbs are visible; antennae 1 and 2 project upwards above the paired eye; thoracic limbs 1 form fused maxillipeds just projecting in front of the head; thoracic legs 2, 3 are subchelate, their segments of origin being fused; thoracic legs 4–6 are slender and some bear gill lamellae (exites); the abdomen, formed of coalesced limbless segments, is the knob at the base of thoracic legs 8. (h) A portunid swimming crab *Macropipus puber*, Crustacea Decapoda, carapace width 50 mm, swimming freely in sea water. The 4 thoracic legs posterior to the chelipeds (the walking legs) are differentiated into walking legs 1–3, sharp tipped and outspread during swimming, and the 4th pair of walking legs which are flattened and used in swimming, much as two twin screws at the back of a boat. At the moment shown they are turned forwards and upwards over the other thoracic legs. Contrast with the bottom-crawling crab in (i) (see Hartnoll 1971). (i) The shore crab *Carcinus maenas*, Crustacea, Decapoda, carapace width 60 mm, running to the right. There is no direct promotor–remotor movement as in most surface runners, see Figs. 2.4, 2.5, 5.2, but a flexor–extensor movement whereby successive legs cause no mechanical interference. The photo is one frame from a cinematograph series. Movements of the right legs: walking leg 1 behind the cheliped is at the end of the propulsive stroke and about to be raised and extended; walking leg 2 is just being put on the ground to start the flexure–propulsive stroke; walking leg 3 is propulsive and pushing on the ground, but near the end of its propulsive phase; walking leg 4 has just been put on the ground and is starting the flexure propulsive stroke. The extensor movements of the left legs are propulsive when the crab is running to the right.

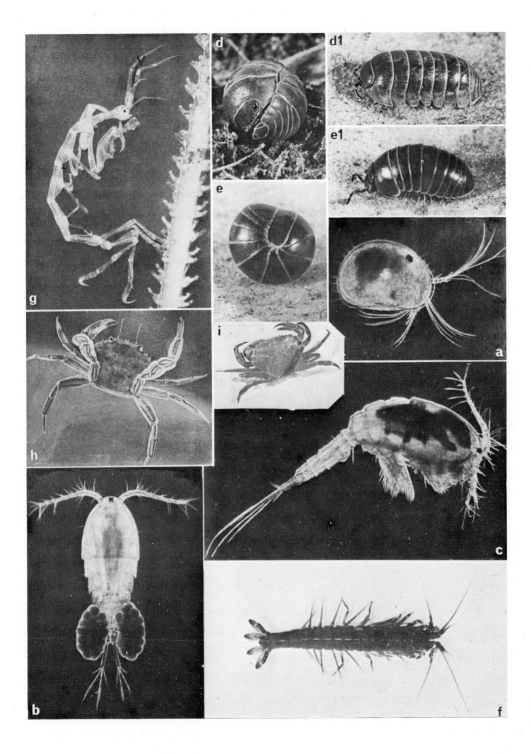

PLATE 4. Uniramia. (a, b) *Peripatopsis moseleyi* The same individual at different body extensions, (a) walking and (b) resting. (c) *Peripatus novaezealandiae* with fewer, larger legs than in *P. moseleyi*. All the round black holes in a piece of card are passable to this animal: the larger holes were passed through rapidly but the small hole took 20 minutes to pass. The incentive for the struggle was attractive moisture on the underside of the card and dry air on the top side. (d, e) The symphylan *Scutigerella immaculata* 4 mm (d) showing trunk flexibility in crawling round a plant stem and (e) walking fast on a flat surface, the antennae rapidly moving and sensing the path. The extra tergites lie over legs 4, 6, and 8. (f) The lysiopetaloid nematophoran diplopod *Callipus longobardius* walking fast towards the top of the page and viewed from below through coverglass. Only the middle region of the body is in the field and a boot has been fixed to a left leg. The sliding sternites are seen, note the distance apart of the bases of the propulsive and forwardly moving legs. (g) *Polyzonium germanicum*, Diplopoda Colobognatha 11 mm, a half-cyclindrical diplopod walking, the legs being entirely covered by the tergal arch. (h) *Craspedosoma rawlinsi* Diplopoda Nematophora 18 mm harnessed to a sledge, gait (3·3:7·7) phase difference 0·04, 26 legs per wave. In normal free fast walking 4–7 legs are off the ground and 4–6 legs on the ground in each metachronal wave and there is no humping as shown here in side view. The white material above is adhesive fixing the harness from the sledge to the posterior tergites. (i) *Polydesmus angustus* Diplopoda Polydesmoidea 22 mm running fast, in side view, gait (6·6:3·4), phase difference 0·1, 9–10 legs per wave, 3–4 propulsive, 4–7 legs off the ground in each metachronal wave. The body is held very close to the ground and the legs flex upwards below keels. (j) *Scutigera coleoptrata* Chilopoda Scutigeromorpha 22 mm, running fast in dorsal view. The steep rise of the legs at the knee makes focus of the entire leg impossible. Legs cross over one another on the forward swing but never in the propulsive phase. Body undulation is almost entirely controlled and much less than in Pl. 1f. Gait (6·3:3·7) approx., phase difference 0·135 speed 420 mm/s, stride 33 mm. Track shown in Fig. 7.5(e). (k) The same, resting, in side view, showing the typical hanging stance of long-legged arthropods, the long antennae capable of sensing a path wide enough to take the leg span and the long antenniform fifteenth pair of legs behind. (l) First instar of *Pauropus gracilis* 0·21 mm, gait (4·0:6·0), phase difference 0·42. (m) *Pauropus gracilis* adult 0·48 mm, walking fast by gait (7·5:2·5) phase difference 0·75. (n) *Craterostigmus tasmanianus* first instar, hatching at 4·4 mm and now 5·5 mm, just starting to walk at a little over 2 days of age. The gait is a slow scolopendromorph gait of near (5·0:5·0), phase difference 0·66, executed with irregularity. (o) A iuligorm diplopod *Blaniulus guttulatus* in the enrolled tight spiral position with legs and antennae protected.

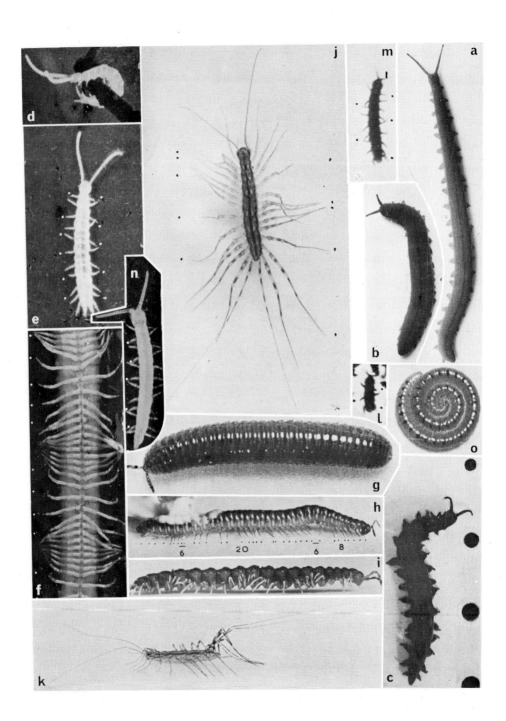

PLATE 5. Chilopoda. (a, b) The geophilomorph centipede *Haplophilus subterraneus*; (a) the anterior end of the body extended (70 mm) and walking forwards towards the top of the page. Gait irregular but near (5·0:5·0), phase difference 0·6; the metachronal waves comprise usually 3 but up to 5 legs and the transverse co-ordination is variable, speed 8 mm/s approx. (b) The same animal longitudinally contracted. The segments when short and wide exert a burrowing heave against the soil. The anterior end of the body is starting to elongate and walk forwards. (c) A large geophilomorph centipede 160 mm long. The animal is progressing forwards under a low glass bridge extending from o anteriorly to o just off the field posteriorly. The segments in front of the bridge (top of page) are advancing slowly; the short, wide segments below the anterior part of the bridge are stationary to the ground and are exerting an upward thrust against the bridge; the longer, narrower segments below the posterior part of the bridge are moving forwards and are fully extended showing long intercalary tergites at the lower edge of the photograph; these segments shorten as they move forwards to join the zone of maximum pushing where the intercalary tergites are invisible and the main tergites, covering the intercalary tergites, are shortened by corrugation (see Figs. 8.1, 8.2, 8.9). As the animal moves forwards the zone of thickening under the bridge moves backwards, to be replaced by another thickening forming under the bridge, the segments from the anterior end of the original thickening walk forwards free from the bridge. In this manner the animal exerts a series of thrusts against the soil, widening a crevice into which the body passes. The anterior legs, holding the animal while it heaves, are stouter than the more posterior legs, cf. the anterior legs, visible clearly on the right, and the posterior legs on the left. (d) The scolopendromorph centipede *Cryptops hortensis* 19 mm, gait (6·0:4·0), phase difference 0·80, see Fig. 6.13. This is the fastest gait recorded for this animal; there are no trunk undulations, which here are entirely controlled with about 4 successive legs swinging forwards between each propulsive leg, cf. the much faster gait patterns shown by Pl. 1f and Fig. 7.1(c), (d) where the wider trunk and better developed anti-undulation mechanisms permit the use of fast gait patterns without crippling undulations. (e) Middle body region of *Lithobius forficatus* 25 mm running fast, showing long and short tergites, the consecutive long tergites being over legs 7 and 8. There is rigidity at the anterior ends of the long tergites, marked X, the body flexures being restricted to the anterior ends of the short tergites. (f) The diplopod *Polydesmus angustus*, 20 mm, male (the female has shorter slender legs), running fast by gait (5·5:4·5), phase difference 0·12, the keels from the tergites being present all along the body but clearly seen anteriorly. (g) *Lithobius variegatus* 25 mm running slowly by gait near (5·5:4·5), phase difference 0·16. The upper slit in the card is freely passable to the animal, while the lower one can be passed only with difficulty by dorso-ventral flattening, but this flattening is much less than is possible to the geophilomorph in (a) which can pass the slit (i). Note the great width of the head and poison claws. The 15th legs are not used for walking. (h) *Lithobius forficatus* 4 mm second instar with 8 leg pairs, gait (5·0:5·0), phase difference 0·25, speed 26 mm/s approx., stride 3·2 mm. (i) Slit in a card which is passable to the specimen of *Haplophilus subterraneus* shown in (a) and (b).

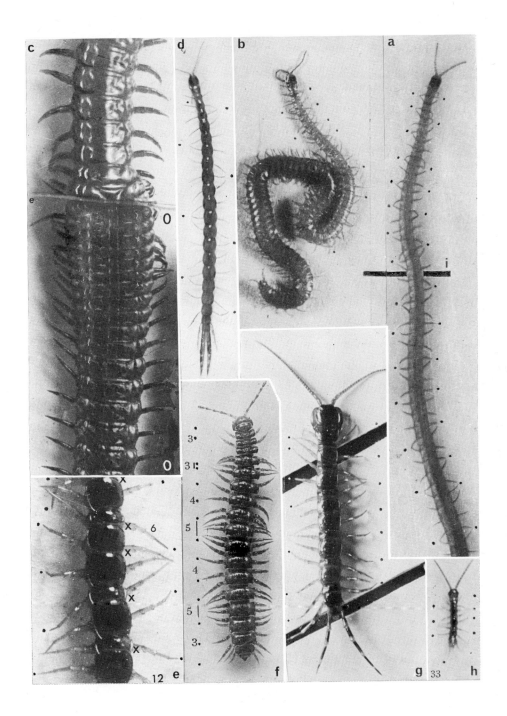

PLATE 6. (a) The collembolan *Tomocerus longicornis* standing, in side view; the springing organ is held up in the ventral abdominal groove, and one antenna shows the extreme terminal mobility. (b) Transverse section of the large hollow spines of the diplopod *Polyxenus lagurus* seen in (j) situated in the lateral spine rosettes. (c) The collembolan *Neanura muscorum* 2 mm displaying the flexibility of body which is absent in *Tomocerus*. (d) The collembolan *Tomocerus longicornis* 5 mm walking over smoked paper by its fastest gait (4·4:5·6), phase difference 0·44, see Fig. 7.3(a), (d), (g). Left leg 1 and right leg 3 are off the ground. The change in successive footfalls is simultaneous and therefore the track is formed by groups of 3 footprints at the same transverse level. The tip of left leg 3 lies in between the footprints of legs 1 and 2. The change in support from one leg to the next is almost simultaneous, cf (i) where right legs 1 and 2 are on the ground simultaneously. (e) The dipluran *Campodea staphylina* 4·5 mm, one antenna with a broken tip, running on smoked paper by gait (3·5:6·5), phase difference 0·4 approx., speed 26 mm/s, pace duration 0·13 s. The footprint groups are forwardly staggered (Fig. 7.12(b)); each leg oversteps the one in front, the differential leg length bringing the footfalls of legs 2 and 3 outside and in front of those of legs 1; left leg 2 and right legs 1 and 3 are on the ground (for gait diagram and gait still see Manton (1972) Fig. 5(a), (h)). (f) The machilid thysanuran *Petrobius brevistylis* 10 mm running over smoked paper, the jumping gait with legs of each pair in similar phase, the footprints on the two sides of the track being level and not alternate as in (d, e, g). The moment in time is unstable, with only legs 1 in contact with the ground; a double claw mark is seen at the tip of right leg 1. The long exposure time (1/60 s), gives a blurred image of legs 2 as they rapidly swing forwards, leaving a single claw mark, showing on the right as a white spot immediately in front of the double claw mark of leg 1 and on the left by a white spot in front and lateral to the footmark by leg 1. Legs 3 have vacated their footprints lateral to those of legs 1 and 2 and are nearing their forward position when they will be put on the ground lateral to the footprints of legs 1 and 2 in front. Legs 3 are not blurred and must be moving forwards more slowly than legs 2. The gait is (5·0:5·0) approx., phase difference 0·33; gait diagram in Fig. 7.13. (g) *Thermobia domestica*, a lepismatid thysanuran, 5 mm, running over smoked paper, exposure time 1/500 s. Paired legs are in opposite phase and the groups of foot prints on opposite sides of the track are staggered and not level as in (f). The gait is (6·7:3·3) approx. Left leg 1 is on the ground but invisible, left leg 2 is moving forwards off the ground, and left leg 3 is about to be put down in front and outside the footprint of leg 2; right legs 1 and 3 are off the ground and right leg 2 is propulsive. The moment is unstable, with only 2 legs in contact with the ground, as in (f), but the paired legs are in opposite, and not in the same, phase relationship. (h) A dipluran japygid sp. from the Solomon Islands, 9·2 mm to show telescopic antenna; the left is 2·4 mm and the right 3·5 mm long with no foreshortening. The thoracic flexibility is well shown, cf. the rigid thorax in (a, d, g, i). (i) Pterygote beetle *Cylindronotus laevioctostriatus* 8·8 mm walking by gait (3·5:6·5), phase difference 0·4, stride length 4·4 mm. Only left leg 3 is off the ground, and right legs 1 and 2 illustrate the long time interval k, see text. Shadows are heavy. (j) The diplopod *Polyxenus lagurus* 3·2 mm in dorsal view; the lateral spine rosettes give cover for the legs. (k) The proturan *Acerentomon nemorale* 1·6 mm, gait (2·0:8·0) approx., phase difference 0·25. Only legs 2 and 3 are used for walking. (l) The diplopod *Polyxenus lagurus* 1·6 mm in lateral view to show the converging groups of propulsive legs with their tips on the ground; legs probably off the ground are 1, 5, 7, 9, 10 but are not easily seen. Gait near (4·0:6·0), phase difference 0·8, stride length 106 mm. (m) The machilid thysanuran *Petrobius brevistylis* 12 mm progressing along a narrow horizontal ledge marked by black lines 2 cm apart. The animal has paused between two small jumps; limb tips 1 and 3 are marked and limb 2 is near 1. The dorsal concavity of the abdomen is increased in the prejump position for the high escape jumps (Fig. 7.15(b)). (n) The diplopod *Glomeris marginata* 14 mm walking, cf. enrolled position in Pl. 3(e).

PLATE 7. (a) The scorpion *Euscorpius carpathicus* walking over smoked paper by gait (2·5:7·5), phase difference 0·45. The animals's track has been an even curve and the groups of footprints on the two sides do not exactly alternate, those on the left being further forward than they would be if the paired phase difference was exactly 0·5. (b) The scorpion *Buthus australis* walking by a slightly irregular gait; but the track shows a phase difference of 0·5 between paired legs and lines of minute dots caused by vibrations of the sensory pectines against the ground. Note the different amounts of the anterior face of each leg visible in the different positions during the pace, indicating the leg rocking; see text. (c) The iuliform diplopod *Ophistreptus guineensis* 175 mm, pushing, but resistance just removed, showing the pushing position of the head and collum, the ventral surface close to the substratum and many propulsive legs in each metachronal wave. (d) The same running freely and fast by gait (5·8:4·2), phase difference 0·08, the number of propulsive legs on the ground and the number in the groups of legs off the ground being marked. Only 5–6 propulsive legs are present in each metachronal wave, cf. (c, e). The ventral surface is held well up off the ground permitting long strides. (e) part of the body of a large iuliform diplopod 295 mm long performing a slow gait, (3·0:7·0), phase difference 0·03 approx., giving a strong push at the head end and long metachronal waves, the two shown possessing 22 consecutive propulsive legs. (f) The colobognathan diplopod *Siphonophora hartii*, walking, ventral view through glass. The longitudinal co-ordination between the legs is stronger than the transverse one. All legs are covered laterally by the keels. The sliding of the sternites can be seen, giving more closely placed coxae on the forward swing of the legs, but it is less extensive than in the millipede *Callipus*, Pl. 4(f). (g) The colobognathan diplopod *Dolistenus savii* 34 mm, walking fast by gait (6·7:3·3) approx. The legs are not entirely covered by the keels during fast walking and the groups of legs performing the forward swing off the ground project beyond the trunk. (h) The diplopod *Polydesmus angustus* first instar, 1·16 mm, gait (3·5:6·5), phase difference 0·4. Only three pairs of legs being present, they have to be used in opposite phase, unlike all later stages where paired legs are moved in the same phase. (i) The centipede *Craterostigmus tasmanianus* 46 mm running slowly by a typical scolopendromorph gait (5·5:4·5), phase difference 0·75 (Fig. 6.13) with converging propulsive legs (cf. *Lithobius* with diverging propulsive legs, Pl. 5(g); the arrows mark the positions of the extra tergite joints dividing the long tergites on segments 3, 5, 7, 8, 10, and 12 which contribute to great secondary trunk flexibility.

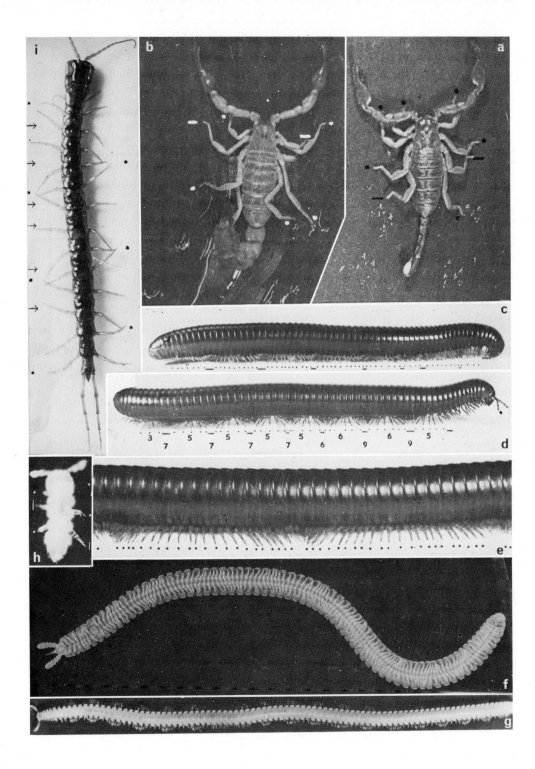

PLATE 8. (a) Arachnida, Pedipalpi, an amblypygid sp. from Jamaica, 10 mm. Stationary but feeling under a ledge with both legs I. (b) The same in posterior view, showing the way in which leg I can swing over the other legs; the knee of left leg IV is folded well over the body; the knee of right leg III is strongly flexed, as used in sideways running or crawling, cf. the outstretched position of leg III of the left. (c) *Galeodes arabs* 25 mm. Left leg I and the tarsus of right leg IV are broken. The animal is running over smoked paper, track not included. (d) The spider *Trochosa ruricola* running by gait (3·7:6·3), phase difference 0·5 approx., speed 76 mm/s, see Fig. 10.7(b). The track, shown by reconstruction from the cinematograph film, reveals two vacated footprints by legs IV, black rings; left leg IV is on its footprint and right leg IV will be placed upon the ground at the ring just in front of leg tip III. (e) The mite *Allothrombium fulginosium* 3 mm walking by irregular gait near (2·5:7·5). (f) The tick *Ixodes hexagonus* 12 mm, female, from hedgehog, walking by a slow irregular gait. (g) British harvester *Opiliones* sp. 10 mm running on smoked paper by gait near (4·0:6·0), phase difference 0·45. (h) The same in side view, walking, showing the hanging stance. (i) Pseudoscorpion *Chelifer* sp. 2·4 mm running slowly forwards in gait (5·0:5·0), phase difference 0·45–0·5. (j) The same running rapidly backwards, the moment chosen showing the limbs in much the same disposition as in (i). The arrows in (i) and (j) show the direction of running. (k) *Anisopus* sp. 9·8 mm larva of dipteran fly, Pterygota. Moving antero-posteriorly by undulations passing along the smooth limbless body, so pressing against viscid cow-pat in which the animal lives, giving forward locomotion. The head is the most anterior unit followed by three thoracic segments marked differently from the rest.

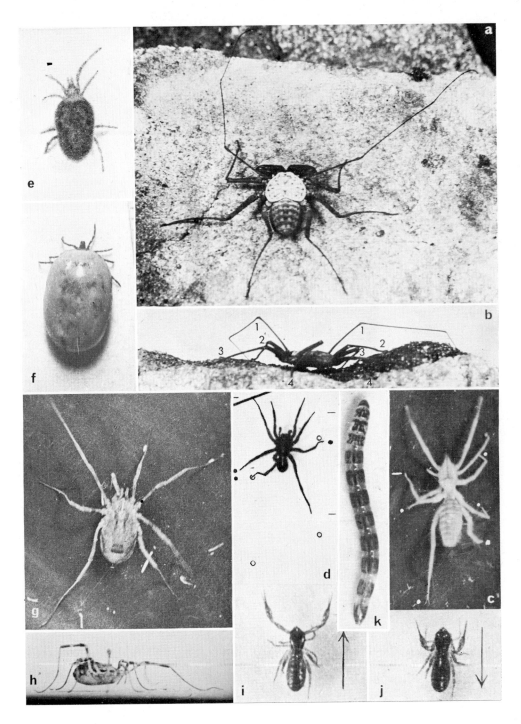

Fig. 10.5(a), (b). A very strong transverse hinge articulation lies across the dorsal face of this joint (Fig. 10.5(c)). The arcuate sclerite forms a stirrup in the ventral arthrodial membrane of the joint and is movably attached to the patella in a marginal notch on either side (Fig. 10.5(a)). Lesser bands of sclerotization also lie in the arthrodial membrane as shown. When the leg flexes the arthrodial membrane at this joint folds up neatly like a concertina.

The main function of the arcuate sclerite concerns the flexor muscles at the femur–patella joint. Some of these arise normally from the ventral rim of the patella (Fig. 10.5(a)), but an additional bulky mass of flexor muscles arises from the arcuate sclerite itself. At strong flexure of the knee, muscles 20b pull the ventral lip of the patella through the space surrounded by the stirrup-like arcuate sclerite. The additional flexor muscles pull on the arcuate sclerite itself, whose passive swinging maintains the tonofibrils of these muscles always at the same angle to the sclerite, a mechanical advantage for the musculature (similarly seen in the swinging origin of the lateral longitudinal muscle 1 of *Tomocerus* which folds up the springing organ after use; see Manton 1972, Fig. 17(a), (c) muscle 1).

The joint elaborations of spiders, together with the musculature and the sense organs within the adjacent cuticle, represent a higher degree of organization than that possessed by any known myriapod or hexapod joint.

(iii) *Leg mechanism of an amblypygid (Pedipalpi)*. In the Amblypygi the coxa–trochanter joint is somewhat as in spiders and scorpions but it provides the only site for levator–depressor movement and probably executes very little rocking. Sclerite f is massive and Y-shaped, the arms uniting with the margin of the trochanter as shown in Figs. 10.1, 10.2(b), sclerite f. The sclerite is complex in form and is flexible to some extent. The joint appears to be a derivative of a rocking joint such as possessed by spiders and scorpions, but enhancing the levator–depressor movements while spiders have eliminated them at this joint.

The slightly oblique position of the axis of levator depressor movements at the trochanter–femur joint in scorpions is in the Amblypygi increased to such an extent as to form an almost vertical axis giving a promotor–remotor swing. The leg beyond this level is turned by a similar amount

Locomotory habits

as far as tarsus 1 (between the black spots on Fig. 10.1(a)); tarsus 1–tarsus 2 articulation is at right angles to that between tibia and tarsus 1, thereby placing the tarsal claws squarely on the substratum. The patella shows some resemblance in shape to that of spiders, but no movement appears to be possible at the patella–tibia joint. The patella is very strongly hinged to the femur; both podomeres are much emarginated ventrally; and a tuft of sclerotization passes into the arthrodial membrane just proximal to a patella notch on the anterior face (Fig. 10.1(a)). There is no notch on the posterior face. It seems that any ancestral mobility of the patella–tibia joint has been sacrificed in providing sufficient strength at the knee which is capable of very acute flexure, a movement needed in the crab-like sideways walking; see the flexed knee of right leg III in Pl. 8(b). A long series of tarsal segments separated by fairly free pivot articulations gives a good grip on the substratum, useful in climbing as in other arthropods such as the plantigrade Pterygota and *Scutigera*, where the intertarsal articulations are often dorsal hinges (Fig. 5.14, Pl. 6(i)).

Thus the structure and movements of the amblypygid legs suggest a derivation from one in which a rocking mechanism has been secondarily almost or entirely abandoned. With the formation of posteriorly directed knees on legs II–IV and a secondary promotor–remotor swing at the trochanter–femur joint, patella-rocking has given place to greater stability of the podomere, an assistance towards acute flexure at the knees.

(iv) *Leg mechanism of Solifugae.* The Solifugae (e.g. *Galeodes*) are said to be the fleetest of all arachnids and their legs are the most remarkable. The proximal segmentation is more complex than in other orders. Three and not one podomere lie between the trochanter and knee and the Solifugae lack a patella (Fig. 10.3). The large fixed coxa amply supports the trochanter; the joint between them is vertical and provides the main stepping promotor–remotor swing, as occurs elsewhere among arachnids only in the Acari. Articulations, barely formed in the Acari, are very well developed in the Solifugae. The basal three podomeres beyond the coxa are stout and rise steeply against the flanks of the body owing to the short dorsal face of trochanter 1, as shown in Fig. 10.3. The position of the coxa–trochanter axis of swing superficially resembles the coxa–body axis giving the same movement in Chilopoda, Symphyla, Pauropoda, and Diplura, but

Locomotory habits

the axis in Solifugae is much more strongly supported. The ventral coxa–trochanter articulation in *Galeodes* is localized and very strong and the dorsal one is more spread out along the coxal rim.

The divided trochanter gives two different movements, a levator–depressor swing at the trochanter 1–trochanter 2 joint places this movement close to the joint giving the promotor–remotor swing, as in *Limulus*, Acari, Myriapoda, and Hexapoda (Figs. 5.15, 10.2). The following trochanter 2–prefemur joint is complex and provides a rock. The anterior face of trochanter 2 is long while its posterior face is extremely short, see right and left sides of Fig. 10.3. The anterior pivot articulation at the trochanter 2–prefemur joint lies so far distal to the posterior articulation of this pivot that the axis of movement comes to be longitudinal on the leg, passing through a posterior articulation which lies against that of trochanter 1–trochater 2. A bulbous prefemur contributes to this most remarkable axis of movement. Counter-rocking is not particularly provided for. Tarsus 1–tarsus 2 joint is held well off the ground; the articulations of the horizontal pivot here are close together, there is a large overhang dorsally by tarsus 1, and sufficient counter-rocking probably occurs here and at the several intertarsal joints which follow. The axis of swing at the prefemur–postfemur joint is displaced (Figs. 10.2(g), 10.3) by the position of the hinge articulation which is not dorsal but slightly anterior in position. The prefemur–femur hinge articulation in the Scolopendromorpha and Lithobiomorpha (the fast-running Chilopoda) is similarly displaced (Figs. 5.3(c), 5.6(a)) and here assists the propulsive extension of the more distal hinge joints and the effectiveness of leg-rocking. It is probable that the position of the prefemur–postfemur articulation in Solifugae has some similar function in promoting speedy movement.

E. *Conclusions concerning leg mechanisms of arachnids*

Some tentative suggestions concerning the evolution of arachnid leg mechanisms emerge from the above data. The legs of *Limulus*, the walking malacostracan Crustacea, the Myriapoda, and the Hexapoda are similar in that a promotor–remotor swing of the leg is sited at the coxa–body junction, however variously this is contrived, and the main levator–depressor movement takes place at the next joint along the leg. It may be assumed that this arrangement of

Locomotory habits

joints is mechanically simple and suitable and it is likely that the ancestors of the Arachnida may have possessed such legs.

The fixation of the coxa on the body in most arachnids may have been a response to the need for stability occasioned by the evolution of long legs in marine ancestors, which could have thereby taken longer strides and contributed greater speeds to locomotion. This immobility may have already started in the Eurypterida. Størmer's (1973) reconstruction of *Parahughmilleria hefteri* (Fig. 10.6) shows the coxae on legs II–V to narrow towards the middle line, the gnathobases converging towards the mouth region, much as in *Limulus* (Fig. 1.9). It is probable that these gnathobases may have been used for feeding as in *Limulus* today (Chapter 3, Fig. 2.6). But the coxae of leg VI are enormous and flat and although they possess a narrow, cusped gnathobasic lobe curving forwards round the metastoma, their shape is suggestive of providing a very solid base housing limb muscles

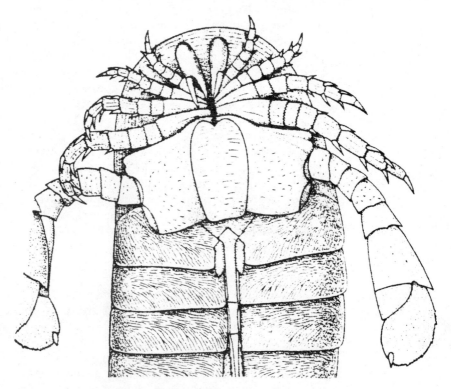

FIG. 10.6. *Parahughmilleria hefteri*, Lower Devonian. Ventral view of the anterior end. (After Størmer 1973.)

used in working the flattened paddle-like 6th limbs. A restriction of chewing to the pedipalps alone in arachnids could have enable legs III–V to become more effective locomotory organs, which then with legs VI gained stability by fixing the coxa on the body. The tracks of the eurypterid *Mixopterus* show shorter strides than those of a modern scorpion (cf. Fig. 1.12 with Pl. 7(a), (b)), which may indicate less effective leg-rocking. A good rocking mechanism is probably incompatible with *Limulus*-like leg movements which probably were used by legs II–V.

The rocking mechanisms of legs of myriapods and hexapods are usually associated with assisting the extension during the propulsive backstroke of a series of hinge joints lacking extensor muscles. Rocking leg-joints of arachnids mainly serve to provide a good stride length, particularly as shown for the scorpion, where the coxa is immobile and there is no direct promotor–remotor swing. Arachnids do not usually possess a long series of distal hinge joints with flexor muscles only as in the Uniramia; but arachnid leg mechanisms and leg jointing are more elaborate than those of the Uniramia and rocking in the scorpion is probably very useful in extending the femur–patella joint, which lacks extensor muscles.

Coxal stability in the larger Crustacea is obtained by a massive and complex apodemal system, the endophragmal skeleton, which carries the extrinsic leg muscles. Smaller Crustacea rely on tendinous endoskeleton or endosternites for extrinsic muscle insertion as well as the tergal cuticle where it can be reached easily. Extrinsic leg muscles insert on the body wall in many groups. Tendinous endoskeleton is also used by Xiphosura, Chilopoda, and Collembola; apodemes alone in Diplopoda; apodemes and tendinous structures in Diplura; and so on. There is plenty of variety. The arachnid mode of obtaining stability differs from the solutions arrived at by other arthropods. It is noteworthy that the Xiphosura, in evolving a massive connective-tissue endosternite, show an alternative mode of gaining coxal stability to that of the Arachnida.

A plausible explanation for the unique arachnid coxa-body joint and the distal variety in leg joints may lie in adaptations to land life taking place at different times. Fig. 10.6 shows the eurypterid prosomal limb-bases to be very similar to those of *Limulus*, where a double movement is used: the promotor–remotor swing and the gnathobasic chewing occurring at right angles to the former. If the ancestral aquatic arachnid groups were similar in this respect, there may have been mechanical difficulties in deriving a coxa with one movement only on the four pairs of post-pedipalpal limbs. It may have been mechanically easier to fix the coxa on the body and evolve a joint giving a promotor–remotor swing elsewhere along the limbs, or even doing without this movement as in scorpions. The crustacean coxa–body junction provides one movement only and that is suitable for both aquatic and terrestrial life. It is possible that the variety in arachnid solutions to the problem of evolving a faster acting limb suitable for the land was due to independent transitions to the land, as is certain for scorpions and trigonotarbids.

No simple scheme of evolution within the Arachnida commends itself to workers in this field, where the great ordinal differences are appreciated (see below, §10). There is far more uniformity within the Myriapoda and Hexapoda although these groups show parallel and different evolutions at the coxa–body joint which produce the basic promotor–remotor stepping movement. The outstanding achievements of at least two of the arachnid orders mentioned above depend, among other things, on their leg mechanism. Bristowe (1947) estimated that spiders eat many times the number of insects consumed by birds. This success of spiders has been promoted by the delicacy, flexibility, and speed of their leg movements used in running, web-building, etc., as well as by other features. This expertise depends on the retention of some of the primitive coxal flexibility on the body, combined with more fully developed rocking mechanisms involving more joints than in other arachnids, and the extreme provision for knee flexure involving the arcuate sclerite and its muscles, the lyriform sensilla in the cuticle, etc. Secondly, the great speed of running in the Solifugae must surely stem from the fine articulations at the joints, namely the formation of the dorso-ventral axis of the promotor–remotor swing at the base of the movable part of the leg followed by the levator–depressor movements of the next joint and by the other remarkable podomere and joint features.

6. Arachnid gaits

The way in which the several arachnid leg mechanisms are combined in each order to form co-ordinated gaits is more

Locomotory habits

uniform and simple than might be expected. Arachnid gaits differ from those of the Uniramia. The principal characteristics of arachnid gaits and their differences from those of the Uniramia are given below. A more detailed account of arachnid gaits can be found in Manton (1973).

As in other arthropods, arachnid legs are well fanned out from their bases, so assisting the separation of the fields of movement of successive legs which is a mechanical advantage, particularly in animals with few pairs of walking legs (Chapter 5, §3, Figs. 5.2, 10.1, 10.7). The movements of successive legs can be very different from one another, although the basic features are the same in each animal.

There is more irregularity in the performance of gaits by arachnids than by Uniramia, but less so in faster running. There is strong unilateral co-ordination between successive legs, as shown by the regular sequence of grouped footprints forming each side of the track (Pls. 7(a), (b), 8(d)). Legs of a pair normally step in opposite phase, but the transverse coordination can easily change, as shown by tracks formed when the path is curved. Turning appears to present some difficulty to arachnids and the elongation of the stride on the outside of a curved path may not at once be accompanied by a corresponding shortening on the other side. Then the grouped footprints on the two sides no longer alternate, and they may be in any position, the paired legs stepping in alternate phase, in the same phase, or in any other relationship. In time the normal bilateral alternation is restored (Pl. 7(a), (b)).

Typical arachnid gaits are illustrated by the plotted leg positions from cinematograph films of a scorpion and a spider in Fig. 10.7 where movements of the left legs are shown by continuous lines and right legs by dotted lines, the thicker parts of the former indicating the propulsive backstroke. The scorpion using a slow pattern of gait (2·5 : 7·5) and the spider a faster one (3·7 : 6·3) show phase differences between successive legs to be much the same in the two gaits. Arachnid gait patterns are usually moderate to slow. No long series of gaits are practised, as by the Uniramia, and changes in speed are largely dependent on changes in pace duration. Common slow patterns of scorpion, mite, and tick gaits are about (2·5 : 7·5), but with a large phase difference between successive legs of 0·45–0·55, in contrast to the 0·2–0·33 used by hexapods for such gaits (Figs. 7.2, 7.9(c)). In the faster

Locomotory habits

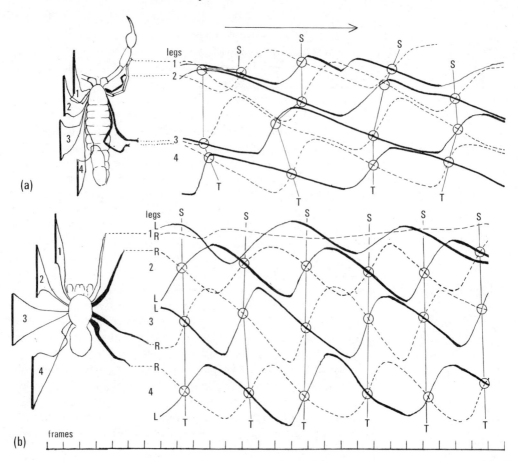

FIG. 10.7. Cinematograph record of (a) scorpion *Buthus australis* (L.) (Pl. 7(b)), using gait (2·5:7·5) approx., phase difference 0·55, speed 102 mm/s and (b) a spider *Trochosa ruricola* (Pl. 8(d)) using gait (3·7:6·3) approx., phase difference 0·5, speed 76 mm/s. Right leg 1 of the spider is idle. The lines S–T, drawn through the circles marking the moments in time when legs of a pair are half-way through the forward and backstrokes, are regular and almost vertical on (b), showing that the phase difference was almost exactly 0·5. In (a) the irregularity in stepping makes the lines S–T crooked. For further description see text. (From Manton 1973b.)

patterns used by spiders (Pl. 8(d)), *Galeodes*, and Opiliones (4·0:6·0) (Pl. 8(c), (g)) the phase difference between successive legs is much the same, not larger than in the slow-pattern arachnid gaits. The more restricted range of arachnid gait patterns is associated with values of phase difference which are quite different from those of hexapods, e.g. gait (2·7:7·5) used by Pterygota, *Campodea*, and *Petrobius* achieves even loading on the legs and footfalls at even intervals of time (Fig. 7.2(b), (g), (i)), but the same gait pattern in scorpions forms a very stable gait with a phase difference of 0·45–0·55 and a long time interval k during which two successive legs are on the ground simultaneously; see the periods of time where two heavy lines lie close together on Fig. 10.7, and compare with Figs. 7.8–10 where the time intervals k are of much shorter duration. The long time interval k in arachnids confers stability on the innately less stable stepping movements, Arachnids usually do not employ the neat ways of gaining stability by even footfall intervals and constant loading on the legs as seen in Uniramia (Chapter 7, §§5, 6C).

The fastest gait pattern recorded for arachnids is only (4·0:6·0) approximately, yet it yields 500 mm/s and more by *Galeodes* with a stride of 30 mm by a specimen 22 mm long. The hexapods can shorten the relative duration of their backstroke much further, even to a shorter duration than the forward stroke and consequent momentary instability, with only two supporting legs. Such fast patterns of gait in hexapods are dependent upon their very exact execution, an accomplishment not conspicuous in most arachnids. The gait (4·0: 6·0) at a phase difference of 0·5 is the best approach to a speedy gait which is possible to an arachnid.

Changes in the pace duration are of much more importance to arachnids in changing speed than are changes in gait pattern. Pace durations are always long in the slower movers, e.g. more than 1·3 s for the hedgehog-tick *Ixodes*. Minimal durations of 0·27 and 0·12 s respectively have been recorded at room temperature for the scorpions *Buthus australis* and *Euscorpius carpathicus*, 0·07 s for the spider *Trochosa ruricola*, and 0·06 s for *Galeodes arabs*, but the latter figure is certainly not minimal for this animal in its natural environment.

Metachronal waves of arachnid limb movement, as in hexapods, can appear to travel either forwards or backwards at different times in the same animal. Forwardly and backwardly directed metachronal waves of limb movement always exist and when their speeds of transmission are equal in either direction, then the wave appears stationary to the eye at a phase difference of 0·5. A wave appears to travel forwards at a phase difference of less than 0·5 and backwards at a phase difference of more than 0·5. The forward or backward transmission of the wave has no direct effect upon speeds of

Locomotory habits

running but the choice of phase difference controls the time interval k.

Both spider and scorpion change the direction of metachronal wave transference just as occurs in some myriapods and in *Campodea* (Chapter 7, §4A, Figs. 7.8(b), 7.10(d), lines r, and Manton 1972, Fig. 5). There is nothing particularly remarkable in this phenomenon, although incredulity has been expressed that any one animal can change the direction of its apparent metachronal wave. The arachnids show easy transfers of this kind. Fig. 10.7(a) shows such a transfer in a scorpion, comprehended most easily by the opposite slopes of line S–T between legs III and IV. When S is to the left of T the wave travels forward and the phase difference is less than 0.5 and when S is to the right of T the wave travels backwards and the phase difference is more than 0.5. When the line S–T is vertical the phase difference is exactly 0.5, the metachronal wave appears to be stationary, and every other leg is in the same phase. The stepping of the spider in Fig. 10.7(b) shows this condition almost exactly; other arachnids appear to approach this but for their irregularities in stepping. This use of a phase difference of approximately 0.5 results in legs I and III on one side of the body stepping in unison as do legs II and IV. Such a gait contrasts with the hexapods where it appears to be avoided, while it is favoured by the Arachnida.

The Arachnida show a number of locomotory features peculiar to certain orders. The pseudoscorpion *Chelifer* habitually walks slowly forwards with pedipalps outstretched at 1–2.5 mm/s or less and runs swiftly backwards with pedipalps folded (Pl. 8(i), (j)). The gaits are apparently the same in both the forward and backward locomotion, pattern about (5.0: 5.0) and phase difference 0.5. Only such a gait could reverse direction without further changes and this type of reversal of direction in running could only develop easily in an arachnid with a habitual tendency to a phase difference of about 0.5 and not in a hexapod which avoids this phase relationship.

Some arachnids are specialists at both fast running and at extremely slow movements barely detectable to an observer. The Amblypygi with their flattened body and lack of upward projection of long legs (Fig. 10.1, Pl. 8(a), (b)) are expert at climbing and at crevice penetration as well as running at great speed for short periods. Leg pair I has become purely sensory, long and very flexible, the coxa being situated dorsal to the level of the coxae of legs II–IV (Fig. 10.1). Thus leg I can feel under a ledge in front, to the side, or behind the animal by swinging over legs II–IV (Fig. 10.8(c), left). Exploration complete, the animal can very slowly move by gaits as slow as (1·8: 8·2), sometimes by crab-like sideways progression, towards the crevice or hide, and when sufficiently close, the imperceptible movements change to ones of extreme rapidity and in a flash the animal has vanished from sight.

7. Hexapodous stepping in Arachnida

The Amblypygi just mentioned are not the only arachnids to have become hexapodous. After the differentiation of a pair of pedipalps took place, ancestral arachnids can never have possessed more than four pairs of walking legs in contrast to the many pairs possessed by ancestral Uniramia, so the change to hexapody must have been less great for arachnids than for ancestral hexapods. Yet hexapody has been independently evolved several times in both groups. The use of only three pairs of legs for walking and running brings the same mechanical advantages to arachnids as to uniramians in an avoidance of overlap of the fields of movement of successive long legs, seen plainly in the amblypygid in Fig. 10.8(c), left, in the uniramian hexapods (Fig. 5.2), and in *Galeodes* which runs only on legs II–IV (Pl. 8(c)).

Varied and presumably secondary, sensory limbs have been evolved from legs II in some Opiliones and gamasid mites as well as from legs I in Pedipalpi and Solifugae (see above) and partially so in some other Acari. This sensory development imposes hexapody on all these examples.

Hexapody in arachnids is not always associated with the ability to run very fast, as in *Galeodes*. The Amblypygi are extremely fleet, but in their very slow movements they employ quite unique stepping as shown in Fig. 10.8. The pace duration is very long and the gait pattern slow (very long relative duration of the backstroke); legs II and IV are almost in similar phase, as is characteristic of arachnids in contrast to uniramian hexapods; and there is a very long time interval k between legs II and III but none at all between legs III and IV. The result, surprisingly, is a fairly close approximation to even loading on the legs (Fig. 10:8(b)). No uniramian hexapod has been found to use a gait as divergent from the usual as this.

Locomotory habits

Thus hexapody in arachnids contrasts with that of the five uniramian hexapod classes in that speed changes are largely achieved by changes in pace duration and little by changes in gait pattern; in that the gaits favoured by Arachnida are those which are avoided by the Uniramia which do not use a phase difference of exactly 0·5 giving legs 1 and 3 stepping in unison. The arachnids gain stability by long time intervals k and not by the uniramian graded series of gaits, often using footfalls at even time intervals, even loading on the legs, or

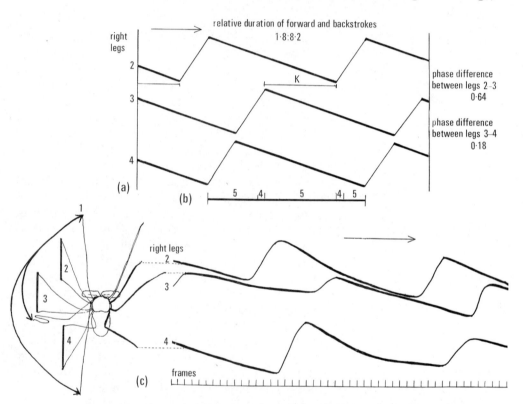

FIG. 10.8. A slow gait of an amblypygid (Pl. 8(a), (b)). (a) Gait diagram showing right legs II–IV performing the gait recorded by cinematography in (c). The gait pattern is constant but the phase difference between legs II and III is 0·64, giving a long time interval k between these legs, while the phase difference between legs III and IV is 0·18, leaving no time interval k between legs III and IV. (b) Number of propulsive legs of successive moments throughout one pace during the gait shown in (a). (c) Plotted positions of the limb tips relative to the head from part of a cinematograph record, conventions and outline of the animal on the left as for Fig. 10.7. The arrows between the three positions for left legs I, given on the left, indicate the range of movement of this leg, used for feeling movements but not for stepping. (After Manton 1973b.)

Locomotory habits

short or non-existent time intervals k. Arachnids do not usually achieve the precision in stepping shown by the uniramian hexapods. Arachnids show an evolution of hexapody parallel to those of the Uniramia but the details of movement in the two groups are not the same. Hexapodous stepping has evolved by different means in the Pedipalpi and Solifugae described above, in the other arachnid orders possessing a pair of antenniform limbs, and in the five hexapod classes. Similar functional advantages are thereby obtained. The Arachnida excel in the variety and perfections of their interpodomere joints while the uniramian hexapods excel in their integrated serial movements performed by more simply constructed legs.

8. Hydrostatic pressure and arachnid jumping

In neither Arachnida nor Uniramia is there any sound evidence indicating that the propulsive backstroke in walking and running is produced by forces other than those generated by the intrinsic and extrinsic leg muscles. Normal hydrostatic pressures in the haemocoel appear to be sufficient to cause certain leg movements, when the tip is off the ground, or exerting little force. By contrast, a sudden and considerable increase in hydrostatic pressures in the trunk is sometimes caused by trunk musculature and leads to jumping by the concerted action of legs IV in jumping spiders (Parry and Brown 1959). The Collembola also jump by means of hydrostatic pressure generated by trunk muscles and these muscles, directly or indirectly, but not the thoracic limbs, produce the high escape jumping of Thysanura (Chapter 9, §§2A, 5B, Figs. 9.3, 9.13, 9.14). The details are entirely different in these three examples.

9. The leg mechanism of Pycnogonida

Since we do not know where the relationships of the Pycnogonida rightly lie, whether they are related to the Arachnida, as seems probable (see Chapter 6, §2E(v)), or whether they are a more independent group, it is of interest to know how their leg mechanism compares with those of the other arthropods described above and in previous chapters. The limbs of the large *Decalopoda antarcticum* and of the much smaller *Nymphon gracile* are shown in Fig. 10.9.

Pycnogonids are found walking about on the sea floor, they climb among hydroids, on sea anemones, on sponges and Alcyonaria on which they feed; and pycnogonids can swim freely by beating the limbs up and down. All their movements are slow and often the legs are very long compared with the slender trunk (Fig. 1.16). The legs are well fanned out owing to the positions of the lateral trunk extensions bearing the coxae. These expansions, stiffened by annulations, are essential for the insertions of extrinsic coxal muscles, which thereby can function without great angular flexures on their tonofibrils passing to the cuticle. The arcuate sclerite of the spider (§5D(ii) above) serves the same need. So narrow a trunk as in the pycnogonids would not be so serviceable for extrinsic leg-muscle insertions.

The Arachnida and Pycnogonida, alone among arthropods, lack a promotor–remotor swing of the leg at the coxa–body union, but the two groups are not alike. The coxa of *Nymphon* and of *Decalopoda* possesses a pivot joint with a horizontal axis between the lateral body extension and the coxa, permitting levator–depressor movements, but, at least in *Decalopoda,* the axis between the anterior and posterior articulations lies above the equator of the joint, so facilitating depressor swimming movements (Fig. 10.9(a), (c)). The antagonistic muscles working the coxa insert on the walls of the body projection with which the leg articulates. The next joint along the leg is a pivot with a dorso-ventral axis between coxa and trochanter 1 (Fig. 10.9(a), (b)). This joint provides the promotor–remotor swing of the leg, used in walking, swaying, and climbing, but not much in swimming. The five more distal joints in *Decalopoda* are pivot joints with articulations on the anterior and posterior faces of the leg and horizontal axes of movement (Fig. 10.9(a), (c)) permitting levator–depressor movements worked by antagonistic pairs of muscles. *Nymphon* is a much smaller animal and according to Morgan (1971) has flexor muscles only from tarsus 1 and presumably the tibia–tarsus joint here is a simple hinge joint. Tarsus 1 and tarsus 2 are relatively shorter in *Nymphon* than in *Decalopoda*. The intertarsal joint is reminiscent of that of

FIG. 10.9 (see opposite). Legs of Pycnogonida (Fig. 1.16). (a)–(c) Leg of *Decalopoda antarcticum*. (a) Lateral face of a leg, the axes of movement at the joints being marked by heavy lines, six of these being longitudinal to the body and transverse to the leg and therefore seen end-on as indicated in (c). Only one joint possesses a dorso-ventral axis of movement giving a promotor–remotor swing, shown separately in (b). All these joints are worked by antagonistic pairs of muscles. (d) Anterior view of the swimming movements of *Nymphon gracile*. The propulsive downstroke shown on the right and the recovery upstroke on the left. (After Morgan 1971.) (e) Anterior view, about 3 mm wide, of the pycnogonid *Ammothea euchelata*, showing the proboscis, chelifores, palps, first walking legs, and hanging stance. (From Hedgpeth 1949).

Locomotory habits

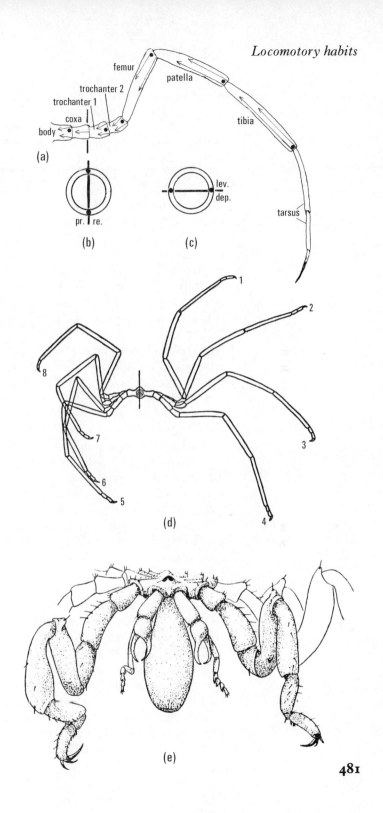

the arachnid *Galeodes* (Fig. 10.3) with a conspicuous dorsal overhang by tarsus 1. The terminal claw is long and strong.

In no other group of arthropods are the first three joints of the leg formed as in pycnogonids. The promotor–remotor swing of the leg is sited at the coxa–body joint in every major group except the arachnids where the coxa is almost or completely immobile (Fig. 10.2, cf. Figs. 2.2, 4.5, 5.3, 5.5–8, 5.14, 5.18). The main levator–depressor movement takes place at the second joint along the leg in arthropod groups other than some orders of arachnids. A secondary promotor–remotor swing is obtained in Solifugae and Acari by a pivot joint with a vertical axis between coxa and trochanter, a similar position to the promotor–remotor axis of swing of pycnogonids. This occurs in no other arthropod but in Opiliones and Amblypygi the movement is situated at the next more distal joint, again uniquely.

A cycle of swimming movements by *Nymphon* in horizontal aspect is shown in Fig. 10.9(d). Positions 1–4 indicate the downward propulsive stroke and positions 5–8 show the upward stroke during which the leg flexures cause least resistance to the water. Presumably slight promotor–remotor movements at the coxa–trochanter joint could cause forwardly directed swimming.

Walking and climbing movements are varied, and as yet no account of the gaits used by pycnogonids has been published; but the function of the unique coxa–body articulation seems to be clear. This joint, and those further along the leg with antagonistic pairs of muscles providing simple levator–depressor movements (Fig. 10.9), facilitate gentle swaying of the body without shifting of the massive tarsal claws. This movement enables the proboscis to engulf coelenterate or bryozoan polyps or sponge tissue without change of footholds, thus keeping the feeding movements inconspicuous; the animal also maintains a camouflage effect by swaying with moving water, in the same way as stick insects, chamelions and others sway in a breeze or in still air. The horizontal axis of coxa–body movement facilitates both feeding in exposed positions and swimming and emphasizes that swimming, climbing, and the maintenance of camouflage are the activities of greatest moment to Pycnogonida.

The coordination between pycnogonid limb movements shows conspicuous irregularities, as in arachnids, in contrast to the myriapods and hexapods where there is greater

Locomotory habits

precision in stepping. The phase difference between successive legs in stepping and swimming in pycnogonids is about 0·5 of a pace, that is, every other leg is in opposite phase. This results in continuous progression during swimming, as opposed to the discontinuous swimming of prawns and medusae. This same phase relationship between successive legs is also found during walking and running in arachnids and in the chewing movements of *Limulus* (Figs. 10.7, 2.6(d)). This resemblance between pycnogonids and arachnids is the more striking because of the avoidance of this phase relationship by myriapods and hexapods, which do not show so long a time interval k (see Chapter 7).

10. Conclusions concerning locomotory habits and the evolution of Arachnida and Pycnogonida

The evidence concerning functional morphology put forward in this chapter is limited in scope but far-reaching in its implications. The feeding mechanisms of chelicerates form a coherent whole differing from those of other arthropods. The transversely chewing gnathobases of five pairs of limbs in *Limulus*, and probably the Eurypterida also, give place to chewing by pedipalps alone in the Arachnida (Chapter 3). The suctorial feeding of pycnogonids could have been derived from the arachnid type of feeding (see Chapter 6, §2E(v)). Pycnogonid locomotory mechanisms, including leg segmentation, the nature of the joints, and the movements they permit, differ from those of all other arthropods. Pycnogonida share with the Arachnida the absence of the promotor–remotor swing of the coxa–body joint in contrast to all other Arthropoda. The pycnogonid leg jointing, although not exactly as in Arachnida, shows the flexibility in joint construction characteristic of this group. The legs of pycnogonids and of arachnids could have had a common origin.

The available evidence of arachnid evolution contains many uncertainties. The uniramous arachnid limbs, suitable for bottom living in the sea or on land, probably represent the endopod of the biramous limbs of merostome-like early arthropods (Chapter 6, Fig. 6.1). But the evolution of this endopod proceeded differently from the prosomal endopod of the Xiphosura and the walking thoracic endopods of Crustacea such as Malacostraca. In gaining stability of long legs by fixing the coxa of the walking legs on the body, instead of by using an endosternite of connective tissue as in the Xiphosura and others or an apodemal endoskeleton, the arachnids are unique. The loss of an original promotor–remotor swing has led to the exploitation of rocking-leg mechanisms, in the precise sense given above, to enhance stride length and at the same time restrict the upward projection of the knees of long legs in burrowing arachnids. A second unique innovation was the evolution of a new promotor–remotor swing at some other leg-joint, but only in some orders and not always in the same position. Thus extraordinary diversity exists in the leg mechanisms of the several arachnid orders and these mechanisms are basic to many habits peculiar to each order. The pycnogonid leg mechanism is associated with habits of particular significance to the group, as in the several arachnid orders. Pycnogonid leg jointing suits the leg movements performed. Long, slender podomeres need firm joints between them which provide precise, minute movements during feeding, when the tarsi remain fixed, but much wider movements during swimming (Fig. 10.9(d), (e)). Good articulations between the podomeres are present (cf. joint construction in Chapter 5 and Fig. 4.27). Joints permitting rocking movements, as described for various terrestrial arachnids and uniramians, would be mechanically undesirable and would serve no useful purpose and no such joints are present. Rocking joints are serviceable on land for some animals, but not in the water.

The Arachnida are basically octopodous in walking, as are the Pycnogonida, and this condition is absent from other large arthropodan groups. There are many examples of hexapodous running in arachnids which have arisen independently and which do not even use the same leg pairs. Hexapodous running in arachnids liberates one leg pair for sensory purposes, which to some extent may compensate for the absence of antennae, particularly in fast runners such as the Solifugae. The hexapodous stepping of arachnids contrasts with the five classes of uniramian hexapods both in morphology and in mechanisms, and is clearly the end term of chelicerate and uniramian invasions of the land, arrived at quite independently.

Both the Uniramia and the Arachnida exhibit marked achievements. The legs of arachnids excel in the variety of their morphological modifications, e.g. their leg-rocking mechanisms and the elaborate knee joints, but it is the long series of gaits and the mouth parts of the Myriapoda and uniramian hexapods which show corresponding versatility in structure and function. However, the general irregularity of arachnid stepping and the almost makeshift or absent provisions for secondary promotor–remotor swing of the leg may well have proved to be a deterrent to a more frequent evolution of hexapody within the group. But very precise leg movements, such as those required by spiders, are executed without difficulty.

The great variety of arachnid leg mechanisms stresses the ordinal differences within the Arachnida (see above, §5E), but gives no clear pointer to the relationships or past history of these orders. More information is provided by comparative external features of development (Anderson 1973). He makes five points of importance:

(a) The Xiphosura retain the most generalized pattern of embryonic development among modern chelicerates. (b) The Scorpionida, in spite of specializations, retain the nearest approximation to this pattern among modern arachnids. (c) All the other arachnids share a derived pattern of development whose basis could lie in that of Xiphosura but not that of scorpions. (d) The derived pattern of development common to arachnids other than scorpions is expressed as two variants (in the opilionids and in Acari, the more generalized of the two, and the araneids, uropygids, amblypygids, solifugids, and pseudoscorpions, with a forwardly flexed opisthosoma). (e) The development of the Palpigradi and Ricinulei is still unknown. In terms of chelicerate phylogeny, these conclusions suggest that the Xiphosura and arachnids diverged from a common ancestral chelicerate stock but that the scorpions diverged early from the other arachnids or perhaps evolved independently of the other arachnids. It can further be suggested that the arachnids other than scorpions are a unitary group with two extant divergent lines whose relationship to each other is not clear, but of which we can perhaps say that the first includes the Opiliones and Acari, while the second has radiated to produce the Solifugae, Uropygi and Amblypygi, the Pseudoscorpionida and the Araneae. How the Palpigradi and Ricinulei fit into this picture remains to be determined. The facts of internal development... insofar as they are available at the present time, do provide the necessary support for the above conclusions.

We know nothing of the ontogeny of the Eurypterida.

Our understanding of the meaning of animal shapes on a functional basis has increased greatly in recent years. The basic organization of the Diplopoda and Chilopoda, for example, and the diverse shapes of their component orders, fit them for performing various burrowing techniques, ceiling living, modes of feeding, running or pushing. The significance of these shapes has been shortly outlined in Chapters 6 and 8 and in greater detail in Manton 1954, 1956, 1958b, 1961, 1965, and in the press. These and many other examples

Locomotory habits

suggest a fresh inspection of the overall shapes and habits of pycnogonids.

The feeding by pycnogonids on prey that does not walk away encourages the evolution of camouflage which is achieved by having a stationary or very slow-moving or swaying body of stick insect-like proportions, movements being often restricted to those of the proboscis which sucks up food, aided sometimes by the chelifores and palpi, the tarsi remaining in place. With this simple habit is correlated pycnogonid morphology suitable less for walking than clinging, swaying, climbing, and swimming among its food organisms. An ancestry diverging from arachnid forbears before they left the sea might be explicable on the evidence of such habits (Chapter 6, §2E(v) and Chapter 10, §9). A suctorial proboscis is found in many crustaceans, tardigrades, and in the gnathosoma of ectoparasitic arachnids and is associated with modification of limbs and with the absence of feeding limbs in Tardigrada. These features, combined with the lack of variation within the Pycnogonida (Hedgpeth 1954), bring us back to the generalization given in Chapter 1, namely that habits which fit an animal to live better in the same or in a variety of environments but fit it to no particular set of environmental circumstances have been a prime factor promoting the advances which have established large taxa. Do we correctly recognize these habits in the Pycnogonida? It is probable that we do. We should not ignore the general similarity between the pycnogonids and the crustacean caprellids: the slender trunk, longer in caprellids than in pycnogonids, with protuberances in both, lodging extrinsic leg muscles; the long caprellid trunk region between head and gnathopods, as there is behind the palpi of *Nymphon* (cf. Pl. 3(g) and Fig. 1.16(a)), in both facilitating reaching different polyps without much movement other than swaying; and what more striking convergence could be found than the tarsi of a caprellid (Pl. 3(g)) and those shown in Fig. 6.6? Somewhat similar tarsi exist in certain harpacticid copepods living on baleen plates of whales and in epizoic and ectoparasitic amphipods and isopods, but in all they enable their owners to resist the swish of the sea.

11 *A further word on polyphyly*

THAT 'arthropodization' must have occurred more than once has been appreciated for over 20 years, although a monophyletic view of arthropod evolution held sway throughout years of discussions of this subject. But it is only recently that sufficient understanding of structure on a functional basis has made it possible to sift out the facts and to show where evolutionary advances do and do not lie. The various groups of arthropods have been functional organisms throughout their long history and any supposed backward marching cannot be entertained. Morphological evolution cannot be understood without an adequate knowledge of function, a counsel of perfection maybe, but modern work is showing the soundness of this outlook. Reliable results are not gained quickly or by guessing or by considering too few examples.

With an admission of independent evolution of the jointed cuticular exoskeleton, parts of which are stiff and firm, on the body and legs, comes the demonstration of much parallel evolution. Modern knowledge of complex organs such as compound eyes, which were once claimed as indisputable evidence of a monophyletic derivation of hexapods and some crustaceans, shows that monophyly is not the explanation of the distribution of compound eyes among those and other arthropods. Similarly, the Mandibulata represent no more than a grade of organization reached independently and by different means in unrelated arthropods. The terms Monognatha, Dignatha, and Trignatha among the Uniramia were originally coined to designate taxonomic groups but now they should be used only to indicate grades of advancement which have been reached independently. The same can be said for the so-called taxon the Entognatha, composed, as can now be recognized, of unrelated parallel evolutions of entognathy among some hexapod classes.

A question now arises concerning the number of times that the arthropod level has been reached. No certain answer can as yet be given, but comparative functional embryology,

A further word on polyphyly

summarized in Chapter 6, points clearly to the independent evolution of: (1) the *Uniramia*, a coherent entity which must have evolved from haemocoelic, lobopodial, metamerically segmented invertebrates, differing from all living annelids and radiating out on land into the extant Onychophora, Myriapoda, and Hexapoda; (2) the *Crustacea*, basically a quite different group with a different mode of embryonic development and with different, as yet unknown, ancestry from that of the Uniramia; the Crustacea radiated out into all types of aquatic environments; (3) the *Chelicerata*, both living and extinct, forming a coherent whole in which the body organization and limb mechanisms differ decisively from those of Crustacea; the marine forms are largely extinct and the terrestrial emergents from the sea parallel the Uniramia in large measure, but all their details are different.

The *Trilobita*, long extinct, with some 2000 described species, had a long reign in geological time; their distinctive morphology contrasts with (1)–(3) above, and perhaps they merit the same status as the Uniramia, Crustacea, and Chelicerata. But it is considered here that the evidence presently available does not warrant a further taxonomic assessment of the trilobites or the many other types of extinct arthropods, some of which have been mentioned in Chapter 1.

A recent attempt to group the arthropods into a diphyletic scheme by accepting the unity of the Uniramia in independence from the rest and lumping together the Crustacea, Chelicerata, and Trilobita rests on spurious arguments. The evidence of comparative and functional embryology is based upon established facts; and although we know only the larval part of trilobite ontogeny, a distortion of the facts cannot support the suggested unity of the three primarily aquatic groups of arthropods. Neither are the functions of the morphological features, so beautifully preserved in certain trilobites and other extinct arthropods, ascertainable by erroneous concepts concerning modern animals which leads to corresponding ill-founded assumptions concerning function in fossil animals. It has taken very many years of diligent work to uncover the known comparative functional embryology of arthropods and the elements of functional morphology summarized in this book. It may be that this essential evidence cannot easily be appreciated, or indeed properly understood, in a very small fraction of the time originally expended in unfolding it and the task will be the

A further word on polyphyly

more difficult where the disciplines are the least familiar. Until more sound evidence is available, little headway will be made with considerations of the probable number of independent evolutions of the arthropod grade of organization, meaning the number of phyla comprised within the arthropod kingdom. There is general consent concerning the independence in evolution of three or perhaps four arthropodan groups and it would be as well to refrain from speculation concerning the rest and not to force the partially understood groups into any hard and fast classificatory scheme.

It is fitting to end with a reference to our limited knowledge of speeds and conditions of evolutionary change in animals. That evolutionary progress has been slow is evident from the geological table of events in Chapter 1. But this has not been uniformly so; evolutionary change has gone on at different rates in various animals living in the same time and place. Bronowski, in discussing *The ascent of man* (1973), drew attention to the rapid human evolution during some 2 million years in one locality while contemporary ungulates remained almost unchanged in the same place. We have information concerning the way in which pressure from the environment controls the persistence of animal variants and how occasional cataclysmal events can have lasting selective effects upon animal populations. We know too how rapid speciation can be when a stock of animals invades a new habitat, such as that of the cichlid fishes in the Rift Valley lakes of Africa and the differentiation from under 20 parent species of *Drosophila* arriving in the Hawaiian Islands into some 700 new species in 1–4 million years since emergence of the islands from the sea and/or 1500 years since the polynesians reached the islands, bringing the cosmopolitan lowland species. These species have remained in the vicinity of human habitations and cultivation, while today the endemic species live in the forests and mountains at higher altitudes.

The arrival of the Uniramia in a new habitat, the land, may similarly have been followed by rapid evolution into a large number of animal types, as suggested here. Later, selection of the various kinds of Uniramia may have been exercised by the environment, those possessing the various habit divergencies already considered persisting to the present day. It is possible, but quite unproven, that a rapid diversification of Precambrian arthropodan fauna occurred and from it selec-

A further word on polyphyly

tion and habit differentiation gave rise to some of the major lines while others died out early. Conspicuous is the absence of intermediates between the many described Cambrian arthropods, both between the lesser-known types and the better-known trilobites and merostomes.

Before passing on from Bronowski's magnificent work just referred to, it must be emphasized that he recognized the enormous importance of habits and their control of human evolutionary progress in the widest sense. He stressed, with his usual modest enthusiasm and awe, the lack of human morphological specialization towards any particular set of environmental conditions, and showed how the evolution of habits has led to modern civilized *Homo sapiens*. The whole of this book on arthropods provides evidence and conclusions of the same type, and Bronowski's introduction to his own account *The Ascent of Man* is in principle equally applicable to the present book. Watson's emphasis on the importance of persistent habits throughout millions of years in understanding the evolution of aquatic reptiles (mentioned in Chapter 1) demonstrates the same thing. What greater support for the 'new look' on the Arthropoda could there be than is provided by the consistency between the work of Bronowski, Watson, and what has been put forward in the previous chapters?

An admission of polyphyly among arthropods spells the end of some other theories, no matter how many arthropodan phyla there may be. The idea that all arthropodan legs possess a basic eight podomeres should be abandoned. This view cannot be substantiated because eight and only eight podomeres are not demonstrable in all taxa. Various numbers of podomeres exist; and the terminal claw or claws have been said to appertain to one podomere or not according to the degree of difficulty in accounting for exactly eight podomeres. Recent work has shown indisputably how podomeres can divide or fuse according to functional needs (see Manton 1958, 1966, and Chapters 5 and 9). The smaller number of podomeres in Collembola than in other uniramian hexapods is correlated with hydrostatic jumping and presumably has directly evolved therewith from soft-bodied lobopodial ancestors.

Redundant too is the frequently copied diagram of a supposed primitive arthropodan leg, arising in flexible pleuron between tergite and sternite, with no close association with either sclerite, the extrinsic leg muscles inserting on

A further word on polyphyly

both. These are unjustifiable assumptions. In most taxa the legs are ventrally and not laterally directed, particularly in aquatic groups, and there is no reason to suppose that these legs have ever been in a different position. Only in Chilopoda and some Hexapoda are the legs ventro-laterally or laterally inserted on the body. There is a coxa–body articulation, or a close union with a sclerite by a short link of arthrodial membrane, except in Collembola where articulations would be functionally disadvantageous, in small Crustacea with wide, flattened limbs, as in most branchiopods and cephalocarids, the shape and size rendering articulation needless, and, insofar as is known, there is no coxa–body articulation in trilobites.

Many misconceptions would not have arisen but for the publication of schematic simplified diagrams, such as the 'primitive' arthropodan transverse section with appended lateral limbs, mentioned above, and the arthropodan heads composed of a row of cylindrical segments behind an acron (Chapter 6, §2C(ii)). Simplified diagrams have been avoided entirely in this book and in the work on which it is based, which has been published during the last 35 years, except for Fig. 9.5(a) which is a functional summary. No proper understanding of either function or morphology can be obtained from simplified diagrams, which thus become a menace and a hindrance. It is the comparative approach and utilization of the fullest morphological and functional details which have led to progress.

Finally, reference should be made to the composition of the anterior end of arthropodan bodies which has been established beyond question by a number of embryologists, the results of whose work is consistent. Nothing new is presented below, the facts having been summarized and discussed most recently by Anderson (1973) and Manton (1960). A table of the accepted composition of the anterior end of the body of arthropods is given below because so many erroneous and irresponsible statements have been made in recent literature without proper consideration of sound published work.

Anderson has shown in some detail how the head end of annelids develops and never again can an annelid be supposed to consist of a series of simple cylindrical segments behind the nonsegmental acron and mouth region. Consequently, anterior external segmentation of polychaetes cannot be compared directly with the different condition in arthropods, and

A further word on polyphyly

the oligochaetes are different again.

Anderson has reconsidered the criteria of the definition of a segment and it becomes more complex in annelid heads than in arthropods. Many of the erroneous statements about head segmentation stem from the lack of appreciation that embryonic mesodermal somites in both vertebrates and arthropods have been demonstrated to evoke ectodermal segmentation, involving ganglia as well as other structures. When a muscle mass is substantial, nerves will grow out to it, but such nerves are no evidence of basic metameric segmentation. Misconceptions also have arisen from not recognizing that the labrum of the Onychophora (Fig. 6.8) is not part of the oral or pleural folds and is a non-segmental structure, as shown in Chapter 6, Figs. 6.3, 6.4. The labrum is functionally analogous from taxon to taxon, but is probably not exactly homologous since the end term of its development lies at different segmental levels (Fig. 6.3). Labral musculature and nerves develop according to functional needs. A second misconception concerns the oral folds (lips) themselves of the Onychophora. They develop exactly as do the oral folds in Collembola and Thysanura (Fig. 6.8), forming in the adult a short pleural fold on the head of *Petrobius* (Fig. 3.23(a)) and the lateral boxing of the entognathous preoral space of adult Collembola and Diplura. These oral folds are no more segmental structures than the labrum.

A basic theme shown by all arthropodan groups is an embryonic back-growth of the mouth and labrum and a lateral forward shift of the rudiments of the anterior segments. Even the trilobites show the effect of such shifts (Fig. 1.5), the mouth lying posterior to the level of the antennae and the oesophagus looping forwards, as in *Limulus* (Fig. 3.13) and other arthropods. It is the polyphyletic progress of adult anterior differentiation which differs in the several phyla. There are some similarities between phyla, as to be expected, because there is a limited number of suitable ways in which cephalic differentiation can proceed. The possible functional reason for a similar number of segments to have shifted to a preoral position in Crustacea and in Myriapoda and Hexapoda has been suggested in Chapter 6, §2C(iv).

In Table 2 the segmental correspondence between trilobites and the other arthropods, indicated above, is conjectural. Entries for legs means walking or swimming legs. The more posterior appendages of the Chelicerata, the

TABLE 2.

The comparative segmental composition of the anterior ends of arthropods and the position of the mouth

ONYCHOPHORA	MYRIAPODA	HEXAPODA	CRUSTACEA	CHELICERATA	TRILOBITA
Antennae **(Mouth)**	*Preantennal* (embryonic, with or without transient limbs)	*Preantennal* (embryonic, with or without transient limbs)	*Preantennulary* (embryonic)	*Precheliceral* (embryonic in some Arachnida)	Antennae **(Mouth)**
Jaws	Antennae	Antennae	Antennae 1	Chelicerae **(Mouth)**	Legs 1
Slime papillae	*Premandibular* (embryonic, with or without transient limbs) **(Mouth)**	*Premandibular* (embryonic, with or without transient limbs) **(Mouth)**	Antennae 2 **(Mouth)**	Pedipalps or legs 1	Legs 2
Legs 1	Mandibles	Mandibles	Mandibles	Legs 2 (or legs I of Arachnida)	Legs 3
Legs 2	Maxillae 1	Maxillae 1	Maxillae 1	Legs 3 (or legs II of Arachnida)	Legs 4
Legs 3	Collum Maxillae 2 Pauropoda Symphyla Diplopoda Chilopoda	Labium	Maxillae 2	Legs 4 (or legs III of Arachnida)	Legs 5
Legs 4	Legs 1 Poison claws Chilopoda	Legs 1	Legs 1 or maxillipeds 1	Legs 5 (or legs IV of Arachnida)	Legs 6
Legs 5 Legs 6 Legs 7	Legs 2 Legs 3 Legs 4	Legs 2 Legs 3 Legs 4	Legs 2 or maxillipeds 2 Legs 3 or maxillipeds 3 Legs 4	Legs 6 (chilaria or metastoma, etc.)	Legs 7 Legs 8 Legs 9

pectines, genital operculum, and respiratory limbs, are not noted. Italic indicates embryonic segments. In myriapods the gnathochilarium behind the mandibles of diplopods is of composite origin and the collum segment, without limbs, behind the gnathochilarium is not simple in ontogeny in Diplopoda (Dohle 1964), but a simple collum segment develops behind the maxilla 1 in Pauropoda. In chilopods legs 1 form poison claws and legs 2, the first walking legs, correspond to the first trunk legs of pauropods and diplopods. The embryonic preantenullary somites of crustaceans are not present in all classes. These features do not detract from the basic characteristics of the anterior end of the body in the several phyla.

The challenge of attempting to understand the past history of the arthropodan phyla cannot be met by studies of palaeontology or of embryology or of functional morphology alone. It is only by a synthesis of sound information from all three sources that progress can be made. There are partially and completely unresolved problems in plenty concerning arthropod evolution.

Classification of the Arthropoda

THE outline classification given below is limited to include those arthropods which feature in this book, although the animals chosen for consideration are as representative of the whole group as was practicable. A full classification is much more complex and in some of the larger groups alternative classificatory schemes exist, no one of them having yet been accorded universal approval. It should be remembered that a classification of living animals is not a system of convenient pigeon-holes as may be suitable for collections of ancient Japanese swords or stamps. On the contrary, living animals are not static and their classification represents a cross-section of evolutionary time; the present-day fauna differs from its antecedents and from its future descendants.

The three arthropodan phyla, for which there appears at present to be substantial evidence for their having reached the arthropodan grade of organization independently, from different origins, are entered. Below each are listed the subphyla and classes into which they appear to have divided, together with a minimal number of other classificatory categories of general usefulness. The trilobites may represent a fourth phylum, but until we discover much more about the early fossil arthropods, it is best to leave the status of trilobites and of many other early arthropods as uncertain.

A classification which embraces every living and fossil arthropod cannot, on present evidence, be considered to be a natural classification; such attempts are to be condemned because doubt and certainty are inextricably mixed up and this does not lead to an understanding of the many zoological problems. Indeed, the classificatory format which suits the several arthropodan taxa cannot be the same in each case because the composition and evolution of the various groups is so different.

Animals which are extinct are marked with a dagger (†).

Nemertine worms (unsegmented no limbs)
    Rhyncodemus sp.

Polychaete worms (segmented with parapodia)
    Hermione hystrix
    Alentia gelatinosa
    Neanthes fucata
    Nereis diversicolor
    Nephthys sp.
    Sphaerosyllis sp.
    Spinther arcticus

(segmented without limbs)
    † *Opabinia regalis*

Classification of the Arthropoda

ARTHROPODA

Relationships uncertain
- † *Marrella splendens*
- † *Burgessia bella*
- † *Cheloniellon calmani*
- † *Leanchoilia superlata*
- † *Emereldella brocki*

Trilobita
- † *Olenoides serratus*
- † *Triarthrus eatoni*
- † *Sao hirsuta*
- † *Shumardia pusilla*
- † *Ceraurus pleurexanthemus*
- † *Naraoia compacta*

Phylum Chelicerata
  Subphylum Merostomata
    Class Aglaspida
- † *Aglaspella eatoni*
- † *Khankaspis Bazhanovi*
- † *Aglaspis spinifer*

    Class Xiphosura
- † *Euproöps thompsoni*
- † *Weinbergina opitzi*
- *Limulus polyphemus*
- *Tachypleus tridentatus*

    Class Eurypterida
- † *Hughmilleria norvegica*
- † *Parahughmilleria hefteri*
- † *Jaekelopterus rhenaniae*
- † *Mixopterus kiaeri*
- † *Baltoeurypterus tetragonophthalmus*
- † *Halipterus excelsior*
- † *Paracarcinosoma scorpionis*

  Subphylum Arachnida
    Order Trigonotarbida
- † *Alkenia mirabilis*

    Order Scorpionida
- *Buthus australis*
- *Euscorpius carpathicus*
- *Androctonus australis*
- *Parabuthus villosus*
- *Butholus alticola*
- † *Waeringoscorpio hefteri*

    Order Pedipalpi
      Suborder Uropygi — Whip scorpions
      Suborder Amblypygi — Amblypygid sp.
    Order Araneae — Spiders
- *Trochosa ruricola*
- *Ciniflo ferox*
- *Xenesthia immanis*

    Order Pseudoscorpionida — *Chelifer* sp.
    Order Solifugae — *Galeodes arabs*
    Order Opiliones — Harvesters

Classification of the Arthropoda

   Order Acari   Ticks and mites

  Class Pycnogonida
   Order Pantopoda   *Nymphon* sps.
              Decalopoda antarcticum
              Pentanymphon antarcticum
              Ascorhynchus castelli

   Order Palaeopantopoda   † *Palaeoisopus problematicus*

Phylum Crustacea
  Class Cephalocarida   *Hutchinsoniella macracantha*

  Class Branchiopoda
   Order Anostraca   *Chirocephalus diaphanus*
              Branchinella australiensis

   Order Lipostraca   † *Lepidocaris rhyniensis*
   Order Notostraca   *Triops cancriformis*
   Order Conchostraca   *Estheria* sp.
   Order Cladocera   *Sida crystallina*
              Daphnia magna
              Moina sp.
              Diaphanosoma sp.

  Class Copepoda   *Cyclops* sps.
              Calanus finmarchicus
              Diaptomus gracilis
              Panjakus hydrophorae
              Amphiascus sp.
              Idya sp.

  Class Cirripedia   Barnacles
              Balanus sp.
              Lithotrya sp.

  Class Ostracoda   Ostracods
              Pionocypris vidua
              Doloria levis
              Cypria ophthalmica

  Class Malacostraca
   Subclass Phyllocarida   † *Canadaspis perfecta*

   Subclass Leptostraca   *Nebalia bipes*
              Paranebalia longipes

   Subclass Syncarida   *Paranaspides lacustris*
              Anaspides tasmaniae
              Koonunga cursor

Classification of the Arthropoda

    Subclass Hoplocarida   *Squilla mantis*

    Subclass Peracarida   Mysids, isopods, amphipods, etc.
                                                                          Lophogaster typicus
                                                                          Gnathophausia sp.
                                                                          Hemimysis lamornae
                                                                          Ligia oceanica
                                                                          Armadillidium sp.
                                                                          Caprella sp.

    Subclass Eucarida   Shrimps, prawns, lobsters, crabs, etc.
        Order Euphausiacea   *Euphausia superba*
        Order Decapoda   *Atya* sp.
                                                                          Astacus (Potamobius Astacus)
                                                                          Homarus americanus
                                                                          Galathea sp.
                                                                          Carcinus maenas
                                                                          Macrocheira sp.
                                                                          Portunas puber

Phylum Uniramia   † *Ayshaeia pedunculata*
    Subphylum Onychophora   *Peripatopsis* sps.
                                                                          Peripatus sps.
                                                                          Macroperipatus sp.

    Subphylum or Class Tardigrada   *Bryodelphax parvulus*
       (possibly should be placed here)   *Macrobiotus* sp.
                                                                          Echiniscus sp.

    Subphylum Arthropleurida   † *Arthropleura armata*
                                                                † *Eoarthropleura devoni*

    Subphylum Myriapoda
       Class Pauropoda   *Pauropus sylvaticus*

       Class Chilopoda   Centipedes
          Subclass Epimorpha (hatch
             with all segments)
              Order Geophilomorpha   *Haplophilus subterraneus*
                                                             Orya barbarica
                                                             Geophilus carpophagus
              Order Scolopendromorpha   *Scolopendra* sps.
                                                             Cormocephalus sps.
                                                             Cryptops anomalans
                                                             Plutonium sp.
                                                             Craterostigmus tasmanianus

          Subclass Anamorpha (hatch
             with few segments)
              Order Lithobiomorpha   *Lithobius forficatus*
                                                             Lithobius variegatus

Classification of the Arthropoda

    Order Scutigeromorpha   *Scutigera coleoptrata*
Class Diplopoda   Millipedes
  Subclass Pselaphognatha   *Polyxenus lagurus*
  Subclass Chilognatha
    Superorder Opisthandria
      (posterior gonopods)   *Glomeris marginata*
                                  Glomeridesmus mexicanus
                                  Sphaerotherium dorsale
                                  Sphaerotherium giganteum

    Superorder Proterandria
      (anterior gonopods)
      Order Polydesmida   *Polydesmus angustus*
                                  Platyrhacus sp.
                                  Oniscodesmus fuhrmanni
      Order Nematophora   *Craspedosoma rawlinsi*
                                  Polymicrodon polydesmoides
                                  Callipus sps.

      Order Iuliformia   *Poratophilus punctatus*
                                  Blaniulus gutulatus
                                  Diopsiulus sp.
                                  Cylindroiulus sps.
                                  Ophistreptus guineensis
      Order Colobognatha   *Polyzonium germanicum*
                                  Brachycybe lecontei
                                  Siphonophora sps.

Class Symphyla   *Scutigerella immaculata*
                                  Hansieniella agilis
                                  Symphylella sps.

Subphylum Hexapoda
  Class Collembola   † *Rhyniella praecursor*
                                  Tomocerus longicornis
                                  Orchesella villosa
                                  Folsomia sp.
                                  Tullbergia sp.
                                  Frisea sp.
                                  Alacma fusca
                                  Sminthurus viridis
                                  Neanura muscorum
                                  Anurida maritima
  Class Diplura   *Campodea staphylina*
                                  Heterojapyx novaezeelandiae
                                  Japygid sp.

Classification of the Arthropoda

Class Protura
Eosentomon sp.
Acerentomon nemorale

Class Monura
† *Dasyleptus brongniarti*

Class Thysanura
Petrobius brevistylis
Lepisma saccharina
Thermobia domestica

Class Pterygota
Blatella germanica
Acridopeza reticulata
Forficula auricularia
Cyclochila australasiae
Erythroneura ix
Aphis fabae
Prylla mali
Locusta migratoria
Cylindronotus laevioc-tostriatus
Anisopus sp.

Class Palaeodyctyoptera
† *Stenodictyon* sp.

Technical Appendix

It may be useful to record brief notes on some simple methods employed in obtaining the data upon which much of this book rests.

Determination of anatomy

The *exoskeleton* can often be inspected from cuticles prepared by boiling in potash or by dissolving away the soft tissues by other less violent means, such as by digestive enzymes. It is helpful to make a small cut in the cuticle so that reagents gain easy access to the interior without disrupting cuticular structures, both external and internal. The staining of cuticles with chlorazol-black is very serviceable in picking out the more from the less sclerotized areas, the sclerites appearing grey in contrast to unstained pleural and arthrodial membranes when the timing is right. The dye can be added to the potash or used later.

Cuticles may be examined in aqueous media, or cleared and mounted in balsam without distortion, or studied floating in containers of suitable size or cavity slides, cleared in benzyl alcohol, the natural shapes of the cuticles being preserved without distortion by outside pressures. The nature of joints and the movements that they permit can be ascertained by manipulation or by using micromanipulators.

Analysis and dissection of the *musculature* present greater difficulties. Well-preserved specimens which have been stored in alcohol for years are the easiest to use because the muscles shrink a little and become brown or yellow. Fresh material can be very difficult to use for muscle examination. An initial examination of musculature is most usefully made from a longitudinal half of an animal, carefully and cleanly cut; one works thereafter from the median plane towards the flanks. If the animal is small it can be measured and embedded in wax. The calculated number of sections can then be removed with a reliably calibrated microtome, checking each section with a binocular microscope when approaching the middle line because although the measurements have been carefully made, the specimen may shrink a little during clearing and embedding. An exact longitudinal half of the animal remains in the wax. The wax is then dissolved away and the specimen returns to its opaque condition in 70 per cent alcohol. Further preparation of the opaque halves must be done with needles or micromanipulators

Technical Appendix

under a suitable binocular dissecting microscope. The viscera and fat body must be removed with the greatest care without damage to the muscles and tendons. This work is often difficult, but it is essential to get rid of these structures, particularly the fat body, this may lie between muscles, and, if not properly removed, will later obscure the muscle detail.

Examination of such muscle preparations can be done most profitably either opaque in ethyl alcohol or cleared in benzyl alcohol from 90 per cent ethyl alcohol, using what ever type of microscope is appropriate to the size of the specimen. Observation of muscles cleared in benzyl alcohol is much easier than after clearing in xylol and balsam. The specimen can be moved about in the benzyl alcohol, propped up on chips of glass to obtain certain aspects, but good ventilation must be maintained because the vapour is somewhat toxic to the worker if many hours are expended in such an investigation. Small specimens in a cavity slide can be sealed by a coverslip.

Polaroids fitted to a dissecting binocular microscope, one below the stage and a revolving polaroid mount on a swinging arm above are most useful, but the muscles under consideration must be brought to a horizontal position if the whole muscle is to be refringent. The tendinous insertions are not refringent and therefore are invisible. Since few muscles are refringent at one time in any one position, the relative levels of muscles and their relationships with one another are not easy to see, but the method is nevertheless very useful. A combination of all methods is usually required for ascertaining the entire musculature, as shown in Figs. 8.5, 8.9, 8.13. Steaks or other pieces are prepared similarly for particular purposes.

Muscle anatomy can also be built up from serial sections stained with Mallory's triple stain, adjusted so that muscles are purple, connective tissue brilliant blue, unsclerotized cuticle pale blue, and sclerotized cuticle red to orange or amber. Series of thick unstained sections prepared in celloidin are also useful.

Determination of limb and body movements

A right-angled prism is useful for observing the movements of an aquatic or terrestrial animal in ventral view. The long face of the prism is set in a horizontal mount and mirror at about 22° illuminates one of the other faces. This piece of apparatus stands on a dissecting microscope stage; the animal walks or feeds on the illuminated half of the prism face and the miscroscope objective descends over the opposite half of the prism. Since the distance travelled by the light within the prism curtails the distance between the objective and the prism face, the size of the prism must be suited to the objective required for viewing animals of particular sizes.

Technical Appendix

Aquatic animals can be held within small containers floored by coverglass.

Photography   The recording of footprints and photography plays a large part in analysing feeding and locomotory movements. But photography alone has its limitations because many arthropods do not behave normally under photographic illumination or show more than some aspects of their capabilities.

Cinematography, at normal or high speeds, may be essential when movements are irregular or jumping or rapid. But the time required for analysis of films is great; some 40 000 leg positions were plotted in a dark room in analysing onychophoran gaits (Manton 1950). Fortunately much more rapid and less laborious methods can be used when movements are regular. Some 40 good-quality still photographs of running animals, with simultaneous recording of footprints made by running over lightly smoked paper, can provide the same information for one animal. Speeds can be recorded by stop-watch or by marking an animal with a spot of white paint and using a fairly long exposure time, such as $1/60$ s; the apparent length of the blurred white spot on the photographic print compared with the real length will give the the speed. In making photographic and speed records it should be remembered that a freshly caught, often hungry and 'terrified' arthropod will turn on its maximum speeds of running only a few times. Thereafter when kept well fed in a terrarium it may perform only at its slower gaits. Other arthropods may walk normally only in dim lighting. Pauropods, for example, need the utmost precision in control of humidity, light, and heat to avoid sudden death. The photographs in Pl. 4(m), (i) were obtained with such controls of animals walking over very small round holes in damp filter paper on a microscope slide.

The use of electron flash illumination keeps the animals cool, but every leg will be sharp and it may not be easy to determine which legs are in the propulsive phase and which are swinging forwards, unless tracks are recorded, or shadows can be seen meeting or not meeting the limb-tip. Note the general resemblance between the body curvatures and limb positions on Pl. 1 Figs. d and f, but the propulsive parapodia of the worm lie on the body convexities and in the centipede in the concavities, marked by black and white spots. Long exposure times and photoflood illumination can give sharp images of tarsi in contact with the ground, while tarsi moving forwards off the ground appear blurred, as in Pl. 6(f) mesothoracic legs.

When legs are black and do not show up against smoked paper, a photograph can be taken of the smoked paper and track with the animal just running off the black paper on to damp white paper. The stepping movements can be interpreted from the stride lengths

Technical Appendix

and the knowledge of the range of movement of each leg shown by the 40 still photographs taken of the animal. Each footprint may be recognizable, as in *Scutigera*, Fig. 7.5(e), or *Lithobius*, Fig. 7.6(d), but when legs are very many the track appears to be a muddle at all gaits. However, the stride lengths can be very different, as can be demonstrated by letting a small drop of 'Durofix' dry on one tarsal claw, so forming a 'boot' which leaves a recognizable mark on smoked paper each time this footfall takes place (Fig. 7.5(a)–(c)).

A rough estimate of the pattern of a gait can be obtained by the numbers of legs of one side of the body in one metachronal wave which are in the propulsive and recovery phases. The phase difference between successive legs can be calculated from the number of segments to each metachronal wave; remembering that a figure of, say, 0·2 or 0·8 depends on the definition adopted for phase difference. Here it is that proportion of a pace by which leg $n+1$ is in advance of leg n.

It is useful to construct on graph paper a series of 'gait diagrams' as shown on Figs. 7.1, 7.3(a)–(c), 7.8–10. A simple series of numerical values for 'gait patterns' and phase difference is easiest to draw, but enough legs should be shown to give a repeat of the metachronal wave, as in Fig. 6.13. Vertical lines through the gait diagrams indicate moments in time which can be drawn as 'gait stills' (see Fig. 6.13). When the range of swing of each leg is known from the still photographs, the position of a particular leg, say, one-half, two-thirds, or three-quarters of the way through the back or forward stroke is known. Photographs of animals can be compared with gait stills. If they fit exactly, then the gait is known. If they do not fit exactly then the necessary changes in gait pattern and phase difference can be made on gait diagrams to give an exact fit with the photograph, so arriving at a correct identification of the gait. Much can be learned about why some gaits are practicable and some are not by making sets of gait diagrams. In fact series of calculated gait diagrams form an essential part of the equipment needed for rapid identification of gaits. If no exact fit can be arrived at, then the animal must have been stepping with irregularity and cinematography is the only way of unravelling the problem, which may actually be fictitious and due to bright photographic illumination.

Hexapods

The gaits of these animals are not so easy to identify as those of the many-legged arthropods by the quick methods just described. Even normal-speed cinematography may yield 2–3 frames per cycle of leg movements and the jumping gaits of *Petrobius* show only one frame per cycle. For the analysis of high jumping escape reactions of *Petrobius* high-speed cinematography has been essential. But the answers to many other problems have been given by still photography combined with simultaneous footprint records.

Technical Appendix

The fields of movements of the 3 pairs of hexapod legs during various speeds of running can be ascertained from some 50 good-quality still photographs of one animal, taken at different times because exposures in quick succession usually depict the same gait. Whether a limb-tip is on or off the ground must be known with certainty, either by a shadow meeting or not meeting the tip or by the position of the leg relative to the stride length shown by the track behind. Limbs do not usually slip on the ground and the limb-tip will be further from the middle line during the forward swing than during the backstroke. The make-up of the groups of footprints on the track must be understood; see Figs. 7.12, 7.5(d), (e), 7.6; a detailed description for *Lithobius* is given in Manton (1952b), p. 140, text-fig. 7.

All the above matters having been attended to, the gaits of hexapods can be ascertained from still photographs. Select a photograph showing, say, right leg 3 at the extreme end of the backstroke and match it against a series of gait stills such as those on Fig. 7.2(a)–(e) which all have right leg 3 in the same position. These gait stills, or a fuller series of them, will lead you to a gait still from your collection of gait diagrams which fits exactly, always making a vertical line through leg 3 at the end of the backstroke on your gait diagram. A check can be made from other photographs taken of the same animal at almost the same time at other moments of time on your gait diagrams. If your photograph will not fit any gait still from any gait diagrams, then the leg movements must have been irregular and if the hexapod in question makes an even track on smoked paper in dim light, then it is probable that its locomotory movements were adversely affected by bright lights used for photography.

Some arthropods use their fast gaits in bright light but will display their other accomplishments only in dim light. The Onychophora may be taken as an example of the way in which a variety of gaits can be worked out from tracks alone.

Fastest onychophoran walking results from longest extensions of the body (Pl. 4(a), (b)) and longest strides. A leg base moves for-

---

FIG. A.1 (see opposite). Variety in footprint groups formed by an onychophoran in dim light. (a) Shows the proportion of the body and legs at a body extension of 67 mm. The arrowed line on Figs. (b)–(t) shows the middle line of the track. Left groups of footprints only are shown, one for each gait, except on (b), each footprint being indicated by a small black spot, the series from legs 1–19 being joined by a line. The footfalls of leg 1 lies at p1 on all diagrams except in (b) where three successive paces are shown at p1–p3. These positions are calculated. Actually each blunt leg forms a wide footprint so that a large common mark, as in Fig. 7.5(d), appears to represent the close rings. The top two rows of diagrams (b)–(j) show footprint groups made at body extensions of 67 mm, values for d decreasing from 3·5 mm to 3·05 mm, d representing the distance travelled by the leg base between the footfalls of leg n and $n+1$. The bottom two rows of diagrams (k)–(t) show slower gaits in which d falls from 3·05 to 2·6 mm, the body extension being 58 mm. The scale of (a) is different from that of (b)–(t). (For further description see Manton 1950.)

Technical Appendix

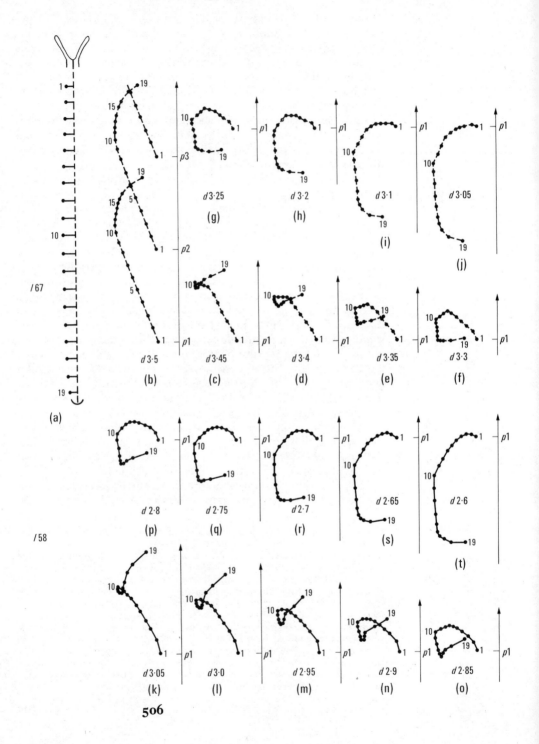

506

Technical Appendix

wards at a constant rate, whether the limb-tip is propulsive or not. The distance the leg base travels forwards during the time interval between the footfall of leg n and $n+1$ may be called d and the distance between the bases of legs n and $n+1$ called t. If $(d-t)$ is positive, the footfall of leg $n+1$ will lie at that distance in front of the footprint of leg n. If $(d-t)$ is negative then the footprint of leg $n+1$ will lie behind that of leg n, leg length and body width determining the distance of each footprint from the middle line; compare the forwardly staggered footprints in Fig. A.1(b) with 1(j). The stride length is shown by the positions of footprint of leg 1, marked p1–p3 on Fig. A.1(b), where d is 3·5 mm and is less for all other groups of footprints shown, the stride lengths and speeds being smaller. Footfalls are calculated and marked as small spots, but on the actual tracks the blunt limb-tip makes a wide mark. Figs. A.1(b)–(j) show how the different patterns of each pace group changes with values for d ranging from 3·5 to 3·05, the animal walking at an extension of 67 mm. The lower two rows of footprint groups result from a body extension of 58 mm, as used during slower walking and values for d ranging from 3·5 to 2·6 mm. The analysis of tracks made in the dim light such as these is rather complex and depends on the availability of further data, but it can be done; see Manton (1950). Some arthropods refuse to walk at all on dry lightly smoked paper, but they will walk over gelatinous culture media. The plates can then be covered and left for the microorganisms to grow and plainly reveal the tracks (Størmer 1975, pers. com.).

References

ALBRECHT, F. O. (1953). *The anatomy of the migratory locust.* Athlone Press, London.
ANDERSON, D. T. (1966). Arthropod arboriculture. *Ann. Mag. nat. Hist.* **9**, 445–56.
—— (1973). *Embryology and phylogeny of annelids and arthropods.* Pergamon Press, Oxford.
BARLET, J. (1951). Morphologie du thorax de *Lepisma saccharina* L. (Apterygote Thysanoure) 1. Squelette externe et endossquelette. *Bull. Annls. Soc. r. ent. Belg.* **87**, 253–71.
BEKKER, E. G. (1926). Zur phylogenetischen Entwicklung des Skeletts und der Musculatur der Ateloceraten (Tracheaten). *Russk. zool. Zh.* **6**, 3–67. [In Russian; German summary.]
BERGSTRÖM, J. (1973). Organization, life, and systematics of trilobites. *Fossils and Strata, Oslo* **2**, 1–69.
BERKLEY, E. (1940). Nahrung und Filterapparat der Walkrebschens. *Euphausia superba* Dana. *Z. Fisch.* **1**, 65–156.
BLOWER, J. G. (1951). A comparative study of chilopod and diplopod cuticles. *Q. Jl microsc. Sci.* **92**, 141–61.
BORRADAILE, L. A. (1922). On the mouth parts of the Shore Crab. *J. Linn. Soc. (Zool.)* **35**, 115–42.
BRIGGS, D. E. G. (1976). The arthropod *Branchiocaris,* n. gen., Middle Cambrian, Burgess Shale, British Columbia. *Geol. Sur. Canada* **264**, 1–28.
—— (in press). The morphology, mode of life, and affinities of *Canadaspis perfecta* (Crustacea Phyllocarida) Middle Cambrian, Burgess Shale, British Columbia.
BRISTOWE, W. S. (1947). *A book of spiders.* Penguin Books Ltd., Harmondsworth.
BROILI, F. (1933a). Ein zweites Exemplar von *Cheloniellon*. *Sber. bayer. Akad. Wiss.* **1**, 11.
—— (1933b). Weitere Beobachtungen an *Palaeoisopus*. *Sber. bayer. Akad. Wiss.* **1**, 33–47.
BRONOWSKI, J. (1973). *The ascent of man.* B.B.C. Publications, London.
BULLOCK, T. H. and HORRIDGE, G. A. (1963). *Structure and function in the nervous system of invertebrates.* Freeman, San Francisco.
BURROWS, M. and HORRIDGE, G. A. (1976). The organization of inputs to motoneurons of the locust metathoracic leg. *Phil. Trans. R. Soc. B* (in press).
CALMAN, W. T. (1909). Crustacea. In *A treatise on zoology* (Ed. E. R. Lankester), Part VII, 1–346. A. and C. Black, London.
—— and GORDON, I. (1933). A dodecapodous pycnogonid. *Proc. R. Soc. B* **113**, 107–15.

References

Cannon, H. G. (1927). On the feeding mechanism of *Nebalia bipes. Trans. R. Soc. Edinb.* **55**, 355–69.

—— (1928). On the feeding mechanism of the Fairy Shrimp, *Chirocephalus diaphanus* Prevost. *Trans. R. Soc. Edinb.* **55**, 807–22.

—— (1929). On the feeding mechanism of the copepods, *Calanus finmarchicus* and *Diaptomus gracilis. Br. J. exp. Biol.* **6**, 131–44.

—— (1931). On the anatomy of a marine ostracod, *Cypridina (Doloria) Levis* Skogsberg. '*Discovery*' *Rep.* **2**, 435–82.

—— (1933a). On the feeding mechanism of the Branchiopoda. *Phil. Trans. R. Soc. B* **222**, 267–352.

—— (1933b). On the feeding mechanism of certain marine ostracods. *Trans. R. Soc. Edinb.* **57**, 739–64.

—— (1947). On the anatomy of the pedunculate barnacle *Lithotrya. Phil. Trans. R. Soc. B* **233**, 89–136.

—— and Manton, S. M. (1927). On the feeding mechanism of a mysid crustacean, *Hemimysis lamornae. Trans. R. Soc. Edinb.* **55**, 219–53.

Cisne, J. L. (1975). Anatomy of *Triarthrus* and the relationships of the Trilobita. *Fossils and Strata, Oslo* **4**, 45–63.

Clark, R. B. and Clark, M. E. (1960). The ligamentary system and the segmental musculature of *Nephthys. Q. Jl microsc. Sci.* **101**, 149–76.

Clarkson, E. N. K. and Levi-Setti, R. (1975). Trilobite eyes and the optics of Des Cartes and Huygens. *Nature, Lond.* **254**, 663.

Daniel, R. J. (1932). The abdominal muscular systems of *Paranspides lacustris* (Smith). *Report for 1931, Lancs. Sea-Fish. Labs.* 26–44.

Dennell, R. (1934). The feeding mechanism of the cumacean crustacean *Diastylis bradyi. Trans. R. Soc. Edinb.* **58**, 125–52.

—— (1937). On the feeding mechanism of *Apseudes talpa*, and the evolution of the peracaridan feeding mechanisms. *Trans. R. Soc. Edinb.* **59**, 57:78.

Dohle, W. (1964). Die Embryonalentwicklung von *Glomeris marginata* (Villiers) im Vergleich zur Entwicklung anderer Diplopoden. *Zool. Jb. (Anat.)* **81**, 241–310.

Dohrn, A. (1881). Die Pantopoden des Golfes von Neapel. *Fauna Flora Golf. Neapel* **3**.

Eastham, L. E. S. (1930). The embryology of *Pieris rapae*. Organogenesis. *Phil. Trans. R. Soc. B* **229**, 1–50.

Evans, M. E. G. (1972). The jump of the click beetle (Coleoptrata, Elateridae)—a preliminary study. *J. Zool. Lond.* **167**, 319–36.

—— (1975). The jump of *Petrobius* (Thusanura, Machilidae). *J. Zool. Lond.* **176**, 49–65.

—— and Blower, J. G. (1973). A jumping millipede. *Nature, Lond.* **246**, 427–8.

Fahrenbach, W. H. (1973). The morphology of the Limulus visual system. V. Protocerebral neurosecretion and ocular innervation. *Z. Zellforsch. mikrosk. Anat.* **144**, 153–66.

Folsom, J. W. (1900). The development of the mouth parts of *Anurida maritima* Guér. *Bull. Mus. Comp. Zool. Harv.* **36**, 87–157.

Fryer, G. (1957). The feeding mechanism of some freshwater cyclopoid copepods. *Proc. zool. Soc. Lond.* **129**, 1–25.

—— (1968). Evolution and adaptive radiation of the Chydoridae (Crustacea: Cladocera): a study in comparative functional morphology. *Phil. Trans. R. Soc. B* **254**, 221–385.

References

—— (1974). Evolution and adaptive radiation in the Macrothricidae (Crustacea: Cladocera): a study in comparative functional morphology and ecology. *Phil. Trans. R. Soc. B* **269,** 137–274.

GRAY, J. (1939). Studies in animal locomotion. VIII. The kinetics of locomotion in *Nereis diversicolor. J. Exp. Biol.* **16,** 9–17.

—— (1968). *Animal locomotion.* Weidenfeld and Nicolson, London.

GOTO, H. E. (1972). On the structure and function of the mouth parts of the soil-inhabiting collembolan *Folsomia candida. J. Linn. Soc. (Biol.)* **4,** 147–68.

GRENACHER, H. (1879). *Untersuchungen über das Sehorgan der Arthropoden inbesondere der Spinnen, Insekten, und Crustaceen.* Vanderhoek and Ruprecht, Göttingen.

HANKEN, N. M. and STØRMER, L. (1975). The trail of a large Silurian eurypterid. *Fossils and Strata, Oslo* **4,** 255–70.

HARTNOLL, R. G. (1971). The occurrence, methods, and significance of swimming in the Brachyura. *Anim. Behav.* **19,** 34–50.

HEDGPETH, J. W. (1947). On the evolutionary significance of the Pycnogonida. *Smithson. misc. Collns.* **106,** 1–53.

—— (1949). A new pycnogonid from Pescardero, California, and distributional notes on other species. *J. Wash. Acad. Sci.* **30,** 84–7.

—— (1954). On the phylogeny of the Pycnogonida. *Acta. zool. Stockh.* **35,** 193–213.

—— (1955). Pycnogonida. In *Treatise on invertebrate paleontology* (Ed. R. C. Moore), Part P: Arthropoda 2, pp. 163–73. Geological Society of America and University of Kansas Press.

HERRICK, F. H. (1895). The American lobster: a study of habits and development. *Bull. U.S. Fish Comm.* for 1895, 1–252.

HESSLER, R. R. (1964). The Cephalocarida, comparative musculature. *Mem. Connecticut Acad. Arts Sci.* **16,** 1–97.

HEYMONDS, R. (1901). Die Entwicklungsgeschichte der Scolopender. *Zoologica, Stuttg.* **13, 33,** 1–244.

HOFFMAN, R. W. (1908). Uber die Mundwerkzeuge und über das Kopfnervensystem von *Tomocerus plumbeus* L. III. Beitrag zur Kenntnis der Collembolen. *Z. wiss. Zool.* **89,** 598–689.

HUGHES, C. P. (1975). Redescription of *Burgessia bella* from the Burgess Shale, Middle Cambrian, British Columbia. *Fossils and Strata, Oslo* **4,** 415–61.

HUGHES, G. M. (1958). The coordination of insect movements. III. Swimming in *Dytiscus, Hydrophilus* and a dragonfly nymph. *J. exp. Biol.* **35,** 567–83.

HUTCHINSON, G. E. (1930). Restudy of some Burgess Shale fossils. *Proc. U.S. natn. Mus.* **78,** 1–24.

IKEDA, K. and WIERSMA, C. A. G. (1964). Autogenic rhythmicity in the abdominal ganglia of the crayfish. The control of swimmeret movements. *Comp. Biochem. Physiol.* **12,** 107–15.

IMMS, A. D. (1939). On the antennal musculature in insects and other arthropods. *Q. Jl microsc. Sci.* **81,** 273–320.

KENNEL, J. (1886). Entwicklungsgeschichte von *Peripatus Edwardsii* Blanch. und *Peripatus torquatus* n. sp. *Arb. zool.-zootom. Inst. Würzburg.* **8,** 1–93.

KÜHNELT, W. (1961). *Soil biology, with special reference to the animal kingdom.* (Transl. N. Walker.) Faber and Faber, London.

References

LEHMANN, W. M. (1959). Neue Entdeckungen an *Palaeoisopus. Paläont. Z.* **33**, 96–103.

LEUZINGER, H. R., WIESMANN, R., and LEHMANN, R. E. (1926). *Zur Kenntnis der Anatomie und Entwicklungsgeschiste der Stabheuschrecke Carausius morosus.* Jena.

MAKI, T. (1938). Studies on the thoracic musculature of insects. *Mem. Fac. Sci. Agric. Taihoku imp. Univ.* **24**, 1–343.

MANTON, S. M. (1928*a*). On the embryology of the mysid crustacean, *Hemimysis lamornae. Phil. Trans. R. Soc. B* **216**, 363–463.

—— (1928*b*). On some points in the anatomy and habits of the Lophogastrid Crustacea. *Trans. R. Soc. Edinb.* **56**, 103–19.

—— (1930). On the habits and feeding mechanisms of *Anaspides* and *Paranspides* (Crustacea, Syncarida). *Proc. zool. Soc. Lond.* **1930**, 791–800.

—— (1931). Photograph of a living *Anaspides tasmaniae. Proc. zool. Soc. Lond.* **1930**, 1079.

—— (1937). Studies on the Onychophora. II. The feeding, digestion, excretion, and food storage of *Peripatopsis. Phil. Trans. R. Soc. B* **227**, 411–64.

—— (1949). Studies on the Onychophora VII. The early embryonic stages of *Peripatopsis*, and some general considerations concerning the morphology and phylogeny of Arthropoda. *Phil. Trans. R. Soc. B* **223**, 483–580.

—— (1950). The evolution of arthropodan locomotory mechanisms. Part 1. The locomotion of *Peripatus. J. Linn. Soc. (Zool.)* **41**, 529–70.

—— (1952*a*). The evolution of arthropodan locomotory mechanisms. Part 2. General introduction to the locomotory mechanisms of the arthropoda. *J. Linn. Soc. (Zool.)* **42**, 93–117.

—— (1952*b*). The evolution of arthropodan locomotory mechanisms. Part 3. The locomotion of the Chilopoda and Pauropoda. *J. Linn. Soc. (Zool.)* **42**, 118–66.

—— (1954). The evolution of arthropodan locomotory mechanisms. Part 4. The structure, habits and evolution of the Diplopoda. *J. Linn. Soc. (Zool.)* **42**, 299–368.

—— (1956). The evolution of arthropodan locomotory mechanisms. Part 5. The structure, habits and evolution of the Pselaphognatha (Diplopoda). *J. Linn. Soc. (Zool.)* **43**, 153–87.

—— (1958*a*). Hydrostatic pressure and leg extension in arthropods with special reference to arachnids. *Ann. Mag. nat. Hist.* ser. 13, **1**, 161–82.

—— (1958*b*). The evolution of arthropodan locomotory mechanisms. Part 6. Habits and evolution of the Lysiopetaloidea (Diplopoda), some principles of the leg design in Diplopoda and Chilopoda, and limb structure in Diplopoda. *J. Linn. Soc. (Zool.)* **43**, 487–556.

—— (1958*c*). Habits of life and evolution of body design in Arthropoda. *J. Linn. Soc. (Zool.)* **44**, 58–72.

—— (1960). Concerning head development in the arthropods. *Biol. Rev.* **35**, 265–82.

—— (1961). The evolution of arthropodan locomotory mechanisms. Part 7. Functional requirements and body design in Colobognatha (Diplopoda), together with a comparative account of diplopod burrowing techniques, trunk musculature and segmentation. *J. Linn. Soc. (Zool.)* **44**, 383–461.

References

—— (1964). Mandibular mechanisms and the evolution of arthropods. *Phil. Trans. R. Soc. B* **247**, 1–183.

—— (1965). The evolution of arthropodan locomotory mechanisms. Part 8. Functional requirements and body design in Chilopoda, together with a comparative account of their skeleto-muscular systems and an Appendix on a comparison between burrowing forces of annelids and chilopods and its bearing upon the evolution of the arthropodan haemocoel. *J. Linn. Soc. (Zool.)* **45**, 251–484.

—— (1966). The evolution of arthropodan locomotory mechanisms. Part 9. Functional requirements and body design in Symphyla and Pauropoda and the relationships between Myriapoda and Pterygota. *J. Linn. Soc. (Zool.)* **46**, 103–41.

—— (1967). The polychaete *Spinther* and the origin of the Arthropoda. *J. nat. Hist.* **1**, 1–22.

—— (1972). The evolution of arthropodan locomotory mechanisms. Part 10. Locomotory habits, morphology and evolution of the hexapod classes. *J. Linn. Soc. (Zool.)* **51**, 203–400.

—— (1973a). Arthropod phylogeny—a modern synthesis. *J. Zool., Lond.* **171**, 111–30.

—— (1973b). The evolution of arthropodan locomotory mechanisms. Part 11. Habits, morphology and evolution of the Uniramia (Onychophora, Myriapoda, Hexapoda) and comparisons with Arachnida, together with a functional review of uniramian musculature. *J. Linn. Soc. (Zool.)* **53**, 257–375.

—— (1974). Segmentation in Symphyla, Chilopoda, and Pauropoda in relation to phylogeny in Myriapoda. *Symp. zool. Soc. Lond.* **32**, 163–98.

—— (1977). Habits, functional morphology and evolution of pycnogonids. *Zool. J. Linn. Soc.* in press.

MARCUS, E. (1972). *In* Ramazzotti (1972).

MENDELSOHN, M. (1971). Oscillatory neurons in crustacean ganglia. *Science, N.Y.* **171**, 1170–73.

METTAM, C. (1967). Segmental musculature and parapodial movement of *Nereis diversicolor* and *Nephthys hombergi* (Annelida: Polychaeta). *J. Zool., Lond.* **153**, 245–75.

MILLER, J. (1973). Aspects of the biology and palaeontology of trilobites. Ph.D. Thesis, University of Manchester.

MITCHELL, R. D. (1955). Anatomy, life history, and evolution of the mites parasitizing fresh-water mussels. *Misc. Publs. Mus. Zool. Univ. Mich.* **89**, 1–28.

MORGAN, E. (1971). The swimming of *Nymphon gracile* (Pycnogonida). The mechanics of the leg-beat cycle. *J. exp. Biol.* **55**, 273–87.

NICOL, E. A. T. (1932). The feeding habits of the Galatheidea. *J. mar. biol. Ass.* **18**, 87–106.

OLROYD, H. (1964). *The natural history of flies.* Weidenfeld and Nicolson, London.

PANTIN, C. F. A. (1950). Locomotion in British terrestrial nemertines and planarians. *Proc. Linn. Soc. Lond.* **162**, 23–37.

—— (1952). The elementary nervous system. *Proc. R. Soc. B* **140**, 147–68.

PARRY, D. A. and BROWN, R. H. J. (1959). The jumping mechanisms of salticid spiders. *J. exp. Biol.* **36**, 654–64.

PAULUS, H. F. (1972). Die Feinstruktur der Stirnaugen einiger Collembolen (Insecta, Entognatha) und ihre Bedeutung für die Stammesgeschichte der Insekten. *Z. Zool. Syst. EvolForsch.* **10**, 81–122.

PEARSON, K. G., FOURTNER, C. R., and WONG, R. K. (1973). Nervous control of walking in the cockroach. In *Control of posture and locomotion* (ed. R. B. Stein, K. G. Pearson, R. S. Smith, and J. B. Redford), pp. 495–514. Plenum Press, New York.

PERRYMAN, J. C. (1961). The functional morphology of the skeleto-muscular system of the larval and adult stages of the copepod *Calanus*, together with an account of the changes undergone by this system during larval development. Thesis, University of London.

PLATE, L. H. (1924). *Allgemeine Zoologie und Abstammungslehre*, Vol. 2: *Die Sinnesorgane der Tiere*. Fischer, Jena.

PRELL, H. (1913). Das Chitinskelett von *Eosentomon*, ein Beitrag zur Morphologie des insekten Körpers. *Zoologica, Stuttg.* **64**, 1–158.

PRINGLE, J. W. S. (1957). *Insect flight*. Cambridge Monographs in Experimental Biology, No. 9. Cambridge University Press.

RAASCH, G. O. (1939). Cambrian Merostomata. *Spec. Pap. geol. Soc. Am.* **19**.

RAMAZZOTTI, G. (1972). Il Phylum Tardigrada. *Mem. Ist. ital. Idrobiol.* **28**, 1–732.

RAYMOND, P. E. (1944). Late Paleozoic Xiphosurans. *Bull. Mus. comp. Zool., Harv.* **94**, 473–508.

REPINA, L. N. and OKUNEVA, O. G. (1969). .Cambrian Arthropoda of Primorye. . *Palaeont. Zh.* **1969**, 106–14.

RICHTER, R. and RICHTER, E. (1929). *Weinbergina opitzi.* n. g. n. sp. ein Schwertiräger (Merost., Xiphos.) aus dem Devon (Rheinland). *Senkenbergiana* **11**, 193–209.

ROLFE, W. D. I. (1969). Phyllocarida. In *Treatise on invertebrate paleontology* (ed. R. C. Moore), Part R: Arthropoda 4, pp. R296–R331. Geological Society of America and University of Kansas Press.

—— and INGHAM, J. K. (1967). Limb structure and diet of the carboniferous centipede. *Arthropleura. Scot. J. Geol.* **3**, 118–24.

ROTHSCHILD, M., PARKER, K., and STERNBERG, S. (1972). Jump of the oriental rat flea *Xenopsylla cheopis* (Roths.) *Nature, Lond.* **239**, 45–7.

RUCK, P. and EDWARDS, G. A. (1964). The structure of the insect dorsal ocellus. 1. General organization of the ocellus in dragonflies. *J. Morph.* **115**, 1–23.

SANDERS, H. L. (1963). The Cephalocarida: functional morphology, larval development and comparative external anatomy. *Mem. Conn. Acad. Arts Sci.* **15**, 1–80.

SARS, G. O. (1901–1903). An account of the Crustacea of Norway 4. Copepoda, Calanoida, Bergen.

SCOURFIELD, D. J. (1926). On a new type of Crustacean from the Old Red Sandstone (Rhynie Chert Bed, Aberdeenshire). *Lepidocaris rhyniensis* gen. et sp. nov. *Phil. Trans. R. Soc. B* **214**, 153–87.

—— (1940). The oldest known fossil insect (*Rhyniella praecursor*) Hirst and Maulik. *Proc. Linn. Soc. Lond.* **152**, 113–31.

SEDGWICK, A. (1909). *A student's textbook of zoology*. George Allen, London.

SERFATY, A. and VACHON, M. (1950). Quelques remarques sur le biologie d'un scorpion de l'Afghanistan: *Butholus altocola* (Pocock). *Bull. Mus. nat. Hist. nat. Paris* **22**, 215–8.

SHAROV, A. G. (1957). Peculiar Palaeozoic wingless insects of the new order Monura (Insecta Apterygota). *Dokl. Akad. Nauk. SSSR* **122**, 733–6.

References

—— (1966). *Basic arthropodan stock with special reference to insects.* Pergamon Press, Oxford.

SNODGRASS, R. E. (1938). The evolution of Annelida, Onychophora and Arthropoda. *Smithson. misc. Collns.* **97**, 1–59.

STEIN, P. S. G. (1971). Intersegmental coordination of swimmeret motoneuron activity in crayfish. *J. Neurophysiol.* **34**, 310–18.

STØRMER, L. (1939, 1941, 1951). Studies on trilobite morphology I, II, III. *Norsk. geol. Tidsskr.* **19**, 143–273; **21**, 49–163; **29**, 108–57.

—— (1944). On the relationships and phylogeny of fossil and recent Arachnomorpha. *Skr. norske Vidensk-Akad.* **5**, 1–158.

—— (1970). Arthropods from the Lower Devonian (Lower Emsian) of Alken am der Mosel, Germany. Part 1: Arachnida. *Senckenberg. leth.* **51**, 335–68.

—— (1973). Arthropoda from the Lower Devonian (Lower Emsian) of Alken an der Mosel, Germany. Part 3: Eurypterida, Hughmilleriidae. *Senckenberg. leth.* **54**, 119–205.

STUBBLEFIELD, C. J. (1926). Notes on the development of a trilobite, *Shumardia pusilla* (Sars). *J. Linn. Soc. (Zool.)* **36**, 345–472.

STÜRMER, W. (1974). X-rays and palaeontology. *Electromedica* **74**, 52–5.

TIEGS, O. W. (1940). The embryology and affinities of the Symphyla, based on a study of *Hanseniella agilis Q. Jl microsc. Sci.* **82**, 1–225.

—— (1945). The post-embryonic development of *Hanseniella agilis* (Symphyla). *Q. Jl microsc. Sci.* **85**, 191–328.

—— (1947). The development and affinities of the Pauropoda, based on a study of *Pauropus sylvaticus. Q. Jl microsc. Sci.* **88**, 165–336.

—— (1955). The flight muscles of insects—their anatomy and histology; with some observations on the structure of striated muscle in general. *Phil. Trans. R. Soc. B* **238**, 221–348.

—— and MANTON, S. M. (1958). The evolution of the Arthropoda. *Biol. Rev.* **33**, 255–337.

TOWE, K. M. (1973). Trilobite eyes: calcified lenses in vivo. *Science, N.Y.* **179**, 1007–9.

VACHON, M. (1953). The biology of scorpions. *Endeavour* **12**, 80–9.

WATERSTON, C. D. (1975). Gill structures in the Lower Devonian eurypterid *Tarsoptella scotia. Fossils& Strata, Oslo* **4**, 241–54.

WATSON, D. M. S. (1949). The mechanism of evolution. *Proc. Linn. Soc. Lond.* **160**, 75–84.

WHITTINGTON, H. B. (1957). The ontogeny of Trilobites. *Biol. Rev.* **32**, 421–69.

—— (1959). Ontogony of Trilobata. In *Treatise of invertebrate Palaeontology* (ed. R. C. Moore), Part O: Arthropoda, pp. O127–O145. Geological Society of America and University of Kansas Press.

—— (1971). Redescription of *Marrella splendens* (Trilobitoidea) from the Burgess Shale, Middle Cambrian, British Columbia. *Bull. geol. Surv. Can.* **209**, 1–24.

—— (1974). *Yohoia* Walcott and *Plenocaris* n. gen., arthropods from the Middle Cambrian Burgess Shale, British Columbia. *Bull. geol. Surv. Can.* **231**, 1–27.

—— (1975a). The enigmatic animal *Opabinia regalis*, Middle Cambrian, Burgess Shale, British Columbia. *Phil. Trans. R. Soc. B* **271**, 1–43.

—— (1975b). Trilobites with appendages from the Burgess Shale, Middle Cambrian, British Columbia. *Fossils& Strata, Oslo* **4**, 97–136.

Index

Bold type indicates a definition and italic denotes a page with illustrations

abdomen, **150**, *21*, *174*, *175*, 240
  use for jumping by *Petrobius*, 337, *338*, *433*, *435*, *436*; by *Tomocerus*, 338, *340*, 402, *405*, *406*, *407*; Pls. 6, 8
abductor–adductor movements and muscles, *see* adductor–abductor movements and muscles
Acari (Arachnida, mites, and ticks), 15
  legs of, 453, 456, *457*; Pl. 8(e), (f)
Acerentomon nemorale (Protura), Pl. 6(h)
Acridopeza reticulata (Pterygota, grasshopper), ambulatory and flight muscles, 442, *457*
acron, **251**, *248*, *249*, *250*
adaptive radiation, **33**
adductor–abductor movements and muscles, **42**, *41*
  of gnathobases of *Limulus*, 50, *91*, *92*, 93
  of legs (coxae) of Collembola, *409*
  of Diplura, *223*; (telopod) of Chilopoda, *211*
  of mandible, Ch. 3 of Crustacea, 73, *78*, 79, *80*, 81
  of Diplopoda, adductors only, *102*, *107*
  of Hexapoda, *119*, *121*, *123*, *127*, *129*, *130*, *131*
  of Myriapoda, *109*, *110*
'advances', **33**, 399
Aglaspella eatoni (Aglaspida), *19*
Aglaspida, 18, *239*
Alacma fusca (Collembola), 401
Alentia gelatinosa (Polychaeta), Pl. 1(d)
Alkenia mirabilis (Arachnida), 266
Amblypygi (Arachnida) habits, 477
  gaits, *478*

Amblypygi (*contd*)
  legs and joints, 454, *455*, *457*
Ammothea euchelata (Pycnogonida), *481*
Amphiascus (Crustacea Copepoda), 69, 153, *155*, *275*
Anamorpha (Chilopoda), gaits and movements, *297*, 298, *323*
  tracks, *307*, *308*
Anaspides tasmaniae (Crustacea, Syncarida), legs, 139, 144, *145*
  mandible, 74–6, *78*; Pl. 3(f)
Androctonus australis (Arachnida) 17
angle of swing of leg, *43*, *46*, 54, 170, *173*, *178*
Anisopus (Diptera), undulations of larva, 266, Pl. 8(k)
annelid, head, 491
  limbs, 176–85, *178*, *180*, *183*, *185*; Pl. 1(b)–(e)
Anostraca (Crustacea), limb movements, 56, *57*
antagonistic muscles, effects of, 348–51 *passim*, 388, 389
antenna, summary, 493, 1–38 *passim* (figs.) Ch. 1, *66*, *78*, *80*, *82*, *243*, Pls. 3–8
Anurida maritima (Collembola), ontogeny, *271*
Aphis fabae (Pterygota), flight muscles, 444
apodeme, **89**, *88*, *287*, 345
  anterior tentorial, of Chilopoda, *110*; of Collembola, *125*, *129*; of Diplopoda, *104*, *106*, *107*; of Diplura, *130*, *131*; of Pterygota, *123*; of Symphyla, *109*; of Thysanura, *119*, *121*
  gnathal lobe apodeme of Diplopoda

515

Index

apodeme *(contd)*
  (partly tendinous), *107*; of symphyla, *109*
  mandibular apodemes of Crustacea, *78, 80, 88*; of Onychophora, *99*; of Pterygota, *117, 123*; of Thysanura, *117, 119, 121*
  phragma of Pterygota, *441*
  posterior tentorial of Collembola, *125, 129*; of Pterygota, *123*; of Thysanura, *119, 121*
  prophragma of Diplopoda, *355, 358, 361, 367*
  spine apodeme, of *Japyx*, 345, *418*; of *Campodea* (equivalent), *419, 420*
Appendiculata, 236
Apterygota, 400
Arachnida, 15, *239*, 262, 452, Pls. 7(a), (b), 8(a)–(j)
  coxae, 453, *455, 457, 459, 463*, 470
  evolution, 266, 483
  gaits, 472, *474, 478*
  locomotory mechanisms, 452–72; opisthosoma, 451
  prosoma, 451
  sternite, 451, 455
Araneae (Arachnida, spiders), habits, 18
  gaits, 474
  legs, 457, 464, *465*, Pl. 8(d)
arcuate sclerite, **465**
Armadillidium (Crustacea, Peracarida), enrolment, 365, Pl. 3(d), (d.1)
arthrodial membrane, 7, 192, *193*, 213, *217, 221, 225, 231*
arthrodial membrane, 7, 192, *193*, 213, *217, 221, 225, 231*
arthropod embryology, 237, 247, *1248, 249, 258, 485, 491*
Arthropoda, origin, 291
Arthropleura armata (Arthropleurida), 27, *28*
  legs, 234, *235*
Articulata, 1, 236
Ascorhynchus castelli (Pycnogonida), *23*
Astacus (Potamobius Astacus), Crayfish, mandible, *80*
Atya (Crustacea, Malacostraca), *168*
axis of swing of leg, positions in crustaceans, 145
  in myriapods, *207, 208*, 209, *213*
  in hexapods, *221, 227*
  in arachnids, *457, 463*
Aysheaia pedunculata (Uniramia) 26, *27*, 288

Baltoeurypterus tetragonophthalmus, Eurypterida, *15*

Balanus (Cirripedia, barnacle), 22, *156, 157, 158, 159*, 266, Pl. 2(c)
benzyl alcohol, use of, 502
biramous limbs, 107
  of *Burgessia*, 53, *10*, 140, 245
  of Crustacea, 142, *145, 146, 149, 152, 158, 159, 172, 173*
  of *Khankaspis*, 245
  of *Limulus*, *143, 152, 463*
  of *Marrella*, *9*, 137
  of Merostomata, 18, *19*, 242
  of Trilobita, *3, 5, 138, 139*
Blaniulus guttulatus (Diplopoda), Pl. 4(o)
blastopore, **250**, *248*
Blatella germanica (Pterygota) 442, *443*, 445
Brachycybe lecontei (Diplopoda), 362
Branchinella (Crustacea, Branchiopoda), 55, *57, 59*
Branchiopoda (Crustacea), 20, 66, Pl. 2(a), (b)
Butholus alticola (Arachnida), *17*
Buthus sps. (Arachnida), *201, 457, 462, 474*, Pl. 7(b)
Bryodelphax parvulus (Tardigrada), *289*
burrowing, 48
  by Colobognatha, 362, *361, 367*, Pls. 4(g), 7(f), (g)
  by Geophilomorpha, 377, *378, 379*, Pl. 5(a)–(c)
  by Nematophora, 360, Pl. 4(h)
  by Polydesmoidea, 356, *358*, Pl. 4(i)

Calanus finmarchicus (Crustacea, Copepoda), 66, *169, 173*, see Pl. 3(b),
Callipus (Diplopoda), *61, 208*, 210, 305, *307, 368*, Pl. 4(f)
Campodea (Diplura), habits, 415, 420, 421
  gaits, *323*
  legs, musculature, and movements, *223, 227*
  mandible, musculature, and movements, *127*, 130, *131, 272*
  skeleto-musculature, *223, 420*
  status, 422
  tracks, *333*
Canadaspis perfecta, 269
Caprellidae (Crustacea, Amphipoda), 264, Pl. 3(g)
carapace, 2, 10, 20, *21, 158, 171*, Pls. 2, 3(a), (h), (i)
cardo, of mandible, *102*
  of maxilla, *121, 125*
Carcinus maenas, the shore crab, mandible, *80*

Index

Carcinus maenus (*contd*)
  limbs, *151*, Pl. 3(i)
carpophagous structures, of Chilopoda, *379*
'cart horses', 357, 365
caudal furca, 11, *21*, Pls. 2(a), (b), 3(b), (c)
caudal spine, *10*
Cephalocarida (Crustacea), *21, 22*, 266
cephalon, **245**, *2, 3, 5, 8, 9, 10, 11, 63–133 passim, 236–93 passim, 241, 253*
cephalothorax, 246
Ceraurus pleurexanthemus (Trilobita), *139*
chelicerae, **12**, *13, 14, 15,* 18, *50, 93, 95,* 238, 244
Chelicerata, 12, *239, 253,* 449
Chelifer (Arachnida), 476, Pl. 8(i), (j)
chelifores, *23,* 24
cheliped, of Crustacea, *151, 152, 154*
Cheloniellon calmani, 8, *11,* 96, 243, 267
Chilaria, *13, 91*
Chilopoda, centipedes, abilities and structure, *378, 379,* 380
  gaits, *283, 322, 323*; legs, *211, 225*
  leg joints, 227
  leg rocking, 212, *214,* 217
  mandible, *100, 110, 111, 272*
Chirocephalus, the fairy shrimps (Crustacea, Branchiopoda), *56, 66*
  legs, *149,* 163, *164*
  mandibles and musculature, *66*–73, *71, 75*
cinematography, use of, 502
Ciniflo ferox, spider, 464, Pl. 8(d)
Cirripedia, barnacles, 22, 157, *158, 159,* 266, Pl. 2(c)
Cladocera (Crustacea) *161, 162,* Pl. 2(b)
Collembola (Hexapoda), 24, 122, 399, Pl. 6(a), (h); coxa–body junction, 403
  gaits, *301, 323*
  habits and evolution, 401, 414
  jumping mechanisms, *340,* 402–15, *405, 406, 407, 409*
  legs, 227, 410, *411,* 412, 413
  mandible, 122, *125, 127, 129, 272*
  pleurite, *405, 409*
Colobognatha (Diplopoda), abilities
compound eyes, 2, 3, 5, 273, 487, Pls. 2(h), 3(a), (g), (h)
Conchostraca (Crustacea, Branchiopoda), *164*
connective tissue, endoskeleton, 345, *347, 385, 435, 436*

convergence, 236, 237, 270–78, 369, 399
coordination of arthropodan movements, 339
Copepoda (Crustacea), 20, *68, 155, 172, 173,* Pl. 3(b), (c)
Cormocephalus (Chilopoda, Scolopendromorpha), gaits, *283,* 322
  habits and facilitating morphology, 380; leg-joints, 227
  leg-rocking, *213, 214*
  mandible, *110*
  musculature, *197, 385*
costa coxalis, of centipedes, *197*
  of arachnids, 455
coupler, of copepods, *170, 173*
coxa (protopod) of trilobite leg *4, 7,* 241
  of other arthropods, 134–235 *passim, 294–486 passim*
coxa–body articulation, 134–235 *passim,* 344–448 *passim,* summary, 225, 227
coxa–body junction, of Arachnida, 453, *457,* 471
  of Collembola, 403, *411*
coxal plate, of Collembola, *411,* **412**
coxal segmentation, of Collembola, *405, 409,* 413
coxo-sternal plate of *Petrobius* abdomen, 433, *435, 436*
crabs, 22
  mandible of, *80*
  swimming crab, Pl. 3(h)
Craspedosoma (Diplopoda, Nematophora), 360–4, Pl. 4(h)
Craterostigmus (Chilopoda), *374,* 388–94, Pls. 4(n), 7(i)
  tracheae, 191
Crustacea, 18, *239, 272*
Cryptops (Chilopoda, Scolopendromorpha), *201, 214,* 380, Pl. 5(d)
cuticle, of Chilopoda, 345, *346, 347*
  of Diplopoda, 345
  of Myriapoda, *346, 347*
  preparation of, 501
Cyclochila (Pterygota), 444
Cyclops (Crustacea, Copepoda), 156, Pl. 3(b), (c); *see also 172, 173*
Cylindroiulus (Diplopoda, Iuliformia), 305, *307*
Cylindronotus (Pterygota), 333
Cypria (Crustacea, Ostracoda), Pl. 3(a)

Daphnia (Crustacea, Cladocera), 160, *161, 162, 164*
Dasyleptus (Hexapoda, Monura), 27, 438

517

Index

Decalopoda (Pycnogonida), leg of, 457, 479, *481*
depressor leg-muscles, 39
  extrinsic, *223*
  intrinsic, *195*, *208*, *211*, *227*, *229*, *232*, *409*, *425*
Diaphanosoma (Crustacea, Branchiopoda), 113, *164*
Diaptomus (Crustacea, Copepoda), 55
Dignatha, 279, *280*
Diopsiulus (Diplopoda), jump of, 335, *336*
Diplopoda (Myriapoda), 27, *285*
  burrowing habits and facilitating morphology, 351–64, *355*, *358*, *361*, *367*
  evolution, *285*, 372
  gaits, *324*, 327
  leg structure and leg rocking, *41*, *195*, *207*, **212**, *227*
  mandibles and mandibular mechanism, 102, 103–8, *104*, *106*, *107*
  musculature, 355–66, *355*, *358*, *361*
  spiralling, 364, *367*, Pl. 4(o)
  tracks, *307*
diplosegments of Diplopoda, 348, 352, 353, *355*, Pl. 7(c)–(g)
Diplura (Hexapoda), 24, 215, Pl. 6(e), (b)
  gaits, *323*
  habits and facilitating morphology, 399, 415–22, *423*, *418*, *420*
  leg structure, *223*, *227*
  status, 415, 422
Dolistenus (Diplopoda, Colobognatha), *361*, *362*, Pl. 7(g)
Doloria (Crustacea, Ostracoda), 168, *171*
dorsal shield, **245**, *9*, *11*
dorsal surface pushing, by Diplopoda, associated musculature, *229*, *231*, Pls. 4(i), 5(r)
Drosophila (Pterygota), 489
Durofix', use of, 504

ecdysis, **236**
Echiniscus (Tardigrada), *289*
embryological evidences of relationships, 258, 287
embryology of arthropods, 237; of mysid crustacean, *248*, *249*
Emeraldella (an early arthropod), *243*
endites, 49, 52, 66, **67**, **136**, 144, *149*, 151, 154, 242
endoderm formation, *248*, *250*
endopod, *3*, 66, **67**, **136**, *145*, *146*, *149*, 151

endoskeletal basement membranes and subectodermal connective tissue, *88*, *187*, *189*, *347*
endoskeletal tendons, of Chelicerata, *91*, *93*
  of Crustacea, *88*
  of Hexapoda, *119*, *125*, *129*, *131*, *406*, *407*, *409*, *410*, *411*, *417*, *420*, *428*
  of Myriapoda, *102*, *109*, *110*, *211*, *347*, *385*, *389*, *393*, *428*, *435*, *436*
  of Trilobita, *7*
endoskeleton, *see* apodemes and endoskeletal tendons
enrolment, 364; *see also* spiralling
Entognatha, 400
entognathy, **122**, **270**, *253*, *272*
  of Chilopoda, *110*, *111*, *393*
  of Collembola, *125*, *127*, *129*, *271*
  of Crustacea, *272*, *273*
  of Diplopoda, Siphonophoridae, *272*
  of Diplura, *130*, *131*
Eoarthropleura (Arthropleurida), 287
Eosentomon (Protura), *221*
Epimorpha (Chilopoda), gaits and movements, *283*, *317*, *322*
  musculature, *347*, *385*, *387*, *389*
  tracks, *308*, Pls. 1(f), 5(a)–(d), 7(i)
epipharynx, **102**, *104*, *109*, *123* (unlabelled)
epipod, an exite, **144**, *151*, *154*, *241*
Erythroneura ix, (Pterygota), 444
escape reactions, 399, 433
Estheria (Crustacea, Branchiopoda), *164*
Euproöps, 18, *19*
Eurypterida, 2, 13, *14*, *15*, *16*, *239*
Euscorpius (Arachnida), *203*, *463*, *474*, Pl. 7(e), (f)
evolution of Arthropoda, 236–93, 487–94
  of Arachnida, 236–93, 449–86
  of Hexapoda, 236–93, 398–448
  of Myriapoda, 236–93, 344–97
exites, 57, 66, **67**, **136**, **144**, *145*, *146*, *149*, *151*, *154*, 241
exoskeleton, 1, most figures in the book, 236
  myriapodan divergencies, 348
eyes, simple and compound, 273, *276*, *3*, *5*, *13*, *15*, *17*, *21*, *27*, Pls. 2(b), 3(a), (b), (c), (g), (h), (i)
  embryonic rudiments of, *248*

fast leg movements, 204
fate maps, 25

Index

fields of leg movements, 201, 203
filter feeding, differentiation of, 166
filter and suspension feeding, 156
filter plates, 157, 161, 164, 169
flagellum, of arachnids, 452
'flat-backed' millipedes, abilities and structure, 356–9, 358, Pls. 4(i), 5(f)
flight, 399, 439, 447; origin of flight mechanisms, 429, 438, 439, 440, 446
flight muscles, 441
   dorsal longitudinal, 444
   dorso-ventral, 444
   pleural, 445
   tergal, 442
   wing adjustor, 456
Folsomia (Collembola), 401
food collection, 49
   transport by currents, 57
   transport mechanical, 49, 50
   hydraulic collection by *Petrobius*, 116
footprints, 303, 307, 308, 333
Forficula (Pterygota, earwig), 201, 203
frontal sclerite, of *Tomocerus* (Collembola), 124, 125, 129
Friesea (Collembola), 401
functional morphology, 30

gaits, 25, 46
   determination of, 502
   gait 'diagrams', **296**
   use of, 504
   gait 'patterns', 48, 298, **300**
   gait 'stills', **296**
gaits of arachnids, 472, 474, 478
   of Chilopoda Anamorpha, 297, 323, 325, Pls. 4(j), 5(e), (g)
   of Chilopoda, Epimorpha, 283, 297, 321, 322, Pls. 1(f), 5(a), (d)
   of Collembola, 301, 401, Pl. 6(d)
   of Diplopoda, 302, 303, 324, 327, Pl. 7(c)–(g)
   of Diplura, 323, 415, Pl. 6(e)
   of Onychophora, 45, 320, 322–4, Pl. 4(a)
   of Pauropoda, 324, 327, Pl. 4(m)
   of Polychaeta, 177–81, 178, Pl. 1(b)–(e)
   of Protura, 422, Pl. 6(k)
   of Pterygota, 299, 323, Pl. 6(i)
   of Symphyla, 321, 322, Pl. 4(d)
   of Thysanura, 331–4, 335, Pl. 6(m)
   of Uniramia, comparative, 282, 283, 322–4
Galathea (Crustacea, Malacostraca), 168

Galeodes (Arachnida), gait and speed, 475
   leg mechanism, 203, 457, 459
gape of mandibles, 76, 79, 117
genital operculum, 12, 13
Geophilus (Chilopoda Epimorpha), 347; see also Pl. 5(a)–(c)
germinal disc, 248
gill, 151, 154
gliding, 428
Glomeridesmus (Diplopoda, Limacomorpha), 367
Glomeris (Diplopoda, Oniscomorpha, 'pill-millipede'), 365; 367, Pl. 3(e), (e.1)
gnathal lobe, of Diplopoda, 102, 104, 107
   of Symphyla, 109
gnathal pouch, of Collembola, 122, 125, 127, 129
   of Chilopoda, 110
gnathobase, **6, 253**
   of Chelicerata, 50, 90, 91, 93, 95, 143
   of *Cheloniellon*, 11
   of Crustacea, 21, 64, 65, 66, 71, 75, 76, 78, 84, 85, 86
   of Eurypterida, 14, 470
   of *Limulus*, 13, 50
   of Trilobita, 4, 5, 7, 8, 188, 189, 256
Gnathophausia (Crustacea Malacostraca), 269
Gnathochilarium, of Diplopoda, 104, 107
grades, of evolution, 279, 280, 285, 487

habits, basic, 31
habit differentiation
   within the Uniramia, 32, 38, 284, 398, 490
   of burrowing techniques within the Diplopoda, 352
   of hunting techniques within the Chilopoda, 326
habit persistence, 34, 286
haemocoel, 186, 187, 189, 198, 287
Halipterus (Eurypterida), 15
hanging stance, 200, 201, Pls. 4(k), 6(a), 8(h)
Hanseniella (Symphyla), 218
Haplophilus (Chilopoda, Epimorpha), 346, Pl. 5(a), (b)
Harvester (Arachnida, Opiliones), 424, Pl. 8(g), (h)
Hawaii Islands, 489
head, 491, 494
   head capsules of Uniramia, 252, 281, 285

519

Index

head (*contd*)
  comparative table, 493
  of Crustacea, 246, 248, 249, 18, 20, 21
  development, 247, 248, 249
  of Hexapoda, 252
  of Myriapoda, 252
  of Onychophora, 252
heart, ostiate, 236
Hemimysis (Crustacea, Malacostraca), 77, 82
  mandible, 85, 86, 139, 248, 249
Hermonia (Polychaeta), 177, Pl. 1(b)
Heterojapyx (Diplura), 215, 221, Pl. 6(h)
Hexapoda, 24, 239, 281, 285
  classes and habits, 398, 447
  coxa–body junction of hexapods, 218–28, 221, 223, 225, 227, 398
  fossils, 448
  gaits, 299, 301, 323, 331, 335
  mandibles, 117, 121, 123, 253
  tentorial apodemes, 116–23
hexapody, occurrence and advantages of, 201, 203, 477
Homarus (Crustacea, Malacostraca, lobster), 149, 152
Homoptera (Hexapoda), 439, 444
Hughmilleria (Eurypterida), 14, 470
Hutchinsoniella (Crustacea, Cephalocarida), 21
  legs, 52, 146, 147, 149, 165
  mandible, 66
hydrostatic pressure, of onychophoran leg mechanism, 187, 188, 189
  of spider jump, 479
hypopharynx, **102**
  of Chilopoda, 110
  of Collembola, 129
  of Diplopoda, 104
  of Diplura, 130, 131
  of locust, 123
  of Symphyla, 109
  of Thysanura, 121

Idya (Crustacea, Copepoda), 69, 155, 275
illumination, for photography, 503
incisor process of mandible, *see* mandibles, cutting incisor process
intercalary sternites of Chilopoda, **346, 347, 378, 379**
intercalary tergites, of Chilopoda, 346, 347
  of Symphyla, 374, 375
insects, 24, 27
intrinsic leg muscles, 194, 195, 207, 208, 211

invasion of land, by arachnida, 18, 450, 451
  by Crustacea, 450
  by Uniramia, 24–7, 284
Isopoda (Crustacea, Malacostraca) large, hard-food feeding and mandibles, 77–9
Iuliformia (Myriapoda, Diplopoda), head, including mandibular mechanism, 100, 102, 103, 104, 106, 107;
  legs, 195, 207, 231; *see also* legs, Pl. 7(c)–(e)
  skeletomusculature and habits, 351–6, 484, 486

jaws and mandibles, 24, 37, 63–133; types, 253
jaw mechanisms, summary table, 65;
  of Onychophora, 97, 98, 99, 247;
  see also mandibles
joints, 236
  body, of Chilopoda, 346, 347, 386, 387, 389, Pls. 1(f), 4(j), 5(a)–(e); of Collembola, 406, 407, 410, Pl. 6(a)–(c); of Diplopoda, 355, 358, 367, Pls. 4(o), 7(c)–(g); of Diplura, 221, 418, Pl. 5(e), (h); of Pauropoda, 219, Pl. 4(m); of Protura, 221, Pl. 6(k); of Symphyla, 373, 375, Pl. 4(d), (e); of Thysanura, Machilidae, abdomen, 435, 436
  of limbs, coxa–body 41; hinge 41; pivot, 41; comparative tables of limb joints, 227, 457
  basic articulations, 191, 193, 195
  specialized articulations or joints, collembolan intra-coxal joints, 405, 409, 411, 412, 413
  coxa–body rocking joints of Chilopoda, 212, 214, 217
  coxa–trochanter anterior pivot of fast-running Chilopoda, 214, 217
  joints providing strength and wide angular movement, in Diplopoda by synovial cavities, 231, 232; in spiders by arcuate sclerite at femur-patella joint, 465; in copepods by the coupler morphology of swimming legs, 173
  leg-rocking joints, 199–235
    passim, 344–448 *passim*
  jumping in water by Crustacea, by Copepoda, 169; by *Paranaspides*, 174, 175; by trilobites, 48

520

jumping (contd)
   on land, by a diplopod, 335, *336*;
   escape jumping reactions on land, Collembola, 286, 338, *340*; other hexapods, 331, 336; Thysanura Machilidae, 336, *338*
   jumping gaits of Thysanura, *335* and associated morphology, 423–33

Khankaspis bazhanovi (Aglaspida), 245
'knee', of arachnids, 456
   of Myriapoda, *207*, *225*
   of Hexapoda, *223*, *225*, *227*
Koonunga cursor (Crustacea, Malacostraca), 83

labium, of *Campodea*, *130*
   of Collembola, *129*
   of *Lepisma*, *121*
   of locust, 123
   of *Petrobius*, *119*
   'labium' of Symphyla, *109*
labral glands, 162
labrum, **102, 250**
   of Chelicerata, 50, 93
   of Crustacea, *21*, *66*, *75*, *76*, *78*, *80*, *82*, 84–6
   of Hexapoda, *117*, *119*, *121*, *123*, *125*, *129*, *130*, *131*
   of Myriapoda, *104*, *106*, *109*, *110*.
   ontogeny of, 248, 249
   of Onychophora, *98*, *271*
lacinia mobilis (Crustacea, Peracarida, on mandible), *86*
lamellae on outer ramus of limb, *3*, *5*, *9*, *138*, *139*, 140, *243*
large food feeding, 77–83
Leanchoilia superlata, 243
legs, adductor–abductor movements, 42
   angles of swing, *46*, 302, *303*; large angles in copepods, *173*, in *Lepisma*, *430*, *431*
   legs of arachnids, 451, 453
   leg-base mechanisms, 36, 199–235, 344–486 *passim*;
   legs, basic movements of, 39, 40, *41*, *43*, *45*, *149*, *151*, *172*
   legs of Crustacea, 56–58, 142–153
   leg dispositions, control of, *303*, 313–14
   fields of movement of legs, *46*, *203*
   hydraulic systems of legs, *53*, *54*, *56*, *57*, *161*, *162*, *164*
   leg jointing, *41*, *43*, *227*, *457*
   leg length, 304
   leg levator–depressor movements, 228

legs (contd)
   leg loading, 314–18, *317*, *299*, *475*
   leg mechanisms, of amblypygids, *455*, *467*; of Chilopoda, 199–235, 294–397 *passim*; of Collembola, 410–15, *408*, *411*; of Colobognatha, 360–64, *209*, *361*; of Crustacea, 38–62, *134*–*98*; of Diplura, 415–22, *223*, *418*; of Hexapoda, 224, 225, 398–448; of Iuliformia, *195*, *207*, 351–56, *355*; of *Limulus*, *50*, *91*, *143*, *454*, *457*, *463*; of Myriapoda, Chs. 5, 7, 8; of Nematophora, *208*, 360; of Polydesmoidea, 229–33, *239*, *231*, *232*, 356–60, *358*; of Protura, 422; of Pterygota, 224, 225, 438–47; of Pycnogonida, 263, 479–83, *481*; of scorpion, 456, *457*, 461–464, *463*; of Solifugae, *459*, *461*, 468–469; of spider, *457*, *464*, 465–469; of Thysanura, 423–38, *425*, *427*, *428*, *429*
   legs, lobopodial, 25, *45*, *182*, *187*, *189*
   legs, metachronal movements, 44, *45*, *46*, *54*, *56*, *161*, *162*, Pls. 1(b)–(f), 2(a), (b), 3(f), 4(a), (e), (f), (h)–(j), 7(a)–(g)
   leg numbers, 201
   legs, order of footfalls, *299*, *301*, 314–19, *317*
   legs, polymerous, 24, Pl. 2(a, in front of leg-less abdomen)
   legs, promotor–remotor movement, **39**, *41*, *43*
   leg-rocking movements, **320**; by Arachnida, *457*, 458–69, *459*, *463*; by Diplura, *215*, *221*, *223*; by Myriapoda, *41*, *212*, *214*, *217*; by Protura, *216*, *221*
   legs, stepping, 42, *43*
   legs, swimming, 144
   legs, types, 241, 236–93 *passim*
   legs, uniramous, 24, *136*, 241, 236–93 *passim*
   legs, walking movements, *178*, 179
Lepidocaris rhyniensis (Crustacea, Lipostraca), *164*, *267*, *268*
Lepisma (Thysanura, Lepismatidae), coxal articulation of, *428*
   coxal movements, 429, 434
   feeding, 116, *117*, *121*
   jumping gaits, 423, Pl. 6(g)
   leg mechanism and musculature, 429, *430*, *431*, *432*
   paratergal lobe, 429

521

Index

pleurites and their movements, 426, 427, 428
track, 333
Leptostraca (Crustacea, Malacostraca), 49, 51, 58
Ligia (Crustacea, Malacostraca), legs, 201
   mandibles, 77, 78
Limulus (Merostomata, Xiphosura), 2, 12, 13, 37, 48, 50, 51; see also legs
Lithobius (Chilopoda Lithobiomorpha), entognathy, feeding, and mandibular mechanism, 110, 111
   hunting technique, 326
   gaits, 297
   legs, 145, 211, 217
   pleuron, 217
   tergite heteronomy, 386, 387
   tracks, 308
lip, of Onychophora, 97, 98
Lithotrya (Crustacea Cirrepedia), 158, 159
lobopodial ancestors of myriapods and hexapods, 396, 399
lobopodial limb mechanism, 182, 187, 189
locust, mandible, 117, 118, 123
Lophogaster typicus (Crustacean Malacostraca), 85, 156, 269

Machilidae (Thysanura), mandibular mechanism, 116, 117, 118, 119
   jumping gaits, 335
   high jumping escape reactions and mechanism, 336, 338, 433, 435, 436
   leg mechanism, 423-6, 425, 427
Macrobiotus (Tardigrada), 29, 289
Macrocheira (Crustacea, Malacostraca, giant crab), 1
Macroperipatus (Onychophora), 299
Malacostraca, see Classification, 497
mammalian locomotion, 294
mandibles, cutting incisor processes and grinding molar processes, 66, 67, 74, 75, 76, 84, 85, 86, 119, 121, 127
   gnathobasic mandibles or jaws, of Chelicerata, 91, 93, 95; of Crustacea, 64-86 including Figures; of Trilobita, 3, 4, 5, 7
   protrusible mandibles, 112, 272, 400
   rolling mandibles, 64-9, 119
   mandibles suiting small food particles, 69, 84; suiting large food particles, 73, 83, 84, 85; transversely biting mandibles, 79, 80, 253

mandibles (contd)
   mandibles of Uniramia, whole-limb, jointed in Myriapoda, 93–133, unjointed in Onychophora and Hexapoda, 98, 99, 116–33
mandibles, and mechanism, 63–133
mandibular mechanism of *Anaspides*, 75, 76, 78
   of *Campodea*, 130, 131
   of Chilopoda, 110, 111, 393
   of *Chirocephalus*, 66-8
   of copepods, 66, 67, 275
   of crab and crayfish, 80
   of Diplopoda, 102, 104, 105, 106
   of *Hemimysis*, 82, 84, 85
   of *Hutchinsoniella*, 66
   of *Ligia*, 78
   of *Lithobius*, 110, 111; 393
   of *Lepisma*, 117, 121
   of locust, 117, 123
   of *Scutigera*, 393
   of *Tomocerus*, 125, 127, 129
mandibular apodemes, 78, 88, 99, 119, 121, 123
mandibular lacinia mobilis, (Crustacea, Peracarida), 85, 86
mandibular spine row (Crustacea, Peracarida), 84, 85, 86
Marrella splendens, 8, 9, 37, 49, 53, 245
maxilla 1 (maxillule), of Crustacea, 78, 82, 84, 85
   of Myriapoda, 109, 110, 393
   of Hexapoda, 117, 119, 121, 123, 125, 129, 130, 131
maxilla 2 (maxilla), of Crustacea, 78, 82, 84, 85, 149, 246
   of Myriapoda, 109, 110, 392
   of Hexapoda (the labium), 117, 121, 123, 129, 130
maxilliped, of Crustacea, 78, 80, 84, 86, 149, 151, 154, 246
Merostomata, 12, 239, 449
mesoderm formation and mesodermal somites, 248, 249, 250
mesosoma, **12**, 13, 14, 15, 240
metachronal waves of limb movement, 44, 45, 283, 310, 475
metacoxal pleurite of Collembola, 405, 409, 413, 414
metanauplius larva of Crustacea, 21
metasoma, 14–17, 19, **240**
millipedes, see Diplopoda
mites (Arachnid, Acari), 453, 454, 457, 458, Pl. 8(e), (f)
Mixopterus kiaeri (Merostomata, Eurypterida), 14, 15, 16, 37, 141, 147

Index

Moina (Crustacea, Branchiopoda), 163, 164
Monognatha, *280, 285*
monophyletic derivation of arthropods, 31, *236–494 passim*
Monura (Hexapoda), 437
mouth, *75, 80, 84, 85, 93, 98, 99, 104, 123, 129, 130, 248, 249*
muscles, basic components of trunk musculature, deep dorso-ventrals, 350, e.g. muscle *dvtr.* on *347, 378, 385*, muscle *dv.* on *407*, muscles dv., dv1., dv2., dv3. on *418*, muscles 20, 21 on *435, 436*
  dorsal longitudinal muscles, 349, e.g. on *187*, long and short dorsal sectors, *dlm., dlm.A., dlm.B., dlm.C.* on *347, 375, 378, 385, 405*, short sectors, muscles *ret. dor.* on *355, 358, 361*
  lateral longitudinal sectors in Onychophora, Diplopoda, epimorphic Chilopoda, and Collembola, *187*, muscle *ret. par.* on *355*, *flexor inferus longus* on *358, 361*, muscle *llm* on *378, 385*, muscles on Fig. 9.2(b), *426*
  sternal longitudinal sectors, e.g. on *187, mus. lac. v.* on *355, retractor ventralis* muscle on *358, slm.,* on *378, 385*, on Fig. 9.3(b), *407*
  deep obliques, 350, e.g. muscles *dvmp., dvma.* on *219, 378, 385*, muscle *ob.* on 407, muscle *dvmd.* on 420, muscles 22, 23, 24, 25A, 25B, 26 on *435, 436*
  superficial obliques, e.g. superficial pleural muscle on *218*, muscles shown on segments 3, 4, Fig. 8.3(a), *355*, muscles *s.ob.a., s.ob.b.* on *378*, pleural muscles on Fig. 9.2(a), *406, 409*
muscles, flexibility promoting in Diplopoda, *355, 358, 361*
  in Geophilomorpha, *347*
  in Japygidae, *418*
  in Symphyla, *218, 375*
  jumping promoting, in Collembola, 403, *405–9*; in Crustacea, 174, *175*; in *Petrobius* leg-muscles for jumping gaits, 423, *425, 427* and abdominal muscles for high jumping escape reactions, 433, 436
  numbers of muscles and functions, 205
  preparation of muscles, 500
  rigidity-promoting muscles in Chilo-

muscles (*contd*)
  poda, Diplopoda, Pauropoda, Collembola, *348–51*, 344–448 *passim*
musculature of legs, of *Campodea, 223*
  of *Chilopoda, 197, 211, 225*
  of Collembola, *409, 411*
  of Diplopoda, *195, 208, 229, 234*
  of Myriapoda, *207*
  of Onychophora, *187, 188*
  of Pauropoda, *219*
  of Pterygota, *225*
  of Thysanura, *425, 427, 431*
musculature of jaws and mandibles, 63–133, 393
Myriapoda, habits and evolution, *236–93 passim*, 285, 281, 344–97, 396

Naraoia compacta (Trilobita), *243, 257*
nauplius larva, of Crustacea, *21,* 153, 267
Neanura muscorum (Collembola), Pl. 6(c), 415
Neunthes fucata (Polychaeta), 176–81, Pl. 1(c)
Nebalia (Crustacea, Leptostraca), 49, 51, 58
Nematophora (Diplopoda), abilities and structure, 360
Nemertine worms, 135, *136*
Nephthys (Polychaeta), 188
Nereis diversicolor (Polychaeta), 178, 185
Notostraca (Crustacea, Branchiopoda), 49
nymph, pterygote developmental stage, 439, 441–6
Nymphon (Pycnogonida), *23,* 479, *481, 383*

ocellus, **273,** **276**
  simple ocelli, of Arachnida, *17*; of Crustacea, *21, 66, 155,* Pl. 3(b), (c); of Eurypterida, *15*; of Myriapoda, *104, 106, 110*; of *Opabinia, 290*; of Onychophora, *98*; of Pycnogonida, *23*
  compound eyes, of Crustacea, 78, 80, 82, 152, Pls. 2(b), 3(a), (g)–(i); of Hexapoda, *117*; of Trilobita, *3, 5*
Olenoides serratus (Trilobita), 2, *3, 4,* 51, 57, 245
Oniscodesmus (Diplopoda, Polydesmida), *376*
ontogeny, of Crustacea, *21, 155, 260, 348, 349*
  of Chilopoda, Pls. 4(n), 5(h)
  of Diplopoda, Pl. 7(h)

523

Index

ontogeny (*contd*)
  of Symphyla Pl. 4(i)
Onychophora (Uniramia), habits, 284, 398
  haemocoel, *187, 188*
  head and jaws, *98, 99*
  evolution, 284, *285*
  limbs (lobopodia), 186, *187, 188*
  ontogeny, 258
  structure, 182
Opabinia regalis, 290
Ophistreptus guiniensis (Myriapoda, Diplopoda), size effects, 329, Pl. 7(c)–(e)
Opiliones (Arachnida), 240, *457,* 474, Pl. 8(g), (h)
opisthosoma, **12,** *13, 14, 15, 16, 17, 19,* 240, 451
Orchesella (Collembola), 401
Orthoptera (Pterygota), 439, 445
Orya barbarica (Chilopoda, Geophilomorpha), 377–80, Pl. 5(a)–(c)
ostia, of heart, of pericardial floor, of muscle sheet, 3, 186, *187, 189*
Ostracoda (Crustacea), 20, 156, 168, *170, 171*
outer ramus, of trilobite and other fossil arthropodan legs, *3, 4, 9, 10, 11, 138, 139,* **241**
ovigers, of Pycnogonida, *23,* 24, 240

Palaeodictyoptera, 438
Palaeoisopus (Pycnogonida, Palaeopantopoda), 262, *263*
palp (Pycnogonida), *23,* 240
palp, mandibular (Crustacea), *66, 76, 78, 80, 82*
Panjakus hydrophorae (Crustacea, Copepoda), *172*
Pantopoda (Pycnogonida), 22, *23,* 239, 457
Paracarcinosoma (Merostomata, Eurypterida), *15*
Parahughmilleria (Merostomata, Eurypterida), *470*
parallel evolution of entognathy, *272,* of hexapody, 398, of eyes, 274
paragnath (Crustacea), *71, 75,* **79,** 80, *82, 84, 85, 275*
Paranaspides (Crustacea, Syncarida), 77, *174, 175*
Paranebalia (Crustacea, Leptostraca), 149
parapodia, 177, *178*
paratergal lobes, of Thysanura, 423, *427*
'pattern' of a gait, 298, **300**

Pauropoda (Myriapoda), *219*
  abilities, 396
  gaits, *324*
  leg-joints, *227*
  leg-rocking, 215
  leg-structure, *219*
  musculature of trunk and leg, *219*
pedipalp, 17, 94, *95*
Pedipalpi (Arachnida), 452
Peripatopsis and *Peripatus* (Uniramia, Onychophora), *45, 271,* Pl. 4(a)–(c)
Petrobius (Thysanura, Machilidae), feeding mechanism, 115–18, *117, 119*
  gaits, 332, *335,* 423
  jumping escape reactions, 433–7, *435, 436*
  mandibular mechanism, 116–19
  movements and muscles of legs, 424, *425, 426, 427,* 432, 434
  musculature of abdomen, *435, 436*
  pleurites, 424, *425, 427*
  tracks, *333, 335*
phase differences between successive legs, **300,** 309, 310, 474
  between paired legs, *46,* 309, 322, 474
photography, use of, 502
phragma, of Pterygota thorax, 345, *440, 443*
Phyllocarida, 269
phyllopodium, 147, *149*
phylum Chelicerata, phylum Crustacea, phylum Uniramia, 238, *239,* Ch. 6
'pill-millipedes', 366, *367,* Pl. 3(e), (e.1)
Pionocypris (Crustacea, Ostracoda), 168, *171; see also* Pl. 3(a)
pivot joint, 41, 192, *193*
plantigrade tarsus, of *Scutigera* and Pterygota, 225
Platyrhacus (Diplopoda, Polydesmida), *229, 231, 357, 358,* 369
pleural apodeme, of Thysanura, **425,** *426, 427*
pleural folds, *117, 271*
pleurites, of Chilopoda, *217, 346, 378, 379*
  of Collembola, *405, 409*
  of Diplopoda, *209, 367*
  of Diplura, *221*
  of Pterygota, *225,* 440
  of Symphyla, *218, 373*
  of Thysanura, 424, *425, 426, 427,* 437, 446
  of Uniramia and homologies, **345,** 437
Plutonium (Myriapoda, Chilopoda), 391
'Polaroid', use of, 502

524

Polychaeta (Annelida), 177–85, *178, 180, 183, 185,* Pl. 1(b)–(e)
Polydesmoidea (Myriapoda, Diplopoda), abilities and structure, 357, *358,* 359
Polydesmus (Myriapoda, Diplopoda), *207, 229, 231, 232, 358,* Pls. 4(i), 5(f)
Polymicrodon (Myriapoda, Diplopoda), 362, Pl. 4(h)
polyphyletic derivation of arthropods, 31, 236–93 *passim,* 487–94
Polyxenus (Diplopoda, Pselaphognatha), 370, Pl. 6(b), (j), (l)
Polyzonium (Diplopoda, Colobognatha), 360–4, 367, Pl. 4(g)
Portunas, swimming crab, 142, 147, Pl. 3(h)
preantennal segment, 247
preantennulary somite, *248, 249,* 250
premandibular segment, 247
preoral cavity of Collembola, *129*
  of Crustacea, *75, 84, 85*
  of Diplopoda, 104
  of Diplura, 131
  of Onychophora, *98*
  of Pterygota, 123
  of Symphyla, 109
prisms, use of, 501
proboscis, of pycnogonida, *23,* 240
procoxal pleurite, of Chilopoda, 345, *346, 378*
promotor–remotor movement of leg or mandible, **39,** 41, 72, *76, 191*
  of Amblypygi, 454, *455*
  of *Callipus* (Diplopoda, Lysiopetaloidea), *209*
  of *Galeodes* (Arachnida), 459
  of other Arachnida, 457
  of Machilidae, 424, *425, 426*
  of Lepismatidae, 426–33, 42, *429*
prophragma, of Diplopoda, **345, 352,** *355, 358, 367*
prosoma, **12,** *14, 15, 16,* 17, 18, *19,* 238
protractor muscles, e.g. *197, 211*
protonymphon larva of pycnogonids, *23,* 24, 262
protopod, 142, 241, *145*
protrusible mandible, *see* mandibles
Protura (Hexapoda), 216, 422
  thorax and leg base, *221*
  entognathy, *272*
  leg joints, *227*
  pleuron, 445, Pl. 6(k)
Pryla (Pterygota), 444
Pselaphognatha (Diplopoda), 370–2

pseudo-limbs, *Rhyncodemus,* 135, *136*
Pterygota, 399, 438; ancestral, 447
  coxa–body joint, 225, 445
  head endoskeleton, *see* tentorial apodemes
  wing movements and flight, 439
  legs and leg-joints, 220, *227; see also* legs
  mandibles, *see* mandibles
  thorax, 440
Pycnogonida, habits and accomplishments, 264, 486
  legs and jointing, 479, *481*
  structure, 22, *23*
  status and evolution, 261, 239, 483
pygidium, *3, 4,* **6**

raptatory habits, 155
relationships of larger arthropodan taxa, *239,* 257, 487
remotor movements, **39,** *45,* 68, 115
retractor muscles, e.g. 39, 72, 76
Rhyncodemus, nemertine flatworm, 135, *136*
Rhyniella praecursor (Collembola), 27, 401
rocking leg movements, 212; *see also* leg
running, fast, in Chilopoda, 298, 376; Lysiopetaloidea (*Callipus*), 368; Pauropoda, 396; Symphyla, 372
  slow, in Diplopoda (for strong movements), 351

Sao hirsuta (Trilobita), *4*
sclerites in arachnid joints, *455,* 464, *465,* 466
sclerotization, 25
Scolopendra spp. (Chilopoda, Scolopendromorpha), Pl. 1(f) gaits, *297, 283*
  mandibles, *100, 110;*
  tracks, *368; see also* Pl. 1(f)
scorpions, 17, 18
  stepping movements, 461, *463*
  leg joints, *457,* Pl. 8(a), (b)
segment, **251**
  determinants of segment numbers, 314, 353
segmental tendons, mandibular, *88,* 87
  trunk, 345, *347, 385, 387,* 389, *406, 407, 409, 435, 436*
Scutigera (Chilopoda, Scutigeromorpha), *203, 217, 225, 297, 303, 393,* Pl. 4(j), (k)
Scutigerella (Myriapoda, Symphyla), 109, *322, 372, 373, 375,* Pl. 4(d), (e)
shapes, significance of, 34

525

Index

Shumardia (Trilobita), *4*
Sida (Crustacea, Branchiopoda), 55, *147*, *160*, *164*, Pl. 2(b)
Siphonophora spp. (Myriapoda, Diplopoda), *209*, *272*, *362*, Pl. 7(f)
size, determinants of, 329; effects of, *328*, *329*
slime papillae, of Onychophora, *98*
Sminthurus (Collembola), 401
Solifugae (Arachnida), *457*, Pl. 8(c)
Sphaerosyllis (Polychaeta), *178*
Sphaerotherium (Diplopoda, Oniscomorpha), *365*, *367*; see also Pl. 3(e), (e.1)
spiders, 18, *464*, *201*, *473*, *474*, *457*, Pl. 8(d)
speed, 298, 300, 399, 402
spine apodeme, of japygids, *417*, *419*
spine row, of peracaridan mandible, 86
Spinther (Polychaeta), *177*, *180*, *183*
spiralling, of Diplopoda, *355*, *364*, *367*, *368*, Pl. 4(o)
Squilla (Crustacea, Malacostraca), 153
staining, use of, 500, 501
status, of Collembola, 414
 of Diplura, 421
 of Protura, 422
 of Pycnogonida, 261
 of Thysanura, 437
 of Trilobita, 261
Stenodictyon, 438
stenopodium, 145, Fig. 4.5(a), **147**
stepping, frequency of, 298
sternites, *41*, *91*, *93*, *95*
 of arachnida, *95*, *451*, *455*
 sliding sternites of Diplopoda, *209*, *362*, Pl. 4(f)
 of Uniramia, **345**
stipes, mandibular (Diplopoda), *102*
stomodoeum, *248*, *249*
striated muscle, 236
stride lengths, 47, 304, *307*, *308*, *333*, 447
strong leg movements, 304
'sub-coxal' segments, of Collembola, 403, *405*, 410
 of *Lepisma*, *436*, 437
swaying movements of Pycnogonida, *482*
swimming, 53, 55
swimming crab, 15, Pl. 3(h)
 by Nymphon, *481*, *482*
Symphyla (Myriapoda), *372*, 445
 abilities and structure, *373*
 grade of organization, 280

Symphyla (*contd*)
 gaits, 322
 mandible, 108, *109*; musculature, *218*, Pl. 4(d), (e)
Symphylella (Symphyla), *373*
Syncarida (Crustacea, Malacostraca), 22, Pl. 3(f)
synovial cavities in diplopodan leg joints, 230, *231*, *232*

Tachypleus (Merostomata, Xiphosura), *91*, *93*
Tardigrada, 29, *239*, 289
teloblasts, ectodermal and mesodermal, *248*, 251
telson, *21*, **150**, 240, *248*, *249*, **252**
tendons, ribbon-like, of Collembola, *406*, *407*, **408**, *409*
tendons, *193*, 345, 347
 endoskeletal, of Collembola, *406*, *407*, *409*, 410, 411
 from gnathal lobe of mandible, Diplopoda, *102*, *107*
 transverse intersegmental tendons and derivatives, embryonic and simple, *7*, *88*
 endosternite of Xiphosura, *91*, *93*
 transverse mandibular, maxilla 1, maxilla 2 tendons, *71*, *72*, *73*, *102*, *109*, *110*, 115, *119*, *125*, 128, *131*
tentorial apodemes, anterior and posterior, *see* apodemes
tentorium of Pterygota, 114, 115, *123*
tergites of Uniramia, 345
tergite heteronomy, of Chilopoda, 381, *385*, *386*, *387*
 of Pauropoda, *219*, 396
 of Symphyla, *218*, *375*, 376
thorax, **3**, 20, *21*, 150, *152*, **240**
Thysanura, 24
 accomplishments and habits, 423
 endoskeletal tendons, *428*, *435*, *436*
 jumping gaits and associated structure, *335*, *423*, *425*, *427*, *428*–*31*
 jumping escape reactions of Machilidae, *336*, *338*, *433*, *436*
 lack of primitiveness, 434, 438
 leg-base, **222**, *424*, *425*, *427*, *428*, *430*, *431*
 leg-joints and muscles, *424*, *425*, *427*, *431*
 pleurites, *424*, *425*, *426*, *427*, 437
 tracks, *333*, *335*, Pl. 6(f), (g)
ticks, see Acari
time interval k, **311**, 322, *323*, *475*, *478*
tracheal systems, of *Craterostigmus*

Index

(Chilopoda, Scolopendromorpha), 391
  of *Plutonium* (Chilopoda, Scolopendromorpha), 391
  of Onychophora, 190
tracks, of beetle, 333
  of *Callipus*, 307
  of *Campodea*, 333, Pl. 6(e)
  of *Cormocephalus*, 308
  of *Cylindroiulus*, 307
  of hemipteran, 333
  of *Japyx*, 333
  of *Lepisma* and *Thermobia*, 333, Pl. 6(g)
  of *Lithobius*, 308, 317
  of *Petrobius*, 333, 335, Pl. 6(f)
  of *Scutigera*, 307
  of *Tomocerus*, 333, Pl. 6(a)
  use of, 504
Trignatha, a grade of organization, 279, 280, 285
Tomocerus longicornis (Collembola), 401
  coxal segmentation, 405, 409, 413
  gaits, 301
  jumping mechanism and facilitating skeleto-musculature, 338, *340*, 401–15
  mandible and musculature, 122, *125, 127, 129*
  legs, 227, 409, 411
  ribbon tendons, 406, 407, 409, 410
  'sub-coxal' segments, non-existent, 403, 405, 410
  tracks, 333
  trunk musculature, 405, 406, 407

Ticks (Acari), 453, *457*, Pl. 8(f)
Triops (Crustacea, Branchiopoda), 24, 49, 51, 163, *164*, Pl. 2(a)
Trochosa, spider, 457, 458–61, 465–7, Pl. 8(c)
Triarthrus (Trilobita), 2, *5, 7, 8*, 51, 57, 245
Trigonotarbida (Arachnida), 266
Trilobita, 1, 2, *3–8*, 138, 239, 258, 261
Tullbergia (Collembola), 401

undulations of trunk in running, 295
unguligrade stance, 447
Uniramia, habits, 284
  evolution, 281, *285*
  locomotory mechanisms, 294–343 *passim*; series of gaits in Uniramia alone, 282

Waeringoscorpio (Arachnida), 18, *19*, 266
Whip scorpions, Uropygi (Arachnida), 452
wings, evolution, 445
  movements and musculature, 441

Xenusion, 26
Xiphosura (Merostromata), 2, 12, *13*
  feeding and walking movements, 50
  legs, gnathobases and jointing, *91, 93, 457, 463*
  respiratory limbs, *143*

yolk sac of embryo, *248, 249*